Making a New Land

Making A New Land

Environmental histories
of New Zealand

NEW EDITION

Edited by
Eric Pawson and
Tom Brooking

Published by Otago University Press
PO Box 56 / Level 1, 398 Cumberland Street
Dunedin, New Zealand
university.press@otago.ac.nz
www.otago.ac.nz/press

First published 2013
Text copyright © the contributors as named 2013
Volume copyright © Otago University Press 2013

The moral rights of the authors have been asserted.

ISBN 978-1-877578-52-6

A catalogue record for this book is available from the National Library of New Zealand. This book is copyright. Except for the purpose of fair review, no part may be stored or transmitted in any form or by any means, electronic or mechanical, including recording or storage in any information retrieval system, without permission in writing from the publisher. No reproduction may be made, whether by photocopying or by any other means, unless a licence has been obtained from the publisher.

Front cover: William Sutton, *Hills and Plains, Waikari 1956*, oil on canvas. 89/143, 1956, Christchurch Art Gallery Te Puna o Waiwhetu

Publisher: Rachel Scott
Editor: Gillian Tewsley
Design/layout: Fiona Moffat
Index: Diane Lowther

Printed in New Zealand by PrintStop Ltd, Wellington

Contents

List of Figures and Tables	7
Contributors	10
Preface	15
1 Introduction Eric Pawson and Tom Brooking	17

Part I Encounters

2 A fragile plenty: pre-European Māori and the New Zealand environment Atholl Anderson	35
3 Contesting resources: Māori, Pākehā and a tenurial revolution Evelyn Stokes	52
4 Resource frontiers, environment and settler capitalism, 1769–1860 Jim McAloon	70

Part II Colonising

5 Settlers transforming the open country Robert Peden and Peter Holland	89
6 Mining the quarry Terry Hearn	106
7 Destruction under the guise of improvement? The forest, 1840–1920 Graeme Wynn	122

Part III Wild places

8 Children of the burnt bush: New Zealanders and the indigenous remnant, 1880–1930 Paul Star and Lynne Lochhead	141
9 The meanings of mountains Eric Pawson	158
10 'Swamps which might doubtless Easily be drained': swamp drainage and its impact on the indigenous Geoff Park	174

Part IV Modernising

11	The grasslands revolution reconsidered Tom Brooking and Vaughan Wood	193
12	An interventionist state: 'wise use' forestry and soil conservation Michael Roche	209
13	On the edge: making urban places Eric Pawson	226
14	The empire of the rhododendron: reorienting New Zealand garden history James Beattie	241

Part V Perspectives

15	Postcolonial environments Katie Pickles	261
16	An updated history of New Zealand environmental law Nicola Wheen	277
17	Ngāi Tahu and the 'nature' of Māori modernity Michael J. Stevens	293
18	Mastering the land: mapping and metrologies in Aotearoa New Zealand Andreas Aagaard Christensen	310
19	Epilogue Eric Pawson and Tom Brooking	328

Notes	337
Glossary of Māori terms	378
Index	381

List of Figures and Tables

Figures

1.1 'Five Mile Ave, Forty Mile Bush', c. 1875 19
1.2 A landscape of 'improvement', looking towards Akaroa, c. 1925 21
1.3 William Fox, *House in Nelson, 1848*, watercolour 24
1.4 A landscape of modernity, South Taranaki, c. 1964 29
2.1 Map of places and districts mentioned in the text 37
2.2 The distribution of Rafter Laboratory radiocarbon dates on rat bone 39
2.3 Windward and leeward provinces in New Zealand, showing Māori forts (pā) and moa-hunting sites 43
2.4 Distribution of forest and open country at approximately AD 1000 and AD 1800 49
2.5 A regionally differentiated ecological model of socioeconomic change in pre-European New Zealand 50
3.1 Pākehā encounter before 1840 55
3.2 Te Hiku o Te Ika: Pākehā land claims and mission stations, 1840 57
3.3 Te Hiku o Te Ika: exploitation of resources before 1840 58
3.4 Te Hiku o Te Ika: cultural relations, 1840 62
3.5 The extinguishment of native title 65
3.6 Alienation of Māori land in the North Island, 1860–1939 68
4.1 Map of New Zealand in the global economy, 1770–1840 73
4.2 Portrait of Governor Philip Gidley King (1758–1808) 74
4.3 Frederic Alonzo Carrington, 'Plan of the Town and Part of the Settlement of New Plymouth, New Zealand, 1842' 80
4.4 Map of Europeans and the south, 1843 83
5.1 Map showing the open country 90
5.2 Sheep in the holding yards at Shag Valley Station, North Otago, during the 1880s 95
5.3 Birch Hill Station at the head of the Tasman Valley in the late nineteenth century 96
5.4 Merino lambs being drafted on a high-country station during summer in the 1920s 100
5.5 A wagon laden with rabbit skins awaiting export from the Port of Dunedin in the 1920s 103
6.1 Goldfields and gold production, 1857–1921, by decade 107
6.2 Damage caused by silting, Waihou River, June 1907 111
6.3 Blue Spur hydraulic sluicing and elevating claim, Gabriel's Gully, Otago 117
6.4 Map of coalfields and production, 1883–1920, by decade 119
7.1 Julius Geissler, *Primeval Bush*, 1919, pen 124
7.2 The structure of the New Zealand forest 125
7.3 Map of the forest, c. 1840 126
7.4 Bullock team and timber workers alongside a kauri tree, 1897 130
7.5 Map of the North Island forest, 1880–1910 131
7.6 Clearing new farms of trees under Mt Egmont, c. 1900 132
8.1 Map of protected areas, 1906–07, showing national parks and scenic reserves 142

8.2	Kennedy's Bush, c. 1920 149
8.3	Plan of Tongariro National Park by Leonard Cockayne, 1908 151
8.4	Charles Blomfield, *The Vaulted Aisles of Nature's Cathedral*, 1921, oil on canvas 157
9.1	Map of the mountain and hill areas of New Zealand 159
9.2	'To the "Hermitage" Mount Cook', c. 1925 164
9.3	Lindis Pass, North Otago, 1926 165
9.4	Graph of alpine fatalities between 1889 and 1959 170
10.1	Excavating a drainage canal, Hauraki Plains, 1910 175
10.2	A plan of swamp drainage, Hauraki Plains, 1908 178
10.3	Hauraki Plains swamp drainage, 1911–12 180
10.4	A swamp plough at work on Mr A. Lusk's farm, Oaonui, Taranaki, c. 1914–15 186
11.1	Graph of changes in the area of sown grassland and the numbers of stock units, 1861–2005 194
11.2	Advertisement: 'Fertiliser helped us grow …', 1977 199
11.3	Cover of the *New Zealand Journal of Agriculture*, July 1946, showing the Four Horsemen of the Apocalypse 201
11.4	Cover of the *New Zealand Journal of Agriculture*, September 1946 202
11.5	Advertisement for Tordon Brushkiller, 1999 204
12.1	Map of state indigenous and exotic forests, 1930 212
12.2	The arrival of the first trainload of logs at Kawerau, site of the Tasman Pulp and Paper mill in 1955 216
12.3	Sketches of 'Nature's revenge', 1939 219
12.4	'Giving wings to soil conservation', 1955 222
12.5	'The land we have we must hold!', advertisement, 1961 223
13.1	Evoking the modern. Cover of the city guide, *Auckland: The Gateway to New Zealand*, 1931 230
13.2	'Civilising' the Waimakariri River since the 1860s: its changing courses and flood-control works 233
13.3	Map of Auckland suburban development, 1915–45 234
13.4	'Salubrious St Clair': auction notice for suburban sections, 1913 236
13.5	'Timaru by the Sea': promotional image from the 1930s 239
14.1	Map showing Cantonese networks of migration, information and plant exchange 242
14.2	Sketch of Horeke, 1840, pencil and ink 247
14.3	*Alfred Ludlam's House and Garden*, 1850, watercolour 249
14.4	Unidentified Chinese man and Rev. Alexander Don outside a dwelling in Waikaia, c. 1900 252
14.5	Well-known market gardeners Ah Sam and Joe Quin with vegetables, in Roxburgh, 1903 253
14.6	Graph of the value of New Zealand exports of fungus, 1880–1920 256
15.1	'Mother Nature', Al Nisbet cartoon, 2011 264
15.2	William Sutton, *Hills and Plains, Waikari 1956*, oil on canvas 267
15.3	Juliet Peter, *Nor'west, 1939*, linocut 268
15.4	The 'New Napier': Napier Carnival poster, 1933 272
15.5	The fallen statue of John Robert Godley, Cathedral Square, Christchurch, 24 February 2011 273
16.1	Map of the location of Lake Manapouri, showing the power scheme and tailrace tunnels 279

16.2   'The view stinks but it's worth millions!', Sid Scales cartoon, 1959   280
16.3   'Mastering the vast energy potential of the Clutha River'. Map of Scheme F, 1976   283
16.4   Government ministers kneel before the graven image of Aluminium, as the Minister of Energy offers a jug of Clutha water into the sacrificial sink, Sid Scales cartoon, 1980   284
16.5   Map of the Project Aqua proposal, showing the location of the canal and six power stations, alongside the Waitaki River   289
17.1   The organisational chart of Te Rūnanga o Ngāi Tahu   294
17.2   TRoNT's 18 regional papatipu rūnanga   295
17.3   Map of places in and around Canterbury mentioned in the text   300
17.4   Map of places in Southland mentioned in the text   304
17.5   Kura-mātakitaki Stevens and his great-grandfather, Tiny Metzger, prepare pōhā using rimurapa from Omaui in advance of the 2012 tītī harvest   305
18.1   Surveyors cutting down the bush to make survey pegs, 1903   315
18.2   A hundred years of New Zealand map making   318
18.3   Key sequences in the production of geographical knowledge of New Zealand environments, 1800–2000   320
18.4   Recording New Zealand, 1930–2010   322
18.5   Production systems under siege in the Wairarapa district, 1959   323
18.6a Metrologies for decency, anno 1956   326
18.6b Normalising the produce: wool classers at work with New Zealand wool, Hornchurch, England, during World War I   326
19.1   'The height of happiness', 1927, colour lithograph   331
19.2   Blasting tree stumps in Taranaki, c. 1900   334

Tables

5.1   Area (in acres) under cultivation, and sheep numbers, for Hawke's Bay, Wellington, Marlborough, Canterbury and Otago provinces   101
7.1   New Zealand sawmills: number, distribution, capacity and output, 1907   129
9.1   High ascents from The Hermitage, 1914/15 to 1938/39   168

Contributors

ATHOLL ANDERSON, CNZM, FAHA, FRSNZ, FSA, is Emeritus Professor at the Australian National University, where he was previously Professor of Prehistory and Director of the Centre for Archaeological Research. He has worked on the prehistory of human–environment relationships for 40 years in New Zealand and throughout the Pacific and Indian Ocean islands, especially on human colonisation, the exploitation of maritime resources and avifaunal extinctions. He is the author of *Prodigious Birds: Moas and Moahunting in Prehistoric New Zealand* (1989) and *The Welcome of Strangers: An Ethnohistory of Southern Maori, AD 1650–1850* (1998).

JAMES BEATTIE is Senior Lecturer in History at the University of Waikato. He has published widely on world, British imperial, garden and environmental histories. His monograph *Empire and Environmental Anxiety, 1800–1920: Health, Science, Art and Conservation in South Asia and Australasia* (2011) presented new perspectives on British imperialism. Forthcoming co-edited and co-written books include *Lan Yuan: A Garden of Distant Longing* (2013); *Networks of Nature in the British Empire* (2014); and *Climate, Science and History in Australasia* (2014). His next major projects examine Chinese–New Zealand environmental connections, and Chinese art collecting.

TOM BROOKING is Professor of History at the University of Otago where he has supervised six PhDs in environmental history to completion. He also works in New Zealand agricultural and rural history as well as political history, the history of ideas, and the Scottish–New Zealand connection. His book with Eric Pawson, *Seeds of Empire: The Transformation of the New Zealand Environment* (2011), was based on research carried out under a Marsden grant. Currently he has a co-authored book on Scottish migration to New Zealand (also funded by Marsden) in press; and his biography of New Zealand's longest-serving Prime Minister, Richard John Seddon, will appear shortly.

ANDREAS AAGAARD CHRISTENSEN has an MSc in Geography and Cultural Studies and a BA in Psychology from Roskilde University. He is a PhD Fellow at the Department of Geosciences and Natural Resource Management, Copenhagen University, Denmark. His research project is the study of land use culture among farmers in European and colonial agrarian landscapes and the effect of such subjectivities on the ecology of rural landscapes. His research interests include landscape ecology, cartography, GIScience, environmental history, environmental anthropology and rural planning.

TERRY HEARN received a PhD from the University of Otago for his thesis on resource policy and resource-use conflict in nineteenth and early twentieth-century New Zealand. He was a contributor to the *New Zealand Historical Atlas* (1997), and spent six years as the Historian of British Immigration in the Ministry for Culture and Heritage, where he co-authored *Settlers: New Zealand Immigrants from England, Scotland, and Ireland 1840 to 1945* (2008, with Jock Phillips). Since 2001 he has worked as a consultant on claims lodged by Māori under the Treaty of Waitangi: he has contributed to nine major regional inquiries with a particular focus on land issues and the social and economic experience of Māori.

PETER HOLLAND was educated at the University of Canterbury and the Australian National University, and has worked in Canada, Kenya, New Zealand and South Africa. He was appointed Professor of Geography at the University of Otago in 1982, and has been Emeritus Professor since 2006. Trained as a biogeographer, his current research interests include environmental change during the early colonial period. His book *Home in the Howling Wilderness: Settlers and Environment in Southern New Zealand* was published in 2013. He is a past President of the New Zealand Geographical Society and was awarded the Society's Distinguished New Zealand Geographer medal in 2008.

LYNNE LOCHHEAD received a PhD from Lincoln University in 1994 for her thesis, 'Preserving the Brownies' Portion: A History of Voluntary Nature Conservation Organisations in New Zealand, 1888–1935'. She lives in Christchurch and is currently involved in a long-term project to revegetate a section of Harts Creek and Birdlings Brook, in the Lake Ellesmere/Te Waihora catchment. She also works to conserve the heritage of built landscapes, something that has become all the more urgent in the aftermath of the Canterbury earthquakes.

JIM McALOON taught history for many years at Lincoln University and since 2009 has been Associate Professor in History at Victoria University of Wellington. He has research interests on the borders of environmental and economic history and has published widely in the history of nineteenth-century New Zealand, including a regional history of Nelson (1997) and *No Idle Rich* (2002), a study of wealthy settlers in Canterbury and Otago. More recently he contributed a chapter on the relationship between British markets and New Zealand grasslands to Brooking and Pawson, *Seeds of Empire* (2011).

TIM NOLAN is the director of Blackant Mapping Solutions, a cartography and geographics company based in Christchurch. With a BSc and MSc in Geography from the University of Canterbury, he has a wide range of interests in geography and mapping, particularly in visualising historical landscapes. His maps have appeared in numerous publications, including the *Historical Atlas of New Zealand* (1997) and the *Macmillan (NZ) World Atlas* (2008). He is currently a member of the Committee of the New Zealand Cartographic Society.

GEOFF PARK was a Wellington-based ecologist with a tendency to history, in particular the effects of colonisation on the indigenous Pacific. He was the author of *Ngā Uruora: The Groves of Life: Ecology and History in a New Zealand Landscape* (1995). He was a Concept Leader at the Museum of New Zealand Te Papa Tongarewa, and worked as a researcher for the Waitangi Tribunal on the flora and fauna claim (Wai 262). A collection of his work was published as *Theatre Country: Essays on Landscape and Whenua* (2006). He died in 2009.

ERIC PAWSON is Professor of Geography at the University of Canterbury. He is a graduate of Hertford and Nuffield Colleges, Oxford. He chaired the advisory committee of the *New Zealand Historical Atlas* (1990–97), and is on the advisory committee for *Te Ara – the Encyclopedia of New Zealand*. He has worked with Tom Brooking on projects in environmental history since the late 1990s, including the first edition of *Environmental Histories of New Zealand* (2002). He received the Distinguished New Zealand Geographer Medal in 2007 and a National Tertiary Teaching Award in 2009, and was Managing Editor of the *New Zealand Geographer* (2004–11).

ROBERT PEDEN has a PhD (Otago) and MA (Canterbury) in History. He specialises in environmental/agricultural/settlement history. He spent 25 years working on sheep and cattle stations in the South Island high country before returning to university in 1999. He was awarded a Claude McCarthy Fellowship in 2009, which gave him the opportunity to write *Making Sheep Country: Mt Peel Station and the Transformation of the Tussock Lands* (2011). He has worked as a researcher/writer for *Te Ara – the Encyclopedia of New Zealand* and for the Waitangi Tribunal. Currently he is working on a land history of the South Island high country.

KATIE PICKLES, BA (Canterbury), MA (UBC), PhD (McGill), is Associate Professor of History at the University of Canterbury. Her research interests are wide-ranging, but often concern the history of imperialism and colonialism. Themes of landscape, memory and colonial identity are important in her monograph *Transnational Outrage: The Death and Commemoration of Edith Cavell* (2007). She has written about the environmental history of Bottle Lake Forest in Christchurch, and is currently working on the theme of Antarctica in New Zealand's imperial imagination for a volume that she is co-editing with Catharine Coleborne, titled *New Zealand's Empire?* (2014).

MICHAEL ROCHE is Professor of Geography in the School of People, Environment and Planning at Massey University, Palmerston North. He completed his PhD at the University of Canterbury on forest policy and forest management in New Zealand. He has published two books on the New Zealand timber industry: *History of Forestry* (1990) and *Land and Water: Water and Soil Conservation and Central Government in New Zealand, 1941–1988* (1994). He contributed several plates to the *New Zealand Historical Atlas* (1997) and more recently has written on colonial foresters and forestry in Australia and New Zealand.

PAUL STAR, MA (Cantab), MA PhD (Otago), was brought up in England. He moved to New Zealand in 1972 and worked for many years as a bookseller before returning to academic study in the 1990s. He lives rurally on the Otago Peninsula, near Dunedin. He was a postdoctoral fellow with the University of Otago's History Department (2004–07) and is now an independent scholar specialising in New Zealand environmental history. He contributed the chapter on this topic to *The New Oxford History of New Zealand* (2009) and has published widely on related themes in local and international journals.

MICHAEL J. STEVENS is Lecturer in History at the University of Otago. He was born into a Kāi Tahu family that participates in the seasonal harvest of tītī/sooty shearwaters ('muttonbirding') from islands adjacent to Rakiura/Stewart Island, and was raised mainly in the port town of Bluff. This gives him a lived understanding of mahika kai activities, which compels much of his research. His PhD dissertation, for instance, examined the syncretic knowledge that underlays the tītī harvest. Some of this research has been reworked in academic journal articles. Building on this work, he has more recently begun to reconnect Māori history with maritime history.

EVELYN STOKES, DNZM, held a Personal Chair in Geography at the University of Waikato. Her research interests included historical geography in New Zealand and North America, Māori land and communities and Treaty of Waitangi issues. She served on a number of government committees, and from 1989 was a member of the New Zealand Geographic Board and of the Waitangi Tribunal, contributing to major reports such as that on the Muriwhenua Land Claim (1997). She also authored many of the Māori-focused plates in the *New Zealand Historical Atlas* (1997). She died in 2005.

NICOLA WHEEN has been a member of the Law Faculty at the University of Otago since 1989. She teaches public law, environmental law and international environmental law. She has co-edited two books with Janine Hayward (Politics, Otago University): *The Waitangi Tribunal/Te Roopu Whakamana i te Tiriti o Waitangi* (2004) and *Treaty of Waitangi Settlements* (2012). Much of her recent research has focused on the law around fishing-related mortality of marine mammals and seabirds, and other conservation law issues.

VAUGHAN WOOD has a BSc (Hons) and an MA from the University of Canterbury, and a PhD from the University of Otago. He is a former research fellow of the Geography Department, University of Canterbury, and a contracted report writer to the Waitangi Tribunal in the hearing of indigenous land claims. His main area of research is agricultural history and environmental modification in nineteenth-century New Zealand. He has published articles in the *Journal of Historical Geography*, *Agricultural History* and *Environment and History*, and he is writing a history of the Akaroa cocksfoot seed industry.

GRAEME WYNN, FRSC, is Professor of Geography in the University of British Columbia where he has also been Brenda and David McLean Chair in Canadian Studies (2011–13). He was co-editor of the *Journal of Historical Geography* (2006–12) and is also General Editor of the Nature|History|Society book series with UBC Press. He is widely published in Canadian historical geography and environmental history, and in 2012 was awarded the Massey Medal of the Royal Canadian Geographical Society. He has maintained an active interest in New Zealand scholarship since he taught at the University of Canterbury in the 1970s.

Preface

THIS BOOK is a new edition of one that originally appeared in 2002 as *Environmental Histories of New Zealand*. It is republished under the fresh title of *Making a New Land*, first in recognition of its revision as a university-level text used widely in New Zealand and overseas, and second to extend its readership among a public keen to engage with good, historically informed writing on the environment. To this end we have assembled a team of over 20 good-humoured and patient contributors, all experts in their field. About a third of the chapters are new and the rest have been revised to reflect advances in knowledge over the last decade. There are two exceptions, Chapters 3 and 10: the authors of these chapters have since died, but their original contributions were highly regarded and not readily replaceable. These have been left unchanged, except for the addition of the paragraph of further reading that has been included in this edition for all chapters.

The book may be approached sequentially, as it is structured into five broad chronological sections; or it may be read thematically, dipping into those parts of interest to the reader. Whatever way it is used, *Making a New Land* is designed to provide an up-to-date, considered series of reflections about important themes in New Zealand's environmental histories. The text is illustrated with about 90 figures, one third of which are new to this edition. We are grateful to all the repositories that have granted us permission to use their photographs and artworks, each of which is acknowledged in the captions. We greatly appreciate the talents and dedication of our cartographer, Tim Nolan, who has translated often sketchy ideas into clear maps and diagrams. We have also enjoyed the wholehearted support of Otago University Press in recreating the book for a new generation of readers. We would especially like to thank Wendy Harrex who, as publisher at the time, backed the initial idea; Gillian Tewsley for her careful attention to detail as editor; and Rachel Scott, the current publisher, who has steered the project through to fruition.

ERIC PAWSON AND TOM BROOKING
Christchurch and Dunedin, August 2013

1.
Introduction

Eric Pawson and Tom Brooking

AOTEAROA NEW ZEALAND holds a special place in global environmental history.[1] It is an isolated land of mountains, beaches, wetlands and forests. According to the foundation stories of its first colonists, the North Island was fished out of the southern ocean by the demigod Māui, while his waka, or canoe, became the South Island. The descendants of those colonists, the Māori, had named and claimed these as their own territories by the time the 'large land, uplifted high' was 'discovered' a few hundred years later by Europeans. In the stories of these newer colonists, named 'Pākehā' by their predecessors, New Zealand was first understood as a remnant of the great southern continent, which at that time was believed to be a necessary counterweight to the landmass of Eurasia. Later it came to be a fragment of the ancient supercontinent Gondwana, cast adrift before the advent of mammals.[2] The Australian ecologist Tim Flannery refers to it as 'a completely different experiment in evolution to the rest of the world', showing 'what the world might have looked like if mammals as well as dinosaurs had become extinct 65 million years ago, leaving the birds to inherit the globe'.[3]

The human experiments of making places for living in these islands are more recent than those conducted elsewhere, considered now to be seven to eight hundred years old at most. The effects of Māori hunting, fire and horticulture were extensive. But they were less dramatic than 200 years of Pākehā transformations, initiated as part of the European imperial drive to incorporate new territories into the capitalist world economy using the panoply of people, animals, plants and, less intentionally, pathogens. In turn, the consequences of such experiments for human livelihoods have had to be borne. There has been the hazardousness of rapid and sometimes extreme weather changes, and the risks from events such as earthquakes and floods. These result from the country's location astride a tectonic plate boundary, with its mountains and ranges intercepting moisture-laden winds from the southern ocean and the subtropics. There have been contests for land between Māori and Pākehā, vigorous throughout the nineteenth century and not forgotten by Māori since, and environmental interventions resulting in resource destruction, soil erosion and the spread of unwanted, and costly, pests and weeds.[4]

Relations between people and environments in New Zealand share many of the characteristics of such interactions elsewhere, when migrants arrive in new lands and seek to come to terms with what they find. But, consistent with geographic theories of place, both the context for these interactions and the manner in which they have intersected with this context and each other have produced very particular outcomes. As Joe Powell, the Australian historical geographer, has suggested, 'the best environmental stories ... recognise locational integrities in so far as they have a very definite resonance in [particular] national and regional contexts.'[5] To this should be added cultural contexts, as there is not nor ever has been but one New Zealand. Rather there is a kaleidoscopic complex of Māori and Pākehā identities in place, and the tensions that go with this.[6]

The purpose of this book is to narrate the ways in which human–environment interactions in the places of Aotearoa New Zealand have been constructed. The Introduction situates the project in the context of current thinking and practice in environmental history. It also introduces the individual chapters and explains how they have been grouped so as to provide the reader with clear signposts for approaching them as a collection.

Making a new land

A recent definition of environmental history suggests that it is 'a kind of history that seeks understanding of human beings as they have lived, worked and thought in relationship to the rest of nature through the changes wrought by time'.[7] The breadth of this statement is an advantage: it opens a very broad canvas. In identifying the everyday matters of living, working and thinking, it focuses on the role of everyone as actors in environmental history, not merely the 'leading figures' approach of conventional historical writing in the past. But in other ways the definition is problematic. It is not 'time' that brings change, but the actions of human and non-human agents. And it is not just time past that environmental history should consider, but the ways in which past and present behaviours shape environmental futures. Perhaps the reason for the growing popularity of the field internationally since the 1970s, and in New Zealand in the last decade or more, is that the ways in which history is written reflect the preoccupations and anxieties of its times.[8]

Consider the photograph in Figure 1.1. This was taken about 1875 by James Bragge, a Wellington photographer born in County Durham. Rather than concentrating on portraiture, the lifeblood that sustained his contemporary practitioners, Bragge fitted out a horse-drawn cart as a portable darkroom that allowed him to capture the development of the new city and its environs. He received an award at the Sydney International Exhibition in 1879 for a series of views commissioned by the Wellington City Council. This was a year after his second expedition across the Rimutaka Ranges, north of the city, where he recorded the extension of the railway and the farming frontier into places previously covered in tussock grass, thick bush and forest.[9] The photograph is of what Bragge called 'Five Mile Avenue' in the Forty Mile Bush in the upper Wairarapa. Outwardly it is a portrayal of the impressiveness of the bush; even, given the small size of the human figure, its oppressiveness.

Figure 1.1 'Five Mile Ave, Forty Mile Bush', photograph by James Bragge, c. 1875.
Source: negative no. D000086, Museum of New Zealand Te Papa Tongarewa

Bragge's photograph, however, is no more innocent as a record of what he saw than any other of the extensive records that survive from this time as documents of environmental change. Scholarship on photography has demonstrated that as a practice it was as central to contemporary discourses of colonisation as were written accounts and maps.[10] Photographers brought to their images, especially in the selection of viewpoints and the manner of framing scenes, an 'imaginative geography' that implicitly betrayed their view of nature and its relation to human endeavour. This photograph highlights not so much the bush but the avenue that draws the invading eye through the trees. And if the smallness of the human figure implies awe of the vegetation, it can equally be suggested that the man is not cowed by it. Rather, he represents the colonial mission to cut down or burn the trees, to convert potentially fruitful land to grass, or to transform external resources into social value. As the editor of the *Manawatu Times* put it in 1877: 'although the smoke may inconvenience us and the charred avenues offend the eye, we must accept all thankfully as a mark of local progress.'[11] The photograph documents culture as much as it does nature.

It has been argued that 'the environment[s] we inhabit [are] inseparable from human culture' – in other words, from the ways in which we see and use environments.'[12] Everything that surrounds us – rural and urban landscapes, coastlines, even the sea – is shaped, traversed and harvested in accordance with cultural imperatives and social needs. Our awareness of these environments, and our representations and interpretations of them, reflect human traditions and expectations. Far from being 'objective', this awareness is shaped by deep-seated assumptions about actual or idealised relations between people and nature.[13] For Māori, kaitiakitanga (guardianship) is 'an obligation to safeguard and care for the environment for future generations ... it is an inherited commitment that links ... the spiritual realm with the human world and both of those with the earth and all that is on it'.[14] In contrast, the world view of Europeans reflects a subject position that constructs nature as something external to the individual. It is this assumption that is inherent in the Treaty of Waitangi drawn up on behalf of the British Crown and signed with Māori tribes in 1840.[15] For both peoples, therefore, nature and people have been deeply intertwined, whether this connection is recognised or covert. The making of a new land has been a social process.

This point was recognised by an earlier generation of European commentators. Edward Gibbon Wakefield, the post-Enlightenment dreamer, wrote in his *Art of Colonization* about the necessity of renaming apparently virgin country, including land occupied by Māori who had not used it in ways recognisable to Europeans, as it 'make[s] part of the moral atmosphere of the country'.[16] To Wakefield, colonisation would produce a new geography, 'improving' that which had been 'wilderness' into the Eden of the Book of Genesis. Improvement was the ideology of colonisation: it applied to both lands and peoples, and one of its most potent symbols was the garden. Henry Sewell, the Canterbury colonist, proclaimed in the 1850s: 'The first creation was a garden, and the nearer we get back to the garden state, the nearer we approach what may be called the true normal state of Nature.'[17] The result was deforested landscapes such as that photographed in the mid-1920s by Jesse Buckland above Akaroa, on Banks Peninsula (Figure 1.2). In the words of a local author, 'True gloomy Rembrandt like shadows have disappeared ... but we cannot help fancying that to the thinking person the present landscape is far more gratifying ... in the stead of the past beauties are smiling slopes of grass ...'[18]

The production of this garden colony was seen as a triumph of progressiveness. It had accomplished in one century, as Kenneth Cumberland wrote in the American Geographical Society's *Geographical Review* in the early 1940s, 'what in Europe took twenty centuries, and in North America four'. Cumberland had only recently arrived in Canterbury from Britain; his was an initial broad-brush analysis of what he saw happening around him. He followed it with many years of more detailed research into New Zealand's landscape development.[19] For a while he worked alongside a Canadian visitor in the Geography Department at Canterbury University College, the historical geographer Andrew Hill Clark. Clark in turn published a book based on his extensive fieldwork in the South Island: *The Invasion of New Zealand by People, Plants and Animals* (1949). This, he wrote, was 'a report on a revolutionary change in the character of a region, which occurred in a period of less than two centuries'.[20] These writers also recognised the considerable environmental costs of the transformation. In part they drew from the great American cultural geographer of the time, Carl Sauer, who had observed in 1938 that 'the growing mastery of man over his environment' must set against 'the revenge of an outraged nature'.[21]

Cumberland devoted much of his early career to recording and analysing the extent of the soil erosion that emerged with the removal of bush in wet hill-country areas, such

Figure 1.2 A landscape of 'improvement' with the forest stripped bare for grass, photo by Jesse Buckland looking towards Akaroa, c. 1925. Source: AK: 2003.18.2, Akaroa Museum

as Bragge's Wairarapa, and the degradation of the South Island hill country from what he interpreted as overstocking, as well as from the invasion of pests like rabbits.[22] In 1986, New Zealand was the subject of a lengthy case study in the American Alfred Crosby's *Ecological Imperialism*, which put such invasions in a wider global frame.[23] In 1999, another American, William Cronon, wrote the preface for a reissue of Herbert Guthrie-Smith's study of environmental change on his Hawke's Bay sheep station, *Tutira*, first published in 1921, which Cronon described as 'one of the great English-language classics of environmental history'. In *Tutira*, Guthrie-Smith discussed the effects of interventions on the land as 'the cumulative result of trivialities', and wondered if, after a lifelong commitment to improvement, it would not have been better to 'admire, conserve, let well alone'.[24]

New narratives

The reissue of Guthrie-Smith's book, and the death of Cumberland in 2011, have brought these key texts in New Zealand environmental history back into wider view. As a field of academic study and popular interest, the contemporary practice of environmental history, however, has relatively recent roots. Its cause has been helped by the interest shown by overseas scholars such as Crosby, Cronon and Thomas Dunlap,[25] but a turning point came with the publication, just before the new millennium, of the *New Zealand Historical Atlas*. This was perhaps the first time that environmental perspectives were incorporated in a mainstream historical project. The *Atlas* does this not just for pre-European Māori, but also for the century and a half after the signing of the Treaty of Waitangi in 1840 between the British Crown and Māori tribes. That it was a product of a team drawn from a range of disciplines, but primarily geography and history, helps to explain this. So too does the context of its making: it was begun in 1990 as a sesquicentennial project to mark 150 years of the signing of the Treaty, at a time when a new environmental politics was prominent (leading in 1991 to the Resource Management Act, for example).[26]

The writing of history itself has changed as well, albeit less evenly. One highly regarded book, James Belich's *Making Peoples* (1996), explored at some length the resource dependence of Polynesian colonists in New Zealand, although once it switched attention to Anglo-Celtic immigrants, environmental transformation faded from the account, as if culture had without question subsumed nature. The same can be said of the second volume, *Paradise Reforged* (2001), as it told the story of New Zealand's move from colony to self-governing dominion, and on to an independent nation state. Michael King's *History of New Zealand* (2003) does however acknowledge the work of making environments by both Māori and Pākehā.[27] The third edition of the *Oxford History of New Zealand*, published in 2009 (the first two editions were 1981 and 1992), included for the first time a chapter specifically devoted to environmental history, written by Paul Star, who had earlier noted that since Crosby had 'painted the broad brushstrokes in his book, no New Zealand historians [had] taken a closer look'.[28]

The first edition of this current book, published in 2002, set out to extend these brushstrokes and to fill in some of the detail between. The Introduction described how its narratives 'incorporate a range of approaches to environmental history, as is characteristic of the field'.[29] In the present edition, these approaches again reproduce the disciplinary allegiances of the authors, which range from history and geography to archaeology, environmental law and landscape ecology. This is one sense in which the subtitle of this book, 'Environmental histories', reflects its diverse contents. The contents also share, however, a number of common points and perspectives about 'nature' and the environment, human agency, and how the interactions between these are represented and interpreted.

First, what is often loosely called the 'environment' is not nearly as 'natural' as it might seem. 'Instead,' as Cronon has written, 'it is a profoundly human construction.'[30] The concept of 'nature' has been prominent in Western thought for more than two thousand years, whereas the term 'environment' has been in regular use only since the 1960s and the widespread realisation of the effects of human activity on 'nature'. The non-human world is of course neither unreal nor imaginary: its materiality cannot be denied. Yet in the sense that nature is a human category, that it is seen, understood and shaped by human actions, people have always 'made' nature. They have been inconsistent in the use of the word: to describe something as 'natural', for example, may mean that it is relatively wild and untouched; or it may mean that it fits an explicit or unspoken ideal about how things should be, as Henry Sewell's description of a garden above illustrates.[31] Human ideas of what constitutes nature also change, sometimes quite rapidly. Swamps and bogs, often demonised in the Pākehā imagination, have over the last 20 years become 'wetlands'. In this way they have been accorded a status representing life more akin to traditional Māori valuations of such landscapes. Nature can be many things at once.

Second, everything in nature is fluidly connected to everything else. Change is constant. Despite their apparent re-evaluation, the pollution of wetlands continues apace precisely because they are part of wider ecosystems that include the human-use systems that generate vast quantities of waste, for example from stock and from chemical additives used on farmland. Recent assessments of the state of waterways and wetlands indicate the extent to which water quality is deteriorating in many places. Cronon writes that the 'natural world is far more dynamic, far more changeable, and far more entangled with human history than popular beliefs ... have typically acknowledged'.[32] The environment was not an empty or neutral stage on which Pākehā could without encumbrance act out their visions, even though it was often seen that way. The hazardous nature of apparently benign environments was rarely understood, or, if it was, this was usually quickly forgotten. Nor was the stage empty, as Figures 1.1 and 1.2 imply. The lands and waters of Aotearoa were Māori, and were named and used as such.[33]

Third, New Zealand did not develop in isolation after European arrival, but rather was connected through evolving webs and complex networks with much of the rest of the globe, and with the British world in particular. It was Marx, in his critique of Wakefield, who portrayed the political economy of colonisation as an international division of labour between the metropole and colonies that were developed to produce surpluses of food and materials. Recent work has both filled in and extended this essential insight.[34] Rollo Arnold in *Settler Kaponga* shows how the making of a small Taranaki settlement could not be understood apart from its wider context; and Felicity Barnes in *New Zealand's London* discusses how New Zealand was remade as a kind of rural hinterland of London, both culturally and economically.[35] But the centre did not only manipulate the edge: colonialism was very much a process of exchange that shaped both coloniser and colonised and, by implication, the environments of both. The networks were multiple. Plant material, for example, arrived in New Zealand from Asia and the Americas as well as from Europe, and left in other directions: notably as part of trans-Pacific networks. Environmental anxieties accompanied change, and ideas about how to respond were shared between parts of South Asia and Australasia.[36] The contexts and content of colonisation were rich and variable but, importantly, more actively interconnected than has often been apparent.

The contributors to this book also share certain conventions about historical writing and representation. The ways in which stories about the past are told depend on careful identification and use of historical sources, and on recognition that an 'archive' is a social construct invariably not made for the purposes for which it is subsequently used. What is left out may be as important as what is included.[37] The framework of interpretation is crucial. How a chapter is written reflects the question or issue that the author wishes to address, and the body of ideas and literature that they use to frame their analysis. Similarly, how a chapter is read is shaped by the interests of the reader: not everyone receives the same message, as each person brings their own frame of reference to the text.

The use of figures has received careful consideration. The photographs reproduced in this book are far more than a documentary record – if indeed they are that at all. They are not just 'illustrations' but are framed by the conventions of their time, in the same way that maps and paintings have been. Each can be regarded as a technology. In this context, the contents of the Pākehā archive sometimes sought to record difference and wonder; often, however, they embody an urge to appropriate so-called Māori 'wastelands'.[38] They reflect a form of panoptical vision, in which the recorder was – literally – master of all surveyed. This was often strongly contested, by both Māori and Pākehā, but it was a

Figure 1.3 William Fox, *House in Nelson, 1848*, watercolour. Source: Hocken Collections, Uare Taoka o Hākena, University of Otago

vision that was both boundless, seeking to draw into its gaze all that lay before it, and bounded, in the sense of desiring to delimit and commodify land and resources. William Fox's watercolours illustrate this. Fox was the New Zealand Company agent in Nelson from 1843 to 1850, and his painting of his own house shows a garden transformed with exotic plants, incorporated as a neat and tidy realm; beyond its boundaries lie apparently empty lands awaiting similar transformation (Figure 1.3).[39] The point is that none of the figures in this book should be approached as if they are in some way neutral.

The chapters

The book is divided into five parts. The first, 'Encounters', concerns Māori and Pākehā responses to a new land, and the interaction and contest between them. This is followed by 'Colonising', in which key themes of transformation between the mid-1800s and 1920 (improving grasslands, developing mining and removing the bush) are considered. 'Wild places', the third part, examines particular initiatives and places – such as nature conservation, wetlands and mountains – that were often marginalised by the processes highlighted in Part II. 'Modernising' takes in the major themes of twentieth-century environmental change: the adoption of science to render economic production and protection of resources more effective, and the making of urban environments. Part V, 'Perspectives', ranges widely to reflect on contemporary environmental themes in historical context.

The three chapters in Part I explore pre-European Māori environmental impacts, the contest for resources between Māori and Pākehā, and the wider context of settler capitalism within which resources were sought and nature and 'native' subjugated. In Chapter 2, Atholl Anderson discusses the effect of pre-European Māori on what he calls 'a fragile plenty', as they expanded 'as rapidly as possible using the richest resources with pitiless energetic efficiency'. In saying this, Anderson places Māori colonisation, in its early stages, firmly within the character of 'colonisation everywhere and at all times'. His essay reaches the conclusion that New Zealand was not colonised till the thirteenth century. Thereafter, however, the environment was culturally and materially appropriated, resulting in widespread vertebrate extinctions and deforestation, before environmental learning brought about the sort of adaptive change that is today often taken – by Māori and non-Māori alike – as somehow inherent to Māori culture. Anderson thus counters the mythology of the environmentally benign noble savage that permeates much everyday thinking, and counters those who imagine (as did many European colonists) that pre-European Māori were not capable of environmental change.

Evelyn Stokes builds on this view in Chapter 3, showing how a strong sense of territoriality and boundedness developed as Māori exploited localised, fine-grained sources of protein and highly productive gardens. Identity became strongly enmeshed in place and ancestry, a world view based on 'a sense of custodial occupation … that the environment should be maintained in a fit state for generations to come'. Such a sense led to land rights defined in terms of use and occupation, not in terms of alienation in the European way. Stokes stresses the complexity of and scope for misunderstanding in transactions with Pākehā, while emphasising the adaptability of Māori in taking advantage of opportunities to trade in environmental resources. Increasing power and numbers of Europeans, however, led to appropriation of the resource that both parties valued the most: land. The imposition of British law resulted in alienation of land, formerly held under customary title, to the Crown and private owners.

Whereas Chapter 3 focuses on the North Island, and particularly the Far North, Jim McAloon concentrates more on the South Island in his discussion of early Pākehā contact and settlement in Chapter 4. His emphasis is to place European colonisation in the perspective of the global economy and cultural politics of the time. New Zealand lay 'on the edge of empire', as part of Australia's Pacific frontier, with a range of commodities of interest to markets in Britain, North America, China and India. McAloon shows how lengthy and involved were the tentacles that began to envelop Māori societies in this contact period. These tentacles embodied what he describes as 'the moral economy of capitalism' built on a 'new international division of labour'. In turn, this moral economy was predicated on the European concept of land as a tradable commodity, and on organised settlement through the establishment of towns.

The three chapters in Part II make up the 'Colonising' section. They examine some of the big themes of environmental change between the mid-nineteenth and early twentieth centuries: the remaking of the grasslands, the effects of mining and quarrying, and the assault on the forest. Robert Peden and Peter Holland in Chapter 5 discuss the transformation of the open country of the eastern flanks of the North and

South Islands by the first two generations of Pākehā settlers, who modified and, in the lowlands, annihilated 'evolutionarily tested native ecosystems ... replacing them with functionally incomplete systems dominated by introduced plant and animal species'. The settlers experimented to match introduced plants and animals to local environments, and although they practised burning with some care, they fought to come to terms with rabbits, weeds and the weather. They created landscapes that functioned as the 'engine room' of New Zealand's economy till the 1920s.

In Chapter 6, Terry Hearn focuses on the environmental impacts of mining, frequently neglected in standard accounts of settlement. Gold boosted nineteenth-century export earnings, and had significance in many other ways, drawing in people and capital from a trans-Pacific circuit of goldrushes, and precipitating conflicts between landholders and miners over the effects of discharge of mine waste in particular. The New Zealand government and its agents were more accommodating of the mining interest than those of either Victoria or California. Mining law discounted 'social and environmental costs in the interests of sustained exploitation'. A detailed case study of the consequences of goldmining in the North Island Waihou–Ohinemuri goldfield reveals how the siltation of rivers and flooding generated disputes between Pākehā and Māori landholders. Coal mining and quarrying were by contrast more limited in their effects.

In Chapter 7, Graeme Wynn describes New Zealand in the nineteenth century as 'a wooden world'. The bush, like the grasslands, might appear of a piece, but varied immensely in composition as well as in potential social value. Timber was used for a wide range of purposes in towns and cities at the same time as it was burned to clear land for farming. Within the prevailing discourse of progress and improvement, the results, however messy, were seen often as achievements. This chapter also enlarges on a theme introduced in the two preceding chapters: the growth of disquiet caused by anxiety about the destruction of valuable resources and concerns for ecological and aesthetic impacts. The New Zealand Forests Act of 1874, albeit ineffective, was one of the earliest state conservation measures in the British Empire.

The eventual resolution to the timber issue lay with an allocation mechanism: good land was to be settled, while those areas with little production potential could remain tree-covered. In Chapter 8, Paul Star and Lynne Lochhead trace the fate of this 'indigenous remnant' in the first of three chapters about 'Wild places' in Part III. After identifying utilitarian moves to conserve some resources as far back as the 1860s, they show that, by 1900, more and more people were valuing the indigenous as part of a growing identification with New Zealand as home. Nonetheless, preservation was usually restricted to small islands – literally offshore – and the growing network of small scenic reserves on the mainland. Star and Lochhead chart the growth of urban-based conservation bodies, the role of tourism, and scientific arguments that scenic distinctiveness was linked to local flora, in developing a counter to the focus on 'improvement'. They show that initiatives to conserve flora, fauna and scenery have longer and more subtle histories than is often assumed.

Most of the more extensive areas conserved in New Zealand have been mountain landscapes, and these have for many years been employed as the face by which the country is projected to the world. In Chapter 9, Eric Pawson explores some of the meanings of mountains, to Māori and to the Pākehā explorers whom they guided; to nineteenth-century artists representing topography with a 'commodifier's gaze'; and to Victorian and early twentieth-century tourists and day-trippers. Many of these people were attracted by mountain magnificence, but generally not with the intimacy that might have led to the creation of an antipodean equivalent of the American wilderness myth. Until the 1950s only a small number of people spent time in the mountains for sport or pleasure. These activities were less male-dominated than mountain discourse has usually implied. Women were active in high climbing and in the urban-based tramping clubs that developed in the 1920s.

Pākehā colonisation of New Zealand coincided with a softening of European attitudes worldwide to mountains: from places that incited fear, mountains were moving to places that could be gazed on or used with pleasure. That the country's conservation estate is so extensive – some 30 per cent of the national land area – reflects the willingness to 'set aside' as national parks areas valued for little else. The estate is therefore quite unrepresentative of other indigenous environments: wetlands in particular are almost wholly absent from the conservation estate. In Chapter 10, Geoff Park portrays the demonising of swamps by Pākehā, reflecting the eighteenth- and nineteenth-century drive to turn these into what he calls 'imperial landscapes', through the desire to convert seeming wasteland to productive lands. Here the 'linear logic' of modernity met and beat 'the chaos of nature'. Drained landscapes are therefore sites of 'amnesia and erasure' of alternative geographies. In the swamp, as in the bush, we can read both sides of the process/progress of transformation.

Part IV contains four essays about the environmental contours of modernity, portrayed here as deserving of understanding in multilayered ways. In Chapter 11, Tom Brooking and Vaughan Wood explore 'the grasslands revolution', that marriage of scientific and technical practices that brought about a concerted intensification of land use and expansion of primary production. It was driven by an 'obsession with the development of grasslands at the expense of other land development strategies'. Although it was important throughout the country, the grasslands revolution was of particular significance in bringing the hill country of the North Island into production. It permitted what are, on the whole, relatively poor soils to underwrite the country's post-World War II prosperity. This chapter discusses the tensions within this progressive view, especially between the differing perspectives of the grassland scientists and the soil scientists, and reveals how, over time, landholders have become more attuned in various ways to the environmental effects of farming. All the same, productivist agriculture and the imagery of 'war' against nature persist (Figure 1.4).

The state was intimately involved in the development of the grasslands revolution. It was also the central player in another complex twentieth-century environmental initiative: that of the 'wise use' of resources such as forests, soil and water. This followed a broader fashion among New World lands, although with clear place-specific characteristics. Wise use strategies in New Zealand drew partly on those voices working to temper the excesses of the grasslands revolution; in part they reflected concern for the future of timber supplies. In Chapter 12, Michael Roche makes the point that for much of the twentieth century the state was therefore the most effective proponent of a kind of conservation, which because of its utilitarianism – designed to protect and maintain production values – then became the target of largely urban-based environmentalists whose interests have lain in fostering amenity and intrinsic values in the landscape. Yet state agencies for many years worked to counter not only resource debasement but also the degradation that unbridled progressivism threatened to wreak.

Above all, however, the twentieth century was an urban century for the majority of New Zealanders. It is in towns and gardens that most have experienced the environment – including Māori, who began to shift out of rural areas in growing numbers in the 1950s and 1960s. Eric Pawson argues in Chapter 13 that despite assumptions to the contrary, towns have always been on the edge of the 'howling wilderness'. Whereas they were founded as bridgeheads of civilisation and assumed many of the attributes of modernity, they have been prone to damage by a suite of environmental hazards such as floods, landslips and – most evident recently – earthquakes. Generally, however, the threats that are more apparent to many urban dwellers have been those produced in towns themselves, such as pollution and congestion. The suburb – where most people live – therefore represents a creative compromise between the unruliness of nature on the one hand and the hazardousness of the town on the other.

The suburb has been intensely consumptive of environmental resources, but it has also been the locus of New Zealanders' most intimate engagements with nature. This is the garden, the subject of Chapter 14 by James Beattie. Today's gardens are full of plants from overseas, and even if we think of them as our own, or as from Europe, many are in fact from Asia. In 'The empire of the rhododendron', Beattie explores the extraordinary mobility of plants, which reaches well back into the nineteenth century, and in particular the extent of the influence of Chinese plants and gardening on the New Zealand environment. He also identifies the importance of Chinese gardeners in providing urban settlements with fresh vegetables. His analysis places New Zealand in wider webs of exchange, and 'the networks of individuals and institutions as well as the technologies and markets' that created these. He thereby shows the inadequacy of Crosby's simpler pattern of 'ecological imperialism' for explaining plant movements.

The four chapters in Part V, 'Perspectives', range widely, each taking a different and fresh approach to contemporary environmental themes and placing these in historical contexts. Each chapter demonstrates how environmental histories have lived meaning in the present. In Chapter 15, Katie Pickles draws on the emerging field of sensory history to 'develop new ways of thinking about and researching the environmental colonial past'. She explores how people apprehend their world through their sensing bodies, using two case studies – of the Canterbury nor'west wind, and the reactions to the Hawke's Bay earthquake of 1931 and the Canterbury earthquakes of 2010 and 2011 – to illustrate this. The senses are 'infused' through Māori knowledge, but have been present only in shadowy ways for Pākeha. Pickles argues that the Canterbury earthquakes – the most prominent environmental event of the last decade – have created a 'postcolonial

Figure 1.4 A landscape of modernity, the neat delineations erasing its origins in the thick bush of South Taranaki, c. 1964. Hedgecutters have been battling to keep the introduced boxthorn hedges in trim. Source: Rodney Hiskens, 'Butler Brothers working on the McGuiness farm, Mangatoki'. PHO2013–0003, Puke Ariki, New Plymouth

moment' in which 'the days of repressing the senses and denying engagement with the local environment are gone'.

Environmental law is the focus of Nicola Wheen's chapter. Her thesis is that the law has always been 'very accommodating' of the promotion of social and economic growth and the protection of associated property rights – as indeed Chapter 6 demonstrates with respect to mining. She identifies the growing politicisation of the environment in the last 40 years as the reason for the gradual extension of legal provision for environmental purposes, including conservation. But central to her argument is that while there have been environmental gains, 'developmentalism' has been persistent. Along with the influence of the Treaty settlements process on environmental law, Wheen shows how libertarian ideology has shaped resource management, marketising resources such as fish quotas and carbon emissions. The central principle of sustainability in the Resource Management Act has again been interpreted in very accommodating ways.

Michael J. Stevens explores the 'nature' of Māori modernity in Chapter 17 through an analysis of the recent history of Ngāi Tahu, the iwi that holds mana whenua over most of the South Island. He shows how water is central to Māori self-definition, and that in this respect the issues and questions faced by Ngāi Tahu are shared by iwi throughout the country. Mahinga kai, primarily the harvesting of uncultivated foodstuffs, depends considerably on waimāori (freshwater lakes and rivers) and waitai (saltwater environments). Through a focus on Canterbury and Southland waters, the chapter demonstrates how colonisation is unfinished business in terms of control of and contamination of these resources. Mahinga kai was a key component of the Ngāi Tahu Treaty claim against the Crown that was settled in the late 1990s. However, Te Rūnanga o Ngāi Tahu, the tribe's mandated iwi authority, has both a social and a business role. It thus now finds itself on both sides of debates about environmental stewardship and development, of which South Island dairy expansion is a clear example.

The final chapter, by Andreas Aagaard Christensen, conceives of New Zealand as a 'history of spaces and of the ability of its inhabitants to cognitively, socially and physically control space and resources'. He discusses the development of the production of spatial knowledge about environments, and explores the close ties between knowledge and practice: that is, between understanding the environment through mapping and metrologies, and changing it. His argument moves from an exploration of how Māori use narrative and performance to conceptualise space, to exposing the power inherent in maps and surveys as Pākehā instruments of commodification and control. He highlights the subsequent shift from the authority of the map to a new way of understanding knowledge through information. Today, modern technologies have enabled land information strategies that have revolutionised accessibility to the data required to manage human–environmental relations.

Conclusion

One of the purposes of this book is to demonstrate that there are many 'environmental histories of New Zealand'. The chapters show how, over time, there have been different interests in the creation of places and landscapes, and different ways of approaching the understanding of change. The human encounter with nature has been marked by change, setbacks, reformulations and sudden ruptures. One tendency within environmental history as a field has been to fragment in the face of such particularity, and study things at the small scale. As Frank Uekoetter has recently observed of northern hemisphere scholarship, 'the field is moving towards a situation where the broad outlines of environmental history are getting buried under the weight of ever more case studies.'[40]

Hence we have set out to amplify 'the broad outlines' with our choice of authors and their topics and approaches. As we suggest, quite intentionally, in Chapter 19, the Epilogue, there are always questions about how things are done and to what ends. But we hope we have managed to live up to the verdict of Richard White on the first edition of this book: that, in its concern 'with the origins, meanings, connections, and consequences of environmental change, and the social, cultural, and economic changes connected to it', it is timely and demonstrates sweep, curiosity and ambition.[41]

Further reading

Tom Brooking & Eric Pawson, *Seeds of Empire: The Environmental Transformation of New Zealand* (I.B. Tauris, London, 2010) is the fruit of a major project in environmental history supported by the Marsden Fund. Two collections of essays edited by Jacinta Ruru, Janet Stephenson & Mick Abbott explore land, place, identity and tension: *Landscape and Identity in Aotearoa New Zealand* (Otago University Press, Dunedin, 2010) and *Making Our Place: Exploring Land-use Tensions in Aotearoa New Zealand* (Otago University Press, Dunedin, 2011). James Beattie places national concerns in a wider imperial context in his *Empire and Environmental Anxiety: Health, Science, Art and Conservation in South Asia and Australasia, 1800–1920* (Palgrave Macmillan, Basingstoke, 2011). The unpredictable nature of the environment is highlighted by Don Garden in *Droughts, Floods and Cyclones: El Niños That Shaped Our Colonial Past* (Australian Scholarly Publishing, North Melbourne, 2009); and Peter Holland, in *Home in the Howling Wilderness: Settlers and Environment in Southern New Zealand* (Auckland University Press, Auckland, 2013) shows how people in one part of the country worked to come to terms with it.

I
Encounters

2.

A fragile plenty: pre-European Māori and the New Zealand environment

Atholl Anderson

THE FIRST SIGNS of the considerable pre-European Māori impact on the New Zealand environment came tumbling out of the Taranaki sand at Te Rangatapu in 1847 – moa bones dug up with great vigour by local Māori under the bemused gaze of amateur palaeontologist Walter Mantell. He was not entirely sure that the bones were from slaughtered birds, but five years later, in North Otago, his suspicions were confirmed. At the place he named Awamoa, his Māori guides, equipped with tent poles and exhibiting such archaeological enthusiasm 'as would make you shudder', recovered not just numerous remains of moa and other extinct birds, but the butchery knives, oven stones and charcoal that put the question of a human association beyond doubt. The disappearance of moa, and of other avifauna, became the first and most enduring topic of debate about anthropogenic change in the New Zealand environment.[1]

Debates and perspectives

For more than a century it remained almost the only topic of significance. The other great issue, deforestation, had been foreshadowed by reports from early travellers and surveyors in the eastern South Island of the remains of former forest strewn among the tussock. The implication of Māori activity in widespread deforestation did not rise to continuing scientific prominence until the mid-twentieth century, and then in rebuttal of a contrary hypothesis. Signs of instability in South Island forest structure had been attributed by forester John Holloway, among others, to climatic changes, beginning about the thirteenth century, that were seen as ushering in cooler, windier and much drier conditions that affected forest regeneration and set a trajectory towards scrubland and grassland in eastern districts. Burning increased commensurately with desiccation, but climatic change was the forcing agent. In an influential review of this argument, geographer Kenneth Cumberland made out a convincing case instead for the primacy of cultural firing. This theory became increasingly accepted as palaeoclimatic data failed to confirm desiccation (indeed somewhat the reverse had occurred, with cooler and wetter conditions prevailing from the late fourteenth century), and as radiocarbon dating documented the coincidence of massive deforestation with initial human colonisation.

Pre-European Māori, it came to be seen, had been involved in the loss of 50 per cent of both the primeval forest area and the late Holocene or postglacial suite of bird species. For some time, the force of this conclusion was blunted by the persistence of an elderly hypothesis that divided the pre-European inhabitants into two populations: early moa hunters and the Māori immigrants who supplanted them in the fourteenth century. Moa hunters, it was argued, had been responsible for the massive environmental changes. In comparison, although Māori increased tenfold in population, 'nowhere did [they] cause a wholesale transformation of the environment or a disastrous disturbance of the ecosystem'. When it became apparent that neither Māori historical tradition nor archaeological evidence supported the idea of a separate earlier population, the historical reality was clearly exposed: Māori alone were responsible.[2]

Reactions to this realisation have varied. One that commands broad popular appeal is to compress Māori environmental behaviour into an ethnographic present defined by historical snapshots of rāhui and other forms of resource management, to which early faunal extinction and deforestation are distant and minor aberrations. Another, almost the opposite, has proposed that Māori were destructive from the beginning, shattering not only the natural environment but also an earlier population, the 'Nation of Waitaha', a multiracial (including European) community of gentle gardeners steeped in ecological knowledge and reverence. The latter is merely a 'new age' fantasy that has coopted a genuine tribal name, but it has served to promote a resentful view of pre-European Māori as environmental mismanagers whose descendants are now the undeserving recipients of state largesse in land and resources, delivered on recommendations of the Waitangi Tribunal.[3]

A third view, which underlies the discussion in this chapter, has developed mainly among archaeologists and ecologists. It originates in evolutionary ecology, notably in the economics of consumer choice, and proposes that within their social rules of resource ownership and use, pre-European Māori operated as optimal foragers, exploiting natural commodities in ways that expended the least effort for the greatest return to promote the comfort and development of individuals and lineages, without consideration for distant communities or the sustainability of any particular resource. Consequently, resources diminished or disappeared in proportion to their desirability and their vulnerability to exploitation.[4] It is axiomatic, in this view of the past, that no substantial, desirable and easily gathered resource, such as moa or seals, would be left untouched for centuries while less efficient subsistence activities were pursued. However, just that argument has been implied as a means of reconciling evidence of moa hunting, beginning in the twelfth or early thirteenth century, with speculation about much earlier human settlement. A small, slow-growing and mainly horticultural population in northern New Zealand, treading environmentally with the delicacy of the mythical Waitaha, is envisaged by some archaeologists as preceding the eventual development of a focus on big-game hunting and other activities in the south.[5]

Such contrasting scenarios of ecological adaptation underscore three points about the delivery of the pre-European impact on the New Zealand environment. First, much depends on when the exploitation began. If the loss of forest and megafauna occurred

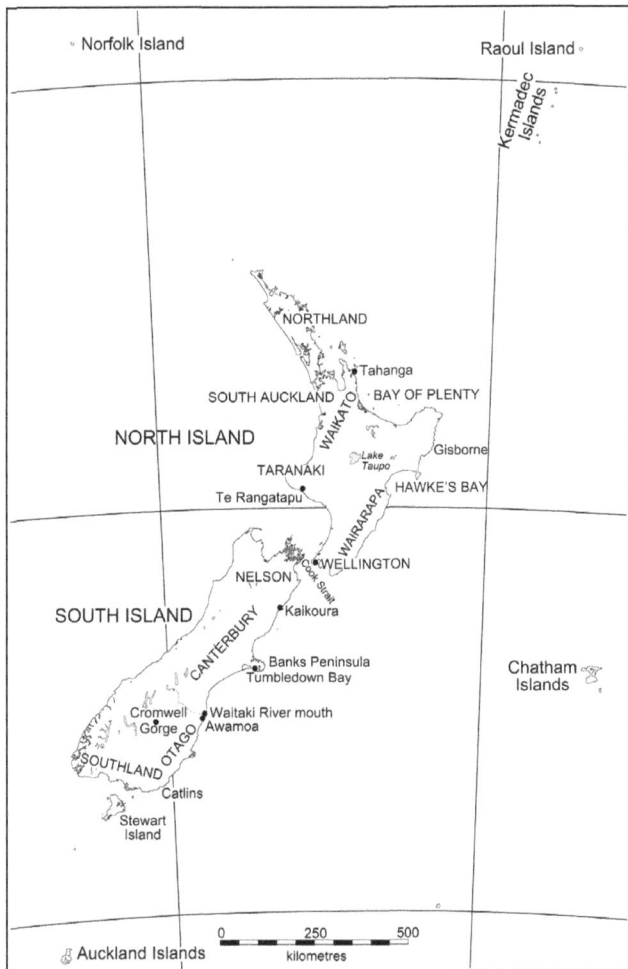

Figure 2.1 Map of places and districts mentioned in the text.

after long occupation had produced conditions of population growth that rapidly accelerated the rate of environmental change, then different processes were at work than if those same changes were the casualties of the first, careless, rapture of overconsumption by newly arrived migrants. Second, the environmental encounter ran in both directions. The growth of Māori society from Polynesian roots owed as much to the constraints and opportunities of the New Zealand environment as the latter was changed by it. Third, considerable environmental variation in New Zealand implies that the land and biota is regionally vulnerable to cultural interference. Out of these elements a broad set of ecological relationships is drawn in order to sketch a hypothesis of socioeconomic change in pre-European New Zealand. The localities mentioned in the text are shown in Figure 2.1.

The chronology of colonisation

Three competing hypotheses of initial colonisation in New Zealand have been developed: a long prehistory beginning about 2000 years ago; an orthodox prehistory beginning 1200–1000 years ago; and a short prehistory beginning 600–800 years ago. The idea of a long prehistory goes back more than 150 years, to conjecture by Mantell and, later, the German geologist Julius von Haast, who was the first Director of the Canterbury Museum, but it has been revived recently in arguments that exploit the methodological vulnerability of sedimentary and pollen analyses, and most especially those that appeal to evidence of early rat colonisation.

In the case of the small Pacific rat, *Rattus exulans*, which was carried – deliberately or stowed away – on large voyaging canoes sailed by the earliest colonists throughout the remote Pacific, it is argued that the people who introduced it either departed more or less immediately or died out, leaving no trace of their settlement. Since rats bred at spectacular rates on islands that previously had no mammalian predators, such as those of New Zealand, they soon became prominent among the prey items of native birds. The earliest radiocarbon dates on the bones of rats from predator roosting sites should provide, therefore, a reasonable proxy for the age of initial human colonisation. Importantly, the accuracy of the method could be tested by comparing dates on rat bone from the later archaeological sites with numerous radiocarbon dates on other materials, such as charcoal, from the same stratigraphic provenances.

Straightforward as the method seems, it ran immediately into controversy. Early results from predator sites suggested that rats had been introduced to the South Island by at least 150 BC, and to the North Island by about AD 200. With East Polynesia probably not settled before AD 900, the first people to reach New Zealand would therefore have been West Polynesian or Melanesian, and not directly ancestral to Māori, who have East Polynesian genetics, language and culture. In addition, the existence of rats in the New Zealand environment for three times longer than previously thought must have had a substantial ecological impact on some plants, many invertebrates, and a wide range of smaller vertebrates. With such profound implications at stake it was important to be sure that the results were accurate.

Comparison of radiocarbon dates on rat bone from predator sites with those from archaeological sites showed a common and peculiar trend: in both cases the dates produced at the Rafter Laboratory, Lower Hutt, were substantially older in samples processed before 1997 than after. As all the archaeological results were on samples from sites that had been independently dated on many other materials to no older than 800 years ago, and as the predator-site results mirrored the archaeological distribution, it was apparent that the quality of radiocarbon dating on rat bone samples had improved with laboratory experience, eventually to produce results exclusively within the expected age range, less than 800 years old (Figure 2.2). Subsequent research has shown conclusively that there is no evidence of rats reaching New Zealand before about 700 years ago.[6]

The pre-European history of vegetation change is quite complex and, in some respects, open to debate. Analyses of pollen spectra, and charcoal abundance in sedimentary cores, enable palynologists to argue that certain kinds and scales of change – typically extensive, unreversed deforestation accompanied by charcoal levels indicative of prolonged burning – were probably anthropogenic in situations where people were present. Reverse that argument, however, and ask how much and what kind of vegetation change could indicate the presence of people, where that is otherwise undocumented, and a difficulty emerges; for clearly almost any kind or scale of change in sedimentary or vegetational conditions *could* be anthropogenic. Given that uncertainty, it has been proposed that brief episodes of charcoal influx or vegetational change extending back several thousand years might represent the presence of a human population that remains unnoticed in other kinds of evidence.

Although lacking any archaeological support, this proposition cannot be ruled out entirely, if only on the logical ground that in appealing to a scarcity of evidence it can hardly be tested; but it is improbable at best and the arguments against it are strong. Vegetational and charcoal perturbations of the same kind are evident in sedimentary records extending into the Pleistocene, long before there were any people in the remote Pacific: even tiny human populations can have a marked environmental effect by fire; it is vanishingly improbable that early settlers would ignore the big game, including moa, for more than a thousand years and then fall on it voraciously; and so on.

Compelling evidence of deforestation associated with fire generally begins after AD 1300 throughout New Zealand. In a few cases, the point of continuous change from

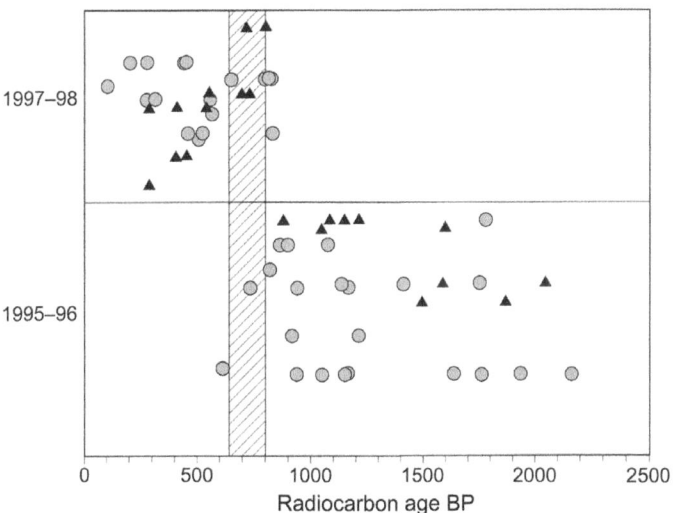

Figure 2.2 The distribution of Rafter Laboratory radiocarbon dates on rat bone. Circles are predator-site samples; triangles are archaeological-site samples. Horizontal axis shows years before present (BP); vertical axis the two periods of sample processing. Vertical shading indicates the expected age of initial rat colonisation. Courtesy of Janet Wilmshurst, Landcare Research

forest to scrub, fern or grass has radiocarbon dates to the first millennium AD, but this seems to reflect difficulties in dating. An important analysis of New Zealand pollen records by palynologists Matt McGlone and Janet Wilmshurst shows that whereas radiocarbon dates on the inception of continuous deforestation recorded in peat-bog strata are characteristically late (AD 1300–1400), much earlier dates on the same horizon came from sediments recovered in lakes and swamps that are susceptible to inwashing of old carbon. Dating of offshore pollen cores on samples from marine organisms that lived in deep-sea carbon reservoirs is similarly open to question. Leaving aside such problematic results, the earliest evidence that might be interpreted as anthropogenic is correlated in 11 pollen records from the northern North Island with deposition of the Kaharoa ash, which is dated to about 600 years ago; and similar ages for widespread deforestation occur in South Island pollen cores.[7]

The direct evidence of archaeology offers no support to speculation about early colonisation. Substantial programmes of archaeological research have failed to locate evidence of any kind that can be dated to earlier than the twelfth or thirteenth centuries. In particular, large-scale analyses of archaeological radiocarbon ages produced by the Rafter Radiocarbon Laboratory and the Waikato Radiocarbon Laboratory, and a wide-ranging study of the calibrated ages, have found no evidence of human habitation before 750 years ago.[8]

On less data, the same story can be read in the records from the outlying islands. The Kermadec Islands were colonised 650 years ago, Norfolk Island perhaps 700 years ago, and the Chatham Islands by at least 450 years ago – although probably somewhat earlier than that, considering the Archaic form of some of the artefacts that have been recovered. Determined voyagers searching in all directions from New Zealand reached even the subantarctic Auckland Islands by 650 years ago, although they did not survive long there.

The rapid discovery of all the outlying islands suggests a substantial early population size, certainly more than the crew of one canoe; and so does the transfer to New Zealand of not just the material supports of a Polynesian existence in plants, animals and technology, but also a rich cross-section of the social structure, customs, legends and traditions. Research on Māori mitochondrial DNA sequences has come to a similar conclusion. An original population 800 years ago would have required at least 50 to 100 women to obtain its genetic diversity, and the retention of rare alleles (the alternative forms of a gene) indicates that the founding population must have been even larger if colonisation occurred earlier than is currently thought.[9] While the evidence of Māori tradition must be approached cautiously, it too suggests that a relatively large founding population reached New Zealand within a fairly brief period. Analysis of whakapapa (genealogies) shows that the preponderance of lines from ancestors who arrived in colonisation canoes is in the range 17–22 generations long up to the early nineteenth century. Calibrated according to modern data on generation length, 19 generations at 29.5 years would imply arrival about AD 1290, which is close to timing by archaeological evidence.[10]

Therefore, it appears that 'South Polynesia' – New Zealand and the outlying islands – was not reached by people until the twelfth or thirteenth century, about the same time that people first reached other islands around the margins of eastern Polynesia, notably Hawai'i and Easter Island.[11] A recent review of the various sources of evidence concluded that 'no human presence in New Zealand is detectable prior to ca. 1250–1300 AD ... Our results strongly accord with the late colonisation, or "short" prehistory model.'[12] The founding population was quite large, probably several hundred people at least, implying the arrival of a number of canoes, although not necessarily together. The early archaeological sites, and other contemporary sources of evidence, thus document not a middle phase of the encounter between Māori and the land, reflecting either a deliberate strategic shift in settlement patterns and economic objectives, nor simply an expedient extension southward of big-game hunting and deforestation driven by population pressure; rather, the evidence illustrates an overriding will by initial migrants to survive and grow in an environment that lay quite beyond their previous experience.

Environmental impact on Māori

The dialectic of human–environmental influence – never a one-way process – opened with the cultural appropriation of the New Zealand environment, marked, we can assume, by the naming of things. This included naming the major topographical features in ways that installed tropical Polynesian claims to ownership. The North Island, for example, became 'Te Ika a Māui', the captive fish of Māui. It also included describing the new biota in familiar terms. The northern palm became 'nīkau' (the coconut frond), recalling the coconut palm; the Polynesian fibre plant *Pandanus* or 'fara' was attached to flax as 'harakeke' (strong pandanus); the New Zealand 'pāua' was named for the black-lipped, iridescent green-eyed pāua or giant clam (*Tridacna* spp.); the barracouta was called 'mangā' after the tropical snake mackerel; 'hāpuku' recalls the coral grouper; and the tiger shark, or 'tanifa', proliferated into a multitude of dangerous water monsters or 'taniwha'. Parrots remained as 'kākā', owls as 'ruru', rails as 'weka', and long-billed waders added to their name the shore-foraging 'kiwi'.[13]

Much of tropical Polynesian technology was able to be transferred directly into exploitation of the New Zealand environment. Adze and blade manufacture, developed on oceanic basalts, translated into the crafting of almost identical tools from Tahanga basalt and Nelson–Marlborough metamorphosed argillite. The bait-hook and trolling-lure kit of tropical Polynesia, deprived of the traditional *Turbo* and pearl shell, found approximate facsimiles in the Cook's turban and the mussel shell. Large bone – hardly available in the tropics – was a bonus, and an abundant source of moa bone enabled a proliferation in the shape and size of hooks and many other small artefacts. The tropical fowling spears, lures and snares worked just as well in New Zealand to take a much wider size range of prey, including moa. Prolonged cooking and soaking to remove the poisons in karaka (*Corynocarpus laevigatus*) and tawa (*Beilschmiedia tawa*) kernels, and to produce sugar from tī (*Cordyline* spp.), were among various ancestral techniques transferred to the processing of New Zealand plant foods.

Yet New Zealand in truth was hardly like anywhere else in Polynesia at all. Massive, cloudy and cold, it stretched to the limit the environmental adaptability of the first settlers. The temperate forest, evolved in the absence of mammals, was poor in large fruits and oil-rich seeds, while the generally humid climate precluded much compensatory development of tubers or starch-rich seeds. Dense forest was consequently a substantial obstacle to human existence. The bulging 'portmanteau biota' that facilitated accessibility to colonisation in the tropics was, in New Zealand, stripped – primarily by climate – of most of its more important contents. No coconut, banana, breadfruit, pandanus, pig or chicken survived, and of the suite of domesticates that did (sweet potato, taro, yam, gourd, tropical *Cordyline*, paper mulberry, dog, rat), only the relatively hardy sweet potato (kūmara) and the faithful but tasty dog (kurī), became substantial aids to subsistence. Tropical horticulture was adapted to the temperate climate by selection of fast-maturing types, the addition of heat-retaining gravel to soils and winter storage of tubers in pits, but even the hardiest kūmara needed a minimum of five to six months to mature, which confined its range to lowland regions north of Banks Peninsula, and its economic prominence to the northern two-thirds of the North Island.

The temperate seasonality in climate was sharpened to almost continental severity in many inland areas, and that, coupled with the rapid succession of frontal systems that characterise the New Zealand climate, reduced the choice of ideal habitats to a few coastal districts, of which Gisborne and Nelson were the warmest overall. But considerations of comfort had to be balanced by those of survival, so the distribution of native resources also moulded the settlement patterns of Māori from the beginning.[14]

One way of looking at this – and it is one with which Polynesians from high islands were profoundly familiar – is the windward–leeward cline. Like tradewinds in the tropics, the dominant westerlies in New Zealand produce contrasting patterns of vegetation and terrestrial life. West to east variations in rainfall (and some rain-shadow reversals) are cross-cut by variations in temperature on islands spread across 13 degrees of latitude. Consequently, the contours of variation are offset to a northwest–southeast pattern (Figure 2.3). At the time of first settlement of New Zealand the mainly northwest windward province was characterised by heavy, humid forest, much of it, in the north, developed on soils of volcanic origin. The great trees were rimu, tōtara, tawa and kauri, and the forest was filled with broadleaved trees and shrubs. In the mainly southeast leeward province, there was a more open forest dominated by mataī, kahikatea and tōtara, with beech forest in the mountains and at the eastern edges. In a few districts, such as the Canterbury Plains and inland Otago, there was a low forest of toatoa, kānuka and kōwhai around open basins covered in light scrub and tussock.

The main biomass of large vertebrates was concentrated in the mosaic vegetation of the leeward province. They comprised, above all, the nine species of moa (Dinornithiformes) – wingless, browsing birds ranging in weight from 25 to 250 kilograms.[15] In addition, there were large flightless geese and a very large rail-like bird, *Aptornis*; a swan; various species of ducks and parrots; and, among other raptors, an immense moa-hunting eagle (Haast's eagle, *Harpagornis moorei*). The absence of turtles was more than made up for by an abundance of the New Zealand fur seal and the sea lion: both of these were most prolific towards the south, although breeding colonies of the fur seal, at least, extended to Northland. Plant foods also were probably more accessible initially in the

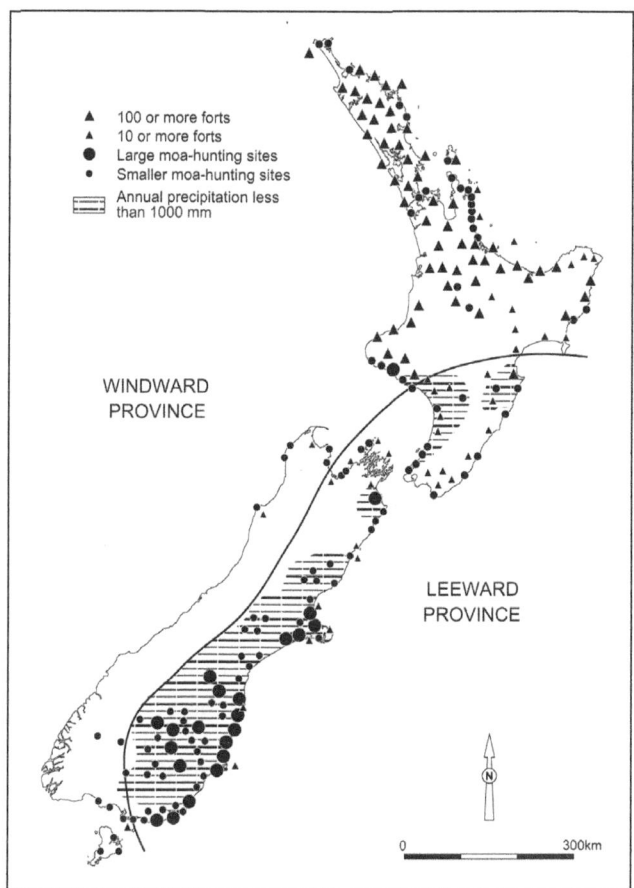

Figure 2.3 Windward and leeward provinces in New Zealand, showing Māori forts (pā) and moa-hunting sites; also the correlation of pā largely with the windward and of moa-hunting sites mainly with the leeward province. Source: author

leeward province, because the most useful of them, such as bracken fernroot (*Pteridium esculentum*) and the roots and stems of young tī, or cabbage trees, were products of the forest edge and open ground.

The distribution of early archaeological sites is not an infallible guide to the way in which the New Zealand environment influenced initial Māori settlement patterns, because it favours the more visible sites of big-game hunting over those where activities that left less obvious traces – such as plant cultivation or processing – may have been carried out. However, given that early bone and shell middens represent the universal foraging aspect of coastal settlement, it is apparent from their size and distribution that many of the colonists favoured the leeward province. In this region of relatively plentiful resources the insular fragility to human intervention of the New Zealand biota was first revealed.[16]

Vertebrate extinctions

The role of people in faunal extinctions is seldom easily documented. In most cases, unavoidable comparison of data from natural sites, where bone accumulated gradually throughout several thousand years or longer, with data from archaeological sites representing selective hunting trips over much less than a century, raises problems of sampling variation that make it virtually impossible to determine whether a species disappeared only after the arrival of people – let alone whether human activity was the critical factor in extinction.

Expanding on the hypothesis of early rat introductions, and assuming no human survival, zoologist Richard Holdaway has proposed that 'many of the extinctions and extirpations occurred when the Pacific rat was the only widespread novel factor in the environment'. For example, he suggests that small petrels, such as the diving petrel *Pelecanoides urinatrix*, were in decline on the mainland through rat predation before continuous Polynesian settlement began. However, this species is well represented in mainland archaeological sites, considering that small petrels lay outside the preferred size range of Māori fowling. Besides, variations in archaeological recovery of small bones subvert any attempt to define underrepresentation of species. Only presence or absence across a wide range of sites constitutes reasonably strong evidence. The data there are quite clear. As palaeontologist Trevor Worthy points out, remains of all prehistorically extinct species of late Holocene birds have been found in at least one archaeological site, apart from two tiny wrens and two recently recognised species (the moa *Pachyornis australis* and a petrel *Puffinus spelaeus*), which are expected to be found in re-examination of archaeological collections. In other words, there is no documented extinction of any New Zealand avian species (indeed of any vertebrate) in the millennium prior to demonstrable Polynesian colonisation.[17]

It was the arrival of people – and rats and dogs – in the thirteenth century or slightly earlier that initiated the great assault on the New Zealand fauna. Much of this remains poorly documented; and it can be assumed that, were the data available, various invertebrate species would be numbered among the extinct. Evolved in the absence of terrestrial mammals, many of them were comparatively large and flightless, just as in the avifauna, and had become 'invertebrate mice' and other small mammal surrogates. They were immensely vulnerable to the introduction of rats, and to forest burning. Much the same is doubtless true of the New Zealand lizards and frogs, several species of which are known to have disappeared in the pre-European era; and others became greatly reduced in range, such as the tuatara, which had been widespread on mainland New Zealand during the late Holocene.

The impact on the avifauna is much better known. The factors of extinction are rather variable, but common factors are: habitats such as light forest, scrubland, or waterside shrubbery that were easily accessible to people and often easily burnt, large body size, flightlessness, ground-nesting, diurnal habits and low breeding rates. Some species, such as the pelican and swan, were vulnerable to any form of human disturbance, especially at their nesting sites. Small ground-nesting birds generally, such as snipe, snipe-rail and wrens, were most at risk of direct predation from 'a grey tide' of rapidly expanding rat populations. Larger flightless or poorly flighted geese, *Aptornis*, and various ducks faced rat predation on eggs, competition with rats for some food sources, disturbance by dogs, habitat destruction and direct human predation, while the moa were largely hunted into extinction, and their natural predator, the great New Zealand eagle, lost its means of existence.[18]

There is general agreement that moa extinction occurred rapidly, but perhaps not as quickly as a century or less after human colonisation, as recent zoological modelling suggests. Predator–prey models oversimplify the complexity of cultural systems within which moa hunting took place. Early colonists, it can be assumed, sought to maintain community integrity, diversity of resources, and external exchange by locating year-round settlements along the coast. These sustained continual predation on local moa populations, but those in less accessible interior districts were probably hunted relatively seldom. Coastal moa populations had probably crashed by the late fourteenth century, especially in the most accessible areas such as the Canterbury Plains, but some archaeological data suggest that they survived rather longer elsewhere. A careful analysis of radiocarbon dates on moa bone by Fiona Petchey, a bone-dating specialist, shows that at least 14 results spread across two North Island and three South Island sites had ranges extending into the fifteenth century and several into the sixteenth century or later; for instance, at Tumbledown Bay on Banks Peninsula, where a moa bone dated in the range late fifteenth to late seventeenth century. None of these data are sufficiently precise to put a date to moa extinction, but they do indicate that while early slaughter was devastating, there were still moa to be found by the fifteenth century, if not later, although survival much beyond AD 1500 seems improbable.[19]

Extinction during the pre-European Māori era claimed 35 species of birds (30 in mainland New Zealand and five in the Chathams), a bat, three to five species of frogs, an unknown number of lizard taxa, and doubtless many invertebrates adapted to dry forest habitats. About 50 per cent of the bird species breeding on the mainland went extinct, most of them predominant in the leeward province. Furthermore, it was mainly the larger species that were lost, either because they had been especially targeted, as seems to have been the case with moa (which are represented by the bones of thousands of individuals in some sites, such as the Waitaki River mouth) or because a generalised hunting strategy impacted more severely on species that were conspicuous but relatively scarce.

The loss of moa was especially important because of the substantial biomass that they represented. This cannot be known precisely: suggested figures range from one to about 30 medium-sized moa per square kilometre, the latter figure close to modern high-country stocking rates for cattle and sheep. In the best conditions, such as in parts of the eastern South Island, this biomass is not improbable. Even at much lower rates overall, the disappearance of moa significantly lowered the human-carrying capacity of forest and open country, while the reduction of fowling opportunities to small birds such as kākā, weka, tūī, and kererū (woodpigeon) completely removed the early comparative advantage of the leeward province in terrestrial protein productivity.

As moa and other terrestrial vertebrates were heading for extinction, New Zealand seals were also declining under human predation. The breeding range of the most common species, the New Zealand fur seal, marked by pup bone in archaeological sites, retreated southward from the Northland coast at AD 1300, to Cook Strait by about AD 1600 and to East Otago by AD 1800. The sea lion breeding range contracted at a similar rate.[20]

The Māori encounter with the land- and shore-breeding New Zealand fauna could hardly have turned out otherwise. The large species had conservative life histories, being long-lived and slow-breeding: moa laid only one or two eggs at a time; seals bore a single pup. This is an efficient long-term strategy for survival in the presence of natural sources of mortality, but it has very little margin for error. The arrival of a superpredator, flanked by dangerous commensals, all with broadly omnivorous habits that released them from the usual swings of predator–prey abundance, that took eggs, chicks, and pups as well as adults, and destroyed vital aspects of the prey ecosystem, could have only one outcome.

Deforestation and its consequences

Having evolved under very low rates of natural ignition, most New Zealand trees and shrubs are killed outright by fire: only a few species, such as mānuka and tī (cabbage tree), are able to withstand it. As the majority of tree species are lowland and slow-growing, it takes centuries for full successional recovery. Tussock grassland, however, profits from frequent firing, which may increase growth, flowering and seed germination. Bracken fern also profits from frequent burning, which eliminates competing species of regenerating shrubs and trees. Consequently, increasing rates of forest firing led to an expansion of fernland and tussock grassland, dotted with cabbage trees and fast-growing shrubs.

It is impossible to say whether these were the desired effects of forest firing, or simply the useful byproducts of it. It is very unlikely that fires had much to do with moa hunting, for which snares set on forest or shrubland trails were probably the most effective means. In northern areas, there would have been some deliberate clearance for horticulture, and it is possible that Māori also burnt forest to facilitate cross-country travel. Underlying these possibilities, however, is the important consideration that closed forest was relatively barren of resources, so that its substantial removal, either deliberately or accidentally, did not represent a material disadvantage. The synchronisation of widespread burning throughout much of New Zealand might suggest either that Māori were indifferent to the substantial loss of forest, or perhaps only that they had no control over the impact of an increased incidence of ignition.

Variations within the general trend of fire-induced vegetation change reflected differences in climate – as had earlier changes based on natural burning and related causes. As the late Holocene climate became cooler and more variable, fire affected some forest in the interior basins of the South Island about 1500–2000 years ago, but devastation in the central North Island wrought by the great Taupo eruption, AD 150, was followed by sustained recovery into a closed forest again after several hundred years.

Māori fires also emphasised regional differentiation. Lower rates of widespread conflagration coupled with relatively rapid regrowth in the humid windward province slowed and fragmented the advance of deforestation, while fires in the drought-prone leeward province often destroyed large areas of forest which, being relatively slow to regenerate, were thereafter suppressed by continued burning. It has been estimated from radiocarbon dates on charcoal that fires occurred in some parts of the North Island at about 80 years apart during the pre-European Māori era. In the eastern South Island, by contrast, a fire interval of 200 years in the mid-Holocene had dropped to about 50 years as late Holocene dry conditions took hold, then dropped again to only 10 years at about 700 years ago, as human settlement began.

The effect was to virtually eliminate the forests of the eastern South Island in a fairly brief pulse of burning, AD 1300–1450, leaving substantial stands only in parts of Southland, the Catlins, Banks Peninsula and some of the eastern ranges, including the Kaikouras. Forest was also burnt out of the lower east coast of the North Island and Hawke's Bay. Throughout the leeward region, tussock grassland was the common result, with fernland prevailing towards the north. At the same time, substantial areas of the windward province (central North Island, inland Waikato, Bay of Plenty, South Auckland and parts of Northland) were reduced to scrubland and fernland, and although the incidence of major fires was relatively low, the incidence of localised burning must have been much higher, perhaps three to five years, to maintain fernlands. Extensive forest remained in the windward province, and some large stands elsewhere, but 'the forests of New Zealand, lacking resistance to fire, slow-growing and without substantial food resources [for people] were doomed from the moment of first human settlement'.[21]

A cascade of ecological consequences ensued from deforestation and the loss of the large browsing birds. As zoologist Graeme Caughley has remarked, 'after AD 1400 the major New Zealand grazing systems ceased to exist … the Polynesians did not just eliminate the moas, they eliminated an ecological and evolutionary process developed over more than 50 million years.' The dispersal of large-fruited seeds (which had relied substantially on moa and other large birds) declined; and fast-growing shrubs (which were no longer browsed) suppressed the growth of podocarp seedlings, perhaps causing a regeneration gap by AD 1400. Such effects, together with the selective loss of the dry lowland forest types, substantially reduced the pre-human forest pattern to its windward and montane types, producing a proportionate increase in beech forest, a relatively barren habitat for birds and people alike.

In the windward province, where fern and scrub colonised the former forest soils, rates of erosion were generally low, although then as now severe episodes of storminess produced substantial slipping in steep hill country. In the leeward province, where tussock grassland was the predominant post-forest cover, rates of soil erosion increased by an order of magnitude. In Canterbury, for example, about three-quarters of radiocarbon dates on charcoal taken from the interface between buried soils and overlying, transported sediments are in the range 500–1000 years ago.

Deforestation and the remobilisation of sediments and soil nutrients were, however, generally beneficial from the perspective of increasing human settlement. Cleared land, especially on rich volcanic soils or alluvium, was needed for gardening and, just as importantly, the better soils also promoted high-quality bracken fernroot, probably the carbohydrate staple of Māori existence in the windward province. Forest edges and open ground also supported the valuable tī, the main carbohydrate source in the leeward province, as well as tutu (a source of fruit juice), wineberry, pōhuehue, flax and other useful plants. Opening up closed forest encouraged many small bird species such as weka and quail, and enhanced light and nutrient levels in lakes and waterways increased production of eels and other freshwater fish.[22]

Adaptive change

The broad imperative of environmental change in pre-European New Zealand was to favour a mid-sequence shift in population growth away from the faunally depleted and southerly leeward province towards the plant-favoured and northerly windward province, a process that was perhaps reinforced by climatic trends. Colonisation, AD 1250–1350, had occurred in generally dry and mild conditions. Conditions became wetter and cooler up to about AD 1500, after which there was a rapid decline into a cold, very wet, and cloudy climate that persisted until the eighteenth century.[23] The historical course of socioeconomic development can be characterised from several perspectives, including those of demographic and societal change, but one that is appropriate here is a simple ecological hypothesis that can be illustrated in archaeological terms by the distribution of moa hunting versus defensive sites (Figures 2.4 and 2.5).

The nature of the pre-European Māori environmental impact was, in its early stages, almost certainly typical of colonisation everywhere and at all times. Migration into new environments releases a powerful instinct to expand as rapidly as possible, using the richest resources with pitiless energetic efficiency. Evolutionary fitness drives lineage competition in the use of unowned resources towards levels of overexploitation described as the 'tragedy of the commons'. In New Zealand this was played out most visibly between the twelfth and fifteenth centuries in the leeward province, although the loss of moa and seals in the windward province shows that it also occurred there.

Once terrestrial protein resources were substantially reduced to a fine-grained uniformity of small birds, fish and shellfish, the means of continuing population growth lay in the gardens and fernlands of the windward province. These combined high productivity of carbohydrates – in an environment where they were otherwise threateningly scarce – with strong spatial definition: everybody could see where they were. Around them – and pulling in nearby resource zones of river, lake, coast and forest – there developed a proprietary sense of boundaries. This became reinforced with population growth, and with increasing climatic circumscription northward of land able to support horticulture, into territorial ownership and defence. It is from there but a small step to the systematic regulation of resource use by allocation among related groups, and seasonal proscription on exploitation. How early this began is hard to say, but since construction of pā goes back to the early sixteenth century, it is possible that such

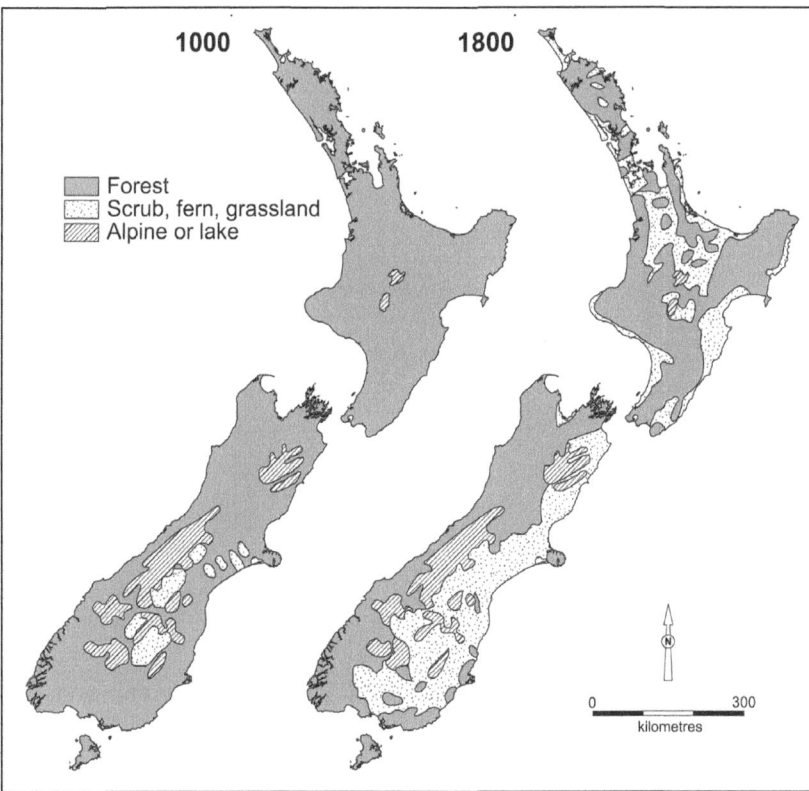

Figure 2.4 Distribution of forest and open country at approximately AD 1000 (left) and AD 1800 (right). Source: modified after M.S. McGlone, 'The Polynesian Settlement of New Zealand in Relation to Environmental and Biotic Changes', *New Zealand Journal of Ecology*, vol. 12 (supplement), 1989, pp. 115–29

regulation was already linked to those developing notions of resource monitoring and conservation that underpin modern Māori perceptions of environmental management.

Continuing population growth and territoriality in the windward province initiated, late in the sequence, a movement of some lineages into the relatively underpopulated leeward province, first in Hawke's Bay and Wairarapa, then Wellington and across to the South Island, in a series of migrations that was continuing as Europeans arrived in the late eighteenth century.[24]

Conclusions

New Zealand was the last major landmass in the world to be colonised by people before the modern era, and its record of prehistoric human–environment relationships is correspondingly diverse and well preserved. It describes colonisation about 800 years

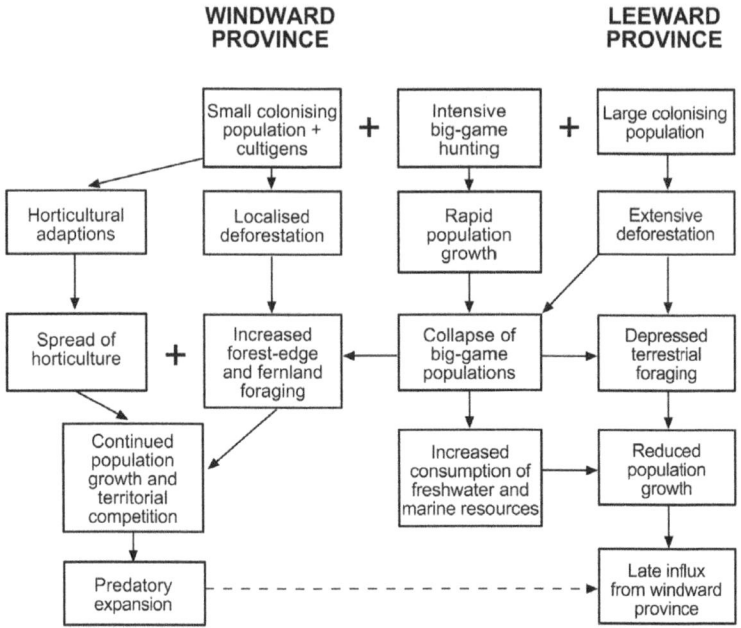

Figure 2.5 A regionally differentiated ecological model of socioeconomic change in pre-European New Zealand. Source: author

ago by East Polynesians, who brought the dog and rat plus a suite of domestic plants. Their subsequent environmental encounter focused initially on the megafauna, most abundant in the relatively dry southeast, and later on fishing, fernroot collection and agriculture, especially in the humid northwest region, where resource competition driven by population growth in deteriorating climatic conditions sharpened territoriality and led to warfare. Sustained predation drove half of the bird species into extinction, and levels of cultural burning devastated the vulnerable forest. Similar changes occurred elsewhere in the Pacific Islands, in Australia and in the Americas. They are characteristic of the entry of people into previously uninhabited and fragile environments.

Further reading

The latest compendium of research results in New Zealand archaeology is Louise Furey & Simon Holdaway (eds), *Change Through Time: 50 Years of New Zealand Archaeology* (Monograph 26, New Zealand Archaeological Association, Auckland, 2004). On pā sites see also Douglas Sutton, Louise Furey & Yvonne Marshall, *The Archaeology of Pouerua* (Auckland University Press, Auckland, 2003); and on the archaeology of undefended settlements, Geoffrey Irwin (ed.), *Kohika: The Archaeology of a Late Māori Lake Village in the Ngāti Awa Rohe, Bay of Plenty, New Zealand* (Auckland University Press, Auckland, 2004). Relevant trends in historical ecology have been discussed by Atholl Anderson, 'Epilogue: Changing Archaeological Perspectives on Historical Ecology in the Pacific Islands' (*Pacific*

Science, vol. 63, no. 4, 2009, pp. 747–57). There is recent discussion of Oceanic demography in Patrick V. Kirch & Jean-Louis Rallu (eds), *The Growth and Collapse of Pacific Island Societies: Archaeological and Demographic Perspectives* (University of Hawaii Press, Honolulu, 2007), and on regional climate change: Andrew Lorrey, Anthony M. Fowler & Jim Salinger, 'Regional Climate Regime Classification as a Qualitative Tool for Interpreting Multi-proxy Paleoclimate Data Spatial Patterns' (*Palaeogeography, Palaeoclimatology, Palaeoecology*, vol. 253, nos. 3–4, 2007, pp. 407–33).

3.

Contesting resources: Māori, Pākehā and a tenurial revolution

Evelyn Stokes

IN MĀORI COSMOGONY the world people live in was shaped by the separation of the original parents, Ranginui, the Sky Father, and Papatūānuku, the Earth Mother. Their children were liberated from Te Pō, the darkness that enveloped them, into Te Ao, the world of light. Some, such as Tāwhirimātea, who was affiliated to atmospheric events, chose to live with Ranginui. Tangaroa chose the sea for his domain. Tānemahuta, who was the prime mover in the separation of his parents, was identified with the forests. Rūaumoko, the youngest and still suckling at his mother's breast, remained with Papatūānuku and is associated with earthquakes and volcanoes.

Māori people are the descendants of the union of Tāne with Hineahuone, the woman created out of earth. Therefore Māori saw themselves as part of their environment, at one with it, not dominating it. This relationship was an intensely practical one of using the resources of land and sea for daily sustenance, but was also deeply spiritual, involving recognition and propitiation of ancestor gods. Such ideas do not always sit easily with Pākehā, accustomed to individual property rights and concepts derived from Judaeo-Christian tradition, and the divine command to Adam and Eve, 'Go out and subdue the earth'.

In the *Muriwhenua Land Report* the Waitangi Tribunal noted:

> The fundamental purpose of Maori law was to maintain appropriate relationships of people to their environment, their history and each other ... Maori law described how people should relate to ancestors as the upholders of old values, to the demigods of the environment as providers of life's necessities, to their hapu, which was the primary support system, and to other peoples as necessary for co-existence ... Maori saw themselves as users of the land rather than its owners ... They were born out of it, for the land was Papatuanuku, the mother earth who conceived the ancestors of the Maori people. Similarly, whenua, or land, meant also the placenta, and the people were tangata whenua, which term captured the view that they came from the earth's womb ... That land descends from ancestors is pivotal to understanding the Maori land-tenure system.[1]

A Māori person's identity is thus linked to both place and ancestry. The tenure rights of a Māori community equated with occupation 'from time immemorial', or at least over many generations – ahi kā, keeping the fires burning on the land. Places were identified with deeds of ancestors, frequently recalled in local placenames, and detailed knowledge of the landscape and resources of the ancestral estate. There was also imbued in this world view a sense of custodial occupation, that the environment should be maintained in a fit state for the generations to come.

Māori customary tenure comprised a complex system of overlapping and interlocking usufructuary rights; that is, rights of use and occupation, but with no right of alienation except in very special circumstances sanctioned by the community. Such a situation might be the allocation of rights to some land or resources on the occasion of a strategic marriage to cement a diplomatic alliance, or to redress an imbalance in relationships (utu), or a form of ceremonial gift exchange of valued resources. Conquest did not provide a right unless it was followed by occupation, usually accompanied by strategic marriages with tangata whenua.

Land rights were thus inextricably bound up into the networks of kinship, ancestry, and a political and social structure that acknowledged leadership in senior lines of descent but with the leader unable to make autocratic decisions. Leadership was based on consultation and consensus. The principal social units for day-to-day sustenance were the whānau (extended family) and hapū (group of whānau). In times of threat to life and landholdings a group of whānau may have joined a loose confederation of hapū, sometimes called iwi or tribe, but the social structure was flexible, and based on kin networks.

Māori customary tenure was therefore much more complex than any single phrase such as 'ancestral estate' might suggest. A fundamental theme in this chapter is that the nature of the tenure of land and resources is the basis of understanding the nature of relationships between people and their environment. Before 1840, although there were already some 2000 Pākehā settlers in New Zealand (Figure 3.1), Māori customary law prevailed and Māori retained control of their environment. The history of colonisation after 1840 for Māori is a revolution in tenurial systems, denial of customary law, and incremental loss of control and alienation of land and resources.

Pākehā encounter

In April 1793 two Māori men, Tuki Tahua and Ngahuruhuru, who were sailing off the Cavalli Islands north of the Bay of Islands, were persuaded to come on board the naval ship *Daedalus*. The commander, Lieutenant James Hanson, had been instructed by Lieutenant-Governor King to bring some Māori back to Norfolk Island to provide expertise in the processing of local flax. The *Daedalus* sailed off with these unwilling passengers, who were taken to Norfolk Island via Sydney. The young men had limited knowledge of processing flax (it was women's work), but they were treated kindly. King personally escorted them back to their home region in November 1793, and recorded his 'gifts', which consisted of:

> Hand axes; a small assortment of carpenters tools; six spades; some hoes; with a few knives, scissors and razers [sic]; two bushels of maize, one of wheat, two of pease, and a quantity of garden seeds ... also ten young sows, and two boars; which Tooke [Tuki] and the Chief promised me should be preserved for breeding.[2]

This was one among many introductions of new plants and animals in the north of New Zealand before the end of the eighteenth century.

The introduced pigs fended for themselves and rapidly spread into scrubland and bush, feeding particularly on fernroot. Potatoes soon replaced fernroot in the Māori diet and were grown as a commercial crop in the Bay of Islands in the early years of Pākehā contact. John Savage, who visited in 1805, commented on the health-giving qualities of potatoes for preventing scurvy among ships' crews, who might be at sea for several months, and was impressed with Māori production of potatoes for trading with visiting ships. By the 1820s, pigs and potatoes had become part of the local economy of most coastal Māori communities, long before Pākehā settlement. American varieties of sweet potato had also been introduced, and from these introductions modern varieties of kūmara have been derived. Fishing, using traditional methods in harbours, along coasts, and often more than 30 kilometres offshore, was also the means to produce a commodity that could be traded. Thus a commercial component was added to established forms of Māori subsistence.[3]

Most European visitors did not really comprehend the nature of Māori society, or iwi and hapū relationships. Those who did record their impressions described what they saw, and interpreted Māori landscape through European eyes. In February 1815, the missionary Samuel Marsden and his companion John Nicholas landed at a village on the coast west of North Cape. Their ship was on its way back to Sydney from the Bay of Islands, but stopped to pick up some baskets of dressed flax. They landed in a canoe in heavy surf and climbed to the top of a nearby hill:

> Below us was a beautiful valley ... laid out in neat enclosures, and planted with the coomera [kūmara] and potatoe [sic]; and at the foot of the hill was built a small village ... the uncommon neatness and regularity of the cultivated lands; the picturesque form of the valley with its clear and beautiful stream winding its way till it lost itself in the sea; the varied and pleasing foliage of the different coppices scattered on the sides of the hills – these gave altogether such an interesting singularity to this view, as must strike the most tasteless observer with admiration.[4]

This idyllic description masks the complexity of the relationship between Māori and Pākehā that would develop through the years leading up to the signing of the Treaty of Waitangi in 1840. In the following sections this relationship is explored more specifically in the Far North, Te Hiku o Te Ika, the tail of the fish, and in Te Ika a Māui, the land of the North Island fished up by the ancestor Māui. In 1840 about 800 of the 2000 Pākehā in New Zealand were settled in the Bay of Islands, Hokianga and further north. The remainder were scattered around the coasts of both islands, mainly in shore-whaling and flax-trading stations (Figure 3.1).

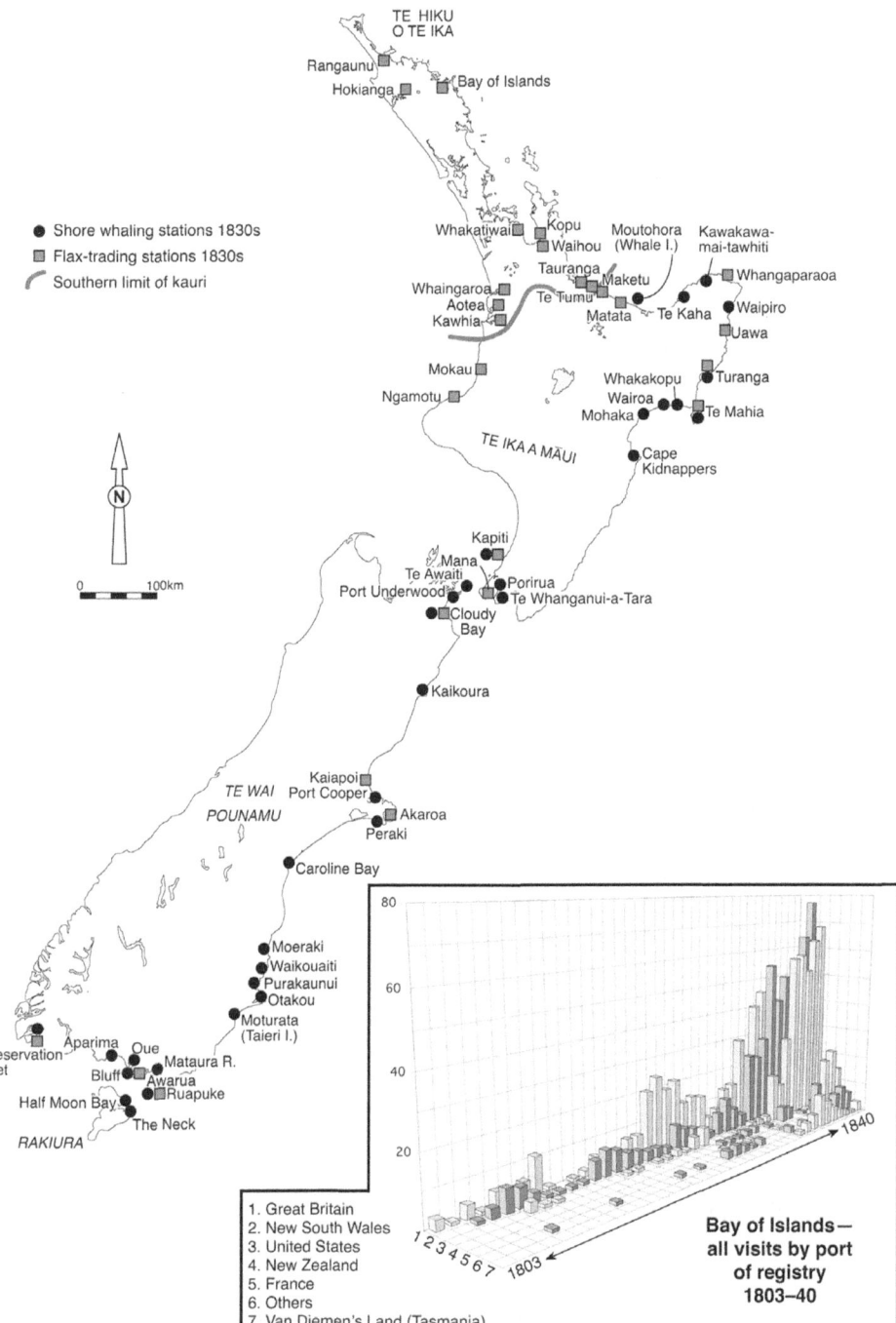

Figure 3.1 Pākehā encounter before 1840. Source: graph adapted from *The Muriwhenua Land Report*, Wai 45, GP Publications, Wellington, 1997, fig. 12, p. 45. Map compiled by author from various sources

Missionaries in the north

The Church Missionary Society (CMS), an Anglican organisation founded in 1799, was persuaded by Samuel Marsden, chaplain to the convict settlement in New South Wales, to sponsor a missionary enterprise in New Zealand in 1814. In planning this enterprise, Marsden relied heavily on one Māori traveller, Ruatara, who had gone to sea in a whaler in 1805. After several voyages he was destitute in London, where Marsden found him, brought him back to his Australian home in Parramatta, and finally in 1812 sent him to the Bay of Islands, equipped with tools and seed wheat. In March 1814, Marsden sent the brig *Active* (which he had just purchased for the New Zealand trade and mission work) with Thomas Kendall and William Hall on board, and more seed wheat, a cock, clothing and other goods destined for Ruatara.

Three Bay of Islands chiefs, Ruatara, Hongi Hika and Korokoro, sailed on the return voyage to Sydney to help arrange with Marsden for the establishment of a CMS mission station. Judith Binney's analysis indicates the complexity and potential for misunderstanding in this transaction:

> ... the missionaries' purposes and theirs were not the same. Each intended to use the other ... Marsden sought to generate a Maori dependency on the European missionaries and their ship, and hoped gradually to persuade them to listen to their evangelical message. But on the voyage home Ruatara revealed his ambivalence ... Marsden knew that the survival of the mission depended on the protection of the chiefs and he assiduously cultivated them.[5]

After Ruatara's death in 1815, the mission of 'godly mechanics' survived uneasily under the protection of Hongi Hika. However, the CMS presence in the Bay of Islands was probably a significant factor in attracting more whaling ships and traders.

From an initial base at Rangihoua, further CMS stations were established at Kerikeri (1819), Paihia (1823), Waimate (1831) and Kaitaia (1834) (Figure 3.2). In each of these places there were negotiations with local leaders involving land to be occupied by the missionaries. There was a public signing of a deed that in English appeared to provide some legal right to the land, but it was obvious to the missionaries that their continued occupation depended on the local chiefs. The fate of the Wesleyan mission at Whangaroa provided an example: the station had to be abandoned when, in 1827, the missionaries were caught up in the rivalries between local chiefs of Ngāti Pou and Ngāti Uru, and Hongi Hika's drive to expand his sphere of influence northward. However, neither the Wesleyans (who were fewer in number) nor the Roman Catholic missionaries (who arrived in the late 1830s) became involved in land transactions in the same way as CMS missionaries did before 1840. The French priests, supported by Irish Catholic traders in the Hokianga, tended to be itinerant. The CMS mission stations developed as settled small communities with a farm to provide subsistence, a church and a school, and a settlement of Māori converts close by: an oasis of Anglican civilisation in the Māori wilderness.

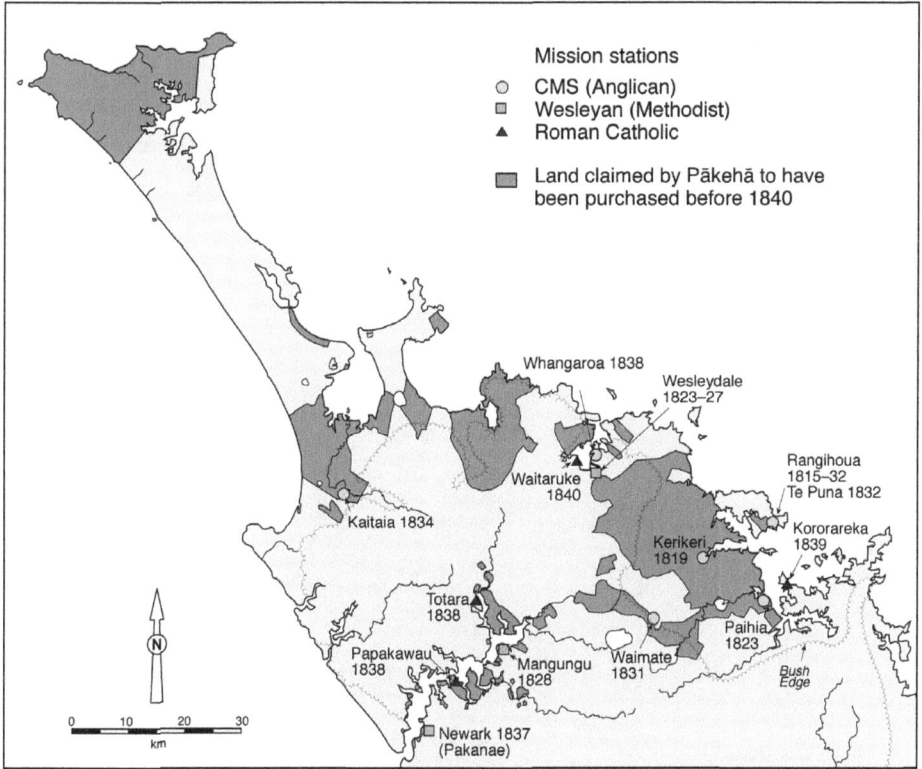

Figure 3.2 Te Hiku o Te Ika: Pākehā land claims and mission stations, 1840. Source: adapted from Malcolm McKinnon (ed.), *New Zealand Historical Atlas*, David Bateman, Auckland & the Department of Internal Affairs, Wellington, 1997, 'Old Land Claims and Mission Stations', plate 28

Exploiting the resources of the north, 1815–40

During the early 1800s, New Zealand came to be regarded as an integral part of the commercial sphere of Sydney and Hobart merchants. A cosmopolitan population on whaling ships from the Australian colonies, Britain, France and North America frequented the New Zealand coast by the 1820s. Traders in flax, kauri spars and timber settled and established trading stations. By the 1830s a number of shore whaling stations had been established (Figure 3.1). The Bay of Islands became the main provisioning place for ships and the focal point for trade with Māori in the early 1800s. By the mid-1820s the kauri forests of the Hokianga attracted sufficient attention to overcome the difficult entrance to that harbour. Pigs, potatoes, vegetables, dressed flax and 'curios' were more portable, and were traded by Māori wherever European sailing vessels made anchorage. Apart from the establishment of mission stations, the principal European activities in the north – and contact with Māori – revolved around the provisioning and repair of ships, trade in processed flax, and extraction of kauri spars and sawn timber (Figure 3.3). In all these

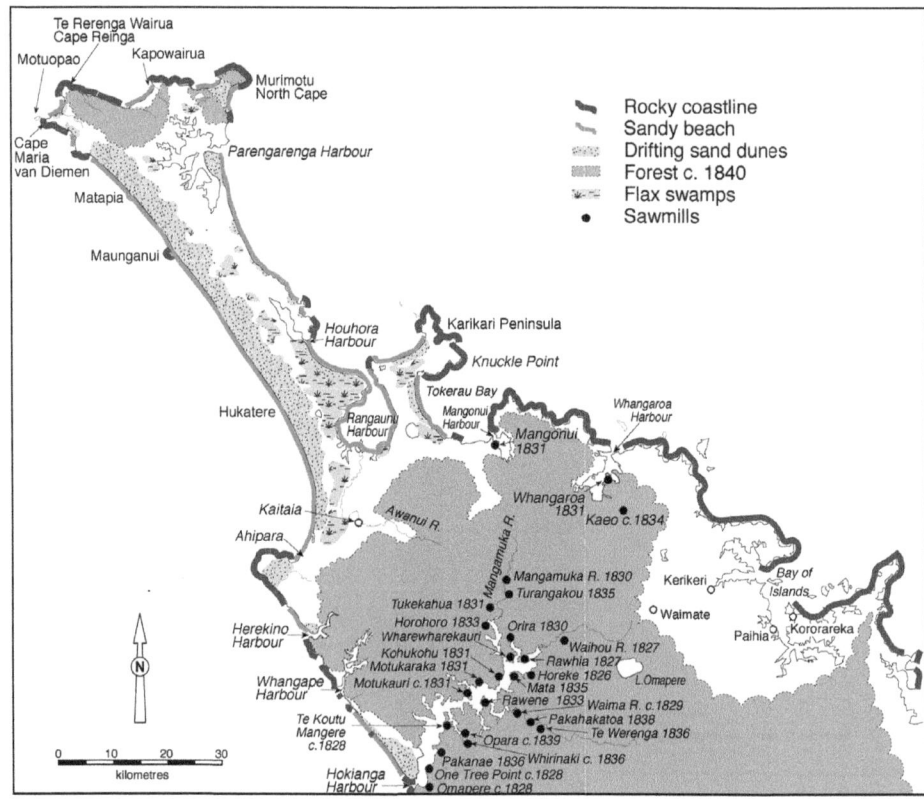

Figure 3.3 Te Hiku o Te Ika: exploitation of resources before 1840. Source: adapted from *The Muriwhenua Land Report*, Wai 45, Government Printer, Wellington, 1997, fig. 13, p. 47

activities Europeans were totally reliant on Māori cooperation for access to resources, assistance in processing, and transport to and loading of ships. Many Māori also worked as crew on whaling and trading ships, found their way to Sydney, London, and other ports, and brought back a knowledge and experience of the European world with which they regaled their kin on their return.

An analysis of 'Shipping Arrivals and Departures' lists compiled for Sydney and other Australian colonies over the period 1803–40 produced a total of 1575 ship visits to the Bay of Islands, including multiple visits by the same ships, although there may have been additional unrecorded visits (Figure 3.1, inset). The number of visits increased through the 1830s to an average of 118 per year. With an estimated 20 to 30 crew and passengers on each vessel, there would have been some 3000 visitors each year. The sperm-whaling grounds to the north of New Zealand were overfished well before 1840, but the southern right whale in coastal waters between Cook Strait and the Chatham Islands had provided an alternative. However, by 1840 this southern fishery was also grossly depleted. American whalers looked elsewhere, deterred by the prospect of British-imposed port fees, and visits to the Bay of Islands declined in the 1840s.[6]

The flax trade

Of all the strange new plants discovered on Cook's first voyage to New Zealand the flax, harakeke (*Phormium tenax*), promised to be the most useful. In a time when sailing ships required ropes and sail canvas, a new source of strong fibre was regarded with considerable interest. It is not surprising, then, that officials and merchants in the colony of New South Wales encouraged the New Zealand flax trade. By 1815 flax was becoming an important item of trade. Between 1815 and 1825 the CMS was actively involved in trading flax and timber, and Marsden sold some flax to the Royal Navy. When Henry Williams took over the CMS in New Zealand there was a change in policy, with less emphasis on trade.

There was a continuing problem of satisfactory processing of flax. European flax dressers found their traditional methods of preparing linen flax were not appropriate. Attempts to mechanise processing of New Zealand flax failed. The flax trade therefore relied heavily on Māori willingness to process flax for supply. During the late 1820s there was an increase in Māori production. The organisation of the trade also changed from sporadic collection by trading ships to the appointment of resident agents of Sydney firms and the establishment of trading stations. This increased the pressure on Māori producers: only best-quality flax would be accepted because of the risk of flax cargoes being rejected in London. The major flax export centres were initially the Bay of Islands and, by the late 1820s, Hokianga. Another important source was the swamps of the Awanui River and around Rangaunu Harbour. By 1830, flax-trading stations were being established further south, into the Thames Valley, Bay of Plenty, and elsewhere (Figure 3.1). Many Māori moved to the swamps to harvest and process flax, which could be traded for European goods, including firearms.

Kauri forest

There are few records of kauri timber extraction before the 1820s. Whalers used kauri timber for ship repairs and, with Māori assistance, cut tall trees for spars. In 1820 HMS *Dromedary*, having landed convicts in Hobart and Port Jackson, was directed to New Zealand to obtain a load of kauri spars. *Dromedary* was accompanied by the 'Colonial schooner' *Prince Regent*. Negotiations for extracting trees from Whangaroa were entered into with local chiefs, and work commenced on construction of a road from the water's edge to where some suitable kauri trees were growing in a ravine, about two kilometres from a tributary that flowed into the harbour.

Removal of the timber took several months. Toward the end of November the job was complete and the crew members working in the forest returned to the ship. Richard Cruise recorded that three local Māori brothers, 'Tippooi, George and Ehoodoo, came on board and received the stipulated payment for the cargo, with which they seemed to be perfectly satisfied'. Cruise also described the difficulties they had encountered in extracting the big logs, which for spar purposes had to be at least 74–84 feet long (22.5–25.6 metres) and 21–23 inches (53–58 centimetres) in diameter and 'perfectly straight' to meet the Royal Navy requirements.[7]

The late 1820s saw the development of a trade in sawn timber destined mainly for the growing Australian colonies. Much of this trade was controlled by Sydney merchants, and timber was re-exported to other colonies. Again, few records were kept and it is not possible to give precise quantities. By the early 1830s the Hokianga and the Coromandel Peninsula were the principal sources of sawn timber. From the mid-1820s through the 1830s the Hokianga district became the centre of a sawmilling industry that drew many more Māori into trade in ships' provisions and work in the bush felling kauri logs (Figure 3.3).

The missionary James Buller arrived in the Hokianga district by ship and on a sunny April morning in 1836 sailed 36 kilometres up the harbour to a point off the Wesleyan mission station at Mangungu. He noted the 'Native villages perched on the hillsides'. Local people came alongside in canoes to barter fish and potatoes in return for pipes, tobacco and other goods:

> At nearly every bend, a rude and lonely hut was standing. This was made of slabs, and thatched with grass. A boat, or a canoe, floated in front of it, or was lying on the beach. It was the home of some white man, living in a semi-barbarous style, with a Maori woman, and surrounded by their half-caste progeny. He was perhaps an escaped convict, or a runaway sailor. About two hundred of these classes were living on the shores of the river. They worked as axe-men, sawyers, etc. for the few traders who were located on their respective establishments …[8]

Buller estimated between 3000 and 4000 Māori living in and around the Hokianga Harbour. By 1838, of the 90 adult male Europeans listed in British Resident James Busby's census as resident in Hokianga, probably 72 were directly involved in the timber trade. It was estimated that between 50 and 60 per cent of all timber exported from New Zealand between 1828 and 1839 came from Hokianga.[9]

Small European settlements associated with production of sawn timber for the Australian market were also established on Whangaroa and Mangonui harbours. Although high-quality spars had been taken from Whangaroa in the 1820s, the timber trade did not develop here immediately. This was partly because of the reputation of local Māori among European timber men, beginning with the burning of the *Boyd* in 1809 and the sacking of the Wesleyan mission station in 1827. By 1836 Busby recorded 23 sawyers and carpenters in residence. The kauri in the Whangaroa district was located well away from the harbour, and Māori labour was needed to get the big logs to a place where they could be cut and loaded.

Exploitation of kauri forests around Mangonui Harbour also began with the spar trade initiated by Sydney merchant Ranulph Dacre. Some of his sawyers – Thomas Ryan, James Berghan, Stephen Wrathall and others – settled permanently, negotiated land transactions and married Māori women. By 1839 Busby's census recorded 11 sawyers at Mangonui, and a trade in sawn timber prepared for the Australian market had been built up. It also became a popular spot in the 1830s and 1840s for whaling ships to refit and obtain water and provisions.

The involvement of Yankee whalers in taking kauri logs to supplement their cargo of whale oil was the means of introducing American methods of extraction already developed in New England for white pine.[10] Like white pine, kauri logs cut in the forests inland of Mangonui and Whangaroa harbours and on the Coromandel Peninsula were floated down streams in floods created when purpose-built dams were tripped. Ernst Dieffenbach visited in 1839 and described the impact of logging in the kauri forests between Mangonui and Whangaroa as 'a melancholy scene of waste and destruction', with the firing of extensive areas of bush 'to make room for the conveyance of logs down to the creek. Noble trees, which had required ages for their perfection, were thus recklessly destroyed in great numbers …' He commented that this destruction of kauri forests was occurring in many places, and noted that, once cleared, 'kauri-land is so exhausted that scarcely anything will grow on it but fern and manuka'.[11]

Impacts on Māori society

By the end of the 1830s many northern Māori had travelled outside New Zealand. Some had worked as crew on whalers. Others had visited Sydney, including several Bay of Islands chiefs who had been entertained at Government House. Some, such as Hongi Hika, had visited Britain. Many had learned English, and some spoke it fluently, although the more usual medium of communication was a jargon language, 'whaler Māori'. Money, often in the form of gold coins, was entering the trading transactions. Elaborate contract arrangements were entered into: some acted directly for themselves and their hapū; others employed intermediaries – usually Pākehā or, less frequently, Māori who had acquired some English.[12]

There is little doubt that Māori participated actively in all aspects of the timber and flax trade. Such participation brought changes in Māori society and economy, not only with the introduction of European goods, but also with the diversion of energy away from customary work patterns into trade. Pākehā commentators showed grudging admiration for Māori proficiency in bartering for goods. Missionaries also noted the amount of energy expended by Māori in trading, and often complained about how tiresome Māori demands could be. The writer of an article in the *Sydney Gazette* in 1829 advised that it required 'no common skill to traffic with the New Zealanders, who are extremely shrewd in making bargains'.[13]

Māori encounters with Europeans sometimes ended in violence. Most of the time people from two very different cultures reached out to each other with a great deal of warmth, hospitality, curiosity and goodwill. But Pākehā were still in the minority and Māori largely controlled the local economy. Busby's 1839 census of the European population for the district from the Bay of Islands–Hokianga north recorded 779 people: 377 male and 109 female adults, and 293 children. Many of the women were associated with mission stations. Most of the 197 European children were of mission families, and another 96 'half-caste' children were the offspring of European men and Māori women 'living as Europeans'. There were probably many more such informal relationships; 'half-caste' children who were brought up in a kāinga (village) were not enumerated.

Markham commented, on Europeans settled in the Hokianga in 1834, that '[a]ll the Sawyers live with the Native women',[14] but he observed that such sawyers and

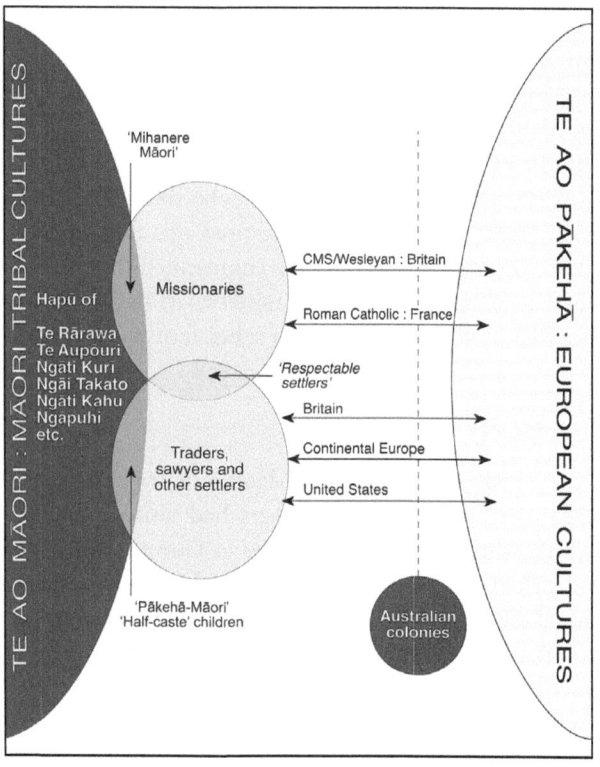

Figure 3.4 Te Hiku o Te Ika: cultural relations, 1840. Source: author

traders remained on the sufferance of the local Māori chief. Settlement was consolidated by marriage – a traditional Māori way of securing alliances. The pattern of emerging cultural relations is shown in diagrammatic form in Figure 3.4; but Māori and Pākehā still lived in different worlds, barely impinging on each other.

There were also more profound changes in Māori society wrought by the expansion of trade and exploitation of resources, as various local chiefs competed to participate in the new economy. Several hapū in the Bay of Islands consolidated their interests to form the strong Ngāpuhi confederation.[15] With the early acquisition of firearms and other trade goods, Ngāpuhi were in a strong position to settle some old scores with tribes to the south. Through the 1820s Ngāpuhi raids into the Kaipara, Waitemata, Hauraki Gulf and Bay of Plenty disrupted large areas. Ngāti Whātua and Hauraki tribes retreated into the Waikato. There was competition for Pākehā traders in other tribal areas, in order to acquire firearms and other trade goods.

In 1830 Philip Tapsell set up a trading station at Maketu in the Bay of Plenty, acting as agent for a Sydney firm. Te Arawa tribes migrated to the swamps of the Kaituna River nearby, reclaiming lands from which Ngāiterangi had dispossessed them several generations earlier. Inevitably there was tension, and in the conflict in 1836, Tapsell's trading station was destroyed. Peace was not made until 1845. In the Waikato the

Hauraki tribes had overstayed their welcome, and after a fight in 1830 at Taumatawiwi on the lower slopes of Maungatautari, Hauraki returned to the shores of the Firth of Thames and lower Waihou River; many were employed there scraping flax for one of Tapsell's agents. Other traders penetrated the Waikato, too.

The ripple effect continued southward. In the 1820s and 1830s Waikato tribes had put pressure on Ngāti Toa around Kawhia Harbour, Taranaki tribes, and Ngāti Raukawa of the eastern Waikato. Te Rauparaha and Ngāti Toa were evicted from Waikato and spent some time in northern Taranaki before moving to Kapiti Island. Several Taranaki tribes joined a heke (migration) south, displacing tribes in the southern North Island, and some moving into the Nelson–Marlborough region. In 1835 Ngāti Tama and Ngāti Mutunga from Taranaki occupied the Chatham Islands. Ngāti Raukawa also migrated to the Horowhenua region. Occasionally, Pākehā settlers were caught up in these conflicts, which represented a massive disruption to Māori tribal and territorial relationships, still fluid in many areas in 1840.

There were other demographic changes. Among the negative impacts of contact with Pākehā must be included the effects of diseases such as measles (with its respiratory complications), influenza, pneumonia, whooping cough, smallpox, typhoid and tuberculosis. Venereal diseases also took their toll and, combined with smoking tobacco, which became a widespread habit, probably contributed to lower fertility rates in women and high infant mortality. There were also changes being wrought by missionary teachings. While the nature of impacts on pre-1840 Māori society and long-term impacts can be debated, it is certain that Pākehā trade and exploitation of resources had profound effects.

The extinguishment of native title

The year 1840 saw the signing of the Treaty of Waitangi and the proclamation of a British colony in New Zealand. Article 1 of the Treaty provided for a kawanatanga or government to make laws for all. Article 2 provided a guarantee to Māori of tino rangatiratanga – full authority – over all lands, forests, fisheries and other property, unless and until Māori willingly chose to sell, and then only to the Crown. Article 3 provided that Māori would share in the rights and privileges of British subjects.

What was not adequately explained to Māori gathered at Waitangi in the Bay of Islands on 5 and 6 February 1840 was that under British law the Crown assumed title to all lands and resources. In this legal fiction the radical title was vested in the Crown, but subject to rights guaranteed in Article 2, and replaced the concept of a communally held Māori ancestral estate. The Crown asserted title to foreshores and seas below high-water mark, to navigable rivers and to other resources, including the royal metals, gold and silver, under Crown prerogative. This was a tenurial revolution that initially was not comprehended by Māori, who continued to live under customary law.

The Crown retained the sole right to negotiate with Māori, protect their interests and extinguish native title to land. This placed in limbo a large number of pre-1840 transactions between Māori and Pākehā settlers because no title could be recognised in British law unless it derived from the Crown. To resolve this, land claims commissioners were appointed to investigate transactions and determine whether such 'sales' were valid. Limits were placed on grants of land to settlers, but the commissioners accepted in many cases that a valid 'sale' had occurred which extinguished native title. This begs the question whether a private citizen could extinguish a native title when the Crown claimed sole right to this function, and the transaction occurred before a colony and British law was proclaimed.

There is a long history of investigation of these 'old land claims', which has been reviewed by the Waitangi Tribunal. The Māori grievance was that when Crown grants were made to settlers who were occupying their lands since before 1840, the 'surplus' reverted not to Māori but to the Crown. The Muriwhenua Tribunal found that these original transactions were not sales because the parties were not of a sufficiently common mind to effect a valid contract: 'Māori contracted with Europeans on the basis of Māori law, which was the only law known to them … As a consequence, the pre-Treaty transactions were not sales but at best conferred a personal right of occupation conditional on acceptance of the norms and authority of the local Māori community as represented in the rangatira [chiefs].'[16] Nevertheless, the Crown recognised such titles as the land commissioners recommended for private individuals, or offered 'scrip' to be redeemed in land elsewhere.

New Zealand Company interests were purchased by the Crown. There were also additional Crown purchases around Auckland, the new colonial capital. For a time Governor FitzRoy waived the Crown right of pre-emption, but this was reimposed by Governor Grey in 1846. By then native title had been extinguished in a number of toeholds of Pākehā settlement (Figure 3.5). Māori frustration erupted into armed conflict in the mid-1840s, in the north around the Bay of Islands, in the Hutt Valley near Wellington, and in the Wairau Valley in Marlborough. Between 1846 and 1865 the Crown right to extinguish native title was assiduously pursued under pressure to provide land for incoming settlers. The Constitution Act 1852, the British statute which established representative government in New Zealand, provided at section 71 for 'native districts' where 'the laws, customs, and usages' of Māori could 'be maintained for the government of themselves'. But the settler Parliament (which had no Māori representatives until 1867) never invoked this provision, in spite of Māori entreaties, including the petition of King Tāwhiao to Queen Victoria in 1884.

The 1850s saw the Crown acquisition of almost the whole of the South Island, with a few small and inadequate reserves left for Māori occupation.[17] In the North Island, where there was a much more numerous Māori population, Crown purchases focused on the Wairarapa, Hawke's Bay and Manawatu regions, small pockets on the west coast, and the hinterland of Auckland. The 1850s also saw a growing resistance by Māori to land sales. In the central North Island, from Waikato to Taranaki, a confederation of tribes evolved in a series of meetings, culminating in the ceremonial installation of a Māori king, Potatau Te Wherowhero, in 1858. The aims of the Kīngitanga movement included retention of a region in Māori control, preservation of Māori customary law and governance, but with coexistence with Pākehā settlers. The settler Parliament was not prepared to accept any form of Māori self-determination, or the possibility of settlers

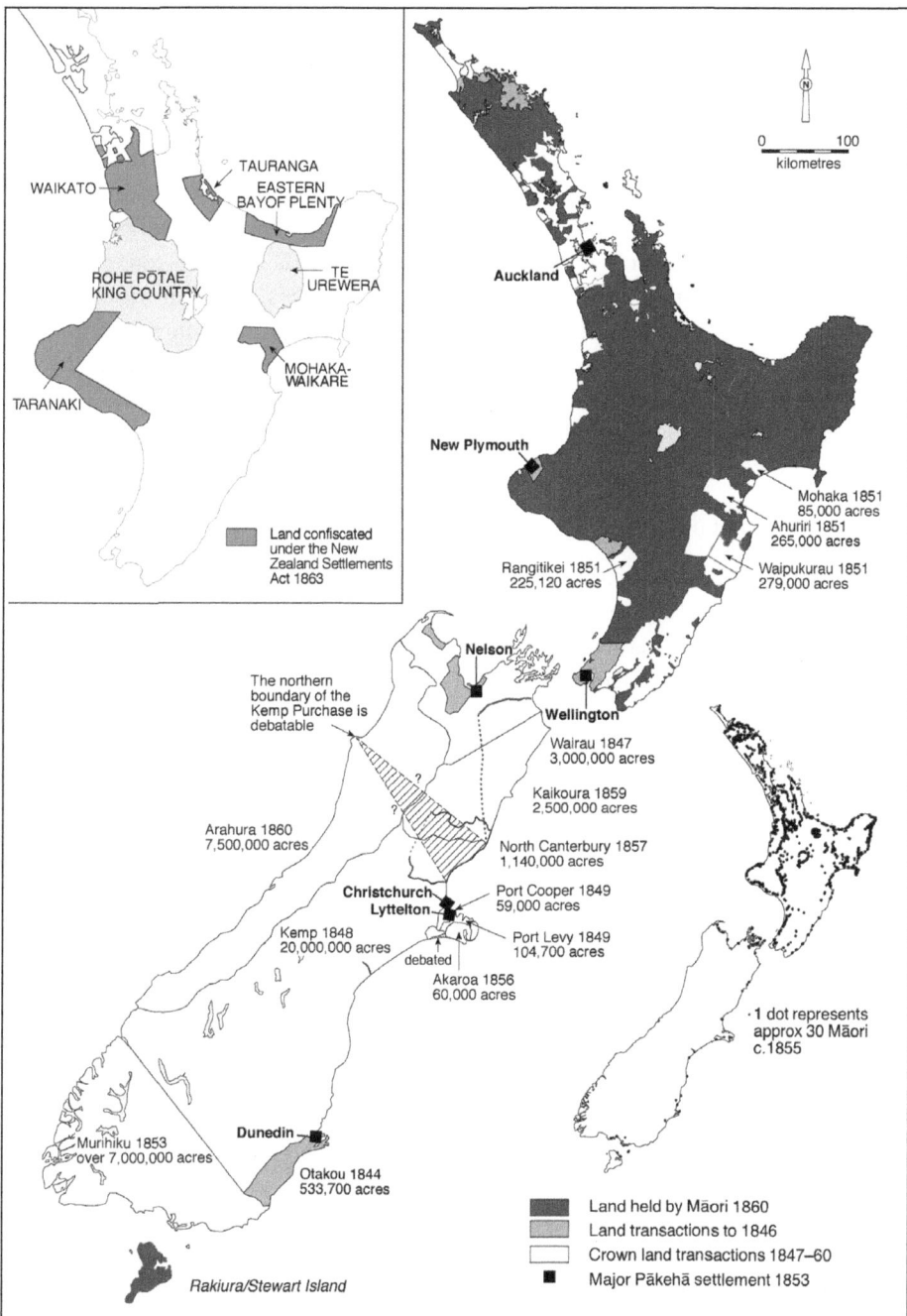

Figure 3.5 The extinguishment of native title. Source: adapted from Alan Ward, *National Overview*, vol. I, Waitangi Tribunal Rangahaua Whanui Series, Waitangi Tribunal/GP Publications, Wellington, 1997, figs 1–5, pp. xviii–xxii; Ward, *National Overview*, vol. III, Figure 23, p. 265; with inset maps compiled by the author from various sources

coming under the control of a Māori king and Māori customary law. There was to be one law for all, British law, and those who did not accept this were deemed to be rebels.

Conflict was inevitable. Fighting began in Taranaki in 1860, and erupted again in 1862. In July 1863, British troops invaded the Waikato, and by early 1864 there were engagements in the Bay of Plenty. Late in 1864 a millennial pacifist movement led by Te Ua Haumene spread from south Taranaki through much of the central North Island to the Bay of Plenty and East Coast. Pai Mārire ('good and peaceful') adherents were labelled fanatical rebels, and more fighting followed. Te Kooti was a 'rebel', too, and so was Titokowaru, and fighting was carried on in the eastern ranges, Taupo district and inland Taranaki into the late 1860s. The lasting result of all this was the punishment of rebels by confiscation (raupatu) of extensive areas of land under the New Zealand Settlements Act 1863 (Figure 3.5, inset). Thus was spawned another set of Māori grievances, some of which are still being investigated by the Waitangi Tribunal.[18]

The Native Land Court

While the North Island was being disrupted by fighting, Parliament developed legislation to provide a more effective mechanism to determine ownership of Māori lands and expedite sales to Pākehā settlers without the complexities of competing claims, which had led to the disastrous transaction at Waitara that sparked war in Taranaki. In the Native Lands Act 1862, Crown pre-emption was waived and a Native Land Court established, with a purpose as set out in the Preamble to the Act:

> And whereas it would greatly promote the peaceful settlement of the Colony and the advancement and civilization of the Natives if their rights to land were ascertained defined and declared and if the ownership of such lands when so ascertained defined and declared were assimilated as nearly as possible to the ownership of land according to British law.

The machinery for the operation of this court was set out in the Native Lands Act 1865, which at section 5 defined it as 'a Court of Record for the investigation of the titles of persons to Native land for the determination of the succession of Natives to Native lands'.

The Act thus ushered in a new framework for extinguishing customary tenure. The legislation was not specific about how this would be done. Under section 21 an investigation of title began with an application in writing by any Māori. In due course a hearing was notified before a Pākehā judge and a Māori assessor: 'At such sitting of the Court the Court shall ascertain by such evidence as it shall think fit the right title estate or interest of the applicant and of all other claimants to or in the land' (section 23). Similarly, the court could 'ascertain by such evidence as it may think fit who according to law as nearly as it can be reconciled with Native custom' the succession of Māori interests in land. The Native Land Court was also made the authoritative body in section 41 in the 'determination of the question or fact of Māori custom or usage' on any matter referred to it by the Supreme Court.

This legislation opened up Māori communities to direct approaches from would-be purchasers of land. Alan Ward described how the operation of the Native Land Court exposed Māori to 'a predatory horde of storekeepers, grog-sellers, surveyors, lawyers, land agents and money-lenders'.[19] The individualisation of title envisaged any owner having an equal right with any other owners in a title to dispose of individual interests in land, thus undermining community political and social structures. Sir Hugh Kawharu called the Native Land Court 'a veritable engine of destruction for any tribe's tenure of land, anywhere'.[20]

Two elements of Native Land Court case law differed significantly from Māori custom. One was the '1840 rule', whereby actual occupation of land at the time British law was imposed constituted a right to title. In reality there were areas where occupation was fluid, and was either shared or still being fought over. There is also some evidence that Pākehā judges placed a greater emphasis on conquest, or the 'might is right' argument, in their codification of customary rights. The second element was the acceptance of succession of interests from either male or female line to all descendants, regardless of usual place of residence. This was a denial of customary occupation based on ahi kā, and later led to multiple ownership of blocks of land, with problems of administration on behalf of numerous absentee owners.

The Native Land Court produced a hybrid title that complicated rather than simplified Māori customary land tenure. This also had 'grievous effects on Māori society'. As Ward explained, 'It set up a body of self-proclaimed experts who had to try, and frequently failed, to interpret Māori custom.' The system invited contention and contestation between parties, and the legalistic nature of the court produced a morass in which Māori floundered for decades, frittering away their estates in ruinous expenses and still all too often not getting equitable awards.[21] In a wide-ranging review of the legislation and operation of the court in the nineteenth century, David Williams remarked on the lack of separation of powers between executive government and the judiciary, and described the court as 'a willing accomplice in promoting government policy'. He also noted 'a clear pattern' of correlation of Native Land Court operations and alienation to Crown or private purchasers.[22] The pattern of land loss is shown in Figure 3.6.

The machinery of the Native Land Court rolled inexorably across the land. In the Rohe Potae (King Country), legislation was enacted to restore Crown pre-emption and allow the Native Land Court process to operate in the 1880s. The Urewera District Native Reserve, which Tūhoe hoped to retain in tribal title, was subject to an investigation in 1899–1900 by commissioners who divided the land into blocks with lists of owners. By 1907 the Native Land Court jurisdiction was extended there, too. In 1891, T.W. Lewis, the long-serving Undersecretary for Native Affairs, told the Commission on Native Land Laws:

> the whole object of appointing a Court for the ascertainment of Native title was to enable alienation for settlement. Unless this object is attained the Court serves no good purpose, and the Natives would be better without it, as, in my opinion, fairer native occupation would be had under the Maoris' own customs and usages without any intervention whatever from outside.[23]

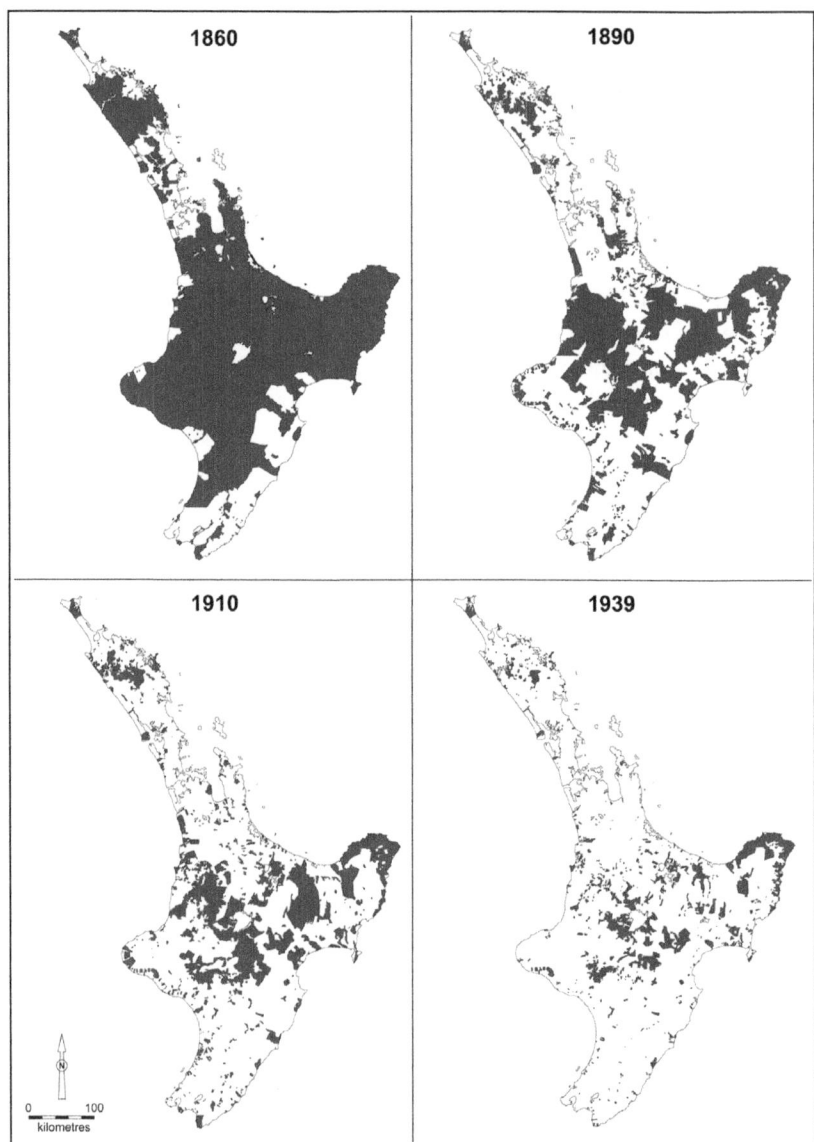

Figure 3.6 Alienation of Māori land in the North Island, 1860–1939. Source: adapted from Alan Ward, *An Unsettled History: Treaty Claims in New Zealand Today*, Bridget Williams Books, Wellington, 1999, pp. 162–66

The commission in its report strongly criticised the 'pernicious consequences of Native-land legislation' and noted the numerous disputes, litigation, and flood of Māori petitions to Parliament – over 1000 in the decade 1880–90. The legislation was complex and voluminous; in 1888 there were eight separate new statutes, and in 1889 nine more. In reviewing complaints about the operation of the Native Land Court, the commission remarked on 'the confusion both in law and practice', which created 'a state of confusion and anarchy in Native-land titles'. The Waitangi Tribunal in the Pouakani Report concluded:

> that Maori paid a disproportionate cost for Pakeha settlement, but little provision was made for Maori participation in the suggested benefits of the introduction of capital and settlers. Indeed, the system of Native Land Court investigation of title and individualisation of interests in land, which could be sold piecemeal, contributed largely to social disruption, dissension over issues of mana and territory, massive debts, costly mistakes in survey boundaries in some cases, and failure to survey in others, and costly litigation ... There is nothing in the Treaty of Waitangi which required the transmuting of traditional Maori forms of tenure into titles cognisable in British law.[24]

Conclusion

The maps in Figures 3.5 and 3.6 illustrate the basis of the piecemeal transformation of the New Zealand environment, a process begun before 1840, as Pākehā settlers acquired land, cut down the trees, drained swamps, developed pasture and constructed roads and towns. Māori lost control of their ancestral lands and resources and became marginalised, relegated to small rural pockets of land or remote hill country unsuitable for farming. The legacy of this failure to acknowledge Māori customary law, of the imposition of British law, and of a policy of assimilation of Māori under the guise of civilisation of the Natives, is the hundreds of Māori claims against the Crown lodged with the Waitangi Tribunal complaining about loss of lands and resources, failure to protect Māori interests, and destruction of environment (see Chapter 17).

Further reading

This essay has been left unchanged out of respect to its author, who died in 2005. Later work extending its themes includes Judith Binney, *Encircled Lands: Te Urewera, 1820–1921* (Bridget Williams Books, Wellington, 2009); Richard Boast, *Buying the Land, Selling the Land: Governments and Māori Land in the North Island 1865–1921* (Victoria University Press, Wellington, 2008); Danny Keenan (ed.), *Huia Histories of Māori: Ngā Tāhuhu Kōrero* (Huia Publishers, Wellington, 2012); Vincent O'Malley, *The Meeting Place: Māori and Pākehā Encounters, 1642–1840* (Auckland University Press, Auckland 2012); Hazel Petrie, *Chiefs of Industry: Māori Tribal Enterprise in Early Colonial New Zealand* (Auckland University Press, Auckland, 2006) and Adrienne Puckey, *Trading Cultures: A History of the Far North* (Huia Publishers, Wellington, 2011). The ongoing reports of the Waitangi Tribunal, available online, are invaluable.

4.

Resource frontiers, environment and settler capitalism, 1769–1860

Jim McAloon

IN THE MIDDLE of 1840 a small schooner, the *Jewess*, sailed from Wellington to Tasman Bay. Part-owned by its master Frederick Moore (the other owners were Wellington and Sydney merchants), the ship had been built in Tahiti. Moore took the opportunity to explore the West Wanganui inlet and quietly examine the northwestern part of the South Island for suitability as a site of European settlement.

This small example highlights some of the connections that placed New Zealand in a Pacific and a world economy. With a range of commodities attractive to Chinese and Indian markets as well as to European and North American, New Zealand was part of Australia's Pacific frontier. While the attractions of New Zealand resources have been frequently noted, the literature has usually focused on relations between Māori and the newcomers. The New Zealand environment, the limits it set on human activity, and the varying ways in which it has been perceived have received less attention.[1]

Nor is the economic context emphasised. Particularly, the revolutionary nature of capitalism is underplayed. As Marx and Engels noted in 1848, 'Modern Industry has established the world market ... this market has given an immense development to commerce, to navigation, to communication by land.' New industries had been set up 'that no longer work up indigenous raw material, but raw material drawn from the remotest zones; industries whose products are consumed, not only at home, but in every quarter of the globe'. The global dimension is fundamental to environmental change in New Zealand after 1769, change that amounts to an ecological revolution, a 'major transformation in human relations with nonhuman nature'.[2]

On the edge of empire

An important theme in European attitudes to nature in the 1760s was the urge to classify and catalogue, an undertaking inseparable from economic use of plants and animals. Britain led the world in systematic natural history by the 1760s. James Cook proceeded in the service of an expanding commercial empire, and his three voyages to New Zealand anticipated much of the pattern that defined the resource economy.[3]

Cook immediately sought to establish New Zealand, like other Polynesian islands, as a source of shipping supplies. Tapa cloth was plentiful in the tropics but highly prized in Aotearoa. At East Coast settlements,

> as every one in the Ship were provided with some of this sort of Cloth, I suffer'd every body to purchase what ever they pleased without limitation, for by this means I knew that the natives would not only sell, but get a good price for every thing they brought; this I thought would induce them to bring to market what ever the Country afforded.

Cook regularly traded for fish, even in excess; the critical thing was 'to incourage them [Māori] in this kind of traffick'. Not that they needed much encouragement: Māori communities would consistently engage with the new world of trade from the 1790s.

The *Endeavour* reports emphasised the commercial dimensions of botany. Joseph Banks became an evangelist for 'the most excellent' New Zealand flax, which produced cord and rope 'of a strength so much superior to hemp as scarce to bear a comparison with it'. The mataī was 'tall streight and thick enough to make Masts for vessels of any size', while hard pōhutukawa wood seemed 'well adapted for the Cogs of Mill wheels &c'. Cook repeatedly emphasised the North Island's potential for European settlement:

> It was the opinion of every body on board that all sorts of European grain fruits Plants &c would thrive here. In short was this Country settled by an Industrus people they would very soon be supply'd not only with the necessarys but many of the luxuries of life …

Cook assessed Te Wai Pounamu (the South Island) as 'Ruged and Mountainous', although he expected it to prove rich in minerals and the middle part of the island had 'much the appearance of fertility'.[4] If the British were overoptimistic, the French dismissed the country as 'less than mediocre … unpromising and sterile'. Like Cook, Marion du Fresne in 1772 looked for fertile land, minerals, and 'other products the value of which might induce the state to set up a branch of commerce'. The French landed chickens as well as pigs, and wheat, corn, pea seed and fruit.[5]

Cook used New Zealand as a base in his two later voyages. The *Resolution* spent six weeks in the autumn of 1773 at Tamatea (Dusky Sound). The 114 men (and the ship's cats and dogs) killed thousands of birds and fish for food, and left other European footprints when they liberated five geese from the Cape of Good Hope. Rats also left the ship. Georg Forster, the expedition's young artist and ethnographer, saw the sojourn as 'demonstrating the superiority of a state of civilization over that of barbarism' by 'the alterations and improvements we had made' – felling trees, taking water, making spruce beer, eating fresh fish. Painting and drawing, 'the polite arts', had appeared, and botany and astronomy marked 'the dawn of science, in a country which had hitherto lain plunged in one long night of ignorance and barbarism'.[6]

Meanwhile the *Resolution*'s sister ship, the *Adventure*, had arrived at Totaranui (Queen Charlotte Sound), where the two ships were reunited. The *Adventure*'s crew had planted potatoes, turnips, carrots and parsnips. Cook added wheat, beans, peas, celery, parsley, garlic, onions and strawberries, as well as more roots. Most flourished. As Forster wrote, 'the most precious plant we left here will be the Potatoe, which in length of time will become one of the finest Substitutes for bread, for future Navigators & the Natives'. The French expedition under Jean François de Surville had given a few breeding pairs of pigs to Northern Māori in 1769. Cook's second expedition landed chickens, goats, more pigs and a ewe and a ram, but these first sheep in New Zealand were also the first to die of tutu poisoning, 'thus all my fine hopes of stocking this Country with a breed of sheep were blasted in a moment'. Cook's pigs were European breeds which, some decades earlier, had been crossed with Chinese and Indonesian subspecies; a few months later Cook returned to the Marlborough Sounds and introduced more pigs, these ones from Tahiti and Tonga. Māori spread the new mammals over both islands, and liberations by whalers and escaped domestic stock later reinforced the populations.[7]

Dusky Sound had no permanent Māori settlements, and the European impact, although noticeable, did not provoke conflict. The French experience in the densely populated Far North was different: 400 men required a lot of fish, and Māori were soon reluctant to trade. De Surville was aware of the problem but remained insistent. In 1772 the pressure on resources from 200 extra men was a factor in the bloody end to Marion du Fresne's visit. Similarly, the killing of some of the *Adventure* crew at Totaranui in 1773 was partly due to pressure on resources and relationships.

From 1779 until 1791 no European appears to have visited New Zealand. George Vancouver, Bruni d'Entrecasteaux and Alessandro Malaspina each made brief visits in 1791 and 1793, but the end of isolation swiftly followed the British invasion of the Australian continent in 1788. Australian historians have debated the motives for the invasion. Some maintain the government was desperate for a penal site; others stress strategic and commercial considerations, including access to maritime commodities and the attraction of a major British base in the southwest Pacific. Joseph Banks, who became a central figure in British imperial science, maintained a lifetime interest in Australasia. In 1783 Banks's disciple James Matra, who had sailed on the *Endeavour*, wrote a 'Proposal for Establishing a Settlement in New South Wales', which conveyed Banks's own views, particularly on flax and timber. Descriptions of New Zealand resources enjoyed some circulation in Britain in the 1780s. Banks himself advocated New South Wales to a Parliamentary Committee on Transportation in 1785. The final plan for the penal colony mentioned both flax and timber.[8]

The resource economy

From the early 1790s New Zealand was brought into a Pacific, indeed a global, economy (Figure 4.1). European interests were specific: timber, flax, seals, whales, and fresh food. Major environmental change resulted, amounting to both plunder and modification. Although Māori society was profoundly transformed, and new diseases wreaked some havoc, the resource economy did not require Māori dispossession; indeed, many Europeans welcomed the integration of Māori into capitalist civilisation.

The Governor of New South Wales, Arthur Phillip, had been instructed to cultivate and prepare flax. The original (1787) plan for William Bligh's ill-fated *Bounty* voyage was for a Sydney-based ship to collect New Zealand flax as well as Tahitian breadfruit.

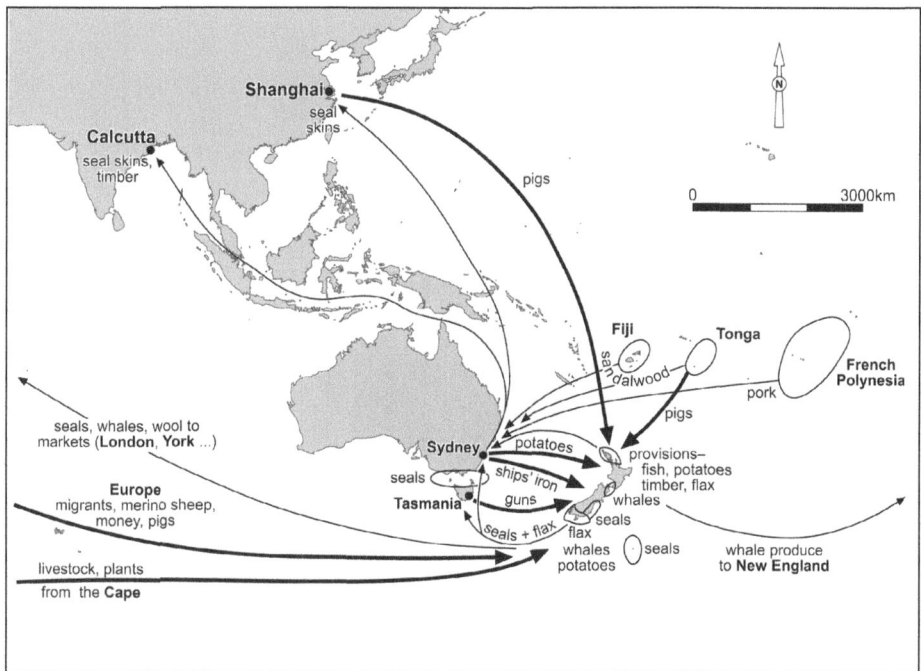

Figure 4.1 Map of New Zealand in the global economy, 1770–1840. This is a schematic representation of the then most important resources and trading links.

Europeans had no idea how to prepare New Zealand flax, which also grew on Norfolk Island. Philip Gidley King, Lieutenant-Governor of Norfolk Island (Figure 4.2), had been sent back to London in 1790 to report on New South Wales. He suggested that Māori be brought over to Sydney to demonstrate flax dressing. In 1791 the Admiralty sent the *Daedalus* to supply George Vancouver's expedition to Nootka Sound, where British and Spanish ships had clashed over the fur trade. The *Daedalus*'s orders were then to pick up Tahitian pigs for New South Wales and to call in at Doubtless Bay and 'endeavour to take a flax-dresser or two'. Exceeding King's intention, the *Daedalus* kidnapped two young men, Tuki Tahua and Ngahuruhuru, but once in Norfolk Island, it was discovered that flax dressing was women's work.[9] The young men gave such assistance as they could once King promised to return them home as soon as possible. King personally escorted them back in November 1793, on the *Britannia*, which was en route from relieving a sealing gang left earlier at Dusky Sound. King gave the two young men pigs, grain, peas and potatoes, the makings of 'a local agricultural revolution'. King justified the modest outlay as 'the means of procuring the goodwill of those who visited us, and the stock, I hope, will prove of publick utility if ever that country should be settled or visited by Europeans'. Firmly convinced of New Zealand's commercial possibilities, King hosted other rangatira in later years.[10]

Figure 4.2 Portrait of Governor Philip Gidley King (1758–1808).
Source: National Library of Australia, Canberra

The *Britannia* was owned by a London firm and had come out intending to seal, bringing a load of convicts to defray expenses. Before the ship could head for the sealing grounds, in one of the first examples of the New South Wales officer corps moonlighting as merchants, she was chartered to bring cargo from the Cape of Good Hope to Sydney. The *Britannia*'s sealers were the first European gang left in New Zealand; both the *Daedalus* and the *Britannia* indicate the extent to which even in the early 1790s New Zealand was one thread in an oceanic fabric of imperial expansion. In ten and a half months the *Britannia* sealers took 4500 skins and built a 40-foot ship. They blamed weather for their low tally, yet perhaps spent more time building their ship – a sensible precaution against abandonment. Sealing in Dusky Sound was sporadic throughout the 1790s, but ships took timber for repairs and spars.[11]

In 1801 George Bass took the *Venus* from Port Jackson to Tahiti, trading for pork, and stopped at Dusky Sound to collect timber for pork barrels. Successful in pork, Bass was emboldened to propose a government-sponsored fishing expedition. He claimed to 'have every proof, short of actual experiment, that fish may be caught in abundance near the South part of the South Island of New Zealand, or at the neighbouring islands. And that a large quantity might be supplied annually to the Public Stores.' Relying on the wide jurisdiction that Britain asserted, Bass petitioned for a monopoly.

> If I can draw up food from the sea in places which are lying useless to the world, I surely am entitled to make an exclusive property of the fruits of my ingenuity, as much as the man who obtains Letters Patent for a corkscrew or a cake of blacking.[12]

Bass disappeared, sailing home to Britain via covert trading speculations in South America.[13]

Northern trading voyages initially concentrated on timber. In April 1793 an East Indiaman, the *Fancy*, visited the Waihou River (Thames) and took 200 spars back to Sydney for India. In November 1794 the *Fancy* called again; and other vessels visited Thames annually between 1798 and 1801. The timber trade was eclipsed between 1801 and 1808 in favour of Tahitian pork, sandalwood, and sealing on both sides of the Tasman.[14]

Sealing became intensive after 1800, and in New Zealand was largely controlled by Sydney mercantile houses. Philip Gidley King, now Governor of New South Wales, had bemoaned the fact that 'we possess no known staple whatever'. By reducing government expenditure in 1801, King forced entrepreneurs out into the Pacific as well as onto the land. Small ships deposited gangs and traded elsewhere while the gangs worked. Wages were minimal, as crews were paid in shares set against provisions. Sealing was an excellent first staple industry. With a cheap process and lucrative commodities, profits were easy. Seal oil was prized. China was the traditional market for furs, but the merchants Simeon Lord and Robert Campbell had 'glutted the Canton market' by 1803. Fortuitously, new techniques made seal fur sought after for hat making in Britain.[15] Sydney commercial developments – and Sydney interests in New Zealand – were also driven from 1800 by British merchants in India, desperate for return cargoes. Campbell and Co. were the first and among the most noteworthy Indian firms in Sydney, either buying skins and oil outright, organising sealing themselves, or selling on commission.[16]

The East India Company monopoly, covering most of the Indian and Pacific Oceans, severely restricted trade and shipbuilding, although it was gradually reduced before being abolished in 1813 for everything but the China tea trade. While Joseph Banks hoped to encourage trade out of Sydney as a way of making the colony self-supporting, the East India Company and British whaling interests sought to maintain punitive duties against 'Australian' produce. Only in 1819 were Australian vessels allowed to trade directly with Britain; and in 1823 the duties were removed. If Sydney and London had not treated New Zealand produce as Australian for customs purposes, the exploitation of New Zealand resources would have been much less.[17]

By 1802 the Bass Strait seal populations had been decimated, and attention shifted across the Tasman. Between 1803 and 1805, Sydney-based ships wrecked the Dusky Sound rookeries. Sealers then moved into Foveaux Strait and further south. In October 1809 a Campbell gang was left on Rakiura/Stewart Island, where its overseer became reasonably fluent in southern Māori – and traded the first iron tools in the south for potatoes. Other gangs experienced much conflict. The most notorious ship was the American *General Gates*, which operated right around the Pacific Rim – Canton, Batavia, Manila, Hobart, Tahiti, Sydney and New Zealand. The *Gates* had little sympathy for Māori sensibilities, and its crews were frequently attacked.[18]

During the 1810s, merchants contemplated the virtues of Murihiku/Southland flax, and trade began in 1822 with the *Snapper*. Having rescued 12 *Gates* sealers who had been on the run from Ngāi Tahu at Chalky Sound, the *Snapper* party came to an arrangement with the chiefs, mediated by James Caddell, a Sydney lad who had earlier been rescued by the high-born young woman Tokitoki when his sealing party was attacked. The ship traded iron and pigs for dressed flax and potatoes. Another ship followed the *Snapper* 'to promote the civilization of the inhabitants of the island by supplying them with British Manufactures in exchange for their flax'.[19] Southern Māori assured this party 'that they did not intend to kill any more white men, now that we had become friends by commencing trade'.[20]

Sealing continued into the 1820s. Gangs – Europeans and Māori alike – killed not only mature seals but also pups and pregnant females. By 1827 the average cargo had declined to 2700 skins per ship; a few years previously it was said to have been 80,000. A Hobart newspaper wished 'that some effective steps were taken to protect and cherish this valuable article'.[21] Capital from sealing allowed Australian entrepreneurs to participate in the London-dominated whaling industry by 1830. Offshore or bay whaling was established by 1807, and ships – British, French, and American – called at the Bay of Islands for fresh food and water. Colonial whalers established shore stations by agreement with Māori, along the Foveaux Strait and Otago coasts, at Cloudy Bay and Port Nicholson, and elsewhere. Whale oil was used for fuel and for lubricating the machines of the industrialising Atlantic world; whalebone was used in corsets and in chairs. In 1840 – just before the industry crashed through over-hunting – half the southern oil came from New Zealand and reached Britain via New South Wales.[22]

The moral economy of capitalism

In 1805 Governor King reported that the seeds he had given to the unwilling flax dressers from the *Daedalus* had become the basis of a thriving trade in provisions. Looking forward to the continued use of New Zealand as a base, King kept up supplies of livestock, both on his own initiative and in response to requests from visiting northern chiefs. The missionary Samuel Marsden had a major role in the agricultural revolution (see Chapter 3). Critical in this was his close relationship with the Ngāpuhi rangatira Ruatara, who had endured a difficult voyage around the world, worked at Marsden's Sydney farm, and finally returned home about 1812 with seed wheat. Marsden himself crossed the Tasman at the end of 1814. Even this voyage had a commercial as well as an evangelical agenda: Marsden had to report on New Zealand's potential for European settlement. All Marsden's voyages would bring back a full cargo of timber, flax, and salted fish for the Sydney markets.[23]

Botanical change was widespread by the 1820s. The potato spread rapidly. Sealing gangs usually brought seed potatoes for their own use; this was probably the means by which potatoes were introduced around Foveaux Strait before 1810.[24] The impact of the potato was revolutionary: it was much easier to grow and gave better yields than the kūmara. In the far south, it encouraged a more settled population because not only did it provide food, but more importantly it was a valuable commodity in trade with visiting ships, as it was further north.[25] By the 1830s, many tons of locally grown potatoes were supplied to visiting ships and to Sydney. Māori traded rather than ate most of the new foods. Fish and potatoes were dietary staples –although younger people were more willing to experiment. A favourite relish combined old and new: tawhara stem, peaches, onions, potatoes, kūmara, fuchsia berries, pig brains, lard and tutu juice.[26]

For both Māori and European, agriculture and iron meant social as well as economic change. Marsden regarded agriculture as the catalyst of Christianity and civilisation, and tried to persuade Ruatara that agriculture would allow Māori access to European goods: 'all these blessings might be obtained by them by cultivating their land and improving themselves in useful knowledge'. He hosted young Māori for considerable periods at his Parramatta farm. Simeon Lord, proposing to trade in flax and timber, declared that employing Māori would introduce them to 'progressive civilization'. In the 1830s Joel Polack, negotiating for a trading position at Kaipara, made similar appeals to the moral economy of capitalism. If his hosts accepted his proposal,

> they would be enabled to compete with their neighbours ... in possessing articles of clothing to protect them from the wintry blasts, and implements of iron to pursue the labours of agriculture ... [and] ammunition to repel an invader; who, aware of their being in possession of such resources, would themselves ... turn their attention to similar pursuits ... and would war no more.

Some 'deplored the innovations Europeans would cause, in lieu of their ancient usages'. One old man, Motarou, regretted 'the warrior [being] obliged to give way to women and slaves, whose utmost ability consisted in paddling canoes, pounding fern-root, or scraping flax'. He regarded muskets as inferior to traditional weapons, and regretted the stigmatisation of cannibalism:

> if we want clothing, we have our women to make them ... if food be our object, we have slaves to plant for us; and of them we shall never be deficient, as long as our enemies exist ... No! let the white man go. Who sent for him? He came from beyond the sea to us – he has seen us. What does he further want? Let him go back.

The opposing case was successfully put by another senior chief, Rapu: iron tools saved time and labour; and pork, potatoes, brassicas, corn, onions, and garlic were 'esteemed edibles presented to the country by the white men'.[27]

The speed with which many Māori communities took up the new trading opportunities testifies to the flexibility and dynamism of Māori society. The motivation was, as always, to increase communal wealth, prestige, power and comfort. While the importance of hospitality and gift exchange in Māori custom is frequently emphasised, 'a tendency for modern scholars to categorise all exchanges of goods as "gift exchange" obscures the economic aspects and ignores the element of negotiation in the more mundane exchanges of commodities'.[28] As the traditional economy changed, entrepreneurial skill became a new part of the chiefly job description alongside diplomatic, military and oratorical ability.[29]

The commodity trade presented threats as well as opportunities and could easily destabilise a competitive society. At Maketu the Danish privateer and trader Philip Tapsell, who had spent 30 years in the Pacific, set up as a flax trader on behalf of a Sydney firm, employing 300. Unwittingly he revived conflict between Ngāi Te Rangi and Te Arawa, with considerable bloodshed. Raukawa Moana/Cook Strait was heavily fought over for its trading opportunities; once there, Te Rauparaha traded flax for guns. Polack's Kaipara hosts were threatened by Waikato raiders and he had to pay muskets for flax.

If Europeans wanted flax, timber and a labour force, they often had to arm their hosts, whose security from attack was the Europeans' security to trade.[30]

Flax exports soared between 1827 and 1830 in a classic boom and bust, with the material often inadequately prepared and of poor quality; it could end up as mattress stuffing. Even so, flax was often more profitable than timber. Kahikatea was easily accessible but not durable; much of it was used in Sydney for building timber. Kauri was preferred for ship's spars, but the requirements were precise as to length, diameter and taper, with one in 10,000 good enough for a main mast. By 1840 considerable damage had been done to the northern and Port Nicholson forests through indiscriminate logging; often, much destruction was caused for the sake of a few prize logs.[31]

Initial colonisation

By the 1820s, whaling and shipping entrepreneurs began to advocate for British colonisation, exaggerating the islands' virtues. In 1823 a proposal for a military colony in the north stressed the 'delightful climate ... uncommon fertility of the soil, [which gives] ... all the necessaries and most of the luxuries of civilized life ... there is no country on earth more favourably circumstanced for the operations of agriculture than New Zealand'. In 1826 another group petitioned for a colony to secure supplies of timber and provisions, allow naval protection of the Pacific trade, and introduce to Māori the 'peaceful pursuits of industry and the blessings of civilisation and of religious and moral instruction'. With economic advantage cloaked in altruism, it is no surprise that, as well as Samuel Enderby and Son, the signatories included Joseph Somes and L. Marjoribanks and Co. – later luminaries in the Wakefields' New Zealand Company.[32]

The colonisation lobby strengthened in the late 1830s. The Wakefield ideals are well known, as is their detachment from reality. On one important issue, however, Edward Gibbon Wakefield anticipated what followed. His prescriptions implied a new international division of labour, with the colonies trading primary produce for metropolitan manufactures. Once in New Zealand, Company agents saw what they wanted to see. William Wakefield described the Hutt Valley as 'of the richest soil, and covered with majestic timber'. Charles Heaphy stressed the climate: summer 'may be said to last for eight months'. Romantic notions of New Zealand were, as a Nelson settler tartly noted, 'written on the strength of a scramble in the Bay of Islands'. In reality the climate and fertility varied widely; European parallels were of very limited applicability, and it took some time for settlers to build up the experience and understanding necessary for successful colonisation.[33]

Systematic colonisation also relied on certain assumptions about land. In 1843 a settler newspaper put the matter baldly:

> Who has the right to most land? He who has cultivated it. This is God's law, and all the chatter about rights of the natives to land, which they have let lie idle and unused for so many centuries, cannot do away with the fact that according to this, God's law, they have established their right to a very small portion of these islands.

Land should no more be considered the property of the uncivilised human beings who find themselves on it, but cannot use or improve it, than of the wild animals of a lower order who roam over it for food or prey, with as much but scarce more ignorance of its beneficial capabilities, and equal inability to turn them to use.

These attitudes conflicted with Governor FitzRoy's ideal of European enclaves in a Māori New Zealand. Māori themselves were often influenced in land sales by conflict over trading opportunities. In controlling the Port Nicholson whaling stations, Ngāti Toa had monopolised trade; Te Āti Awa wanted a European settlement to confirm their own hold on the district. Sales often overrode the rights of others living on the ground; and surveying was highly contentious.[34]

Colonisers were also highly optimistic about commercial prospects. Ernst Dieffenbach, the German radical and the New Zealand Company's dissident naturalist, sardonically criticised the mania for building towns:

To found a dozen capitals and commercial ports, and more than two score of villages, before any population is in the island, any produce raised to support a population, or any article of commerce ready to be exported, is subverting the natural order of things, and would have raised a smile on the lips of William Penn, who is often regarded as the father of modern colonization.

Dieffenbach advocated something like what developed in Nelson: allowing labouring families to squat for 15 years on 10-acre blocks. That would create a class of peasants; only then would capitalists have a reason for coming. Dispersed settlements, he believed, meant weak centralised government and, conversely, strong and democratic local communities. This yeoman view saw land as the passport to independence and self-respect.[35]

Nevertheless, the towns quickly emerged as trading centres and beachheads (Figure 4.3). Auckland became a major entrepôt with an aggressive mercantile community. The young Scotsmen John Logan Campbell and William Brown built a boat and began at Waitemata as 'merchants, commission-agents, pig and potato brokers, anything and everything in both a large and a small way', for Māori were keen to sell their produce and buy clothing, tools and tobacco. Auckland's merchants relied on Māori supply and demand for most of the 1840s.[36] The trading economy differed only in scale from that of the 1830s.

Agricultural progress was slow. Many settlers preferred fernland to swamp; although the latter was often more fertile, draining proved particularly arduous. Fern was difficult enough. First it was burnt, then grubbed before ploughing and liming. At the end of 1843 Nelson counted only 83 farmers, not including subsistence squatters. Five hundred and forty acres was cultivated, with another 50 acres cleared, and 133 acres in gardens. No farm had more than 30 acres ploughed or in cultivation, and most had less than 20 acres. There were only 1130 sheep, and fewer than 500 grazing cattle. Domestic pigs outnumbered sheep until May 1844.[37]

European exploration of the North Island was slow. The central North Island was trisected in the 1840s: Dieffenbach travelled from Kawhia to Taupo; Richard Taylor

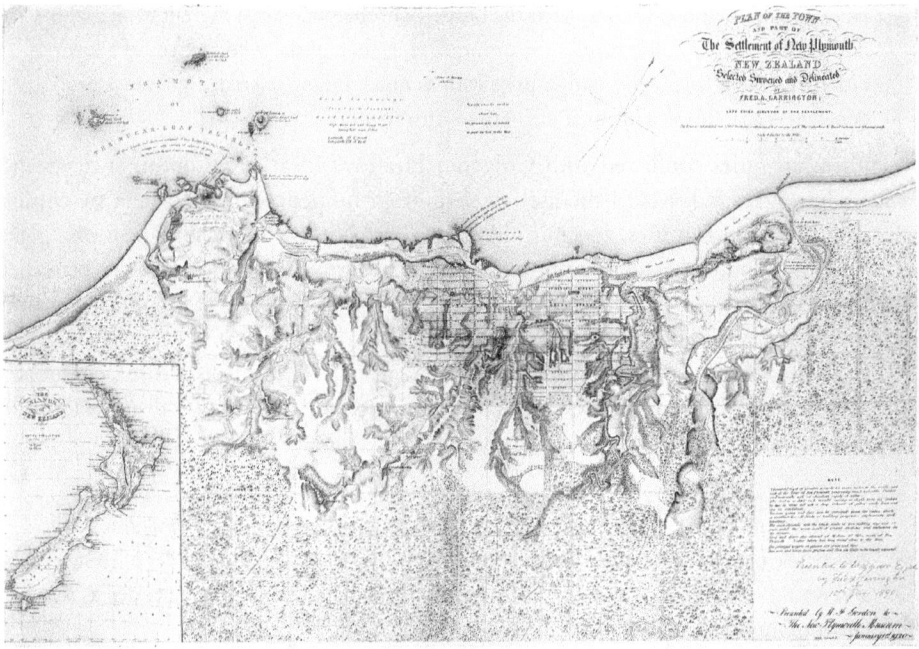

Figure 4.3 Frederic Alonzo Carrington, 'Plan of the Town and Part of the Settlement of New Plymouth, New Zealand, 1842' (1841–42). Source: ARC2004-287, Puke Ariki, New Plymouth

crossed from Whanganui to Taupo in 1845; and two years later Colenso reached Taupo from Hawke's Bay. Explorers were avid botanical collectors and forwarded many specimens to the Royal Botanic Gardens at Kew, which received a consignment from somewhere on the globe for every day of the Victorian era, and distributed plants and seeds to governments, firms and individuals throughout Britain and beyond. Pākehā exploration was overwhelmingly driven by utilitarian considerations, but utilitarianism did not imply efficiency.[38]

New Zealand Company surveys were superintended by a surveyor who had never left Britain – a circumstance much in line with the Company's predilection for theorising over practical experience. In Wellington the Company's principal agent, William Wakefield, a former convict and mercenary noted for 'capriciousness, ruthlessness … vindictiveness', autocracy, laziness, and lack of imagination, constantly second-guessed his principal surveyor, William Mein Smith. Nor did the rectangular perfection of the Company's idealised plans suit the topography. For a long time Wakefield would not allow more practical subdivisions.[39]

When Company surveyors crossed the Rimutaka Range at the end of 1841 they found level land suitable for Wellington's pastoral hinterland. In April 1844 the first pastoralists negotiated leases with Māori and established a station at Wharekaka, near Lake Wairarapa. The merino sheep came from Australia, and the station manager from the Scottish Borders. By March 1847 there were 1300 cattle and 13,000 sheep in the

Wairarapa, and the North Island pastoral frontier began to move into Hawke's Bay. Wairarapa flocks stocked much of Marlborough and Canterbury as well, with New South Wales constantly reinforcing the New Zealand gene pool.

The government – contrary to the Waitangi Treaty – forbad Māori from directly leasing land to settlers (Chapter 3). This edict proved essential in the purchase of the Wairarapa and Hawke's Bay, for it threatened Ngāti Kahungunu income and left them with little choice but to sell land in order to raise capital, despite the obviously low price. Pastoralists supported Crown purchase, preferring the familiarity of English tenure. From 1853 Lands Purchase Commissioner Donald McLean secured large blocks. Ngāti Kahungunu had prospered by leasing; in 1849 they were 'to a very great extent in the enjoyment of European comforts … in every village is to be seen the wheat-field, the stack, and mill, and what is still more gratifying, the use of bread is now becoming universal, and is an article of daily consumption'.[40] Where European colonisation had initially been partly justified by the scope for inspiring such technological change among Māori, once settlers demanded the land, Māori adoption of wheat and flour milling was ignored or dismissed as irrelevant; Māori were characterised as not using the land and were remorselessly pushed aside.[41] Inculcating the moral economy of capitalism was now less important than imposing state control; but after the Wairarapa and Hawke's Bay, European survey and purchase accounted for little else in the North Island before 1860.

Exploration from Nelson was likewise driven by the desire for grazing land. The Wairau grasslands were disputed, with bloody results at Tuamarina in 1843. Misleading or misheard accounts of inland grassy plains prompted various explorations by William Fox, Charles Heaphy and Thomas Brunner between 1846 and 1849. Heaphy, having travelled as far south as the Arahura River, reported that 'the Port Cooper country [Canterbury] appears to be decidedly the most appropriate locality for the next settlement'. Brunner and his guide Kehu's epic journey of 1846–48 confirmed that view. A decade later the Nelson government, expecting mineral wealth, commissioned further exploration of the West Coast. Confirmation of the heavy forest did nothing to dent boosters' optimism. Some suggested that axemen could migrate from eastern Canada to clear the land and let 'wilderness become a fruitful field'. Julius Haast wrote enthusiastically about inland West Coast valleys, particularly the Maruia and the upper Grey, and reported the Grey Valley coalfields, 'a vast store of mineral wealth'.[42]

The pastoral south

The South Island east coast, Wairarapa, and Hawke's Bay quickly became the core of the colonial economy, with New Zealand supplying raw materials for the British textile industry. Having learnt from northern embarrassments, the church associations that organised the colonisation of Otago and Canterbury ensured that the land was 'purchased' well in advance of colonisation. In 1844 the New Edinburgh (Otago) purchasing party travelled extensively.[43]

Wakefield regarded pastoralism as inherently uncivilised, but the suitability of wool as the principal export staple rapidly became clear. David Monro described the New

Edinburgh expedition for local readers – such accounts were crucial in informing settler efforts. He left his readers in no doubt as to the pastoral potential of the east coast: he was 'convinced that [wool] can be grown with greater profit there than in any part of Australia'. The sparse Ngāi Tahu population was not expected to present an obstacle.[44] South Island pastoralism spread rapidly from its Marlborough toehold; after the largely fraudulent Wairau purchase in 1847, Nelson capitalists quickly took up the northeastern valleys. Kemp's massive 1848 'purchase' opened the way for Canterbury and further pastoral expansion (Figure 3.5, Chapter 3).

Pastoralism depended largely on Australia for stock, capital and expertise. Although sheep had been introduced into Australia in 1803, it was not until the early 1830s that wool became the dominant Australian export. For a long time mutton was more lucrative; and British tariff policies hampered colonial wool trade as much as maritime produce. From the early 1820s power machinery in the Yorkshire woollen industry enabled fine merino wool to supplant coarser fibre, and expert Australian wool growers ran sheep to suit. From the late 1830s, with wool prices booming and land prices rising, there was a rush to the easy open country of Port Phillip.

In similar fashion Australian pastoralists moved across the Tasman a decade later. Australian pastoralism had reached the end of first expansion, land tenure was becoming complicated, small farmers assertive, environmental degradation apparent, and New Zealand was cooler and less dry. A few pastoralists crossed to Canterbury early in 1851; their descriptions attracted a number of others. Australian pastoralists were typically from the rural lower middle classes in Britain. Some worked first as station managers; some profited from the Victorian goldrush; others traded on their skill to form useful connections with richer partners. Some New Zealand pastoral fortunes can be traced back to whaling and shipping – like that of the Rhodes family, who arrived in 1839. Robert Campbell of Otekaieke, although he was born in England, was related to the Sydney merchant house. George Moore of Glenmark, who may have been the richest man in New Zealand at the time, began as managing partner for his Quayle inlaws and their Kermode kin; both had been Hobart whalers. Such families bridged the resource and settler economies.[45]

During the 1840s and 1850s British demand for wool grew rapidly, and woollen merchants and manufacturers, particularly in the West Riding of Yorkshire, devoted much time to identifying possible sources of wool around the globe and advising and encouraging local pastoralists to produce what was required. Broadly speaking, Bradford manufacturers required long-staple and relatively fine wool for worsted cloth, while Scottish Borders manufacturers wanted shorter stapled wool for woollen cloth. British demand coincided with New Zealand colonisation, but the economic relationship that drove Australasian environmental change was not inevitable; it was created.[46]

Pastoral occupation contributed unevenly to settler geographical knowledge. In Canterbury, except for the search for an inland route to Nelson and the Wairau – successful only in 1852 – exploration was largely concerned with finding grazing country. J.B.A. Acland and Charles Tripp explored much of the upper Rangitata and Ashburton, and Samuel Butler the Rakaia, Waimakariri and other major catchments.[47] The Main

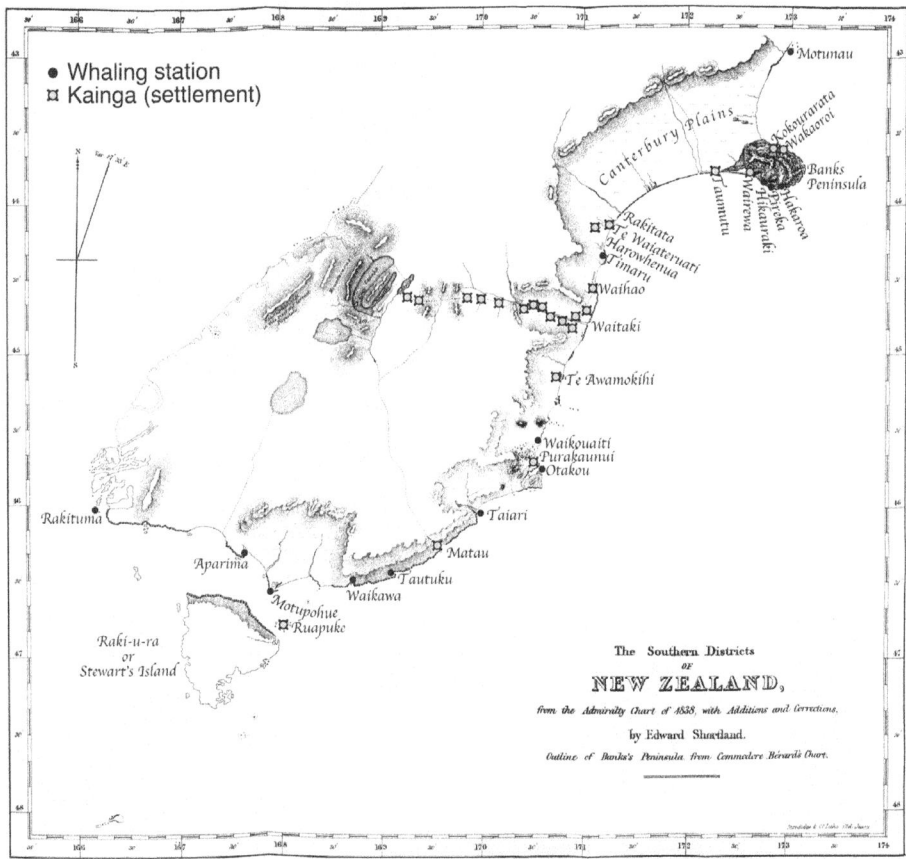

Figure 4.4 Map of Europeans and the south, 1843. Source: scanned, to retain key details, from the map in the inside cover of Edward Shortland, *The Southern Districts of New Zealand*, Longman, Brown, Green & Longman, London, 1851

Divide passes of Canterbury were all assessed in 1863–64 in terms of their suitability for a road to the West Coast goldfields.

Prior to the 1848–51 *Acheron* expedition, European knowledge of much of the New Zealand coastline was little advanced from Cook's time. The French ship *Astrolabe* had charted Cook Strait and the Hauraki Gulf in 1826–27. Edward Shortland's long journey along the South Island east coast in 1843–44 was not published until 1851, and Shortland was more an ethnographer than a cartographer (Figure 4.4). Accurate mapping of the coast was essential for the growing shipping trade and for the infant Otago colony and the imminent Canterbury settlement. *Acheron* parties also explored areas such as the Kaipara Harbour and (guided variously by Māori and by Pākehā whalers) the Kaikoura Ranges, and the terrain between Foveaux Strait and Dunedin.

The southern interior was opened to European eyes by John Turnbull Thomson, who walked 2500 kilometres through South Otago in 1857, effecting a reconnaissance survey of a million hectares. Later that year he repeated the exercise in North Otago and inland as far as Wanaka and north to Pukaki. With the map filled in, pastoral occupation proceeded rapidly; 'instead of [the interior] being a dismal wilderness of snowy mountains, as popularly believed, it was found to be one of the finest sheep countries in the world'.[48] Like many other explorers, Thomson had the rough outline of major rivers provided by Māori – in his case Reko of Tuturau.

More detailed surveys swiftly followed in Otago, but in most other provinces the system was much more lax, with very basic surveys only following selection and serious and accumulated error persisting beyond 1870. Surveyors themselves had to persuade their political masters to allocate funds necessary to go beyond reconnaissance. The rapidity with which the European footprint was placed on the landscape can be easily overestimated.[49]

Conclusion

The environmental transformation of New Zealand between 1769 and 1860 encompassed European discovery, trade in maritime commodities, biological transformations resulting from the new crops, and the beginnings of organised British settlement and the pastoral economy. Transformation was intimately linked to London, Sydney, Hobart, Calcutta, Canton, New England and Yorkshire, and was a consequence of the revolutionary expansion of capitalism.

Cook's voyages had profoundly utilitarian agendas, despite the romance of discovery admitted by participants and emphasised many times since. Cook and his scientists mapped, listed, assessed and positioned New Zealand as a base for maritime exploration and commerce over the next 70 years. The penal colony in New South Wales was a direct consequence of British maritime expansion and created the commercial base from which New Zealand maritime resources were exploited, even pillaged, from 1792 for European, Asian and American markets. Cook's reports and species introductions foreshadowed the way in which New Zealand would service expanding British and American maritime fleets. Potatoes and pigs spread extremely rapidly, and Māori adopted new crops for trade. Cook assessed New Zealand as a site for a settlement colony, and in time that agenda was also adopted.

Yet there were fundamental differences between the resource economy before 1840 and the settler economy thereafter. Both were driven by expanding capitalism, but the resource economy could accept tribal autonomy and – as it developed – the settler economy could not, and sought systematically to relieve Māori of as much land as possible. Although much attention has focused on the weaknesses and downright failure of the Wakefield schemes, the New Zealand Company shared one critical agenda with yeomen and pastoralists, and the imperial government from 1846: that of the wholesale implanting of capitalist social relations in New Zealand, beginning with the commodification of land. The commodification of land rather than simply its produce led swiftly to the expropriation of Māori and the beginning of New Zealand's development as Britain's outlying farm.

Further reading

Recent work situating New Zealand in a wider network of trade in this period includes Alan Frost, *The Global Reach of Empire: Britain's Maritime Expansion in the Indian and Pacific Oceans, 1764–1815* (Miegunyah Press, Melbourne, 2003) and Tony Ballantyne, 'Sealers, Whalers and the Entanglements of Empire' in his *Webs of Empire: Locating New Zealand's Colonial Past* (Bridget Williams Books, Wellington, 2012, pp. 124–36). Hazel Petrie's *Chiefs of Industry: Māori Tribal Enterprise in Early Colonial New Zealand* (Auckland University Press, Auckland, 2006) provides a window on the moral economy of capitalism and the enthusiasm and rationality of Maori engagement with the new order; and Mike Johnston & Sascha Nolden's *Travels of Hochstetter and Haast in New Zealand, 1858–60* (Nikau Press, Nelson, 2011) focuses on an aspect of initial colonisation. Jim McAloon, 'Mobilising Capital and Trade' in Tom Brooking & Eric Pawson, *Seeds of Empire: The Environmental Transformation of New Zealand* (I.B. Tauris, London, 2010, pp. 94–115) emphasises how the relationship between economy and environment was created not given.

II

Colonising

5.
Settlers transforming the open country

Robert Peden and Peter Holland

THE FIRST GENERATION of European settlers in the open country of New Zealand moved to a different social milieu and physical environment, located almost as far as it was possible to be from the British Isles and situated at the heart of the ocean hemisphere. Their stage was the extensive tussock grass and shrub lands, fernland, wetlands and patchy forests of the eastern flanks of both main islands, which they aspired to transform into productive agricultural and pastoral properties. They worked to realise their ambitions with match, plough, introduced animals and plants. In short order they learned about the seasons and how the weather changed from year to year. With experience they were able to interpret signals of change, using that information to manage their operations and select northern hemisphere animal and plant species likely to do well and yield a satisfactory economic return. Structures were instituted to govern the country and regulate the economy, enabling flows of essential services, materials, investment and information. In so doing, the first generation of European settlers were creating their own places in areas that had supported tangata whenua during eight centuries of exclusive occupation of the land.

Not all their activities proved beneficial and not all their expectations were fulfilled. Although introduced pasture plants and grazing animals did well, many well-known species from Western Europe and North America – plants like gorse, broom and thistles – became major weeds. Animals such as red deer, trout and the rabbit were imported for recreational purposes but soon had an adverse impact on native animals, plants and ecosystems. In effect, settlers were modifying and, in the lowlands, annihilating evolutionarily tested native ecosystems and replacing them with functionally incomplete systems dominated by introduced plant and animal species. This was to trigger significant environmental problems within a generation.

In this chapter we highlight the development of the open country from 1840 to 1920 through the transformative activities of the first two generations of settlers. At the beginning of settlement, pastoralism (extensive stock grazing) was the main system of land use in the open country, but in the 1870s it gave way to more intensive pastoral farming or mixed crop and livestock farming at lower altitudes when landholders began to fence, plough and drain their land.[1] Isolation, environmental constraints and rabbits prevented landholders in the drier and mountain lands from following their lowland counterparts,

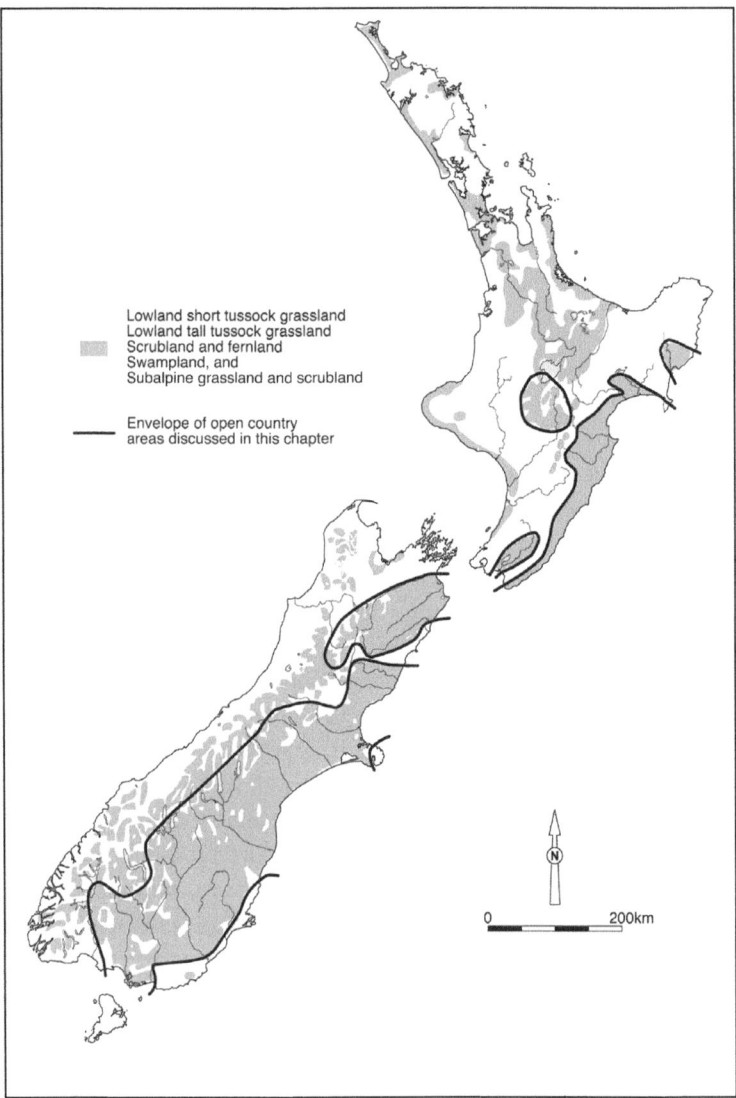

Figure 5.1 Map showing the open country. The parts discussed in this chapter are delineated by thick lines. Source: redrafted from maps in Ian Wards (ed.), *New Zealand Atlas*, Government Printer, Wellington, 1976, pp. 104, 105

and they remained pastoralists. Until the late 1890s, landholders were effectively creating rural landscapes to their own design, but by 1920 most of the formative work was complete and the state was playing an increasing role through regulation, economic direction and publicly funded research and advisory services.

The open country in 1840

In 1840, the open country comprised tracts of lowland short and tall tussock grassland, low shrubs and scattered grasses, large and small areas of wetland, and subalpine grassland. It extended from the shrub and fernlands of Poverty Bay in the north to the far south of the South Island, with tussock, low shrub and swampy land stretching from the Pacific coast inland to the mountains (Figure 5.1). The chief characteristics of the open country are its diverse topography, climate, soils and vegetation.

Some parts of the open country are almost frost-free, whereas others receive annual snow and experience over 200 ground frosts a year. Annual precipitation ranges from 300 mm in the semi-arid parts of Central Otago to at least 1500 mm closer to the Main Divide. In their diaries early settlers regularly referred to the windy weather. Foehn winds east of the Main Divide are common over much of the region and at times these can be extreme. Soils vary considerably and often over small distances. For example, on the Canterbury Plains expanses of shallow, coarse-textured soils may occur a few kilometres from friable agricultural soils in areas previously occupied by marshy ground or where deposits of airborne silt had accumulated.

Early European settlers surmised that the region had been an open landscape for only a few centuries. Throughout the lowlands and low hill country of Canterbury and Otago, scorched logs, root plates and pieces of charcoal were preserved in damp depressions and have been dated to between 600 and 800 years ago.[2] It is thought that much of the region was forested until Polynesian settlers used fire to hunt moa. In the central North Island, episodic ash showers sustained tracts of fern and shrub land, as well as the open tussock grassland that had developed after fires set by Māori and by lightning strike, climatic cooling and drying, or combinations of all three (Chapter 2).

From Banks Peninsula south, European settlers were struck by the small numbers of Māori, although they became aware that Māori trails criss-crossed the whole island. In summer Ngāi Tahu had migrated into the interior to collect food and had routes that crossed the Main Divide to the West Coast, where they had traded for greenstone with Poutini Ngāi Tahu. That Māori had a deep knowledge of the land was demonstrated by their naming of apparently minor topographic features.[3]

Until guidebooks published in Auckland by Brett's became available, the first of them appearing in 1883, settlers planning to occupy land in the open country relied on information in manuals produced for settlement companies or published by the New Zealand government.[4] The earliest of those guidebooks contained glowing accounts of what settlers could expect, and suggested a congenial landscape with fertile soils and a uniformly temperate climate.[5] A frequently drawn comparison was with the weather of northern Italy, something that the administrator of the Canterbury settlement, Henry Sewell, clearly found hard to accept when he wrote the following entry in his diary on 10 February 1853: 'A cold wet afternoon [in Christchurch] by no means answering to the idea of an Italian climate.'[6] Spells of calm, mild weather would often end with a day or two of high winds, heavy rain and snow in winter. This variable, often very windy weather came as a surprise to the first generation of European settlers, and few were prepared for stormy years like 1868 and 1869, widespread flooding in September 1878, or prolonged periods of below average precipitation during the early 1880s. Against that, however, they enjoyed advantages over farmers in Britain in not having to make as much hay, set aside as much of their arable land for turnips and swedes to sustain livestock over the coldest months, or house animals during winter.

Early written accounts, drawings and paintings of the South Island open country portray an almost treeless landscape stretching from Marlborough to the Southland plains, as well as in Central Otago and the Mackenzie Country.[7] This made it an easy point of entry for settlers into pastoralism and farming, as it was much cheaper and far simpler to clear tussock, fern and low shrubs than heavily forested land. Using their knowledge of farming on the open hills of Britain, pastoralists knew that by burning standing vegetation they could quickly establish grazing lands for their livestock.

The majority of the early pastoralists and farmers in the open country were British, although a minority came by way of the Australian colonies. Much has been made of an Australian influence in the establishment of pastoralism in New Zealand, particularly in Canterbury.[8] This argument ignores the leading role played by settlers who came directly from Britain – people like Frederick Weld and Charles Clifford, who established Wharekaka Station in the Wairarapa in 1844, one of the earliest sheep runs in the country. In 1847 these two men set up one of the first sheep stations in the South Island, at Flaxbourne on the Marlborough coast, and in 1850 they established Stoneyhurst Station, one of the first runs in North Canterbury.[9] Arguments for the significance of the Australian influence often overlook the fact that the management of sheep in New Zealand in the early days was more akin to that of the hill districts of Britain than it was to the Australian colonies.[10]

The intellectual, economic and social environments of settlement

If we are to appreciate how and why the settlement and transformation of the open country took place as it did, it is necessary to understand the intellectual, cultural and economic environments within which European settlers operated. Colonisation was predicated on the sweeping economic and social changes then taking place in Britain. The industrial revolution, which was made possible by a host of technological advances, had transformed Britain's economy, and the machine age gave Europeans and North Americans a new conception of human power over the physical environment. Underpinning these changes was the belief that through the application of labour, capital, technology and science people could make the world a better place.[11] The ideology of progress and improvement was a tenet of Victorian society: in a changing world, the Victorians thought that they must move purposefully ahead, otherwise their society would stagnate or perhaps even decline into barbarism. Imbued with the improvement ethos, settlers saw the landscape as a blank canvas that could be transformed according to their wishes.[12] The search for resources to meet the demands of industrial expansion and the desire to ease the pressure of an expanding population were two key forces in the colonisation of the 'new' lands outside Europe.

The European settlement of New Zealand coincided with the spread of technical and scientific inputs into farming. The founding of the Royal Agricultural Society of England (1838), the publication of von Liebig's *Organic Chemistry in its Application to Agriculture and Physiology* (1840), and the commencement of systematic field experiments at Rothamsted research station near London (1843) have been described as major turning points in agricultural science.[13] The opening of the Royal Agricultural College at Cirencester in 1845 established formal agricultural education in Britain, and in 1851 Cyrus McCormick displayed his mechanical reaper to the world at the Great Exhibition. Agricultural historians regard the period from 1850 to 1875 as the golden age of 'High Farming' in Britain.[14]

It was against this backdrop that early British settlers first explored and occupied the open country of New Zealand and turned their attention to wool growing. In the Australian colonies pastoralists had already shown that wool production was not only economically viable, but at times also very profitable. The needs of British mills could not be satisfied without imports, so there was a ready market for wool from South Africa, New Zealand and the Australian colonies. Demand for wool was strong through the 1850s and early 1860s, especially during the Crimean War (1853 to 1856) and the American Civil War (1861 to 1865). Fine wool fetched high prices because it could be made into fabrics that were soft and comfortable. The expansion of the urban middle class in Britain, Europe and the United States also created a large market for worsted cloth made from medium wools; military uniforms, blankets and rough textured cloth were made from coarser wool. Wool had the advantage that it could be stored and transported without loss of quantity or quality, which was crucial when growers and manufacturers were half a world away from each other (Chapter 4).

The entrepreneurial spirit of Victorian society was reflected in the attitudes of the first two generations of pastoralists and farmers in the South Island open country. Jim McAloon, in his analysis of wealth creation in Canterbury and Otago before 1914, noted that in southern New Zealand farming was very much a business proposition, farmers of the time believing that success was based on sound training and hard work.[15] The social commentator and novelist Samuel Butler, who in the early 1860s held the lease of Mesopotamia, a sheep station in the upper Rangitata Valley, observed those values in Canterbury when he wrote that 'people here are busy making money; that is the inducement which led them to coming in the first instance, and they show their sense by devoting their energies to work'.[16]

Despite their distance from Britain and Europe, settlers kept themselves abreast of advances in agriculture and quickly developed formal and informal ways to disseminate this information. Newspapers played an important role in this.[17] In Canterbury the *Lyttelton Times* was established in January 1851, a few weeks after the arrival of the first colonists. The *Otago Witness* first appeared in Dunedin in the same year, and by the mid-1860s newspapers had been established in Hawke's Bay, Wellington, Timaru, Oamaru, Milton and Invercargill. They published original pieces and reprinted articles on agriculture and the markets from the numerous farming periodicals and newspapers published in Britain, Europe, North America and the Australian colonies and imported into New Zealand.[18]

The first generation of settlers also established agricultural societies to promote improved pastoral and farming practices. The societies sponsored lectures, demonstrations and competitions; but in time their main purpose became the organisation of agricultural shows. The first was held in the Bay of Islands in 1842, followed the next year by the Agricultural and Horticultural Society of Auckland. The Christchurch Agricultural, Horticultural and Botanical Society was formed in July 1853 and held a cattle show in October that year. These societies eventually became agricultural and pastoral associations, and others were established throughout the country.[19] The Canterbury Agricultural and Pastoral Association was one of the more active. It was formed in 1863 and held its first show later that year. In 1877 it began publishing *The New Zealand Country Journal*, which was distributed around the country and included material written by leading local farmers and articles reprinted from journals and newspapers from across the world.[20] It kept farmers up to date with wool, meat and grain prices in the local market, and market information from Australia and Britain.

Farmers' clubs were another mechanism for disseminating knowledge. These were usually run by and for small farmers, as opposed to the large open-country farming operations. Farmers' clubs organised demonstrations, ploughing matches and social events. They also held regular meetings where speakers gave lectures on topics such as crop rotation, the use of manure and the selection of sheep.[21] But perhaps the most remarkable achievement in the domain of promoting agricultural improvement was the establishment of an agricultural college, with a 500-acre experimental farm, at Lincoln on the outskirts of Christchurch. This was the first agricultural teaching institution in Australasia and one of the earliest agricultural colleges in the British world. Instruction began at Lincoln in 1880.[22]

Initially, open-country farming was dominated by pastoralists on large properties growing wool for export to Britain (Figure 5.2). Wool returns were greatest where large flocks were grazed in a low-cost system with minimal labour. Pastoralists grazed their livestock on land leased cheaply from the Crown, although when they started to develop their properties, landholders on the plains and better lands began to purchase the freehold of large parts of their runs. Farmers on freehold blocks near townships, with their small flocks, could not survive on wool alone so they fattened stock, grew grain and milked dairy cows, mainly for the local market. It was not until the mid-1890s, when the refrigerated meat trade to Britain had become firmly established, that small- and medium-sized farms were fully integrated into the export trade.[23]

From the 1870s there was growing pressure to subdivide the big estates on the better lands for closer settlement. The success of the frozen meat trade made this possible as small farmers now had two streams of income: wool and meat. The Liberal government actively encouraged the break-up of the great estates to establish smaller runs and farms that would be owned and operated by individual farmers and their families. The Liberals began the process in 1892, but most of the subdivision was done voluntarily over the following two decades as large landowners sold off their surplus land for substantial profits.[24]

Figure 5.2 Sheep in the holding yards at Shag Valley Station, North Otago, during the 1880s. Chemicals to treat sick sheep were stored in the small shed near the drafting race. Source: Hart Campbell Collection, Box 144, ISS, HC026, Hocken Collections, Uare Taoka o Hākena, University of Otago

Although wool growing was occasionally very profitable, returns were volatile and between 1870 and 1905 they fell precipitously. Wool production increased during the second half of the nineteenth century, especially in Australia but also in New Zealand, South America, South Africa and India, which served to undermine prices in the international marketplace. In 1887 John Barton Acland, who owned Mt Peel Station in South Canterbury, compiled statistics showing that the quantity of wool imported into Britain had increased from 151 million pounds (68,640 tonnes) in 1860 to 615 million pounds (279,550 tonnes) in 1886.[25] Increasingly, wool producers were facing competition from other fibres. Cotton production rose markedly after 1865: British manufacturers had encouraged cotton production in India and Egypt during the 'cotton famine' caused by the American Civil War, and when the war ended American cotton flooded back on to the market and at the same time production increased in the new cotton-growing regions.

The decline in wool prices initiated a massive change in land use on the plains and downlands of the open country, where pastoralists could diversify into large-scale cropping, growing mostly wheat, oats and barley for domestic consumption and export. Once the frozen meat trade became profitable, interest in cropping declined as large landholders and small farmers turned to permanent pasture, turnips and swedes to fatten breeds of sheep selected for meat and wool production. In contrast, in the high country (a term in use by the 1890s that remains current today), aridity, rabbits, altitude and isolation meant that landholders did not have these options and they continued to rely on fine wool as their only source of income. As wool prices fell through the 1880s and early 1890s, these large pastoral properties became economically marginal (Figure 5.3). The success of the frozen meat trade entrenched the divide between those low-country

Figure 5.3 Birch Hill Station at the head of the Tasman Valley in the late nineteenth century. This is an example of a remote and difficult high-country run. It was incorporated into Mount Cook National Park in 1953. Source: c/nF74/20, Hocken Collections, Uare Taoka o Hākena, University of Otago

farms that were able to breed and fatten sheep for export and the hill- and high-country stations that could not. High-country stations benefited from the frozen meat trade only in good years, when they had surplus sheep to sell to low-country farmers who used them as breeding stock or fattened them for export.[26]

Fire, tooth and hoof

Transformation of the open country began with the simple practices of burning, grazing and trampling. Burning was used universally to clear heavy vegetation and encourage the growth of palatable fine grasses and broadleaf herbs. Plants like anise, blue tussock, holy grass, plume grass and wheatgrass sustained the rapid growth in sheep numbers during the early years of pastoral expansion on the open country, but these favoured plants were soon grazed out and pastoralists were compelled to introduce pasture species to replace them.[27]

Regular burning was practised to prevent tall tussocks and woody shrubs from invading the short tussock grassland created by the early fires, but the frequency of burning depended on environmental conditions. In the fern country of Hawke's Bay heavy trampling was required to crush bracken before a burn, and the combination of trampling and grazing afterwards checked emerging fern fronds.[28] In the early years there was insufficient stock to do either task efficiently, so the cycle had to be repeated every few years.

Tussock burning has been the most contentious land management practice of the early pastoralists: criticism of it was particularly vociferous during the first 60 years of the twentieth century. Critics have claimed that pastoralists fired the country all year round, used fire indiscriminately and burned for no reason.[29] In the 1980s scientists used pollen analysis, carbon dating, vegetation transects and repeat photography to reassess this criticism and concluded that burning did not result in the wholesale destruction of the native vegetation, nor did it precipitate phases of erosion in the scree-covered mountain slopes of Marlborough and Canterbury.[30] More recently, historians studying station records and diaries have found that early pastoralists were generally very careful about the practice of burning, which they used to enhance or maintain pasture quality for sheep.[31] Pastoralists burned in spring when plants recovered best and when fires were easily put out with the onset of dew in the early evening. In dry areas settlers burned less frequently, as grazing alone could maintain the desired grassland environment. Burning took place every year in moist regions, but the same piece of the country was not burned annually, as some critics of tussock-burning have claimed.[32]

Matching plants to environment

After burning tussock and clearing low shrubs, landholders broadcast grass and clover seed in the ashes. A few months later they turned out their livestock on the young sward. Widespread ploughing of the open country began in the late 1860s and was almost complete by the 1890s. The best of the land was planted in wheat and oats, and several successive grain crops were often harvested from the one field without the addition of a mineral fertiliser.[33] The remaining land was converted to grass. For much of the nineteenth century a common aspiration was a species-rich, multilayered artificial pasture that could provide nutritious and palatable herbage for as long as possible throughout the growing season, would last for a decade or longer before resowing, would persist by self-seeding and would resist the establishment of less palatable species and weedy plants. Experience showed the first generation of settlers that this goal might be achieved in more temperate parts of the country, such as western Southland, but was almost unattainable elsewhere.

Between the 1850s and 1920 advice to farmers about seed mixtures and sowing weights for permanent pasture changed considerably, although for much of the period species-rich mixtures were popular. In 1851, for example, farmers in Canterbury were advised to sow each acre of cultivated land with the following mixture:[34]

4 lb perennial ryegrass	2 lb cocksfoot
2 lb meadow fescue	2 lb meadow foxtail
2 lb rough stalked meadow grass	2 lb timothy
2 lb wood meadow grass	1 lb crested dog's tail
1 lb fiorin	1 lb golden oat grass
1 lb hard fescue	1 lb smooth meadow grass
0.25 lb sweet scented vernal	3 lb perennial red clover
2 lb white clover	

This mixture, based on agronomical research and practice in Britain, lacks a clear dominant species. In southern New Zealand, the first generation of landholders and seed merchants discovered that perennial ryegrass, cocksfoot and white clover were the chief elements of a palatable, nutritious and reasonably long-lasting pasture, and one, two or all three species quickly came to dominate commercial seed mixtures. In 1918, agricultural scientist Alfred Cockayne wrote that New Zealand farmers could select from many more commercially available species than farmers elsewhere.[35] The average diversity of seed in sown mixtures fell after World War I as a result of research carried out by government scientists and technicians. Breeding and varietal selection, supported by certification of purity and germination rate, strengthened the dominance of cocksfoot, Italian and perennial ryegrass, and the importance of red and white clover throughout the open country.

Nineteenth-century farmers knew that seed from some areas was superior – for example, browntop from Waipu for the few who appreciated it, Chewings fescue from western Southland, cocksfoot from Banks Peninsula and Sanson, perennial ryegrass from Poverty Bay and white clover from Manawatu – and some were willing to pay a premium for it.[36] They also recognised that different physical environments called for different mixtures of species. In 1899 a recommended seed mixture for clay soils in pumice country was dominated by Chewings fescue, with abundant danthonia and white clover.[37] A decade later, North Canterbury farmers who wanted a temporary pasture suitable for raising lambs were advised to sow only Italian ryegrass and white clover, while farmers on drought-prone land were advised to sow a mixture of cocksfoot and other deep-rooting grasses, as well as red and white clover, with minor amounts of burnet and chicory to ensure palatable and nutritious herbage for stock during hot dry spells in summer.[38]

Despite experimenting with seed mixtures, landholders often experienced difficulty establishing permanent pastures. Many farmers wrongly ascribed the spontaneous replacement of palatable species by weedy plants to some innate deficiency of introduced pasture plants, whereas, as agricultural scientist E.B. Levy proposed in an article in 1921, the more likely cause was that desirable pasture species required more mineral nutrients than the soil could supply.[39] Even in the 1880s it was clear that when frozen meat and dairy products were exported, essential mineral nutrients such as calcium, nitrogen and phosphorus that plants take from the soil also left the country. Until the advent of cheap artificial fertilisers (for example, superphosphate, ammonium nitrate and urea), guano – imported from the western Pacific and Chile – enabled interested New Zealand farmers to replace phosphorus lost in the meat trade, but it was not used widely.

From the outset some farmers sought advice from diverse sources about the seed mixtures of pasture plant species best suited to their properties and the regional climate. Seed merchants were one such source, but settlers also found guidance in newspapers, farming magazines and books like the widely read *Brett's Colonist's Guide*. We have traced more than 1000 recommended 'recipes' for seed mixtures for the seven decades to 1920, and undoubtedly there were many more.[40] Some included at least 10 species, while others had as few as two or three, and even similar mixtures differed in the weight of seeds for each recommended species.

After the establishment of the Department of Agriculture in 1892 farmers could seek advice from experts employed in regional field stations, and their responses were published in issues of the *New Zealand Journal of Agriculture*.[41] A farmer would usually describe the conditions in which he planned to establish a pasture – altitude and exposure, prior vegetation cover, seasonal water retention and natural drainage, soil organic content and friability – and would seek advice on the best seed mixture. Some farmers even requested details of seed mixtures for different landforms on their properties, such as river flood plain, river terraces, gently sloping downs and the like, and received often subtly varied lists of species for each of those habitats. By 1914 almost 30 pounds of seed on average was being sown to the acre.[42] Perennial ryegrass tended to be the largest component of seed mixtures for well-watered, fertile soils, whereas cocksfoot was usually recommended where there was a risk of two or three months of fairly dry soils in summer.

Even though ecology as a recognised scientific discipline was only nascent until the 1880s, ecological thinking was evident in the advice that pioneer farmers could read in guidebooks produced by settlement companies, the publications of state employees such as Alfred Cockayne, or newspaper articles. They were advised to buy only the best quality seed, select pasture plants suited to the environmental conditions of their properties, sow species with complementary growth surges and rooting patterns to ensure nutritious and palatable stockfeed for much of the year, manage grazing especially carefully until the sown pasture became established and be attentive to problems thereafter, and aim for a mixture of species that would persist for several years, would resist establishment of weedy plants and less palatable grasses, and would not require frequent resowing. Ideally, one or two coarse grasses would form a fairly open canopy, a similar number of fineleaf grasses would form a more uniform subcanopy, and five or more small grasses and broadleaf herbs like plantain and chicory would cover the topsoil and provide variety for sheep and cattle. The poor quality of seed generally available from merchants, and the prevalence of weedy plant seeds in what they supplied, meant that this ecological ideal was rarely, if ever, achieved in colonial New Zealand. Until the advent of government-backed seed certification in the 1930s, the low germination rates and weedy contaminants of commercially available seed were severe trials to grassland farmers.

Matching sheep to the environments of the open country

Raising sheep has been the primary land use in the open country since the start of European settlement, although many pastoral stations also grazed cattle and there were small dairy farms on heavier soils near townships. However, until establishment of the frozen meat trade in the 1880s, pastoralists generally ran cattle in places where they could not graze sheep, such as swamp country, or as an adjunct to sheep grazing to control growth of coarse pasturage.

From the outset, the pastoral system was very simple. Sheep were imported from the Australian colonies and grazed on native pastures, which were modified by burning. Over time this system became more intensive and complex as pastoralists and farmers selected more productive livestock and transformed large areas of the open country by tussock-burning, fencing, cultivation, oversowing and drainage. In the late nineteenth and early

Figure 5.4 Merino lambs being drafted off their mothers on a high-country station immediately before shearing during summer in the 1920s. Source: c/nF460/14, Hocken Collections, Uare Taoka o Hākena, University of Otago

twentieth centuries sheep breeders responded to changes in demand for different types of wool, development of the export meat industry, intensification in land use on the plains and downlands, and an overarching desire to have the right sheep breed in the right place.

Britain had numerous local breeds of sheep adapted to environmental differences across the country, so settlers were mindful of the importance of matching livestock to a property's environment. By the 1850s selective breeding techniques were well established and breeds like the Leicester, Lincoln, Cheviot and Romney had been improved to thrive in particular environments. In the early 1860s the *Lyttelton Times* reprinted articles from British journals expressing the importance of matching sheep with their environment, with observations such as 'the nature of the pasture ... ought to determine the breed of sheep kept on it' and 'sheep should always be adapted to the peculiarities of the country'.[43]

Most sheep owners found that merino were well suited to the system of large-scale pastoralism that dominated sheep production in the early years of settlement (Figure 5.4). Fine wool was in demand, and with low stocking rates merino thrived on natural pastures. However, by the mid-1870s the combination of declining wool prices and intensified farming methods encouraged the development of breeds better suited to the new economic and environmental conditions of the open country.[44] Intensive farming, by way of cultivation, fencing, drainage and oversowing, changed the environment

Table 5.1 Area (in acres) under cultivation, and sheep numbers, for Hawke's Bay, Wellington, Marlborough, Canterbury and Otago provinces.

	Under cultivation	Sheep numbers
1871	865,450	8,920,464
1881	3,895,040	11,682,787
1891	6,711,930	15,318,895
1901	9,326,793	16,656,669
1911	9,781,844	18,727,693
1921	11,149,919	18,830,496

Source: B.L. Evans (compiler), *Agricultural and Pastoral Statistics of New Zealand 1861–1954*, Government Printer, Wellington, 1956

for sheep, particularly on the plains and down lands.[45] In 1871, 865,450 acres was under cultivation in the provinces that made up the open country, but by 1881 that had increased to 3,895,040 acres and by 1891 it was 6,711,930 acres.[46] Cultivation, oversowing and other practices raised stocking rates from one sheep to 3 acres to at least one sheep to the acre on better land. However, higher stocking rates led to the spread of footrot through merino flocks, causing their productivity to decline.

Land improvement presented a conundrum. Many pastoralists embarked on programmes to boost production when fine-wool prices began to decline, but the costs involved meant that they needed higher returns to justify their investment. A light in the gloom was that by the 1870s Bradford mills began paying a premium for longer, stronger wools.[47] Developments in mechanical combing and spinning had given rise to a huge increase in the production and popularity of worsted fabrics, which were stronger, smoother and cheaper than woollen cloth produced from fine merino wool. Halfbred wool (from crossing merino with English long-wool breeds) proved ideal for this trade. Halfbred sheep are also bigger and fatten more quickly than merino, making them suitable for the frozen meat trade as well as wool production.

In response to these changing environmental and economic conditions, more and more pastoralists began mating long-wool sires with their merino ewes to breed a first cross halfbred. Selection of a long-wool sire for crossbreeding often depended on environmental conditions: the Lincoln was favoured on drained swamplands and the heavier soils of the plains; the Leicester on lighter soils and dry, easy hills; the Border Leicester (bred in the Scottish Borders by crossing the Cheviot and Leicester breeds to make a hardy type) in the harder hills; and the Romney Marsh in higher rainfall districts. Inevitably, as crossbreeding increased in popularity there were fewer merino ewes to breed from. In 1880 merinos were still the most numerous sheep breed in the South Island, but by 1895, at 4 million they made up just over one-third of the region's total sheep flock. Over the following decade that number halved to 2 million and by 1912 it had dropped to 1.5 million.[48] In the face of this decline, leading breeders intensified their experiments to fix new breeds of sheep, without having to go through the process of continual crossbreeding.

By 1920 the merino had been replaced everywhere except in the high country and semi-arid lands where grazing remained largely unimproved and stocking rates low. Elsewhere in the open country, sheep breeders had developed three new breeds that were adapted to local environmental conditions as well as to the requirements of the wool and frozen meat trades. The Corriedale, an inbred halfbred, became popular on the lighter plains and dry easy hill country. Sheep owners on harder hill country and easier high country grazed the 'Colonial' halfbred, developed by crossing merino with English long-wool breeds. And in higher rainfall areas, as well as on intensively farmed plains and downs, the New Zealand Romney became the most popular breed because it is less susceptible to footrot and copes with wet rank pasture better than other breeds.

Rabbits

During the nineteenth and early twentieth centuries European settlers brought in many animal and plant species, some of which proved environmentally and economically disastrous. Hedge plants such as boxthorn, broom, gorse and hawthorn required close annual maintenance if they were not to spread into 'waste' land and riverbeds. In lower and better watered country nearer the coast, sparrows and finches, imported to control insects that attacked grain crops, became pests in their own right, and county councils carried out numerous poisoning programmes to deal with the 'small birds nuisance' in the last three decades of the nineteenth century.[49] Individuals and acclimatisation societies released many small and large vertebrates, including the European rabbit *Oryctolagus cuniculus*, which turned out to be an environmental catastrophe for large tracts of the open country.

Rabbits were released in many parts of New Zealand in the earliest period of settlement for sport and also as a source of food. At first they struggled to become established, but by the early 1870s rabbit populations in the Wairarapa, Marlborough, Kaikoura and Southland had begun to increase and spread. It is difficult now to appreciate the vast number of rabbits and their rapid spread between the mid-1870s and mid-1890s. It is also hard to comprehend the devastation they caused, especially in semi-arid parts of the open country. In Central Otago, the Upper Waitaki Valley, the Mackenzie Country, the Amuri and inland Marlborough, overgrazing by rabbits destroyed the vegetation cover and triggered soil erosion.

A powerful way to gauge the effect of the rabbit irruption is through the collapse in stock numbers and production on pastoral stations. A report to Parliament in 1887 noted that 545,000 hectares had been abandoned in Otago because of the rabbit plague.[50] The sheep flock on Castle Rock Station in northern Southland declined from 50,000 to 20,000 as rabbits ate out the country, and it was estimated that this loss in production cost the owners £5000 a year.[51] Rabbiters took 244,000 skins off Kawarau

Figure 5.5 A wagon laden with bales of cleaned and graded rabbit skins awaiting export from the Port of Dunedin in the 1920s. Despite the problems rabbits caused, the sale of rabbit skins supported many rural families and kept some marginal properties stations economically afloat during the 1880s and 1890s. Source: c/nE719/5&6, Hocken Collections, Uare Taoka o Hākena, University of Otago

Station near Cromwell in 1884 and 283,000 the following year. The death rate of the station's sheep flock rose from 3.5 to 10.5 per cent and the number of lambs born per 100 ewes fell from 70 to 45.[52] Earnscleugh Station, a 32,400 hectare property in Central Otago, carried 24,000 sheep in 1879. In 1897, with his flock reduced to 12,500 sheep, the runholder abandoned his lease and the Crown took over the land. In 1902 Stephen Thomas Spain purchased the lease and in his first five months on the property employed 32 rabbiters, who killed over 250,000 rabbits. Under Spain's care the Earnscleugh flock stabilised at around 12,000 sheep, half the size of that before the influx of rabbits.[53] In 1917 W. Nosworthy purchased the lease of Mesopotamia Station, which had carried 23,000 sheep annually since about 1880. By 1920 rabbits had forced him to reduce his flock to 5000.[54] Entries in the cash book for Black Forest Station in the Mackenzie Country during the 10 months from 25 April 1916 show that rabbit extermination accounted for 49 per cent of all listed expenditures.[55]

How to control the rabbit irruption was the major preoccupation among runholders and farmers in the open country. Landholders adopted different methods, including shooting, trapping, dogging, digging out burrows, mass poisoning using grain treated with phosphorus, arsenic or strychnine, and fumigation of rabbit warrens (Figure

5.5). Government agencies and a small number of private individuals imported many thousands of ferrets, weasels and stoats in the belief that they would provide a cheap and effective measure of control. They also bred these animals in captivity, then released them in large numbers. Mustelids were largely ineffective in rabbit control but took a heavy toll on New Zealand's native birds and lizards. The government and private individuals also constructed rabbit-proof fences in an attempt to keep rabbits from spreading, but they too were ineffective.

From the outset of the first rabbit irruption the government played an active role in rabbit control and passed the first of many Rabbit Nuisance Acts in 1876. The 1881 Act established a system of rabbit inspectors, and another Act in 1882 increased their powers. They had the right to inspect properties at any time, could take landholders before a magistrate, and could oblige landholders to carry out control measures. Inspectors came under the authority of the Department of Agriculture after it was set up in 1892 and rabbit control became one of the department's major functions, accounting for a quarter of its budget in 1895–96.[56]

The impact of the rabbit plague had far-reaching consequences for affected runs, especially those in semi-arid areas. Lost production was a huge cost, as was the ongoing expense of rabbit control. As a result, runholders lacked surplus capital to invest in improvements and became reluctant to spend money in development because a sudden lift in rabbit numbers could wipe out oversown or drilled pastures within a season. In semi-arid areas, the only improvements were on the best soils of the lower slopes and flats that could be protected by rabbit-proof fencing and by control measures such as shooting and poisoning. Outside these pockets, the open hills and mountains remained undeveloped and at the mercy of the next rabbit irruption. Consequently, production in the semi-arid lands declined from pre-rabbit levels, then remained static until implementation of the technological and scientific advances that followed World War II.

Conclusion

By 1920 the open country of New Zealand was a highly modified landscape. Indeed, the plant cover of most low country was almost unrecognisable from what it had been in 1840. Settlers had ploughed the tussock and low shrublands, cleared fern, drained swamps, subdivided the country with fences and hedges, and planted blocks of trees which were beginning to provide shelter from the ever-present winds. Leading sheep breeders had been successful in developing new breeds that were better adapted to the new environmental and economic conditions. They had made progress in selecting seed mixes suited to the different environmental zones found across the open country, and by 1920 government scientists and technicians had begun breeding grasses and legumes that would suit local conditions better than imported species.

Stock loss and property damage during episodes of adverse weather, the high cost of establishing and maintaining infrastructure, fluctuating returns for primary products on the English market, declining soil fertility and trace element deficiencies led to a reduction in economic production and a decline in optimism among settlers. Farmers struggled to maintain permanent pastures and had to plough and resow every few years. A range of weed species invaded low fertility sites: gorse and broom spread across hill districts and along riverbeds; sheep sorrel and several varieties of thistle invaded pasturelands. Trace element deficiencies showed up in poor stock health and reduced productivity, and internal and external parasites caused high death rates in young stock and reduced the productivity of older animals.

The problems of the low country were magnified in the semi-arid lands and some hill- and high-country areas where the rabbit irruption that followed World War I wreaked further damage on lands that had not recovered from the first rabbit plague. By 1920 these lands were in a sorry state. In places the tussock country was grazed bare and landholders struggled to remain on their properties. In the worst areas, without vegetation to hold it, the soil blew away in nor'west gales.

The transformation of the open country between 1840 and 1920 had been crucial in the development of the economy of New Zealand. The area had been the engine room for the country's economic growth from the 1850s, producing fine wool and grain for export, and for several decades after 1882 it produced most of the meat for the frozen meat trade. However, all that came at a cost to the land. By 1920 the productivity of parts of the open country had tapered off and farms in the recently cleared forest lands of the North Island, utilising the untapped soil fertility, had taken over in terms of the number of stock carried and in wool and meat production. The problems that constrained production in the open country were later to emerge in the former bushlands as soil fertility declined there. It was not until after World War II that farmers had access to the scientific knowledge and technology needed to solve these problems. Some problems, like rabbits and weeds, remain with us still.

Further reading

Three books which emerged from the 'Empires of grass' Marsden-funded project discuss the themes of this chapter at a range of scales: Tom Brooking & Eric Pawson, *Seeds of Empire: the Environmental Transformation of New Zealand* (I.B. Tauris, London, 2010); Peter Holland, *Home in the Howling Wilderness: Settlers and Environment in Southern New Zealand* (Auckland University Press, Auckland, 2013) and Robert Peden, *Making Sheep Country: Mt Peel and the Transformation of the Tussock Lands* (Auckland University Press, Auckland, 2011). Robert Peden also explores a central theme in '"The Exceeding Joy of Burning" – Pastoralists and the Lucifer Match: Burning the Rangelands of the South Island of New Zealand in the Nineteenth Century, 1850 to 1890' (*Agricultural History*, vol. 80, no. 1, 2006, pp. 17–34). Michele Dominy adopts an anthropological perspective in *Calling the Station Home: Place and Identity in New Zealand's High Country* (Rowman & Littlefield, Lanham, 2001) and Jim McAloon writes as an economic historian in *No Idle Rich: The Wealthy in Canterbury and Otago, 1840–1914*, (Otago University Press, Dunedin, 2002).

6.
Mining the quarry

Terry Hearn

MINING PLAYED A KEY role in the New Zealand economy through to 1921, encouraging exploration, stimulating immigration and settlement, promoting internal commerce and fostering technological innovation and industrial development. A wide range of minerals was mined at various times and in various places. These included the metalliferous (especially gold), non-metalliferous (limestone, clays, and aggregates), and fuel minerals (mainly coal).[1] From 1861 to 1865, the industry generated 65.2 per cent of New Zealand's export earnings. It still accounted for 15.3 per cent for 1901–05, before declining to 3.2 per cent in 1915–20. Mining also employed a significant proportion of the European male workforce: 14.5 per cent in 1874, although that fell to just 2.0 per cent in 1921, with a pronounced shift during that time from goldmining to coalmining and quarrying.

Mining, in spite of its benefits to the economy, was recognised as environmentally damaging. In 1890, W.N. Blair attributed the rapidity and scale of landscape change in New Zealand in part to 'the washing down of mountains for gold'.[2] The environmental impact of mining has not, in fact, previously been systematically investigated.[3] This chapter will explore three major issues involving goldmining: the disposal of mining debris, land settlement and the preservation of agricultural land. It will also consider, more briefly, quarrying and coalmining.

The goldrushes

Gold was discovered in New Zealand during the 1850s, but the first large-scale rushes took place to Otago in 1861, followed by Marlborough, Nelson–West Coast, and Hauraki–Thames–Ohinemuri (Figure 6.1). They were part of a nineteenth-century New World series of rushes that began in California in 1849 and extended to New South Wales and Victoria in 1851; Colorado and the Fraser, near Vancouver, in 1858; Western Australia and South Africa in 1886; and the Klondike (Alaska) in 1898. The circulation of capital, participants and practices linked these discontinuous series of events together, as did experiences of excitement and privation, riches and disappointment. Behind the rushes lay a burgeoning demand for gold to finance the expansion of world trade, to sustain the growth of metropolitan – especially British – industry and commerce,

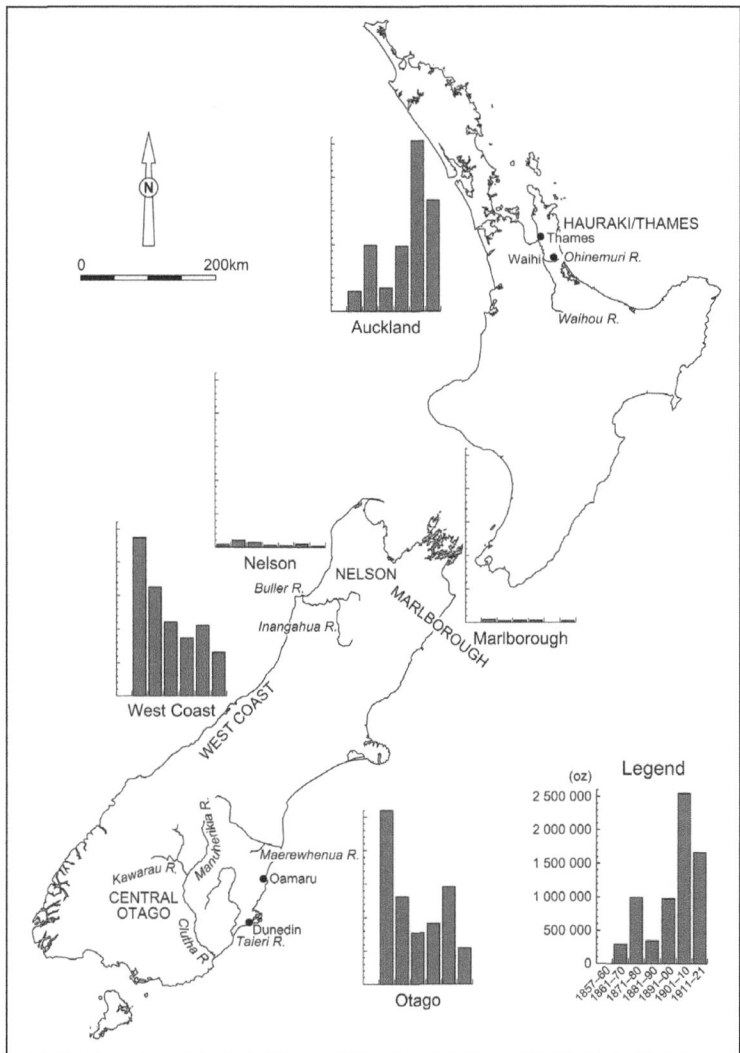

Figure 6.1 Goldfields and gold production, 1857–1921, by decade. Source: Mines Statement, *Appendices to the Journals of the House of Representatives (AJHR)*, C2, 1916 and 1917–22

and to stimulate the settlement and development of the colonial periphery. The scale and character of the rushes also reflected the distribution of the gold. Initially, at least, this offered ease of entry, required minimal capital outlay and held out the prospect of sudden wealth.

In order to promote and regulate goldmining, the New Zealand Parliament initially adopted the policy and principles of Victorian mining law. Simultaneously, it laid the basis for conflicts between the private user rights created by that law and the common-law rights of private property. As a result, Parliament was compelled to try to reconcile the competing ideals of the freedom of the mines and private property rights. Those efforts offer insights into resource developmental priorities, resource conservation and environmental protection and management, and also into the forces that encouraged policy reappraisal and redirection. The policy of goldmining law was clear: to allow the private mining of the public domain with minimal restrictions and at minimum expense, and to accord primacy to miners' requirements. It employed the law of capture to allocate resources among miners, created user rights and established an institution to apply mining law, resolve disputes and collect taxes. These private user rights differed sharply from private property rights, which included rights of possession, use, management, income, security, capital, transmission and absence of term.[4]

User rights and private property rights thus constituted two different rights structures. One conferred exclusive enjoyment and freedom of disposition for an unlimited time; the other, occupation and usage for limited periods contingent on specified performance. The application of these two structures to the same resources of land and water created on- and offsite conflicts between miners and landowners. Miners sought freedom of access to all lands for mining purposes, and the unrestricted right to discharge mining debris into watercourses. Landowners sought security of tenure and protection for investment. Simultaneously, it became apparent that nineteenth-century mining law had been devised for 'pick and shovel' mining, which created relatively minor environmental effects. It contemplated neither the major changes that took place in mining technology – especially in hydraulic elevating, gold dredging, and hard-rock mining – nor the growing scale of environmental impact.

Discharging mining waste

It was not long before anxiety was being expressed over the discharge of mining debris into rivers, the elevation of riverbeds, increased susceptibility to flooding, and the dangers posed to farmland. Miners denied both actual responsibility and legal liability, and rejected suggestions that they stack tailings or construct debris dams and settling ponds.[5] Concern that a successful appeal to common law riparianism (that is, the flow of water be undiminished in quality and quantity for the owners of land on both sides of a river) would mean 'the ... abandonment of goldmining' induced miners to ask Parliament to enact legislation to confirm the right to dump, and to ask that riparian lands be withheld from sale.[6] Two landowner challenges in Otago in 1875 (*Borton v. Howe* and *Glassford v. Read*) established three things: riparian rights of freeholders outside the proclaimed goldfields were deemed not to have been affected by mining legislation; the riparian rights of freeholders within the goldfields were deemed to have been affected only in so far as mining law authorised the diversion of water from natural courses; and mining legislation did not authorise miners to pollute streams.[7]

The implications for mining were serious, and hence Parliament enacted measures to extinguish existing riparian rights and prevent the acquisition of new riparian rights. The Gold Fields Act Amendment Act 1875 (No. 1) empowered the governor to proclaim any watercourse to be a sludge channel open for the dumping of mining debris. Affected riparian proprietors were entitled to compensation, payable by the Crown. The governor could issue regulations 'prescribing the mode and times at which such tailings mining debris or waste waters may be discharged' and impose restrictions on the right to discharge. Miners were not satisfied: they claimed that mining legislation had created rights that transcended those of common law, and that Parliament could abrogate the latter without paying compensation.

Further, the Act implied that miners were not entitled to discharge into watercourses that had not been proclaimed and for which compensation had not been paid. Continued mining was thus subject to the forbearance of riparian proprietors, or the willingness of the government to utilise the Act's provisions. The miners sought to have the Act replaced by a measure that would 'afford relief … by securing them in all the rights and privileges which they have fairly and legitimately acquired'. The Legislative Council declined to approve a repealing measure, and the Act remained the basis for the settlement of riparian rights disputes. It also provided that no person who purchased land after 21 October 1875 (the date the Act came into force) would have 'any right … to the flow of any watercourse which shall have been at any time proclaimed … [a sludge channel] running through or on such land'. Additional provisions limited the acquisition of riparian rights, absolved miners from any responsibility for damage resulting from dumping, and limited compensation claims.[8]

Different approaches to debris dumping

Very different approaches were employed in Victoria and California. In Victoria miners – claiming a prescriptive right – freely dumped debris. The government appointed a board of inquiry in 1886, when confronted with demands from miners for protection against legal action, and from landowners for the preservation of their lands. The board of inquiry's recommendation – that miners be required to restrain debris – was finally acted on following the expansion of gold dredging. The Mines Act 1904 defined precise discharge standards and established a Sludge Abatement Board to regulate any form of mining 'by which mineralised or impure water, sludge, or mining debris is discharged into any river stream watercourse lake or reservoir'. The board could inspect, impose and enforce appropriate orders, including the construction of restraining works, and investigate and prosecute complaints.[9]

In California, prolonged conflict involving the Hydraulic Miners' Association and the Anti-Debris Association culminated in *Woodruff v. North Bloomfield Gravel Mining Company* (1884). The Supreme Court declared that it had 'never recognised the validity of any custom to mine in such a manner as to destroy or injure the property of others' and issued an injunction that effectively suspended alluvial mining.[10] Claims that the state had indirectly authorised the dumping of debris by according mining a special place in the order of policy priorities were rejected. Various solutions were canvassed. These included that miners pay for the right to discharge debris or purchase affected lands, the construction of river-control works, and legislation to authorise discharge. Neither the

Californian legislature nor Congress ever attempted to authorise debris dumping. In 1888 a report ordered by Congress concluded that mining could be resumed if restraining dams were erected in the mountains and wing dams built to constrict rivers on the debris plains. In 1893 an Act created the California Debris Commission and required miners to construct restraining works and fund the commission.

In New Zealand legislative solutions continued to be followed. The expansion of gold dredging after 1895 engendered further conflicts, especially on the Clutha River in Central Otago. It was claimed that 'The whole of the gold duty obtained would not pay for the compensation that would be claimed if the river were proclaimed': an acknowledgement of the magnitude of the interests involved.[11] Further litigation established that miners had no prescriptive right to foul rivers and that legislation enacted since *Borton v. Howe* had authorised neither the discharge of tailings nor the fouling of water. A notice to proclaim the Waipori River produced compensation claims amounting to £18,000. The government declined to act, lest it encouraged riparian proprietors elsewhere to demand proclamation and compensation. The position became increasingly serious from about 1900, with dredges winning some 20 per cent (by value) of the gold produced each year.

This pressure forced government to appoint a commission to establish the basis for a comprehensive settlement. In fact, the Rivers Commission of 1900–01 simply investigated all watercourses required by miners and recommended proclamation where mining was established or likely to expand. It recommended proclamation of the Inangahua River, into which annually 120,000 tons of debris was being discharged from Reefton's quartz mines.[12] Conversely, it did not recommend proclamation where rivers – including the Clutha, Kawarau, Manuherikia and Buller – had been long used to carry off mining waste, or where the auriferous character of the ground was uncertain, compensation likely to be substantial, interested parties could reach agreement or state land-settlement schemes were likely to be adversely affected.

In all, the Rivers Commission considered claims amounting to £150,000. By 31 March 1907, 51 watercourses had been proclaimed and over £51,000 paid in compensation.[13] In so doing, the commission discounted the implications of proclamation for river management, public and private property, the protection of agricultural lands, and riverine ecology. It had interpreted its terms of reference as requiring resolution of disputes through the existing mechanism of proclamation, and at the least possible cost to the state. The limitations of that approach became apparent in the major dispute involving the Waihou and Ohinemuri rivers, which drain the North Island's Ohinemuri goldfield.

The Waihou–Ohinemuri dispute

Although the Ohinemuri goldfield was proclaimed in 1875, hard-rock mining awaited the introduction of cyanide extraction in the late 1880s. Then, inflows of English capital, and the amalgamation of small companies into large enterprises, stimulated a major expansion of the industry. The numbers of men employed, stampers installed, tons of ore crushed, bullion recovered and dividends paid all rose sharply.[14] Mining towns, especially

Figure 6.2 Damage caused by silting, Waihou River, June 1907. *Top:* A silt bank at Pereniki Bend: 'Settlers used to tie up their boats at willow where figure standing, in about 6 ft. of water.' *Bottom:* View of silt deposit on Barrett's paddock. Source: Report of Commission Appointed to Inquire into Silting of Waihou and Ohinemuri Rivers, *AJHR*, C14, 1910

Waihi, boomed. Extensive areas of bush were cleared to satisfy demand for fuel and construction materials. Continuing improvements in ore treatment and recovery rates reduced treatment costs and brought millions of tons of low-grade ore within the range of economic working. From 1890 onwards, hundreds of thousands of tons of tailings, including cyanide-injected waste, were tipped into the Waihou and Ohinemuri rivers (Figure 6.2).

In 1895 the government, acting in advance of land settlement, declared the two rivers to be sludge channels. Major floods occurred in 1898, 1907 and 1910. In 1901 the Ohinemuri County Council initiated a series of investigations. The council and the Mines and Public Works Departments reached contrasting conclusions concerning the extent of environmental impacts.[15] Miners rejected proposals that they stack tailings clear of watercourses, or construct California-type timber retaining weirs and backfill stopes (underground excavations), as too expensive and unnecessary. The Minister of Mines noted that there had been no objections to proclamation, that miners were legally entitled to dispose of their tailings into the rivers and that settlement had followed proclamation. Any silting, he claimed, was attributable to bush clearance in the watersheds.

Petitions calling for remedial measures were ignored, including that of Ngāti Tamaterā, who had settled along the Ohinemuri. Ngāti Tamaterā acknowledged that in 1875 they had ceded to the Crown the right to mine over certain lands, but claimed that contamination of the Ohinemuri had rendered the water 'unfit for use by man or beast'. In addition, cultivated land and burial grounds had been covered with tailings, and fish, 'an important part of their sustenance', had been destroyed by cyanide in contravention of the Treaty of Waitangi under which 'the fisheries of the Natives were specially reserved'. Ngāti Tamaterā insisted that they had never been advised of government's intention to proclaim the two rivers as sludge channels. The iwi petitioned Parliament again in 1907 and presented evidence to the Goldfields Select Committee, which was particularly interested in satisfying itself that the iwi had received a monetary benefit arising from the 1875 cession.[16]

Other petitioners claimed that 200,000 acres were 'more or less injured', outlet drains filled up, navigation impeded, the Thames fishing industry damaged. They also complained that any scheme for draining the Piako lands near Thames would be 'seriously retarded' – all to benefit 'wealthy outside companies'. Further investigations by the Mines Department failed to produce a definitive assessment of the actual and potential damage and practical remedies. The flood of 1907 and further petitions led to investigation by the Goldfields and Mines Committee. This committee decided, on the basis that the district's local authorities contribute to any control scheme, to refer the petitions to government for 'favourable consideration'. The short debate in Parliament that followed was remarkable only for the Minister of Mines' insistence that the petitions be rejected, as the area was naturally prone to flooding, and that mined ore was so finely crushed that waste was carried directly out to sea. In any case, the miners would want huge levels of compensation.[17]

Anti-dumping interests formed the Ohinemuri Silting Association, which, familiar with overseas developments, claimed that the miners could stack tailings at minimal cost. Miners called for a commission of inquiry, certain that it would 'explode the contentions of the farmers that the mining tailings are proving a menace, and a source of great danger and loss to them'. The Waihi Gold Mining Company insisted that not more than 50 acres had been affected, and that the Ohinemuri floods were no greater than in the past. The admittedly higher Waihou River floods reflected the 'immense number of artificial drains emptying into the river' and the growth of willows, although, it claimed, the river would remain navigable. The company noted that the county had benefited from mining through roads, bridges and drains, mostly funded by revenue derived from the gold duty. Stacking was not practical and finely ground residues were carried out to sea. Revocation would render government liable to provide both secure storage sites and the means of conveyance.[18]

In 1910 the government appointed a Waihou and Ohinemuri Rivers Commission to investigate these conflicting claims and to propose remedial measures that could be adopted 'without injury to any other persons, corporations, or interests'. This commission found that the annual dumping of 558,000 tons of debris had destroyed 371 acres and affected 'deleteriously' a further 7507 acres of agricultural land.[19] As landowners could not have foreseen the 'wide-reaching and disastrous effect' of proclamation, it recommended that, though not legally entitled to compensation, they should be recompensed. It noted that the adoption of tube mills meant that slimes rather than coarse sands were being discharged into the rivers, and recommended that 'grinding to slimes should be compulsory'. Any restriction to the use of the river as a sludge channel for slimes only would be in the mining companies' interests.

The problems to be remedied related largely to tailings already discharged. While it also noted that cyanides were present in the rivers and that fish were no longer found there, the commission concluded that farmers had not been affected. Navigation, on the other hand, had been, and 'seriously' so. This it attributed in part to 'the large volume' of light sands discharged into the upper Waihou following settlement. It also predicted future shoaling problems in Thames Harbour, which would require the dredging of both the harbour and the Waihou River. Various schemes for the disposal of tailings were considered. None was felt to be in the state's interests because of the likely impact on the cost of gold production, which could well render processing of lower-grade ores uneconomic.

Hence the 1910 commission recommended that the Ohinemuri River proclamation be amended to allow only the discharge of 'slimes', in the belief that, once the rivers had been straightened and dredged, these would be carried out to sea. Tailings should be stacked or, where that was not possible, discharged under special permit. It recommended revocation of the Waihou River proclamation, although existing mines could continue dumping, and 'that if any other mines in its watershed should desire to utilize the river or its tributaries, permission be specially given, after careful investigation and on special conditions, particularly as to fine grinding'. Otherwise the commission proposed extensive river control works, estimated to cost £150,000, to be raised by loan. This was to be administered by a new river board funded by the miners, rates levied on lands in the river board district, and by the state. The miners' contribution, at the rate of three-quarters of a penny per tonne of waste, was 'a tax which cannot affect the working of any low-grade ore, and an amount which it is believed the mining industry can well afford'.

An integral part of this proposal was that Māori riparian lands should be purchased or 'brought into line with lands owned by Europeans in the matters of taxation and contribution towards the construction and upkeep of the new works'. Ngāti Tamaterā interests were accorded no particular recognition, while the proposed membership of the board did not provide explicitly for their representation. The commission's concern was that the large area of land in Ohinemuri County 'held and occupied' by Māori should be rated. No reference was made to the destruction of Māori river fisheries, and it considered that the Thames fishery had not been seriously affected. Were it to become so, 'the industry must give way to the general benefit to the Dominion caused by the continuance of the mines'.

The Waihou and Ohinemuri Rivers Improvement Act 1910 embodied most of the commission's recommendations but did not permit amendments to be made to the 1895 proclamation. That proclamation remained in force, the government not prepared to contemplate action that would have rendered it liable to claims for compensation from miners. The Waihou River was the subject of another inquiry in 1919, but it focused on flood control, while a commission in 1921 investigated the allocation of burgeoning control costs. Miners, insisting that their industry was 'waning and wasting', asserted that they had contributed more than sufficient to remedy any damage from dumping. The Public Works Department claimed that the right to discharge debris 'was a benefit of incalculable value to the gold-mining companies', obviating 'enormous' costs in stacking, and that the companies 'could well afford to pay their share of the remedial works required'. The commission itself criticised government's refusal to issue amended proclamations, recommending revocation and reissue in a form that would permit dumping 'only on such conditions as will obviate all liability to damage of either the rivers or adjacent agricultural lands, and also on payment of ... charges'.[20]

The fact that in 1921 quartz mining yielded 79.2 per cent (by value) of New Zealand's total gold output, and that 77.1 per cent of that came from the mines of the Thames–Coromandel, ensured that the government would respond cautiously. Again the drive for development had triumphed over any impulse to conserve or preserve landscape, ecology or traditional Māori access to food resources.

The national Rivers Commission 1919–20

The findings of the 1921 Waihou and Ohinemuri Rivers Commission marked a shift that was earlier apparent in the findings of a national commission appointed in 1919 to investigate river problems. While it cited dumping as a major cause, this commission noted, with respect to the Maerewhenua in North Otago, that suspension of mining 'would not mitigate the trouble'. It also judged the value of the gold being won as insufficient 'to justify our recommending any contribution towards the cost of remedial measures being demanded from this source'. On the other hand, the government, having proclaimed the river a sludge channel, 'could not escape liability to contribute a substantial sum towards the costs of the works recommended'.

Dumping had damaged Oamaru's water supply, roads, railways, bridges and 800 hectares of land.[21] Dumping in the Clutha River – which had never been proclaimed – had resulted in damage of 'a manifold nature', while the cost of protective works was estimated at £165,000. The commission concluded that agricultural and pastoral operations, burrowing by rabbits and burning of the indigenous plant cover had accelerated the rate of 'natural denudation' as much as had mining, 'and therefore the settlers and miners are equally responsible'. Mining operations had displaced some 300 million cubic metres of material, which was then being swept out to sea at an annual rate of 670,000 cubic metres.[22] The similar use of the Taieri River had resulted in damage to some 12,000 hectares and to roads, bridges and railways.[23]

In each instance, the costs were to be borne by local authorities, the state and landowners. One important outcome of this investigation was section 48 of the Finance Act 1921–22, which provided that the exercise of the right to discharge mining debris into watercourses or to mine in or on their banks was to 'be subject to such conditions and restrictions as may be set out in the Proclamation'. The state, thereby, assumed more coercive powers to limit damage caused to river systems by mining and dredging.

Mining or land settlement?

The priority accorded mining was apparent in the manner in which the government resolved the conflict between miner and settler over land as well as water. As small-farm settlement was promoted on the goldfields, miners became anxious about the implications for their industry. The goldfields press insisted that 'the miner must not be encroached on or curtailed in his operations'.[24] In Central Otago the resumption and close subdivision of large areas of pastoral lease land from 1878 onwards aroused alarm. Much of that land, miners claimed, was auriferous and mining was 'capable of almost unlimited extension'. But miners and settlers were in dispute over the priorities to be accorded each group's natural resource requirements; the miners were apprehensive that litigation as the result of the fouling of watercourses 'would smother the mining industry'.[25]

At the heart of the issue lay the application of the two different rights structures to the same resources of land and water.[26] Resolving the inherent conflict was a complex matter, but the major approach was to establish new forms of land tenure. Between 1860 and 1907, eight tenures were introduced. These were agricultural leases (in 1860), deferred payments licences (1872), perpetual leases (1882), small grazing runs (1885), leases-in-perpetuity (1892), occupation with right of purchase (1892), occupation leases (1894) and renewable leases (1907). All, with the exception of deferred payments, were forms of leasehold. The term varied considerably, but all were renewable, apart from the lease-in-perpetuity, while all eventually offered lessees the right to acquire the freehold.

The desire to promote close settlement was embodied in acreage limitations and mandatory residence and improvement conditions. To meet miner demands, the right of entry to all leasehold lands was reserved, all leases could be terminated if the land was required for mining, and, in the case of four of the tenures, owners or occupiers were not to have any right of action for any damage resulting from the diversion or fouling of watercourses. Thus selected incidents of ownership were modified in the interests of miners, but none of the tenures met their demands for unrestricted and unqualified entry and liability to termination and resumption without compensation. Government also refused to deny would-be settlers right of purchase in goldmining areas.

Freehold lands were not open to miners. Although the Crown retained the ownership of gold or silver found in either Crown or private lands, it had neither the right to enter private lands nor the power to authorise others to enter for mining purposes. An alternative was the sale of all lands, including those sought for mining. Proponents claimed that determining resource allocation priorities by legislation precluded optimal allocation and use. They alleged that mining was facilitated by the transfer of external costs to the community, while it secured the (largely) untaxed profits. Authorising miners to enter private property meant that Parliament was being asked to pass 'exceptional legislation for the protection of an industry which was not nationally profitable nor nationally desirable to encourage'.[27] Finally, proponents insisted that mining freeholds, by offering greater security of title, would encourage investment and increased production, while land sales generally would effect the rapid settlement of lands closed 'in deference to the wish of goldfields "demagogues"'.[28]

Miners opposed the replacement of the legislatively determined user regime by private property and the competitive market as the allocative mechanism. The sale of mining freeholds would create monopolies and render entry into the industry difficult, while the sale of land for settlement would impede all mining through the multiplication of private property rights. Rejecting claims that their demands amounted to confiscation, miners sought legislation that would allow entry on private lands for mining. The outcome was the Resumption of Land for Mining Purposes Act 1873, by which all lands thereafter alienated were liable to resumption without compensation. Further measures followed: the Mining Act 1891 provided that all lands held under lease or licence from the Crown with the right to acquire the freehold were open for prospecting and liable to resumption. But it also established rules to preclude indiscriminate entry and destruction as well as the frustration of legitimate mining enterprise.

Preserving agricultural land

Both the measures relating to land tenures and to entry to private land applied irrespective of the land's primary production potential.[29] As early as 1871, Dunedin's *Evening Star* declared that everyone had an interest in the preservation of the soil resource. Parliament should therefore devise some means of restoring the soil in areas once mined.[30] Few listened, until the introduction of hydraulic elevating and gold dredging (Figure 6.3). This allowed the profitable working of low-lying, low-value ground. Vast dredging schemes were proposed. Opponents insisted that the state should preserve 'the staple productive resources of the country' by withdrawing land from the operation of mining law or, in the case of freehold, expropriating it for resale under conditions that prohibited mining.[31] Miners in turn insisted that only rich land would be mined, affecting only a minute proportion of the national land resource. Dredging would anyway improve swampy, cold and sour land, and advance stripping could be employed, although mandatory restoration conditions could render mining unprofitable.

Accordingly, governments up to 1908 dismissed claims over the area being destroyed and insisted that attempts to restrict the sale of land for mining were politically impossible. Landowners believed themselves to be the best judges of the most profitable use of their land.[32] The debate involved the familiar conflict between those who felt that private interest would effect the optimal allocation of resources, and those who wished Parliament to impose controls to ensure resource conservation and sustained yield.

Figure 6.3 Blue Spur hydraulic sluicing and elevating claim, Gabriel's Gully, Otago. 'At this claim about 50 acres of auriferous fluvio-glacial conglomerate has been removed by sluicing and the remarkable fault-face to the right has been laid bare for a height of about 400 ft.' Source: Report on the Goldfields of New Zealand, *AJHR*, C3, 1911

Victoria adopted a different approach. Covenants inserted in mining leases issued after 1906 required advance stripping, redistribution of soil, and sowing of grass seed; and in 1909 the Mines Department was instructed not to issue leases over any land the value of which exceeded £3 per acre. In practice, the covenants were often ignored: an inquiry in 1913 concluded that the sole remedy was to withdraw all agricultural land. In California, the dredging of agricultural land proceeded largely unopposed: miners claimed that the land involved was largely marginal or had been mined, that overgrazing and soil erosion were more destructive, and that good land could be reclaimed by resoiling.[33]

In New Zealand the controversy intensified during the 1900s. Arguments were advanced that mining was part of a larger destruction being wrought on indigenous fauna and flora. Land of high agricultural value, it was proposed, should be preserved irrespective of its gold content. Government noted that such land was largely freehold and that owners had 'the perfect right to [do with it] as they wished', rejecting repurchase as a means of ensuring preservation. Objections to the destruction of orchard lands, and a petition seeking to halt mining of agricultural land,[34] persuaded Parliament in 1910 to include a clause in the Mining Amendment Act 1910. This allowed wardens to impose conditions on any licence to obviate the destruction of agricultural and pastoral land. It was prospective and discretionary in application, and failed to provide for inspection and enforcement.

In a report to the National Efficiency Board in 1918, the Mines Department conceded that considerable areas of agricultural land had been destroyed, and that little had been done to restore it.[35] The outcome was section 12 of the Mining Amendment Act 1919. This required submission of an application for a dredging claim to the local Commissioner of Crown Lands for assessment of the agricultural value of the land, and the warden to impose such conditions as were necessary 'to prevent, so far as practicable, the destruction of the surface of the land or the rendering of it unfit for pastoral or agricultural purposes'. The section did not provide for the reservation or other protection of such land. It remained prospective in application and did not apply to freehold land for which a mining licence was not required. Once more use rights triumphed over an attempts at conservation, and control of damage resulting from goldmining remained limited. Given its by then relative lack of economic importance compared with farming, goldmining continued to receive preferential treatment.

Mining for coal

Coal was mined in New Zealand from the 1830s, when whalers in Otago exploited sea-beach outcrops. Annual production grew slowly through to 1895, and more rapidly thereafter. The Grey Valley emerged by 1880 as the most important field, although by 1914 the Waikato, responding to the growth of dairy processing and the expansion of Auckland's domestic and industrial markets, produced a quarter of the country's annual output (Figure 6.4).[36] Most of the coal was extracted from underground mines with apparently limited environmental consequences. There were few slagheaps, while land stability and subsidence, accelerated leaching from the surface soil, acid drainage to other areas, soil conservation, site rehabilitation and grading of spoil dumps were not obvious matters of concern. Investigations into the industry focused on such matters as production and delivery costs; mine management and inspection; health, safety, ventilation, and sanitation; and nationalisation, labour relations and prices.[37]

While the legislation under which coalmining leases were issued did not require mine owners to assess and make provision for environmental protection, the leases themselves did impose conditions. Those issued over Crown lands, initially under the Land Acts and later under the Coal Mines Acts, required lessees to work the mines 'in the most approved manner', and to inflict as little injury as possible to the land surface. On the expiry of leases, lessees were required to 'fill up, level, or substantially cover' all pits, shafts and open works; clear surface slack and stone; and 'restore the surface to its original state, or as near thereto as reasonably may be or be required'. Such conditions were of a standard rather than site-specific character, and concerned safety and the property's 'tenantable condition' rather than environmental matters.

The same was true of licences issued for the raising of lignite, while the conditions included in leases issued in respect of other minerals matched those relating to coal.[38] Parliament occasionally imposed other conditions, notably with respect to transport from mine to port.[39] Otherwise, concerns related to the clearance of bush around mining settlements, and the often unsanitary condition of towns constructed in bush-clad, hilly and wet areas. However, the isolation of many coalmining townships, especially on the West Coast, the perils of mining, and the fact that many mining families had been drawn from the English coalfields, with their traditions of social and political activism and solidarity, helped to create strongly integrated mining communities.

While coalmining legislation reserved the surface, water, watercourses and trees on all land leased for mining, it also provided that these could be used for mining purposes

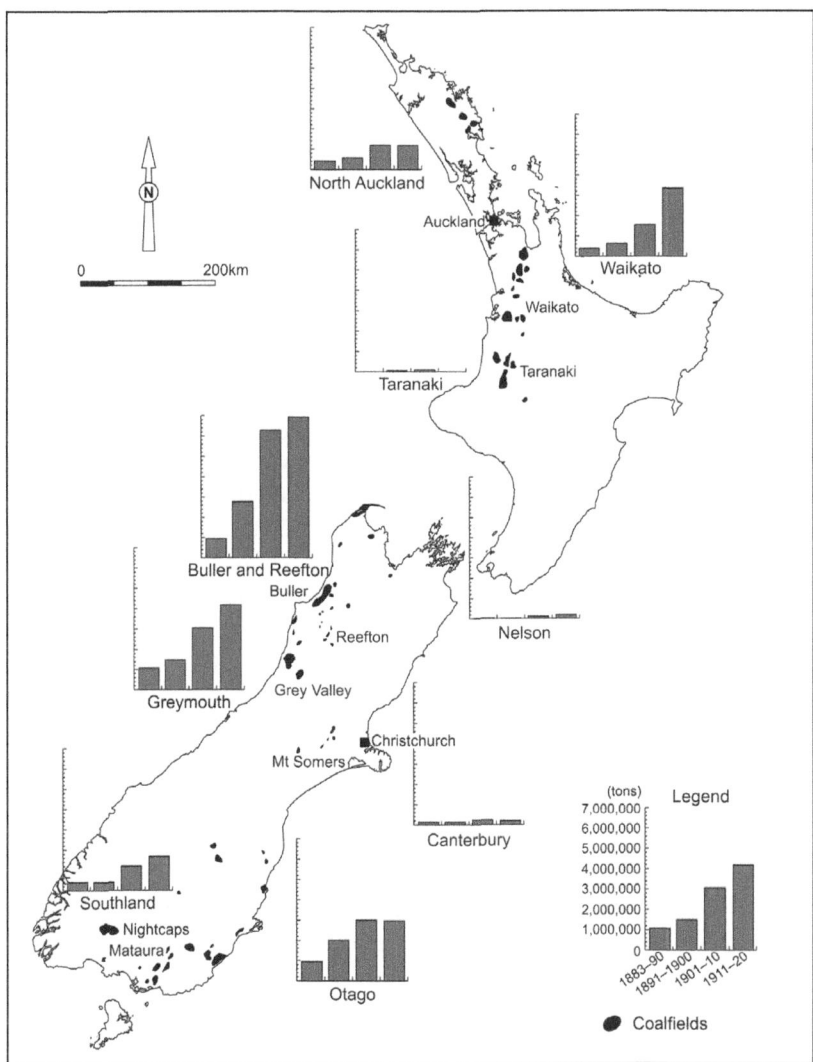

Figure 6.4 Map of coalfields and production, 1883–1920, by decade. Source: Mines Statement, *AJHR*, C2

without further consents having to be obtained. Opencast mining, with its potentially greater environmental impact, was conducted on a very small scale until 1945 and the introduction of machinery to facilitate the stripping of overburden.[40] Up to 1921 almost all opencast mining was confined to the Mt Somers (Canterbury), Mataura and Nightcaps (Southland), and Central Otago fields, with small pits opening and closing as demand fluctuated. Opencast coal did not exceed 5 per cent of annual production before 1921.[41]

Quarrying

A widespread form of mining was quarrying – of stone for building, aggregates for roading and reclamation, limestone and phosphate for agriculture, and claystone for bricks and tiles. Quarry sites were reserved throughout New Zealand.[42] The few statistics available suggest that between 1911 and 1921 the number of quarries in operation, men employed and production fluctuated considerably from year to year. In 1921, 257 quarries employed 1561 men and produced 1.19 million tons. Gravel for macadamising and ballast accounted for 56 per cent of this total, stone for harbour works 16.7 per cent, limestone for cement 15.2, limestone for agriculture 10.4, stone for building 1, phosphate for agriculture 0.5, and claystone for bricks and tiles 0.2 per cent.[43]

The first Quarries Act, passed in 1910, dealt with the certification of managers, safety and inspection, and made no reference to the discharge of waste and storm water, the control of dust and noise, or site rehabilitation. Moreover, it excluded state quarries; those in which the height of the rockface was under 20 feet; and those in which explosives were not employed. There seem to have been few problems with rural quarries. The Halswell Quarry outside Christchurch, which developed into one of New Zealand's largest, did not attract complaints; nor did the opening of a quarry on Mt Taranaki.[44] Quarries in urban areas, however, did cause some concern.

In 1902 the Heathcote Road Board sought – unsuccessfully – an amendment to the Public Works Act to prohibit the operation of quarries within three miles of a city or borough: the object was to preserve property values. Similarly unsuccessful efforts were made to control drainage from quarries alleged to be causing silting in the Heathcote River. Abandoned quarries could also cause anxiety to residents as city suburbs expanded to encompass areas where quarries and industries had formerly operated.[45] From about 1910 onwards there was growing concern in Auckland about the quarrying of the area's volcanic cones. The Reserves and Other Lands Disposal and Public Bodies and Empowering Act 1915 imposed some restrictions and required quarry owners to plant approved trees and shrubs. Not satisfied, the Auckland Town Planning Association campaigned to prevent further despoliation, attracting the support of Māori and tourist interests, while many of Auckland's local authorities opted to use scoria obtained from alternative sources.[46]

Conclusion

New Zealand's colonial development involved the conversion of public natural resources into sources of private output and wealth, a process facilitated by legislation and supported by state-funded programmes for the construction of roads, railways and communications. Mining law and land law, in particular, imposed few restrictions on the use of resources, while discounting social and environmental costs in the interests of sustained exploitation. Goldminers interpreted mining law as offering unlimited licence for their enterprise and sought appropriate measures whenever such licence appeared to be under threat. Parliament largely acceded to their demands, proving much less imaginative than its Victorian and Californian counterparts. Controls were only gradually introduced after the costs of unrestrained mining and dumping for water quality and flood control, and the extent of damage to infrastructure and wildlife habitats, became apparent, and only after the community began to realise that it would have to pay for the damage inflicted by companies concerned solely with their own profits. Even then, government attempts at control were intended to resolve conflicts among competing interests rather than to ensure wise resource use.

The first signs of a shift in attitudes and values are apparent from about 1900, in the findings of the Rivers Commissions as well as inquiries into pastoral and hill lands, forests and sea fisheries. This shift reflected a growing debate over the social and environmental costs of untrammelled exploitation, technological change that constantly redefined resources and offered an enhanced capacity for environmental change, and the roles of the state and the 'market' in resource assessment and allocation. Further, it reflected concern over the approaching closure of the frontier, the challenge to the notion of 'inexhaustibility', and the shift from 'physical' to chemical wastes. Although legislative controls were reluctantly enacted, often loosely formulated, and prospectively and sparingly applied, such changes recognised, albeit in qualified ways, the principles of 'efficient development', 'wise use', and active management, and did embody roles for the state beyond those of promoter and facilitator. The general public acceptance of those principles ensured, a century later, strong opposition to government proposals to open certain categories of conservation land to mining – sufficiently so that they were abandoned.

Further reading

Little has been written about mining and the environment in New Zealand in the last decade, with the exception of the recent debate over mining on the conservation estate: Geoff Bertram, 'Mining in the New Zealand Economy' (*Policy Quarterly*, vol. 7, no. 1, 2011, pp. 13–19); Gundars Rudzitis & Kenton Bird, 'The Myth and Reality of Sustainable New Zealand: Mining in a Pristine Environment' (*Environment: Science and Policy for Sustainable Development*, vol. 53, no. 6, 2011, pp. 16–28); and Parliamentary Commissioner for the Environment, *Making Difficult Decisions. Mining the Conservation Estate* (Wellington, 2010).

7.

Destruction under the guise of improvement? The forest, 1840–1920

Graeme Wynn

LATE IN LIFE, novelist Frank Sargeson recalled his uncle's farm, high in the King Country beyond Taumarunui. 'Tucked away at the narrow top end of a valley', it was so crowded with 'thick and heavy bush' that its new owner could only imagine its future as a farm when he occupied it in 1913. By 1920, only a few patches of bush remained. On these broken hills, seven years was sufficient to reprise a major theme in the short European history of New Zealand.

A little sawmill, 'not much more than a roof supported on upended logs, with much bigger logs bedded into the ground to take the weight of the machinery', had been built below the farmer's hut. A hauling engine fed the mill, dragging the biggest of the farm's trees roughly across the earth and along a wooden tramline that snaked through the valley. The farmer cut what the millers had left, setting the debris to burn at the end of each summer. Grass was sown, fences were built and sheep introduced. With remarkable speed, it seemed that 'the natural order of things … had been obliged to yield to intelligent human guidance'. An 'appearance of settled and civilised pastoral order' was imposed on the once-forested landscape.[1]

This 'making over' of the North Island ranked, in the estimation of geographer George Jobberns writing in 1956, as one of 'the outstanding achievements of our people'.[2] The historian G.H. Scholefield measured things rather differently. In *New Zealand in Evolution*, published in 1909, he described the assault on the forest as a 'pitiful war'. The rights given sawmillers were 'simply and solely … executioner's warrant[s]', allowing greedy, selfish individuals to 'pick out the eyes of the forest, to slay and ruin the rest, and then go elsewhere'. The countryside was being devastated. Hundreds of mills were 'killing and slaying and burning and wasting'. The prevailing 'doctrine of progress' had unleashed 'unbridled rapine and licence'. The forest was being destroyed under the guise of improvement.[3]

The drama

Achievement or disaster? Why were opinions so different? Such questions lie at the heart of this chapter, which charts what happened to the forests between 1840 and 1920, seeks to understand why those who led the assault on the forest acted as they did, and reflects on contemporary and later assessments of these events. The story is complicated because neither forest nor society was of a piece. Between Auckland and Southland, 'the New Zealand forest' varied enormously in species composition, appearance, utility, susceptibility to fire, and so on. Personal and family histories and location differentiated people; their engagements with, and attitudes toward, the indigenous vegetation of the country varied enormously. There was no unanimity of opinion about the processes under way. Important gains as well as distressing losses were recognised.

In this, New Zealanders were late participants in a long-running drama. Like actors in a touring theatre company, they spoke lines adapted to local audiences and comported themselves according to the configurations of the particular stage on which they played, but their performance was shaped by a well-used script. Exploration, encounter, exploitation, adaptation and transformation were its common acts, and it had been rehearsed for centuries before it reached New Zealand. Those who rendered the South Pacific version of the show felt the burden of this experience.

From the fifteenth century onwards, trade and politics drew European interests outward. Plants, animals, diseases and ideas accompanied European voyagers abroad: simultaneously, elements of New World environments became resources for use and commodities for exchange. Place after place was transformed. Progress – from 'wilderness to fruited plain', as one of dozens of works on this theme has it – was widely applauded. But its costs came to be counted as signs that all was not right with the ways in which humans were using the earth.[4]

Incorporated into the New Zealand version of the global drama of colonial settlement, these experiences imparted a distinctive accent to the late local version of the play, which cannot be comprehended without an appreciation of this international context. By the same token, the significance of New Zealand's transformation after 1840 is reduced unless it is understood against the backdrop of changes induced in New World lands by the overseas expansion of Europe. Just as Sargeson saw what happened to the hills of Taumarunui as a 'small-scale repetition of … [the] country's history', so New Zealand epitomises processes that shaped the modern world.[5]

A 'tangled and impervious forest': 1840

Distinctive, impenetrable and extensive, the forest captured the attention and the imaginations of Europeans in early New Zealand. From Cook and Banks, who noted 'imence woods', 'lofty Trees' and 'the finest timber', through descriptions of views that encompassed 'little else than a succession of steep irregular hills, clothed with dense forests', to claims that these were 'the stateliest forests and the densest underwood in the world', diarists, journal keepers and others recorded their impressions of New Zealand's vegetation. Some marvelled at its remarkable trees. Others were fascinated by the luxuriant growth of 'grand broadleaved ferns palmated like the horns of the elk', mistletoes, the clematis that 'spangled' the shade with 'silver stars', 'innumerable parasites and climbing plants, vegetable boa-constrictors in appearance', and a 'carpet of lichens

Figure 7.1 Julius Geissler, *Primeval Bush*, 1919, pen. Source: gift of Mr John Leech and the Old Colonist's Museum, 1926, Auckland Art Gallery Toi o Tāmaki

and mosses and fungi'. With the help of several luminous portrayals by early artists (Figure 7.1), this 'forest' is relatively easily imagined – but far from precisely rendered.[6]

Typically, it forms a complex multilayered ecosystem, including almost 400 plant species, in five more or less distinct strata (Figure 7.2). The giants of the forest are podocarps: rimu, kahikatea, tōtara, mataī and miro. Below them, hardwoods – taraire, tawa, kāmahi, tōwai, pūriri, black maire – form a dense, diverse canopy. A third layer of tree ferns and small trees – the nīkau palm, māhoe, wineberry – erect a subcanopy. Comprising, in the main, relatively fast-growing pioneer species adapted to colonising open ground, this stratum is most evident in the high-light environments of the forest margins; with time, these trees are overtopped and shaded out by the slower-growing but taller hardwoods and podocarps. Shrubs of the *Coprosma* species, horopito, tutu, rangiora and the tree nettle ongaonga make up a fourth storey; and the ferns, mosses and lichens, the lilies and the grasses that grow on the damp, dark forest floor a fifth. Vines such as supplejack, bush lawyer, passionflower, clematis and rātā climb through the forest layers; and epiphytes – orchids, ferns, lilies – form often spectacular hanging gardens high in the trees.

Variations from this archetype are best understood in biogeoclimatic terms. The forest was richer in species and more luxuriant in appearance in the north than the south, and on lower rather than higher ground. Differences in humidity and rainfall, soil and drainage conditions, and aspect and exposure altered the vegetation cover. Among the

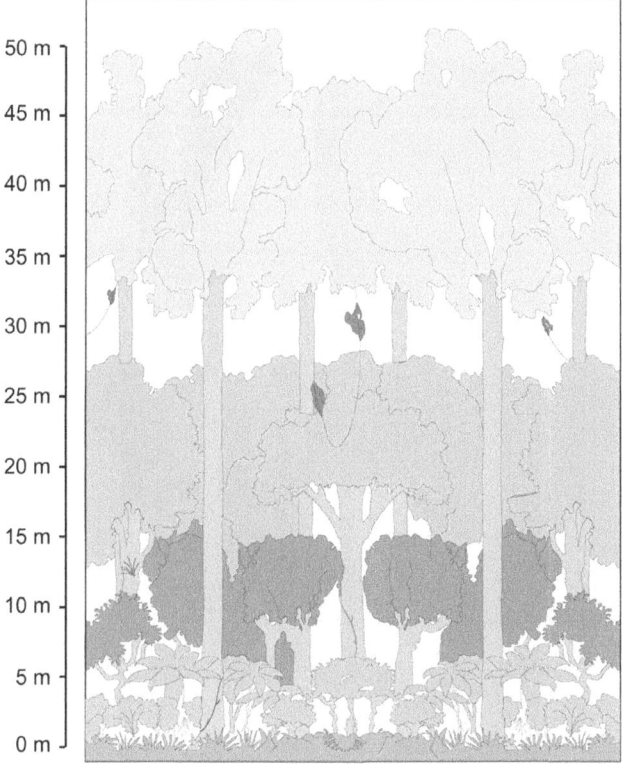

Figure 7.2 The structure of the New Zealand forest. Source: author

podocarps, matai and kahikatea favoured fertile soils; tōtara flourished on good, dry soil; and kahikatea – the 'white pine' of early European parlance – did best in the swampy ground of tidal estuaries and river banks. In the hardwood canopy, taraire gradually yielded to tawa south of Auckland, and tawa gave way, in turn, to kāmahi in the South Island. On coastal sandplains, a great range of forest species grew as low windswept shrubs. On volcanic soils, 'the bush' was generally lighter; on heavier non-volcanic soils, it more closely approached the form that surveyors and settlers came to regard as the quintessential 'dense primeval New Zealand Forest'.[7]

Beyond this, two arbitrary latitudinal boundaries help clarify the form of the forest. North of 38°S (and below the 800-metre contour), that most magnificent – and most commercially valuable – of New Zealand trees, the kauri, rose above all others in the early forest. Typically growing some distance apart, in smaller or larger groups, on slightly elevated sites, the kauri towered over other species in the mixed podocarp–hardwood forest. South of 42°, common North Island species – nīkau, karaka, tawa, kohekohe, rewarewa, pukatea and tānekaha, rangiora, kanono, and wharangi – disappear. Here, too, subantarctic rainforest, dominated by southern beech (*Nothofagus*), becomes more prominent. Although *Nothofagus* grow singly in the podocarp–hardwood forest and are

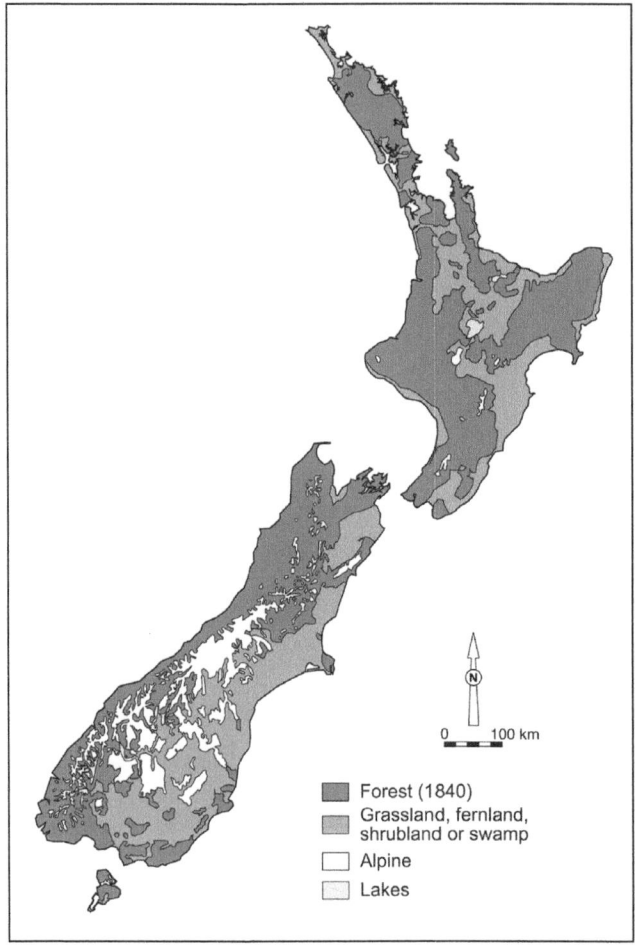

Figure 7.3 Map of the forest, c. 1840. Source: adapted from Malcolm McKinnon (ed.), *New Zealand Historical Atlas*, David Bateman, Auckland and Department of Internal Affairs, Wellington, 1997, plate 12

relatively common in lowland Taranaki and the East Cape, as well as in the northern and southwestern parts of the South Island, they dominate the poorer soils and steeper slopes of the cooler south.[8] In the filtered half-light beneath the *Nothofagus* canopy, there are fewer lianes, epiphytes and tree ferns than elsewhere, and beech seedlings often dominate the understorey, which also includes shrubs such as mingimingi and weeping matipo.[9]

Volcanic eruptions and burning by Māori had reduced the area of this highly varied forest by half well before 1800, and the activities of timber-getters – reflected in the rising quantities of timber shipped from Hokianga and Whangaroa early in the nineteenth century (discussed in Chapter 3) – pushed back its edges on the northern peninsula.[10] James Hector, Director of the Colonial Museum, made the earliest systematic estimate of New Zealand's forest cover in 1874. Using information collected from provincial superintendents, he suggested that 31 per cent of New Zealand was forested in 1830.

Later, more scientific estimates, derived from ecology as well as the historical record, have consistently revised Hector's figures upward. Several concur broadly with geographer Kenneth Cumberland's claim that 43 per cent of the colony was forested in the 1840s. Regional proportions and lines on maps are more difficult to establish. Perhaps the best that can be said is that 'bush' covered some two-thirds of the North Island and rather less (25–30 per cent) of the South (Figure 7.3).

Destruction

Writing to her father and sisters in England, shortly after her arrival in New Plymouth in 1841, Sarah Harris reflected: 'The forest is behind me and the sea in front.' Like countless other nineteenth-century immigrants, she had little choice but to confront the interior, into which she initially dared 'not venture … for fear of losing' herself. Enormous effort was spent in this confrontation, as settlers sought to realise their 'earthly paradise' in the 'future Britain of the South', especially after relatively open country, such as the South Island tussock and the shrub and fern land of Hawke's Bay, was occupied (as described in Chapter 5) by runholders and fortunate farmers who found ready markets in the growing towns.

As New Zealand's Pākehā population rose, farms had to be carved from the forest, and the replacement of existing ecosystems by new forms of land use was taken for granted in the march of 'improvement'. The task of 'clearing' was so central that it measured everyday existence. For some it even offered a light to live by. Preaching to his congregation in northern Taranaki in 1844, the Wesleyan minister Cort Schnackenberg urged them to treat their mortal souls as they did their farms: 'if you find your mind, your heart to be a wilderness, cultivate it in the same manner as you do your fields, cut down the bush, great and small, spare no sin'. Even those who escaped active service in the war on the woods were somehow caught up in the onslaught. Hardly a soul was free from dependence on the bush and, more pointedly, its exploitation.[11]

Nineteenth-century New Zealand was a wooden world. Its citizens' very existence rested on the forest's bounty. In 1881, seven of every eight New Zealanders lived in a wooden dwelling; 30 years later, the proportion was even higher. People used wood to warm themselves, and to cook and heat water. Families burned between two and four tons of firewood per person each year. Wooden barrels and boxes were ubiquitous. Ships that plied the coastal trade, and the carts and carriages that linked farms to markets and ports to hinterlands, were built of New Zealand wood.

Bridges were made of wood, and logs were laid in low-lying areas to form corduroy roads. Wooden sleepers supported the tracks of the expanding railway network. Wooden poles supported telegraph lines the length of the country, and wooden posts carried thousands of miles of fencing wire around the fields of growing numbers of farms. Barns, woolsheds and sheep pens, much furniture, and many implements were made of wood.

Countless colonial industries depended on wood. Without exploitation of the bush, New Zealand could not have developed as it did.[12]

Although settlers generally sought to remove bush from their lands, the settings and circumstances of their assault on the forest differed. Wood was produced for different purposes – with kauri, rīmu, kahikatea and mataī favoured for house-building, kahikatea for boxes, tawa for barrels and tubs, and tōtara for telegraph poles and railway sleepers – and sent to different markets. Yet the main features of the story are clear. Millers and settlers were the prime movers of change. Their assault on the forest was directed to personal ends – accumulation of profit, establishment of home and family – but it lay squarely within the prevailing discourse of progress and improvement. 'The sawmiller,' said one of their number, revealingly, 'cannot be taxed with the destruction of forest, since he merely converts the standing timber into a marketable product.'[13]

For 30 years after 1840, sawmilling was concentrated in coastal districts. Its main centres were Auckland and the Coromandel Peninsula, the northern reaches and west coast of the South Island, Banks Peninsula, and the Catlins River area of Otago, with a smaller concentration in Wellington.[14] Here most of the colony's early mills – six in 1843, 12 in 1847, 15 in 1855, and 93 in 1868 – provided timber for growing Pākehā settlements (a population of 32,000 in 1855, and 10 times this 15 years later), as well as trans-Tasman markets. Flotillas of small craft ferried wood from ports near the mills to adjacent settlements – from the Catlins River to Dunedin, for example. Slightly larger vessels ran the coast – from Pelorus Sound to Wellington and Lyttelton; from Auckland to ports south. Others carried exports, particularly kauri and kahikatea, overseas.

Generally the early mills were small; few were capable of producing even 1 million superficial feet a year. Nonetheless, a combination of wasteful exploitation and the concentration of production (in 1868 one-third of New Zealand's mills and many of its largest were in Auckland) quickly left its mark on the landscape. Ferdinand von Hochstetter lamented that 'extensive districts … formerly … covered with kauri woods are now totally destitute of such'. Meanwhile, James Hector urged millers to practise more careful and selective logging rather than continue their wholesale and rapid destruction of forests.[15]

With population growth and the quickening of economic activity after 1870, the sawmillers' assault on the forest magnified. Railways opened up new areas for exploitation and new markets for consumption. In the mid-1880s, timber and firewood accounted for half the freight carried on the Nelson, Picton and Napier lines, and approximately the same proportion of all goods shipped in the province of Wellington. Railways facilitated sawmilling and settlement in the southern reaches of both islands. Botanist and forester Thomas Kirk, noting the rise of Southland as a supplier of timber to markets 'from Invercargill to Ashburton', Lyttelton and elsewhere, wrote: 'The rapid development of the Southland trade has closed the mills in Catlins River, annihilated the coastal timber export of Westland, and greatly restricted that of Marlborough and Nelson.'[16]

In the hill country of the North Island, mills were established at 'successive railheads' as tracks and settlers pressed into the bush. Although many mills operated for less than a decade, others underpinned more permanent settlement. By creating markets

Table 7.1 New Zealand sawmills: number, distribution, capacity and output, 1907. Figures in parentheses indicate percentage of total.

Land district	No. of mills	Horsepower	Employment	Annual capacity superficial feet	Annual output superficial feet
Auckland	59 (14)	2847 (28)	2367 (33)	256,325,000 (36)	190,543,000 (44)
Hawke's Bay	38 (9)	1059 (11)	721 (10)	70,804,000 (10)	40,868,118 (9)
Taranaki	29 (7)	737 (7)	453 (6)	32,158,000 (4)	16,824,281 (4)
Wellington	83 (20)	1458 (15)	1388 (19)	118,440,000 (16)	70,138,000 (16)
Marlborough	14 (3)	494 (5)	195 (3)	15,770,000 (2)	9,869,000 (2)
Nelson	71 (17)	852 (9)	403 (6)	61,127,000 (9)	16,594,399 (4)
Westland	49 (12)	912 (9)	645 (9)	94,526,000 (13)	44,933,813 (10)
Canterbury	9 (2)	102 (1)	54 (1)	3,000,000 (0.4)	1,164,000 (0.2)
Otago	9 (2)	131 (1)	82 (1)	5,690,000 (0.8)	3,190,000 (0.7)
Southland	50 (12)	1408 (14)	831 (12)	61,100,000 (8)	38,087,000 (9)
TOTAL	**411**	**10,000**	**7139**	**718,940,000**	**432,031,611**

Note: Employment total includes sawmill hands only; bush workers not enumerated.
Source: 'The Timber Industry in New Zealand in 1907', *AJHR*, C4, 1907

for supplies as well as timber, and by employing local men as casual labour to build tramways or cut the bush, they helped settlers find their feet in their first years on the land. In many localities, such as Feilding, where eight mills were said to support 1268 men and their dependents, it is fair to say – as H.C.D. Somerset did of 'Littledene' (Oxford in Canterbury) – that 'the mill made the township'.[17]

There were 150 mills in 1876, 204 in 1881, 243 in 1891, and 414 in 1905. Early in the twentieth century, New Zealand's largest mill could produce 18 million feet a year; six others were capable of 10 million, 20 more of 5 million feet plus. Twenty-two of these large mills were in Auckland, and all were in the North Island. Elsewhere, portable mills able to cut no more than 200,000 feet a year, and more permanent structures with capacities of 500,000 to 2–3 million feet, consumed their varying shares of bush (Table 7.1). Skid roads, chutes and tramways, dams built to flush logs downstream, and the shanties and cookhouses of the workers marked the landscape. Logging was never a tidy operation. Stumps, sawdust and vast quantities of debris were left behind; the earth was gouged and disturbed; soils were exposed to rain and wind (Figure 7.4).

The consequences of this onslaught were most evident in the kauri districts. Sawn kauri production peaked early in the twentieth century, at approximately 120 million superficial feet, when even the most optimistic contemporary estimates indicated that barely 1112 million feet remained standing, on Crown, Māori and private lands

Figure 7.4 Bullock team and timber workers alongside a kauri tree, 1897, probably in the Whangarei District. Source: Mrs B. Jones Collection, Alexander Turnbull Library collections, National Library of New Zealand Te Puna Mātauranga o Aotearoa

(Figure 7.5). A forest once among 'the grandest sights to be found in the entire range of the vegetable kingdom' had been reduced, by the turn of the twentieth century, to the point where trained foresters anticipated its likely extinction. All recognised that the kauri timber used in building 'prosperous provincial cities' (in Australia as well as New Zealand) represented an 'almost fabulous sum of money', but the bonanza had run its course. Despite technological improvements that gave loggers access to formerly uneconomic trees, the annual kauri harvest fell during the 1920s to a tenth of its peak level; 30 years later, output was less than a million feet.[18]

This landscape suffered a double indignity: it was further ravaged by a small army of prospectors in search of the gum exuded by kauri. From the 1850s through the 1920s and beyond, 'gumdiggers' collected resin from living trees, removing it from cracks in their bark or 'bleeding' them with purposeful incisions. Most, however, simply fossicked for fallen lumps of dried gum deposited on and in the forest floor over thousands of years. Armed with makeshift tools, they searched the swamps and hills north of Auckland for the substance, which was used in making varnish and linoleum. In peak years around the turn of the century, their efforts yielded over 10,000 tons of gum for export. They left behind extensive areas pocked with excavations and tailings, the surface consisting of bare clay of no value for agriculture.

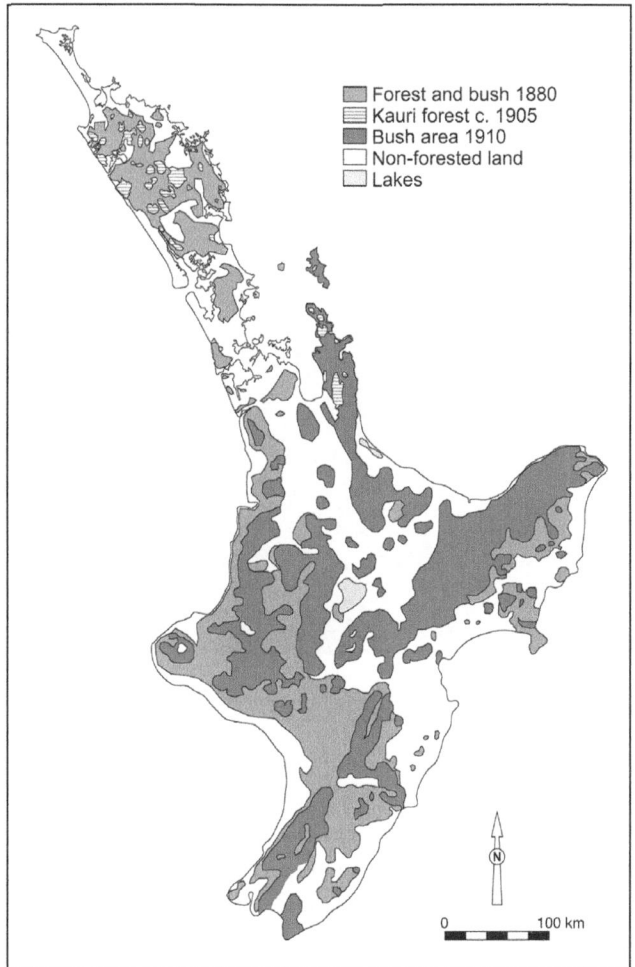

Figure 7.5 Map of the North Island forest, 1880–1910. Source: adapted from Malcolm McKinnon (ed.), *New Zealand Historical Atlas*, David Bateman, Auckland and Department of Internal Affairs, Wellington, 1997, plates 47, 48

Farmers had their due elsewhere. For 40 years or so after 1870, settlers pressed into the bush districts of Taranaki, Manawatu, Hawke's Bay and, later, the King Country.[19] Each farm created from the bush was a triumph of hard work and resilience in daunting circumstances. Hemmed in by forest, living in a rough hut made of wood slabs or tree ferns, those who pioneered these areas endured a 'Crusoe-like existence'. Their survival rested on the physically demanding, often dangerous work of bush felling, and a thorough late-summer burning of the dried debris (Figure 7.6).

The surveyor and scholar (and later bureaucrat) Edward Tregear, who watched the process unfold in Taranaki, captured its wider essence. There, he wrote,

Figure 7.6 Clearing new farms of trees under Mt Egmont, c. 1900. Source: PHO2007–301, Puke Ariki, New Plymouth

the first settlers … emulated each other in a struggle of dire economy and stubborn endurance, getting enough perhaps from the grass seed sown on the rich ashes of their first burn to keep life and soul together against another winter of reeking moisture and mud … The women worked side-by-side with the men from dawn to dark in the work of 'logging up' the charred and smutty timber and hewing out the massy stumps. Little by little, year by year, their reward became visible; a cow was bought, poultry increased, a better roof was over their heads; schoolhouses began to be built near improved roads …[20]

These fruits of 'heroic self denial … [and] patient courage' transformed the interior of the North Island. By 1900 the Seventy Mile Bush, 300,000 acres of forest so intractable that Scandinavian axemen were recruited to lead the charge against it, was no more.[21] Across the drainage divide, sheep and dairy farms covered the Rangitikei–Manawatu area, once among 'the best-timbered districts in the colony' (Figure 7.5). In Waipawa County, sown grassland covered an eighth of the land area in 1878; 30 years later it accounted for almost two-thirds. In Manawatu County the proportions rose from less than 5 per cent to more than 50 per cent over the same period. By one account, 1,684,231 acres of land leased or sold by the Crown was cleared between 1895 and 1914. Approximately 60 per cent of this lay in the Wellington and Hawke's Bay districts.[22]

None of this would have been possible without the settler's most essential – and ferocious – tool: the firestick. Burning was the only practicable method of opening up bushland to cultivation. Each new farm was edged by 'blackened prostrate logs', emblems of the 'ever-widening circle' that settlers strove, relentlessly, to expand in the 'seemingly impervious bush'. Travelling through the North Island by train in the 1880s, James Inglis noted that clouds of smoke darkened the sun and that the 'deepening gloom' of the evening was broken 'at fitful intervals by the lurid glare of ... forest fires'. In the Manawatu, he wrote in *Our New Zealand Cousins*, 'Burning is going on all around. The air is dense with smoke. Our clothes get white with falling ashes, and our eyes smart with the pungent reek.'[23]

But fire is notoriously difficult to control. A rising wind, a shift in the breeze, and away it might go. Sparks from the fireboxes of trains and traction engines, smouldering cigarette butts, scrub fires set by gumdiggers – often disparaged as 'fire raisers in chief' – could also set extensive tracts of forest ablaze. In dry years, with kindling in abundance wherever timber-getters had worked, 'terrible bush fires ... raged throughout the country'. The summer of 1885–86 was especially bad. After months with lower than normal rainfall, it seemed that there was 'scarcely a newspaper in the North Island', that did not 'narrate tales of disaster and peril by reason of the spread of fire'.[24]

Disquiet

Was this a bad thing? Inglis, a widely travelled, well-educated man with a dozen years' experience developing and operating an indigo plantation in India, was by no means certain. Although revolted at what he regarded as the destruction of New Zealand's natural endowment, he recognised that 'circumstances alter cases'. Perhaps the profligate haste with which the country was being developed was 'really the cheapest and the best' course. Most nineteenth-century New Zealanders, caught up in the immediacy of getting on, saw little reason to equivocate about the future they believed they were bringing into being. The bush was abundant. A yeoman empire was the goal. Reduction of the former was necessary to realise the latter. The 'invariable rule in a new country', observed a prominent member of the New Zealand Parliament in the 1860s, 'was the lavish use of native forest'. Inglis was not alone in his doubts.[25]

Concerned about future supplies of timber for its navy, the British government grew anxious about the destruction of the kauri forests of northern New Zealand even in the 1840s. A quarter-century later, the sight of Canterbury's Port Hills 'covered, for weeks together, with thick and lurid smoke' led Thomas Potts, an amateur botanist resident on Banks Peninsula, to lament the 'barbarous improvidence' of contemporary attitudes toward the forest. He and a handful of other educated and influential individuals pointed to the ways in which deforestation worked to disturb the 'equilibrium' arrived at by nature. Removal of the forest was seen to be 'a primary cause of excessive inundations'. Desiccation, too, could result from the 'destructive processes' unleashed in 'attempts to bring new countries under the domain of ... [human] wants'. In the minds of this minority, New Zealanders had a 'duty to preserve' as well as to develop their islands.[26]

A few years later, economic terms were added to the equation of unease. Some parts of the colony were already short of wood; in a few years New Zealand's progress might be hindered by lack of timber. Depleting the resource without thought for the future was akin to a man marooned on an uninhabited island consuming his limited supply of food in a few weeks though months might pass before his rescue.[27]

Debate about the depletion of New Zealand's forests continued, but the balance between waste and improvement was hard to find. The argument was not simply, as forester George Perrin alleged, over 'the destruction annually of thousands of pounds worth of timber – all for the sake of a few sheep'. People, farms, domestic markets were vital to the development of the country and the prosperity of its industries. Opening a major conference for leading members of the timber industry, convened by the government in 1896, Prime Minister Seddon noted that the timberlands of the colony were 'getting smaller day by day', and lamented that little benefit had fallen to the country or its people from their exploitation. Settlers were encouraged to colonise the backblocks, but by burning bush they destroyed millions of feet of timber 'more valuable than the land' eventually fenced and sown. When interrupted with the question, 'Why put them there at all?', however, Seddon was quick to respond: 'If we did not encourage settlers, there would be no sawmills, and there would be very few of you here today.'[28] The problem was that the first business in every new country was 'to clear the forest, and open up the best land for farming and settlement' but allowing this to go 'too far' threatened to make wood scarce, increase its price and harm the economy.[29]

In 1913 the Royal Commission on Forestry sought a solution by acknowledging the fundamental principle that good land should be settled, while insisting that inaccessible areas, and those with little agricultural potential, should remain forested for watershed protection, scenic benefit or timber supply. Recognising the slow growth rates of New Zealand's indigenous trees, they also suggested that once-forested areas of limited agricultural value might be afforested with introduced species to meet future timber needs.[30]

Restraint

This resolution was a long time coming, even though the doctrine of 'improvement' had not gone unchallenged through the decades after 1840. From the beginnings of European settlement, colonial officials and some settlers sought to regulate and restrain use of the forest. Their efforts were more often intended to control and direct rather than prevent exploitation, and even at that they were more often flawed than effective. They were, however, not insignificant. They defined an alternative, and sometimes tempered the ethos of exploitation. In their ineffectiveness they revealed much about prevailing attitudes to both the forest and regulation.

Early steps toward forest management in New Zealand came in 1849, with the introduction of licences permitting the cutting of timber on Crown land. Adapted and extended by some provincial governments after 1852, these provisions also saw the establishment of forest reserves, intended to secure timber supplies for the future, in Canterbury and Otago. But inadequate surveillance of reserves and licence holders allowed 'great waste and destruction' of the bush. Where demand was high and authority weak, it was nigh impossible to deflect sawyers and settlers from exploitation of the forest by fiat alone.[31]

In 1874 the New Zealand Forests Act, introduced by Prime Minister Julius Vogel, allowed for the creation of state forests and the appointment of a conservator. It gave Parliament authority to regulate access to, punish trespasses against, and protect these areas from fire. Although many of the Act's provisions were emasculated two years later, Inches Campbell Walker as the first Conservator of State Forests articulated cogent arguments against 'mining' the resource, and for the gradual rather than wholesale opening of the public domain for settlement and timber supply. He also asserted the need for forest reserves on upland slopes beyond the wise reach of settlement. By 1881 half a million acres had been set aside under the 1877 Land Act for the 'growth and preservation of timber' or 'climatic forest conservancy', Little effort, however, was given to managing these or other Crown forest lands. A few years on, the editors of the *New Zealand Times* claimed that regulations governing Crown forests had 'been ill-observed' and that there had been very great 'waste of trees by reckless and extravagant chopping and by fires'.[32]

Under the provisions of the State Forests Act (1885), the newly appointed Chief Conservator of Forests, Thomas Kirk, insisted that the boundaries of state forests be marked before adjoining land was sold, and imposed penalties for the destruction or damage of the forests by fire, felling or grazing. New mountain reserves were established to protect rivers, streams and climate, and several new regulations were imposed on timber-getters and sawmillers. Still, the effect of such measures depended on local officials, who worked in difficult circumstances and in the face of widespread sentiment that the wealth of New Zealand nature was for the settlers' benefit.

Kirk was often exasperated by the reports he received. One forest ranger charged with responsibility for the kauri district 'resided in Auckland and had only paid two or at most three visits to the Puhi Puhi Forest, although he had spent many days in the Whangerei hotels'. In midwinter Southland, 'everlastingly overhung' with cloud, physical conditions made the work of even the most dedicated ranger difficult. Every step in the often 'scrubby tangled broken' bush seemed to bring down 'a drenching shower'. Sleeping on bare boards, or on a bed and in clothes that were rarely dry, Thomas Hickson captured the harshness of this life when he wrote: 'One would require the constitution, hide and coat of a seal or polar bear to stand it.'[33]

In 1888 the Forests Branch in the Crown Lands Department was abolished and the rangers were discharged. In the next three decades large areas of land were withdrawn from state forests to promote settlement. In 1897 alone, over 20,000 acres of bush was cleared from former state forest land. Still, some influential people clung to the principles Campbell Walker had introduced in the 1870s. Attempts to open bush in Southland to settlement brought a reminder that 'the preservation of a reasonable share of the forest lands of the colony was a matter of great importance'. In the Lands Department, where settlement was always a priority, officials acknowledged the importance of forest cover for watershed protection, and some grew concerned about the ecological consequences of turning marginal land into farms.[34]

Early in the twentieth century, as timber production climbed toward a 1907 peak of 432 million superficial feet, growing anxiety about future timber supplies also tempered spendthrift attitudes toward the forest. There was no reliable inventory of the resource: of one estimate, made in 1909, it was said that 'no-one can say whether the amount is too much or too little'. The notion of an impending timber famine was powerfully underwritten by the sharp fall in estimates of how long timber supplies would last – from 70 years in 1905 to 30 years in 1913. Coupled with a growing sense that the battle to settle the land had been won, and with rising interest in the conservation of New Zealand's indigenous flora and fauna, this was enough to initiate an important shift in the terms on which New Zealanders treated the bush. In this respect the Surveyor General's turn-of-the-century recommendations that forests be conserved to maintain water supplies and climatic equilibrium, to prevent degradation of the high country and deposition in the lowlands, and to protect flora, fauna and natural beauty, were a prescient blend of the utilitarian and aesthetic concerns that would shape the future.[35]

Reflection

Speaking in the parliamentary debate on the Forests Bill of 1874, Mr Hunter of Wellington City challenged the value of Julius Vogel's arguments for the proposed legislation. Vogel, known as an impassioned advocate of economic expansion who considered colonisation the 'heroic' work of New Zealanders, had come to his understanding of the forest question by combining local knowledge with book learning. Travels in the South Island alerted him to the rapidity with which domestic demands were reducing forests and altering the landscape. From American and European literature dealing with human impacts on nature he gained broader insight into the environmental consequences of forest destruction.

Vogel imagined a future blighted by climatic changes induced by deforestation – by the drying-up of streams, the erosion of soils, the desertification that had overtaken once-forested North Africa and Asia Minor. It was time, he insisted, to curb the heedless and irresponsible destruction of New Zealand's forests. Hunter, though, would have none of this. In his view, the Prime Minister's examples from Mediterranean Europe and the United States were irrelevant. They were 'a mere history of the past' of little consequence to people 'in a young country and in a new land'.[36]

There were other arguments against the bill, but Hunter's intervention went to the heart of the matter. Few of those who opposed the bill were prepared to accept the collective, ecological and long-term viewpoints on which it rested. The forest question, said Vogel, was 'something for New Zealand to cling to for generations'; it was about preserving the 'intrinsic value' of the country. Most who spoke against the bill expressed immediate concerns and specific anxieties. Individualism, self-interest and material progress were the bedrocks of their understanding. They preferred 'the gospel of trade' to official regulation; they railed against the government's 'desire to become paternal' and argued that settlers should be 'left alone' to conduct their lives.

Knowledge derived from everyday local experience – common sense – was valued above expert opinion from overseas. Formal efforts to conserve forests before real shortages emerged were alarmist, and probably completely unnecessary: 'Nature', after all, had been an effective 'conserver of forests ages before the science of forestry was known'. Nelson was a 'sleepy hollow' compared with those 'hives of industry' Christchurch and Dunedin, precisely because it stood amid 'almost interminable forest', while the other cities were surrounded by 'grass-covered plains'; clear the forest, make waste lands fit for settlement, and Nelson, too, would enjoy prosperity. All the people of his province wanted, said Mr Bunny of Wellington, was to 'disforest' their land, foster development and produce a large profit.

These arguments echoed the central principles of mid-nineteenth-century liberalism – laissez-faire, localism, economy, and progress – but they owed at least as much to colonial circumstances as they did to the intellectual climate of Victorian Britain. New Zealand *was* a 'young country'. Having abandoned the places of their pasts in anticipation of better futures, its Pākehā settlers faced the daunting task of establishing homes in a new territory. Houses, barns, fences, roads, bridges, farms and cities had to be developed, and exploitation of the colony's natural capital offered the most direct means to this end. From this perspective, the onslaught on the New Zealand forest after 1840 was a constructive rather than a destructive process, yielding valuable export revenues, providing essential commodities and bringing 'unproductive' bush into 'higher use'.[37]

Only slowly were most of those caught up in the work of colonisation able to look differently at the world they were making, to consider the wider consequences of their actions on the environment and future generations. In this context, expressions of disquiet at the onslaught on the forest, and efforts to restrain it, were more remarkable than the assault on the bush itself. But they, too, are attributable to the particular circumstances of colonisation. Such developments owed much to the timing of European settlement in New Zealand and the composition of the colony's migrant stream. Without the insights derived from the hillsides of Vermont and the fringes of the Mediterranean by George Perkins Marsh; without the voyage of the *Beagle* and all that Darwin meant to understanding of the natural world after 1859; without the experience of European-trained foresters and the work of the Forest Conservancy in India; and without the presence in New Zealand of a group of settlers spared, by their wealth, from the general struggle to gain a sufficiency from the land, the assault on the bush might have been more unfettered and protracted than it was.[38]

Conclusion

New Zealanders, said Captain Campbell Walker, with reference to the prospects for New Zealand's forests in 1877, 'can not be wise too soon'. But it would be a mistake to simply indict the attitudes and actions of late nineteenth-century New Zealanders. Even as they watched (as Sargeson's uncle did in the 1920s), the creeks through their new-cleared land rising quickly after rain and running discoloured where once they had been clear, even as they saw landslides scar once-forested slopes and pondered the invasion of briefly lush pastures by biddy-bid and bracken, there were reasons – in the sheep, the taut fencelines, orchards and dwellings that had replaced the bush – to take pride in the transformation of the landscape. Setbacks such as declining soil fertility could be addressed, it seemed, by clearing more land or, later, by using chemical fertilisers.[39]

Yet the ambivalence James Inglis felt about the course of New Zealand development was not unwarranted. Consider the heartfelt cry offered by Herbert Guthrie-Smith near the end of his remarkable life as a Hawke's Bay runholder, natural historian and author. No one in New Zealand between 1880 and 1940 had a closer knowledge of or love for the land. Yet for years he had followed the common 'stream of tendency' in his development of Tutira station, valuing it for its 'ability to grow meat and wool' and failing, he said, to afford his acres the respect that the long view demanded. In search of pastoral perfection, the hills had been remade, and left largely 'bare of trees, of woodland, of sedge'. He harboured some misgivings about this along the way, but not until 1940, looking back on a lifetime's work, did he come to ask: 'Have I for sixty years desecrated God's earth and dubbed it improvement?'[40]

Further reading

Much of the recent work relevant to this chapter has come from James Beattie and Paul Star and is referenced in their joint essay, 'Global Influences and Local Environments: Forestry and Forest Conservation in New Zealand, 1850s–1925' (*British Scholar*, vol. 3, no. 2, 2010, pp. 191–218). See also Michael Roche, 'Sir David Hutchins and Kauri in New Zealand', in John Dargavel (ed.), *Araucarian Forests* (Australian Forest History Society Inc, Kingston, 2005, pp. 33–40), and his essay, 'Edward Phillips Turner: The Development of a "Forest Consciousness" in New Zealand 1890s to 1930s' (*Proceedings* of the 6th National Australian Forest History Conference, Augusta, WA, 12–17 September 2004); Catherine Knight, 'The Paradox of Discourse Concerning Deforestation in New Zealand: A Historical Survey' (*Environment and History*, vol. 15, no. 3, 2009, pp. 323–42); and, for a wide-ranging discussion, Geoff Park, *Theatre Country: Essays on Landscape and Whenua* (Victoria University Press, Wellington, 2006).

III

Wild places

8.
Children of the burnt bush: New Zealanders and the indigenous remnant, 1880–1930

Paul Star and Lynne Lochhead

Naked, denuded,
Forestless, fernless,
Mute, now, and songless,
Sharp on sheer sky gape the lips of the gully;
Burden'd with black is the grass of its pasture;
On whose long slopes
The sheep in their browsing
Must leap o'er a million,
Strewn, helter-skelter, headlong and helpless
Burnt bones of the Bush ...[1]

THE SCENE DESCRIBED by Blanche Baughan was familiar to New Zealanders at the dawning of the twentieth century. They lived as people still very much engaged in the rapid transformation of the environment. Baughan was one of a growing number in that generation who questioned the direction of settler society, who challenged the assumptions on which the rate and extent of the transformation were based, and who considered it vital to preserve an indigenous remnant.

The forest area of New Zealand, in 1886 still 22 million acres or about a third of the total land surface, had been reduced to 17 million by 1909 (Figure 7.5). Correspondingly, the total area 'under occupation' increased from 35 million acres in 1901 to 40 million in 1910. This reflected the Liberal government's settlement policy and its accelerated purchase of remaining Māori bushland after 1892.[2] At the same time, efforts were made to reserve some of the pre-European natural heritage. By 1909, over 2 million acres had been set aside under the State Forests Act 1885, but timber on this land was intended for eventual milling. More significantly, the Land Act 1892 was the first to cater for scenic reservation, and by 1907 nearly 3 million acres had some such classification, much of it as national park (Figure 8.1). Our purpose in this chapter is to describe and analyse the movement to protect a part of the native environment during the decades of its greatest destruction.

141

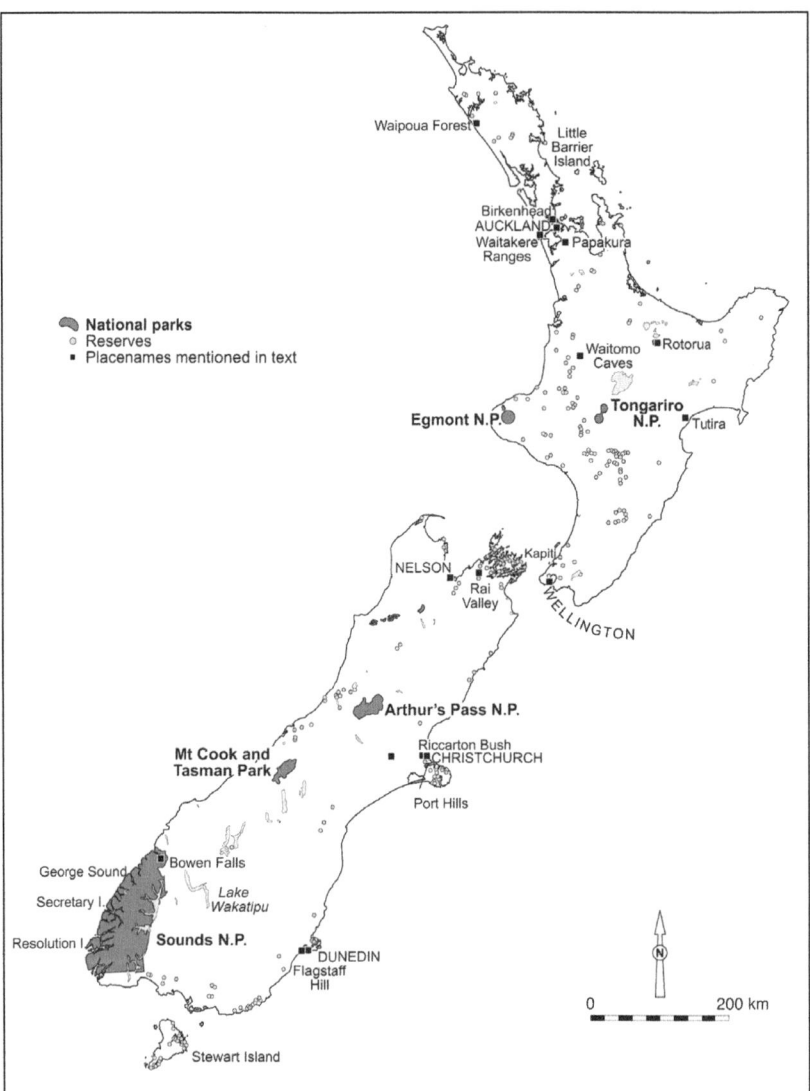

Figure 8.1 Map of protected areas, 1906–07, showing national parks and scenic reserves and places mentioned in Chapter 8. Source: Department of Lands: Scenery Preservation, *AJHR*, C6, 1907

The rapidity of change has led to the assumption that the settlers feared and hated the bush, but the reality was rather more complex.[3] The grandeur and luxuriance of the forest were widely appreciated, and though it was often described as 'gloomy', this needs to be understood within the context of nineteenth-century canons of taste. Whereas it may suggest negativity to us, any literate Victorian would recognise this as a conventional expression of the sublimity of nature.[4] Actual experience of travel on foot was apt to change appreciation into a feeling that the forest was interminable, and experience of the difficulties in removing it could lead to a negative assessment.

Victorians were as ready as we are to suppress aesthetic feelings in the name of utility. They were helped by a belief that the progress of civilisation depended on their efforts to 'subdue and replenish' nature. Whether or not they appreciated its beauty, most felt morally bound to turn uncultivated nature ('wilderness', 'barren' or 'waste' land) to domesticated, civilised use.[5] Any regret felt for the passing of indigenous flora and fauna diminished with the conviction that it was merely hastening a natural process. Until the 1880s the apparently inevitable displacement of 'weak' southern hemisphere species by more vigorous northern hemisphere species that had evolved with greater competition was perceived as a universal law by most scientists and laymen alike.

Economic motives for conservation

Given this frame of mind, it is not surprising that the earliest conservation measures were utilitarian in motive. Government designed the Forests Act 1874, and its successors, to ensure future supply in the interests of long-term development of the colony, prompted by growing recognition that timber was a limited resource. The legislation also reflected anxiety that loss of forests contributed to droughts and increased downstream flooding, both of which imposed economic costs on settlement.

There was concern, too, about the depletion of marine resources. In 1870 the Otago Institute investigated coastal whaling stations and the resulting slaughter of young whales. A close season was introduced for oysters (1866) and seals (1872), and the Fish Protection Act 1877 allowed for protection of fish through close seasons and control of the mesh size of nets. Fears that a burgeoning flounder export trade might be 'nipped in the bud' through flouting of the law led to a strengthened Fisheries Conservation Act 1884, with provision for the protection of breeding areas. It also contained a first anti-pollution measure, against the dumping of sawmill waste into streams.[6]

From the 1860s to the 1880s, economic arguments for conservation at times carried the day over the protests of many who believed nothing should stand in the way of settlement. For a few, concern went beyond the purely utilitarian. In 1874 Premier Julius Vogel contended that forests also affected the 'beauty, healthfulness, and pleasure-bestowing qualities' of the nation as well as 'its character and intrinsic value'. Discussion of 'forest influences' signalled some realignment in thinking about the human relationship with nature. Science had been linked to the goal of dominating the environment since the seventeenth century, but just when the rapid advance of technology and the spread of 'civilised and civilising' Europeans throughout the globe seemed to bring that dream close to fulfilment, some scientists began to sound warnings. An American, George Perkins Marsh, wrote of limits to the control that could be exercised over nature. No one yet spoke of sustainability, but the recognition of limits was an essential step towards acceptance of this concept.[7]

Island reserves and native bird protection

The need for preservation remained a minority viewpoint throughout the nineteenth century, but the pace quickened in its final decade. The year 1891 proved a landmark for fauna preservation with the establishment of Resolution Island in Fiordland as a sanctuary, in response to a recommendation from the Australasian Association for the Advancement of Science. Others followed – Secretary Island (1893), Little Barrier (1894–97) and Kapiti (1897). The specific idea of *island* reserves was probably a 'world first' for New Zealand, a nation of islands. An appreciation of the frequent incompatibility of indigenous fauna and European settlement had led T.H. Potts, a Canterbury runholder and keen ornithologist, to suggest in 1872 that Resolution Island 'might be placed under tapu from molestation of dog and gun' as a refuge for 'wingless species'. The Auckland-based botanist Thomas Kirk envisaged a similar future for Little Barrier Island[8] (see Figure 8.1 for locations in text).

Implementation of this idea seemed increasingly urgent after 1884 when, in a misguided attempt to reduce the vast population of rabbits that ravaged pasture intended for sheep, the government introduced stoats and weasels. While these creatures were the 'natural enemy' of rabbits in England, in New Zealand the flightless birds made easier prey. Native bird populations decreased rapidly at the same time as interest in them increased, and stoats became the scapegoats for a decline that probably resulted more from destruction of habitat.

In Europe, zoological gardens and private collectors paid good money for unusual antipodean specimens. Lord Rothschild of Tring amassed nearly 200 kiwi skins and skeletons. As its scientific name (*Apteryx*, or 'no-wing') indicated, the bird challenged European taxonomists as profoundly as had Australia's duck-billed platypus. International fascination with such evolutionary oddities as the kiwi, the takahē and the lizard-like tuatara helped the government to appreciate that, in these species, New Zealand had something worthy of protection. Reserving one or two island refuges where at-risk species might be spared from extinction was quite easily done. It proved much harder to gain government support for action on the mainland, where protection of stoats and weasels, even in areas without rabbits, remained until 1903.

The declining numbers of species of native birds proved unquantifiable, but it was a fact no one doubted. It went with the removal of native vegetation and the introduction of exotic species of flora and fauna, and was recorded in broad outline by many early settlers.[9] While the best accounts of the process of ecological transformation did not appear until the 1920s, these were written by two men – Herbert Guthrie-Smith and G.M. Thomson[10] – who had themselves participated in biotic change for half a century and had published earlier, less extensive commentaries. First efforts to protect native birds therefore occurred against a background of quite widespread awareness of cause and effect.

These early measures were half-hearted at best, occurring on an ad hoc basis under the Animals Protection Acts. Based on British game laws, they were primarily concerned with protecting acclimatised species. An 1864 provision for a close season for native ducks and pigeon – species that settlers recognised as worth shooting for sport – was clearly utilitarian. The list of game species expanded over the years, including kākā and morepork in 1887, but since the law was not strictly enforced it afforded little protection.

Protection of tūī proceeded from a different motive. Canterbury's Superintendent William Rolleston advocated legislation to counter the destruction of native birds 'characteristic of the country'. He observed that one saw 'bundles of tuis hanging up for sale in shops and heard of people dining off kiwis', which he considered 'gross abuse of the present privilege to kill'.[11] Sale of tūī was prohibited in 1872, but kiwi, which would become a synonym for New Zealanders from around 1916, were not protected. An 1873 amendment added tūī to the schedule of native game, effectively giving them permanent protection, as an open season was never declared. Crested grebe and white heron were similarly treated from 1888, huia in 1892, tuatara in 1895, and finally kiwi, kōkako, kākāpō, saddleback, stitchbird and bellbird in 1897.

While this gradual addition of native species was in part a colonial response to Britain's Wild Birds Protection Bill 1872 and its subsequent legislation, there was a growing, locally inspired feeling for native and especially endemic species. According to the Minister of Internal Affairs, by the 1900s 'most of our older settlers [had] practically disassociated themselves from the bird-life of the older countries'. The occasional one even acknowledged that 'we are sadly inferior to the old Native chiefs in this matter', describing rāhui (Māori protection orders), which prevented unseasonal slaughter of tūī, kiwi and muttonbirds. The Christchurch politician Harry Ell sensed within Parliament, in 1900, a 'unanimity of opinion that every effort should be made to save our native birds from destruction', but only during debate in 1910 does this appear a genuine concern for most MPs. The result was an important amendment to legislation which, rather than listing favoured species, protected them *all* unless, like the predatory harrier and kea, they were specifically exempted.[12] Comparable recognition of the special value of native plants – simply by the fact of indigeneity – did not follow until the Native Plants Protection Act of 1934. Despite gradual legislative improvement, protection of native fauna remained haphazard until the formation of the Wildlife Service in 1931.

Early scenery preservation and tourism

In 1874 ex-Premier William Fox introduced the national park idea to New Zealand in a letter urging acquisition of the Rotorua thermal region to ensure its controlled development for the public benefit. He cited as a precedent the recent protection of Yellowstone's volcanic wonders within a public park. But the resulting Thermal Springs Districts Act 1881 did not emulate the Yellowstone model. Rather, it was designed to promote settlement to take advantage of the revenue-producing potential of sanatoriums. However, with the example of Yellowstone in mind, the 1885 Land Act made provision for designating both mineral springs and 'natural curiosities' as reserves. This was amended in 1892 specifically to include scenery, but the provision was seldom used before 1900. Nevertheless, there was a growing realisation that New Zealand's finest scenery could attract tourists and that, by failing to protect it, they risked killing the goose that would lay the golden egg.

The Union Steam Ship Company began cruises to Fiordland in 1877, though the area did not gain protection until 1905. Following a private initiative that established the Hermitage in response to growing international interest in mountaineering, the Department of Crown Lands designated Hooker Valley near Mt Cook a recreation reserve in 1885. The Tasman Valley was added in 1887. That same year the government also finally turned to the national park idea after Te Heuheu Tukino IV, the leader of Ngāti Tūwharetoa, gifted the three volcanic peaks in the central North Island to the nation.[13] The Tongariro National Park Act 1894 formally constituted New Zealand's first national park. General acceptance of the park owed much to the fact that the land was 'almost useless as far as grazing was concerned', and that as a tourist attraction it was likely to be 'worth ten times more to the country than if it was agricultural or pastoral land'.[14] In contrast, a debate on scenery preservation in 1891, prompted by Clutha Member of the House of Representatives Thomas Mackenzie's call to improve tourist access, revealed a widely held view that roads should not be opened solely for tourism at the expense of the needs of settlers. For many in the community, tourism was only acceptable provided it did not hinder the dominant goal of settlement. Despite this perception, the government remained committed to tourism, as signalled by the creation of the Department of Tourist and Health Resorts in 1901. The department's annual reports regularly stressed the importance of protecting scenery.

The growing support for scenery preservation throughout the 1890s was also manifested in the formation of scenery preservation societies. The Dunedin and Suburban Reserves Conservation Society was established in 1888 to improve and preserve the natural attractions of that city. Other societies followed between 1891 and 1899 in Taranaki, Nelson, Wellington, Christchurch, Auckland and Birkenhead. Most chose to use the name Scenery Preservation Society, and, though independent, they felt they were part of a nationwide movement. They were not slow to recognise the tourist potential of scenery. The Taranaki society published a guide to Mt Egmont in 1892, indicating routes and accommodation and describing the flora and fauna. Participants in these groups felt a genuine love of nature but recognised 'that the faculty of comprehending, reverencing and loving natural beauty grows up slowly and gradually'. Promoting utilitarian benefits offered the best hope of achieving protection until others shared their sense of reverence. Members knew their contemporaries well enough to recognise that they would be regarded as 'fanatics not untouched with lunacy' if they stressed preservation for its own sake.[15]

Participants were mainly middle-class males – lawyers, politicians at national and local level, newspapermen and businessmen – while the most active members of the groups were typically scientists or men with surveying experience. Historian Ross Galbreath has said in explanation of the movement, which was predominantly urban-based, that distance from the bush lent enchantment to the view, but supporters were conscious that continued indifference on the part of most settlers was the more likely result.[16] They stressed the need to encourage contact with nature. This was one reason behind publication of the guidebook. It also led to an emphasis on protection of accessible areas. Although designation of Egmont National Park (1900), including the valuable lowland forest of the Kaitake Range, and the Otira/Arthur's Pass reserves (1901), which formed the nucleus of the later Arthur's Pass National Park (1929), were important results of this activity, protection of forest remnants in or near the city – such as at Riccarton Bush, Birkenhead, Papakura and the Waitakere Ranges, or in the town belts of Dunedin and Wellington – was more characteristic (Figure 8.1).

Portrayal of an early conservation movement preoccupied with offshore islands, remote mountains or 'ferny glades, waterfalls and forest clearings for bush picnics'[17] does not present a rounded view. Certainly, ferny glades and other favoured picnic sites were protected, which was consistent with the goal of encouraging contact with nature. But far from focusing on offshore island reserves (however great their significance in retrospect), the preservation groups advocated reserves where birds were actually breeding, as with the 1893 campaign by the Taranaki Scenery Preservation Society to protect coastal bush at Whitecliffs.

There were boundaries, however, beyond which indigenous protection could not at first extend, as clearly signalled by the attempt to preserve the Ronga and Opouri valleys near Rai Valley township on the Nelson–Blenheim highway.[18] The area was rich in millable forest, but also provided a major lowland breeding and feeding ground for species such as pigeon and kākā. Local millers sought revocation of its existing forest reserve status in 1897. The Nelson Scenery Preservation Society responded in 1898 with a counter-petition for a national park of some 20,000 acres, claiming that birds should be preserved in a mainland sanctuary where they were thriving, rather than removed to places where they might perish. In addition, they argued, the area contained some of the densest, grandest, most diversified forest in New Zealand, and had the advantage of low fire risk and ready accessibility. Reversing the usual settler equation of undeveloped land with wilderness, a key supporter of the park claimed that if the valleys were 'subjected to the usual ravages of milling and fire, nothing but a howling wilderness would be the result'.[19]

Never before had conservationists sought to preserve in perpetuity such a large area of demonstrably high-value timber on land suitable for settlement. Its very suitability for those purposes meant that it was an increasingly scarce example of the finest forestland, some of which, supporters argued, should be preserved for posterity. They reasoned that, if future generations were to experience not just the trees but also a forest teeming with life, small blocks would not suffice because they were 'absolutely useless for birds'. Opponents, arguing that a national park would mean 'complete stagnation' for the district, pointed out that other parts were far more picturesque. Yet having dismissed its scenic value, they had to concede that the 'one unique peculiarity' about the Rai was that 'you can see grand forestry'. In fact, Charles Adams of the Land Board admitted he had 'never seen better forest … anywhere in New Zealand'.[20] The campaign highlighted a difficulty with the term 'scenery preservation', soon to be enshrined in legislation. In the popular mind it was associated with particular types of scenery and did not sit comfortably with the ecological approach towards which some campaign participants were already feeling their way.

The Waste Lands Committee recommended that the government carefully consider how best to conserve the forests of the Rai, but milling rights were granted in 1906, and in 1914 it was opened for settlement. This conclusively answered the question raised during the conflict, whether 'in this young country, the fringe of which is not yet tilled, land is in such demand that 20,000 acres cannot be spared for a National Park'. They could not if they contained good timber or were suitable for settlement. New Zealand did not prove ready to save a major commercially valuable lowland forest until Waipoua was set aside in 1952 (see Chapter 12).[21]

The significance of the scenery preservation societies went beyond campaigns to reserve land. They sought effective management of reserves at a time when there was scarcely any political awareness of the need. They drew attention to damage caused by vandalism, browsing animals, weed invasion and fire. The Taranaki society undertook fencing of reserves, and a campaign launched by it in 1898 resulted in the Noxious Weeds Act 1901. This group also successfully lobbied for a Board of Conservators for Egmont Reserve. Appointed in 1892, it set a precedent for Tongariro and later national parks; and curators were also appointed for the island sanctuaries. Approval was given for the first appointment – Richard Henry on Resolution Island – in 1893, but for the majority of reserves there was no improvement, and there was no official inspector of scenic reserves before the appointment of E. Phillips Turner in 1907.

Arguably the greatest achievement of the early groups was the Scenery Preservation Act 1903, which in turn provided the groundwork for the current system of parks and reserves. Scenic preservation in the 1890s had been the responsibility of the Department of Lands and Survey and could apply only to Crown land. The new Act paved the way for compulsory purchase of private or Māori land if the land's features were of sufficient note. Between 1904 and 1906 a commission, reporting to the Tourism Department, identified 380 scenic and historic places throughout New Zealand worth preservation. Over three-quarters were recommended for their scenic attributes; fewer than a quarter for their historic associations.[22] The first reserve made under the Act was the Waitomo Caves, on Māori land, and Flagstaff Hill behind Dunedin was repossessed from its European owners soon after, but most reserves were on Crown land, which involved the government in no additional expense. About 214,000 acres had been preserved under the Act by 1914, in 363 different locations (Figure 8.1). Scenic reserves covered a lesser total area than national parks, but they resulted from a more continuous effort.[23]

Harry Ell, who ardently campaigned for it in Parliament, is usually given credit for the Scenery Preservation Act, but its promotion by Prime Minister Richard Seddon was also a key factor.[24] Furthermore, Parliament's willingness to listen resulted from a decade of lobbying on a range of preservation issues by the scenery preservation societies. Paradoxically, the Act then seems to have contributed to their demise. Having achieved government action, some societies lapsed, and others turned increasingly to the provision of amenities. In Christchurch the nature protection role was taken over by Ell's Summit Road Association, formed in 1909 to promote his far-sighted schemes for the Port Hills (Figure 8.2).[25] A few groups modelled on these early societies continued to form after 1903, but there was a temporary loss of vigour prior to the emergence of nationally based groups in the second decade of the century.

An eye to the tourist dollar continued to motivate scenery preservation at the official level. Particular emphasis was given to major travel routes like the Main Trunk Line between Auckland and Wellington. Thomas Donne, the Superintendent of Tourism, believed that 'magnificent areas of forest … preserved in its primeval beauty contiguous

Figure 8.2 Kennedy's Bush, c. 1920. It was the first of the scenic reserves gazetted (in 1906) in association with Harry Ell's Summit Road. Source: CCL-PhotoCD01-IMG0079, Christchurch City Libraries Ngā Kete Wānanga o Ōtautahi

to the railway, will afford a great attraction to travellers ... whereas miles of burnt and blackened logs would prove a weariness to the spirit'. As long as there was no interference with settlement, Donne considered it a 'national obligation' to protect such patches of bush, which would 'not only agreeably adorn the routes of travel, but also serve as a last home for the rarer New Zealand birds'. Consistent with this attitude, the government's Scenery Preservation Board bent over backwards not to alienate settlement interests, stressing that they never withdrew farmland for 'aesthetic purposes'.[26]

A sense of scarcity

Before protection could extend beyond the parameters so vividly shown in the case of the Rai Valley, new factors had to emerge. Undoubtedly, the increasing scarcity of some native bird species and of certain kinds of native forest was a stimulus towards their preservation. In 1907, when the Department of Lands and Survey expressed concern that the quantity of millable native timber had fallen from 43,000 million superficial feet to 36,000 million in just two years, one of the most striking facts was that less than 650 million superficial feet of kauri, the most sought-after timber, remained. There was general acceptance that kauri would soon be milled out, and this focused attention on the need to preserve at least one representative sample of original kauri forest. The decline in overall area of bush was harder to appreciate, and for many the impression remained that there was ample left.

Nor did calls for preservation necessarily follow from increasing scarcity. Many argued, rather, that acclimatisation of introduced species such as insectivorous birds and exotic timbers was all the more important as native species grew rarer. The 1913 Royal Commission on Forestry (which included botanist Leonard Cockayne) was frequently urged to recommend planting of indigenous trees to help prepare for a forthcoming 'timber famine' (see Chapter 12), but concluded that no native timber, save perhaps beech, would grow or regenerate quickly enough.[27] The solution lay with the continued one-off utilisation of native forest until sufficient exotic timber became available. While a regard for native species was increasing, few of the requirements of settler society seemed likely to be met by them. Basic need favoured the exotic – which did not improve the case for preservation of native forests.

On the other hand, the rejection of a crude displacement theory, which had been used in support of suppression of native species, increased respect for the indigenous remnant. Already in the 1880s New Zealand-based botanists and some other colonists, trusting more to their own eyes than to the generalisations of overseas theorists, questioned the notion.[28] By the 1920s, Cockayne had published evidence that native plants could hold their own unless humans (directly or through the grazing of their beasts) continued to modify the environment in favour of exotics. Displacement theory was disproved – which meant that preservation was not, intrinsically, a lost cause – and, in its place, understandings based on ecological science gained credence.

Leonard Cockayne and ecological motives for preservation

Much as Marsh's *Man and Nature* had crystallised the forest conservation ideas of Potts and Vogel in the 1860s and 1870s, so the ecological studies of the Dane Eugene Warming provided a focus in the 1890s for the botanical conservation work of Cockayne. His 'oecological' survey of the Chatham Islands (1901) described a vegetation destined for extinction through the combined influence of exotic plants, introduced animals and fire – all initiated by human action. Cockayne's government-financed surveys of Kapiti, Tongariro, Waipoua and Stewart Island (1907–09) then familiarised him with a series of remarkable environments that appeared capable of protection. He seized this opportunity to develop the arguments for protection of the indigenous remnant, and his persuasive introductions to the resulting botanical reports provided an influential rationale for conservation in New Zealand over the ensuing decades.

Cockayne adduced all the usual arguments for preservation based on tourism and protection of lowland farms from erosion and flooding. His particular insight was to link the distinctiveness of New Zealand's scenery with its flora, an argument expressed most cogently in his recommendations for extensions to Tongariro National Park (eventually achieved in 1922). Progressing beyond the standard argument that Tongariro was a scenic wonder, he reasoned that similar geological features could be found elsewhere:

> … the special features of any landscape depend on the combinations of plants which
> form its garment, otherwise a monotonous uniformity would mark the whole earth.
> Therefore the more special the vegetation, the more distinctive the scenery. And nowhere

does this dictum carry more weight than in New Zealand, where the vegetation is unique ... Nor is it merely the individual species which are interesting, but equally important and of greater moment to the scenery is the manner in which they are associated together.[29]

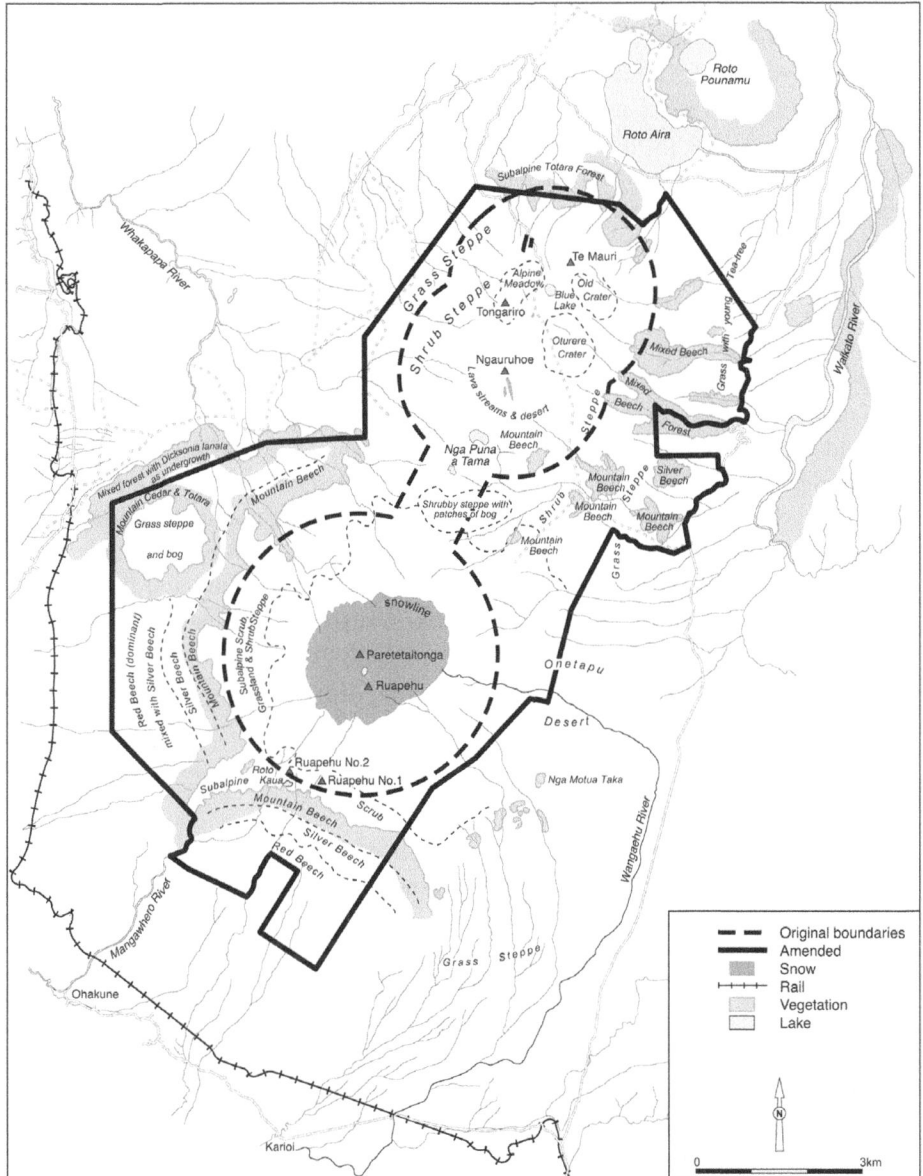

Figure 8.3 Plan of Tongariro National Park, from Leonard Cockayne's report on his botanical survey of the park. Source: Department of Lands, Report on a Botanical Survey of Tongariro National Park, *AJHR*, C11, 1908

Extensions to the park should not be into the forest alone, since 'no area gives an accurate picture of the district ... if it does not contain typical examples of all those combinations of species called scientifically "plant associations"'. Here was a specifically ecological argument for the preservation of the *range* of habitats, one that could be assimilated into the popular and legislative emphasis on scenery. It opened the way for protection of areas that were not conventionally scenic, such as tussock, low shrub and bog associations of the Volcanic Plateau (Figure 8.3).

The corollary of Cockayne's emphasis on plant associations was stress on the importance of reserves as natural museums. Individual species could be preserved in botanic gardens, but plant communities could not. Once destroyed, they were lost forever. If necessary, conservation should take priority over access, as on Kapiti, which was not for the 'merely curious' and 'no place on which to picnic'. Cockayne believed that Kapiti and the kauri forests of Waipoua were examples of 'actual primitive formations'. Since nothing comparable remained in the Old World, they were 'of extreme botanical interest to foreign scientists'. He further remarked that preservation of forests such as Waipoua, by facilitating the study of ecological succession, not only advanced science but also had a 'distinct economic bearing'.[30]

Cockayne claimed that his recommendations for the extension of Tongariro included 'no land which could finally become of economic value'. He consistently endeavoured to accommodate the needs of settlement as well as preservation, a position that aided his public employment as a botanist and commissioner, and prevented his dismissal as a crank. It also facilitated a gradual acceptance of his ecological approach to conservation. Yet he was surely aware that, at a time when even Campbell Island and the Auckland Islands in the subantarctic were leased for grazing, few areas were considered devoid of economic value. When unique remnants such as Riccarton Bush, Waipoua, and those on remote islands or around Tongariro itself were threatened, Cockayne pushed for preservation ahead of economic gain.

Nature conservation and management after 1914

European botanists held Cockayne in the highest esteem for his work in plant ecology, but his significance to us lies more in his readiness to utilise his science in the cause of preservation. He became a commanding presence in the New Zealand Institute (later the Royal Society), which, under his influence – and that of like minds such as the Dunedin scientist G.M. Thomson – became increasingly active in this respect. Cockayne also participated in the popular preservation movement, both through the Christchurch Beautifying Association and the shortlived New Zealand Forest and Bird Protection Society (1914 to about 1918), the first conservation organisation with a national constituency – not to be confused with the still-extant Royal Forest and Bird Protection Society, which began as the Native Bird Protection Society in 1923.

This latter group, together with its companion the New Zealand Forestry League (1916), marked a new departure. Nationally based, it developed a network of active branches and regional representatives. It also formed working relationships with locally based conservation groups, such as the Nelson Bush and Bird Society founded by Pérrine

Moncrieff in 1928.[31] While more and more New Zealanders were becoming urban-based in the early twentieth century, the presence of nature study as a school subject from 1905 encouraged new generations to remain in touch with the land. An address by Thomson to the Dunedin Naturalists' Field Club in 1929 suggests that such clubs could play a similar role with adults. For some, 'going bush' with the tramping clubs that emerged in the 1920s, or going hunting, perpetuated direct experience of native environments through outdoor recreation.[32]

Confident of a wide range of supporters and building on the work of the earlier groups, Forest and Bird brought a new depth of purpose to the cause. Agencies that administered the indigenous remnant were exposed to a high degree of scrutiny and public criticism. They could not afford to ignore relentless pressure through the press, direct political lobbying and the groups' own publications. This ensured that existing laws were adequately implemented and improved. Particular emphasis was placed on the need for active management. With the support of the Royal Society, the groups helped to shape policy on the control of browsing animals and of erosion and flooding. They also influenced the creation of an administrative structure for protection of native flora and fauna.

In 1919 the Forestry League secured the establishment of an independent Forestry Branch which, initially at least, seemed to offer better hope of protection than Lands and Survey or Internal Affairs. Challenging the dominance of agriculture was no easy task, however, and little came of the movement's proposals for an expert 'demarcation' of remaining forest land into areas best suited for forestry or farming. Conflict increased as the State Forest Service gradually moved away from its initial commitment to sustained-yield management of indigenous forests (see Chapter 12). It was only with the much later establishment of a Department of Conservation in 1987 that the 'unity of control' the groups had long advocated was fully achieved.

Most importantly, the preservation movement continued to shape the definition of parks and reserves, which had begun with no concept of indigenous integrity. Thomas Donne believed that, though New Zealand was extraordinarily beautiful, wealthy tourists would only visit if there were things to shoot as well as to see. Through the Tourist Department he had promoted both scenery preservation and deer acclimatisation, even though attention was already being drawn to the harm that browsing animals caused. In 1904, while supporting a successful move by Thomas Mackenzie to make much of Fiordland a national park, Donne 'enhanced' the area by releasing wapiti at George Sound and white-tailed deer at Lake Wakatipu.

The close link between scenery and preservation had also created ambiguities. For instance, in an 1897 report on the conservation of New Zealand forests, George Perrin, Tasmanian Conservator of State Forests, suggested that mānuka scrub should be replaced with 'more ornamental and useful trees' to render thermal areas attractive.[33] Cockayne's identification of the relationship between vegetation and the uniqueness of scenery helped to alter perceptions. By 1905 for Riccarton Bush, and again in 1907 for Kapiti, he argued that no exotic species – even those that were apparently harmless – had a place in such reserves.

The rapid spread of heather, introduced at Tongariro National Park from 1914, helped to crystallise views on the need to keep reserves 'inviolate'. As a result, by the mid-1920s the Forestry League, the Royal Society and the Native Bird Protection Society all adopted the principle that national parks and reserves should be preserved in a natural state. In 1925 the Native Bird Protection Society, influenced by the entomologist J.G. Myers, went further and argued that reserves must also be kept free of indigenous species that had not formerly been present in the area. This lay behind a successful campaign in 1927 to prevent transfer of South Island robins to Kapiti, where it was feared they would have hybridised with the northern species.

Preservation and nationhood

Increasingly, native flora, fauna and scenery were not merely appreciated, but also fulfilled a psychological need. Native birds, scenic wonders and mountains first appeared on New Zealand postage stamps in 1898.[34] Four years later, Harry Ell wanted schools to be provided with pictures that would make the new generation 'more familiar with our beautiful flowers, and the more interesting specimens of the New Zealand native birds, instead of having only as an object lesson the cards of English birds'.[35] Indigenous nature became linked to both individual and national identity.[36]

Scientists supportive of conservation tapped into a gathering sense of nationhood. G.M. Thomson, in 1901, regretted that in towns 'the native flora had been more or less displaced by introduced species, so that there was nothing in the vegetation to tell a visitor he was in New Zealand'. To counteract this trend, he suggested plantings by the Natives' Association, which comprised a growing number of New Zealand-born Europeans who felt pride in the country of their birth. The following year Cockayne called for 'a national garden in the centre of the colony' (Wellington) to protect native plants. Such remarks fell short of any direct statement linking native flora with national identity, but they demonstrate the interplay of developing ideas. Protection of scenery, flora and fauna was justified on the grounds of their New Zealandness, while, in parallel, landscape and biota helped to define the values a New Zealander would hold.[37]

By the early 1920s the link with national identity was being made explicit. In 1924 J.G. Myers associated exotic-free reserves with patriotism. He argued against the 'abominable practice' of introducing foreign species, considering it a 'sin against posterity' to allow them in parks and reserves. He was scathing of those in authority 'whose patriotism, scorning those natural glories which embody the very spirit of our country, rises no higher than a desire to create in New Zealand a paltry replica of other lands'. In 1925 James Cowan, opposing proposals to generate power and manufacture fertiliser at Bowen Falls in Fiordland, argued that 'landscape beauty is bound up with the soul of a country'. National parks should be treasured not so much for tourists as for New Zealanders, since 'they make a definite impress on our spirit of nationhood'. In a slightly different vein, another well-known writer, Johannes Andersen, argued that the sparsity of New Zealand's human history, compared with European countries, was more than compensated for 'by the rich offering from the hands of Mother Nature'.[38]

Both to the natural scientist and to the nationalist, the indigenous remnant was greater in value the less it had been subjected to external influences. For contemporaries with ethnological interests such as Andersen and Cowan, evidence of the indigenous human impact had tremendous value in the creation of a national identity, but for a botanist like Cockayne it was an external factor interfering with the pristine environment preceding *any* human occupation. In common with most of his contemporaries, Cockayne underestimated Māori influence on the indigenous environment, but this did not negate his insight that the endemic quality of vegetation (and indeed birds) mattered most. Of all the plants that hold a place in our affections, he believed 'none can ever rank quite as high as those which slowly took their shape on New Zealand soil in the far-distant past'. For naturalists, personal identity was particularly likely to find expression through the indigenous biota; but, increasingly, all Europeans who settled became 'Kiwis'.[39]

Preservation and appropriation

Ross Galbreath, in 1989, analysed the development of a protective attitude towards the indigenous remnant within the wider context of a society engaged in its destruction. He found that 'as British colonists became settled New Zealanders they began to develop a sentimental attachment to native bush scenery and birds, and to elevate images of them into symbols of their new country and their new identity'.[40] More recent studies have supported this analysis, but whereas Galbreath made no suggestion that the process he described was anything other than wholesome, some writers have viewed it as what might be called '*environmental* appropriation'. For historian Paul Hamer, 'the active incorporation of nature into the realm of settlement – to create a national *park*, for example, was to lessen nature's threat and make it seem accessible to settlers'.[41]

In 1995 the Canadian scholar James Muir described settlement as a process of 'ecological colonialism', the final stage characterised by 'the beginning of a celebration of the new world's environment'. Reservation of Little Barrier Island, an example of the final stage, involved the eviction of Māori – an 'anti-Māori position'. Similarly the Animals Protection Act 1907, by protecting over 20 native bird species, effectively outlawed traditional Māori bird hunting.[42]

Pursuing this line of thought, ecologist Geoff Park argued that the park and reserve system started from 'the Western, imperialist principle that people and nature are irreconcilably opposed, and that the victory of one necessarily entailed the vanquishing of the other'. Thus, for him, 'today's conservationists are the obvious descendants of the pioneers' in that both deny the impact or relevance of indigenous people on a land deemed 'original' prior to the advent of the European. Park stressed that preservation often entailed the 'effective exclusion' of Māori from their land and their customary rights.[43]

The reservation of Little Barrier was important, but the reservation of Resolution Island was equally significant. Here there was no conflict with Māori land claims, only an interest in the preservation of native species. Further, the Animals Protection Act listed for protection some species rarely hunted by Māori, and others, such as the huia, that had already been the subject of rāhui. Arguably, it was as much an extension of Māori concerns as a denial of them. Conflict occurred, but there was also cooperation. In Taranaki, the scenery preservation society worked with Māori to preserve pā sites. And when Te Heuheu gave Tongariro and adjacent peaks to the nation, he was in effect utilising European concepts of sanctity and conservation to ensure that his tribal land remained sacrosanct.

As contemporary newspapers make clear, the New Zealand situation was most similar to, and most influenced by, parallel situations in Australia and America, where indigenous rights were also being eroded, along with the assertion of nationhood.[44] But a variety of motivations and influences lay behind conservation in this period, including a shift in Western thought away from the idea of opposition between people and nature. By 1900 this was strengthening the protection movement not only in colonies and ex-colonies, but also in long-settled European countries, where subjugation of a native race was clearly not a major factor.

Is protection of the indigenous remnant best viewed as part of an evolving process whereby European New Zealanders embraced and dominated a land and its people, or did preservation genuinely challenge the vision held by most of settler society around 1900? Our analysis favours the latter approach. There might have been a degree of cynicism and calculation in the extent to which Parliament and society as a whole yielded to the preservationist lobby, but for many the concern was sincere, at times even noble.

Conclusion

The society that reduced so much of the country's landscape to the 'burnt bush' of Blanche Baughan's poem did so consciously, but for many individuals this was accompanied by an increasing sense of sadness. To historian James Beattie, the artwork and journalism of Alfred Sharpe, who lived in the Auckland region from 1866 to 1887 and portrayed its remaking, are expressive of 'environmental anxiety'.[45] Charles Blomfield thought it impossible to exaggerate 'the exquisite beauty' of the bush that he had known from the 1860s and celebrated in his paintings: for him, it was 'a crime to destroy so much'.[46] Like Baughan in her poetry, Blomfield developed a reverential attitude towards 'the vaulted aisles of Nature's cathedral', powerfully evident in his 1921 painting with that title (Figure 8.4).

Baughan, Sharpe and Blomfield were exceptional individuals whose art gave heightened expression to the stirrings felt by many around 1900. The increasing sense of both response and responsibility towards the indigenous remnant cannot readily be quantified, but that sense was there, and the attempts towards environmental preservation that we have analysed were its clear manifestation.

Further reading

Paul Star, 'Humans and the Environment in New Zealand, c. 1800 to 2000', in Giselle Byrnes (ed.), *The New Oxford History of New Zealand* (Oxford University Press, Melbourne, 2009, pp. 47–70)

Figure 8.4 Charles Blomfield, *The Vaulted Aisles of Nature's Cathedral*, 1921, oil on canvas.
Source: Auckland War Memorial Museum Tāmaki Paenga Hira

provides overall context, with a focus on conservation in David Young, *Our Islands, Our Selves: A History of Conservation in New Zealand* (Otago University Press, Dunedin, 2004). Kirstie Ross, *Going Bush: New Zealanders and Nature in the Twentieth Century* (Auckland University Press, 2008) explores how the public related to the indigenous environment. Peter Gibbons, 'Cultural Colonisation and National Identity', (*New Zealand Journal of History*, vol. 36, no. 1, 2002, pp. 5–17) argues the case for European settler activity as cultural appropriation, a view applied to New Zealand's environmental history by Geoff Park, *Theatre Country: Essays on Landscape and Whenua* (Victoria University Press, Wellington, 2006). Tom Brooking, *Richard John Seddon* (Penguin Books, Auckland, 2014) portrays the critical moment in scenery preservation as one part of the Seddon government's enactment on 'nearly everything'.

9.
The meanings of mountains

Eric Pawson[1]

THE FIRST SET of 'pictorial' New Zealand postage stamps was issued in 1898. There were 13 stamps in this set, the earliest to depict anything other than one-colour portraits of Queen Victoria. Three featured native birds, and two, thermal terraces at Rotorua. But no fewer than seven displayed mountain scenes. Mount Cook, today known by its Māori and European names as Aoraki/Mt Cook, framed the set. It was on both the lowest and highest denominations – the halfpenny and the five-shilling stamps. Milford Sound (twice), Lake Wakatipu and Mt Earnslaw, Lake Taupo and Tongariro, and Otira Gorge with an inset of Ruapehu, also appeared.[2]

If Mt Egmont (later officially renamed Taranaki) were to be added, this list would be a good illustration of those mountains then best known, within and beyond New Zealand. That a set of pictorial stamps with a strong alpine theme should be issued at all at the end of the nineteenth century is perhaps indicative of how such landscapes had become almost sacralised in European societies by this time.[3] Mountains have always been an inescapable part of New Zealand's topography, to Pākehā and Māori alike, in a land where a considerable portion of the surface area lies above 600 metres, and the ranges are visible from many towns (Figure 9.1). The backbone of the South Island is the extensive chain of the Southern Alps, its highest peak, Aoraki/Mt Cook (3754 metres), surrounded by permanent icefields and glaciers. If the North Island is less alpine in nature, it is still characterised by rugged hills, and by the volcanic peaks of Tongariro and Egmont National Parks. The ways in which these mountains and ranges assumed social and national significance is the focus of this chapter. Four themes are examined in turn.

The first theme outlines the meanings of mountains both to Pākehā and to Māori, and the extent to which the former relied on the knowledge of the latter in their explorations of hill and mountain country. This is followed by discussion of the relationship between the mountains, their discovery and use by tourists, and an emerging sense of Pākehā national identity. Third, the growing popularity of mountain-based sports, such as mountaineering, hunting, tramping and skiing, is traced, from the founding of the first New Zealand Alpine Club (NZAC) in 1891 until the 1950s. The fourth theme explores the relationship between mountains and the creation of national parks over a similar timeframe. Whether or not the landscapes of national parks naturalised a particular idea

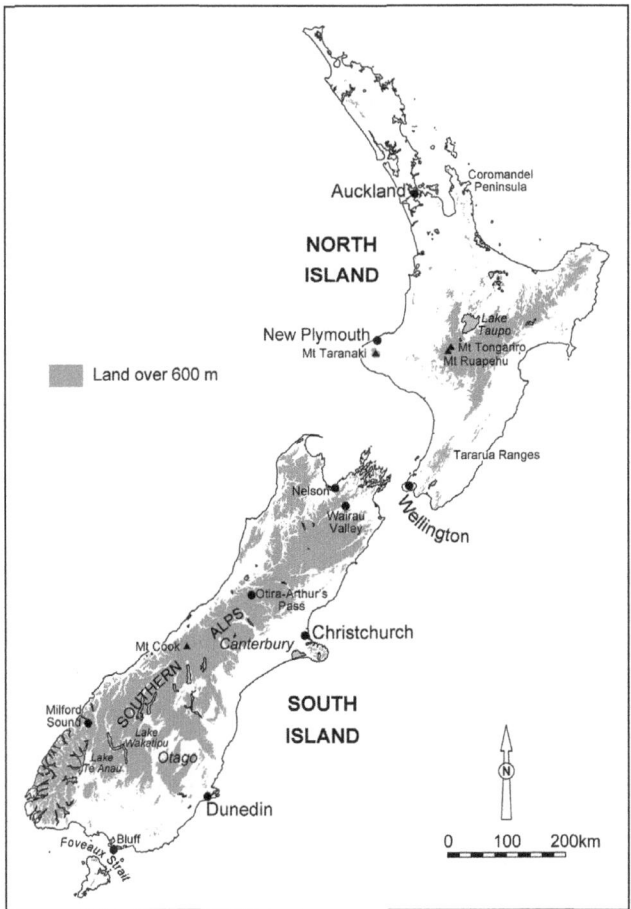

Figure 9.1 Map of the mountain and hill areas of New Zealand, showing also the proximity of main towns.

of New Zealand, in the process 'silencing' other environmental meanings, is questioned in the chapter's conclusion.

Cultural meanings

Mountains have not always had an unequivocally positive hold on the European imagination. When John Ruskin wrote *Modern Painters* in the 1860s, he felt obliged to distinguish between 'mountain gloom' and 'mountain glory' as Western responses to alpine regions. He was very much a proponent of the latter, but dread or death had been associated for hundreds of years with mountains in Europe. Anxiety of the unknown, about the apparent wildness of unfamiliar environments, and often about their animal and human inhabitants as well, heightened fears of the realm that Simon Schama characterised as the meeting point of the terrestrial and the celestial. Mountains were then, as now, places of the imagination, lying beyond the borders of social control.[4]

Some accounts from early colonial New Zealand share this alienation from what was frequently termed the 'wilderness'. '[T]he summits of the mountain-chain are ... barren and rude, where jetting precipices ... are disjoined ... by deep gorges and ravines ... and deeper chasms that, Erebus-like, appear awfully fearful.' Other observers, however, experienced awe and exhilaration in the face of 'sublime' vistas. Julius Haast's accounts of his explorations in the Southern Alps are riddled with superlatives. Many painters embellished their portrayals of mountains – Charles Heaphy, for instance, exaggerating the prominence of peaks such as Taranaki. But other colonists were dismayed by the apparent uselessness of the uplands, at a time when, employing the Eurocentric convention of the pastoral, godliness was seen as synonymous with transformed landscapes of tidy productiveness.[5]

The Māori understanding of mountains has long been rather different, with the formation of ranges, peaks and lakes written into the stories of the ancestors. Tribal territory was often linked to one mountain in particular. What Hong-key Yoon has called 'motto maxims' were, and are, used by speakers to identify themselves before an audience, reciting their whakapapa, or genealogy, in relation to mountain, river and locality. In this way, prominent features of the land anchor people to place (see Chapter 18). The peaks were the homes of atua, or spiritual beings, and often also the burial places of high chiefs. They were simultaneously places of attachment and apprehension. The explorer and scientist Ernst Dieffenbach made two attempts to climb Mt Taranaki in 1839, on both occasions with Māori guides. The second attempt succeeded, although the guides refused to go further than the snowline because, he recorded, of their superstitions and the cold affecting their uncovered feet. In the Tararua Ranges, guides interpreted poor weather as indicative of the displeasure of the atua.[6]

Such accounts were inevitably framed by Eurocentric suppositions. Māori were not unfamiliar with hill or mountain terrain. Archaeological finds suggest they explored the Tararuas early, and travelled along the tops. There was a network of trails across the Southern Alps, connecting population centres in the east with sites of greenstone in the west, and with inland food sources. The naturalist Walter Mantell observed how his guide Te Wharekorari could recite a long list of important sites, such as wetlands, up the Waitaki Valley. The surveyor John Barnicoat recorded that the Māori of Foveaux Strait 'sometimes make excursions to the Snowy mountains and catch 300 woodhens per night'. Ngāi Tuahuriri went birding and trapping native rats in the ranges behind Kaiapoi, especially from April to July when prey had overfed on the tawai berry. The rat was the most prized food, and rat runs were strictly controlled by family groups against poaching.[7]

The Waitangi Tribunal has recorded how:

> European names [in the hills and mountains] ... celebrate the deeds of the nineteenth century explorers ... These later adventurers were not the first, neither were the paths they took untrodden. Māori guides often accompanied such men, using trails with landmarks long familiar to them and places named after their ancestors.

When Heaphy and Thomas Brunner undertook their epic journey from Nelson down the South Island's West Coast in 1846, they went with Kehu as their guide, and as they progressed southwards, they 'recognised mountains, hills, rivers, streams, headlands and other natural features from [his] prior description'. Europeans first found a route over the Main Divide of the Southern Alps when Leonard Harper and a Mr Loch, guided by four unnamed Ngāi Tahu from Kaiapoi, 'discovered' the Harper Pass from Canterbury to the West Coast. But when John Henry Whitcombe, the Canterbury Provincial Surveyor, attempted the apparently shorter pass that bears his name with a Swiss companion alone, he ended up drowning in the Taramakau River after a badly miscalculated journey over treacherous terrain.[8]

These early European travellers in the mountains were not out to experience scenery, although they often recorded their reactions to it in detail. Rather, their motives were utilitarian: the search for routeways, resources or land (Chapter 4). From this perspective, mountains were obstacles. The Tararua and Rimutaka ranges were barriers between the infant settlement of Wellington and the pastoral lands of the Wairarapa, which the New Zealand Company's surveyor had painted as 'a vast English park'. Heaphy and Brunner were looking for places into which the sheepmen of the hemmed-in colony of Nelson could expand.[9] On another journey south, earlier in 1846, they had been accompanied by William Fox, New Zealand Company agent in Nelson, who subsequently became a prominent politician and four times prime minister. Fox was also a prolific artist, adopting a Claudian style of perspective in his views of valleys and lakes that they encountered. European eyes appropriated the land for European use. It has been said of Fox that he recorded 'not so much unspoiled primeval nature, but potential productivity'.[10]

Mountains, tourism and nationalism

Colonial artists often applied this 'commodifier's gaze' to mountain landscapes. Fox enclosed copies of his watercolours with the written dispatches he sent to the New Zealand Company in London. In 1858, the surveyor and painter John Turnbull Thomson wrote a glowing description of the country of inland Otago and the 'magnificent spectacle' of Mt Aspiring, concluding that, within a lifetime, one might yet 'see these beautiful lakes frequented as summer retreats for a change of air and recreation from business'. In 1865–66, the well-known Australian-based artist Nicholas Chevalier visited the South Island, supported to the tune of £200 apiece by both the Otago and Canterbury provincial councils. Their interest in his trip was that he would produce a series of his spacious, atmospheric paintings that would, by being shown overseas, encourage immigration and tourism. Chevalier obliged, exhibiting in Melbourne on his return, in Paris in 1867 and at the Crystal Palace in 1871.[11]

The best-known late-Victorian artist resident in New Zealand was John Gully, who – like Fox – worked for some time out of Nelson, where from 1863 he was a draughtsman in the Provincial Surveyor's office. Etchings of his drawings appeared regularly in publications such as the *Illustrated New Zealand Herald*, a monthly newspaper 'for Home readers', published in Dunedin 'in good time for transmission home by the San Francisco mail'. In 1874 this paper used several of Gully's drawings of the mountains in the upper Wairau and Aorere valleys. When it printed a picture of his of the Remarkables range above Queenstown in 1878 (the year after he resigned to paint full time), the *Herald* asserted that 'the chief characteristics of New Zealand scenery have become by this time familiar to the Australian public, and that they have done so is to a large extent due to the pictures of Mr Gully'.[12]

Gully produced some famous paintings of the Fiordland sounds, which by this time was one of the better-known alpine districts. Initially, the only way to see the sounds was on a summer excursion by steamer from Melbourne, such as aboard the *Alhambra*, which visited in 1874, although there appear to have been Australian tourist ship visits before this. Soon the intercolonial steamers between Melbourne and New Zealand were calling at one or other of the sounds, usually Milford, in the summer months. In 1877 the Union Steam Ship Company started its 'alpine excursion by steamer' from Port Chalmers, Dunedin. A traveller on one of these, in 1883, noted that the vessel carried 250 passengers and crew to the sounds, but that another 50 hopefuls had been turned away. By 1884 there were two such excursions each January.[13]

Other visitors went overland. There were two popular alpine tourist routes in the South Island. The first was from Bluff north to the Otago lakes – often called the 'Lakes District' at the turn of the twentieth century – and back to Dunedin. The second was the road from Christchurch, over Arthur's Pass, to the West Coast. Overseas visitors on both routes were prone to all sorts of comparisons as they struggled to fit the scenery into a European frame of reference. Anthony Trollope, here in 1872, found a steamer at Kingston, at the lower end of the lake from Queenstown, even at that early date. 'I do not know that lake scenery can be finer than that of the upper ten miles of Wakatip,' he wrote, adding: 'In New Zealand everything is ... very like to that with which we are familiar in the west of Ireland and the highlands of Scotland.' A decade later, the Reverend W. S. Green was more effusive: 'Wakatipu is amazingly beautiful; the only lake in Europe which can surpass it is Lucerne.'[14]

The Fiordland Sounds, however, were more likely to be compared with Norway, whereas the Otira road was described by a visitor writing in *The English Illustrated Magazine* as 'one of the sights of the world ... compared with which in dizzy boldness of engineering and road-making, those of which I have had experience – whether in the Alps, the Carpathians, the Balkans, or the Himalayas, are tame and prosaic'. This was quoted with approval in *Maoriland*, the Union Steam Ship Company's guide to New Zealand.[15] One description that has stuck through time was that added as the headline by the editor of the London *Spectator* to an article by Blanche Baughan on the Milford Track: 'The finest walk in the world'. The track had been opened in the late 1880s, with government encouragement. 'Everything is Titanic in scale,' said the Department of Tourist and Health Resorts in *Overland to Milford Sound*, published in 1903 – including, it seemed, the cost. This was nearly £15, covering first-class rail travel from Dunedin to Lumsden, coach to Te Anau, then boat to the start of the track.[16]

Despite the hyperbole, the actual numbers of people making such journeys were small. Only 287 people walked the track in the 1905/06 season, although this had risen to 484 three seasons later, of whom 137 were women. In 1905/06, 185 guests stayed at the Hermitage at Mt Cook, which had been opened in 1885 before being taken over by the government 11 years later. In 1908/09, numbers rose to 309, with the replacement of the coach by a motor service, which reduced travel time from two days to one from the railhead at Fairlie; but there were still only 539 guests in 1913/14. The most readily reached mountains at this time were in the North Island. Mt Taranaki was close to the town of New Plymouth, which had been connected by rail to the capital, Wellington, by 1885. In the 'record' season of 1902/03, the combined number of visitors to the three visitor houses on the mountain – at North Egmont, East Egmont and Dawson Falls – was about 2500. In 1910, over 1000 people climbed Mt Holdsworth in the Tararuas, accessible from Wellington.[17]

In 1922, the Mount Cook Company, which had been providing a road service from Fairlie, took over the Hermitage. After this it was promoted vigorously, in conjunction with the Railways Department: 'To reach this alpine resort, – which is not excelled by any in the world, either in its appointments or its setting, – we travel 40 miles by rail to Fairlie which is 1000 feet above sea level and the terminus of the branch railway from Timaru ... At Fairlie, powerful motor cars are requisitioned to complete the journey.'[18] A description adopted by the resort was that it was 'Thousands of Feet Above Worry Level', and advertising regularly contrasted the virtues of alpine freedoms with the confining nature of office work (Figure 9.2). There were some clever plays on New Zealanders' underlying fears (see Chapter 13) of the polluted nature of towns:

The air is so dry, pure and invigorating –
As to make the winter climate at the Hermitage –
The most delightful, cheering and healthful in the Dominion –
GET AWAY FROM THE FOGGY, STUFFY TOWNS[19]

Two themes that emerge as important in the growing use of the mountains are therefore the accessibility provided by transport links, and the recreation and health benefits to be gained by taking time among the peaks: 'The body and the mind is rejuvenated in this land of enchantment which modern transport now places within the reach of all', as *The New Zealand Railways Magazine* put it.[20]

The extension of the railway system popularised other alpine districts, such as the peaks of Tongariro National Park (the Main Trunk railway was completed near Ohakune in 1908), Arthur's Pass (the Otira tunnel was opened in 1923) and Franz Josef Glacier (reached via the train to Hokitika and 'thence 95 miles by connecting motors').[21] The Railways Department was soon running day excursions to 'the alpine wonderland' of Arthur's Pass from Christchurch. With a 10-shilling second-class return fare, about 10,000 people a year were making this trip in the 1930s. It was claimed that nine-tenths of them accompanied a 'well-informed railway officer' on the 'World-Famed Otira Walk', a 10-mile trek from Arthur's Pass station 'across the ridge of the Alps into Westland'.[22]

Such initiatives matched the views of the Minister of Works in the first Labour government, who was 'firmly of the opinion that the natural beauties of our country shall be within easy reach of all our citizens and not merely those who visit us from overseas'. This was consistent with programmes of physical welfare then being adopted by governments

Figure 9.2 'To the "Hermitage" Mount Cook', c. 1925. Source: pamphlet collection, Hocken Collections, Uare Taoka o Hākena, University of Otago

Figure 9.3 Lindis Pass, North Otago, 1926. Today this is the main tourist highway from Canterbury to Central Otago; then it was little more than a track. Source: c/nE500/6A and c/n766/1, Hocken Collections, Uare Taoka o Hākena, University of Otago

in the British world. The Ministry of Internal Affairs undertook limited development of mountain tracks to encourage hiking, known in New Zealand as 'tramping'.[23] The Ministry of Works invested increasingly in highway development. The Lewis Pass road across the Alps from North Canterbury was opened in 1937. Unemployed workers were placed on public works projects in South Westland, in the Haast Pass and on the road to Milford Sound (although the Homer tunnel was not completed until 1953). The summer-only road (today's State Highway 1) from Waiouru to Taupo, on the eastern flank of the central North Island volcanoes, was upgraded, across what a contemporary motorists' guide referred to as 'the Gobi Desert'.[24] Access to Whakapapa on the western flank had improved sufficiently for a hotel, the Chateau Tongariro, to be opened in 1929, in an area becoming popular for skiing. But despite high levels of car ownership in New Zealand by international standards, a traffic census in 1937–38 revealed that daily use by motor vehicles of what were often rough mountain routes was low (Figure 9.3).[25]

Given the praise, the pictures and the growing, if small, number of visitors to accessible alpine districts, to what extent were the mountains important in an emerging national consciousness? In America, Alfred Runte has claimed that wilderness has been an integral aspect of American nationalism, in a country with spectacular scenery but without the 'great art, literary attainments and time honoured traditions' of Europe. He dates its emergence to the popularity of the Hudson River school of painters in the 1820s–30s, with its later flowering in the writings of Walt Whitman, Henry David Thoreau, John Muir and Aldo Leopold. New Zealand, however, which long idealised the ruralism of productive farm life, had no equivalent wilderness ethic; it has even been claimed that 'as a people we are afraid of the "wilderness myth"'.[26]

At the time of the New Zealand Centennial in 1940, M.H. Holcroft wrote of the absence of the untamed environment in the imagination of Pākehā: 'The general crowd of New Zealanders likes to read the laudatory remarks of tourists ... If there is a sentimental interest in the country for its own sake, it remains more or less inarticulate.'[27] Rather, it was framed for its utility, if not for tourism and recreation, then for the popular activity of hunting in the bush – for wild pigs or intentionally introduced species such as deer, tahr and chamois.[28] But there had been earlier writers (see Chapter 8), such as Blanche Baughan, who urged enjoyment of scenery in itself. Another was Leonard Cockayne, at a time when he was campaigning for national park status for the Arthur's Pass and Otira landscapes. 'Mountains are the noblest recreation ground,' he wrote, 'the finest school for physical and moral training, a source of perfect health for those that visit them, and the place of all places for enlarging our minds by the study of nature in Nature's greatest laboratory.'[29] His words appeared in 1900 as part of a publication marking Canterbury's fiftieth jubilee. Nonetheless, every other essay in the volume celebrated the virtues not of nature but of urban culture.

Mountains and sport

In the late nineteenth century small numbers of overseas and local climbers had started to discover the potential of New Zealand peaks for mountaineering. The NZAC was founded in 1891 by a group of Canterbury and Otago enthusiasts, who consciously modelled it on the Alpine Club of London. On Christmas Day 1894, Tom Fyfe, George Graham and Jack Clarke climbed Mt Cook to the summit for the first time, described a century later as 'a symbolic event not to be underestimated in its contribution to the self-respect of a [country], painfully emerging from 19th century colonialism'.[30] This statement is well worth examination.

The first attempt on Mt Cook had been by the Reverend W.S. Green in 1882. An Irishman, he had climbed in the European Alps since the late 1860s, and had seen some photographs of the peak, taken 'by a lady recently returned from New Zealand', at a British Association meeting in York in 1881. 'They showed me enough to convince me that Mount Cook was a splendid peak, and the conquest well worth the trouble of the long journey.'[31] If the wilderness could not be transformed, it could still be humanised by individual skill and effort. However, in reaching the summit in 1894, the three New Zealanders deliberately forestalled an attempt by a visiting Englishman, Edward FitzGerald, and his Swiss guide.

Fyfe, Graham and Clarke were not moneyed and leisured as were the overseas climbers who came to the Southern Alps, nor were they middle class like those who had founded the NZAC. They were young labouring men from the small South Canterbury towns of Timaru, Waimate and Temuka. They had come to the Hermitage, combining paid guiding – usually of tourists onto the glaciers – with self-motivated climbing. This was not the European model, where guides served clients alone; but then New Zealand climbers – both middle class and working class – usually had to be more self-reliant. Compared with Switzerland – which A.P. Harper, a New Zealander who had climbed in both countries, thought full of 'feather bedded mountaineers' – high-altitude guides were initially scarce, porters hard to get, tracks and accommodation rudimentary, all of which gave colonial mountaineering a more expeditionary feel.[32]

But if the mountaineers saw themselves partly in terms of colonial pride, and if their early feats have entered New Zealand mountain lore, their achievements seem not to have had wider meaning at the time. The Mt Cook climb was not widely reported in the press. The *Otago Witness* was enthusiastic, but the *Otago Daily Times*, the *Oamaru Herald* and Auckland's *New Zealand Herald* ignored it. The Christchurch *Press* was curmudgeonly: why hadn't the New Zealand climbers 'played the game' and waited to take FitzGerald with them? A recent historian considers that 'the ascent led to no outpouring of patriotic pride'.[33] In fact, the NZAC went into recess not long after, and did not re-emerge until after World War I. The weather at Mt Cook in the summer of 1895/96 was poor. Then five of the leading lights of the club moved away, to the North Island or overseas. Instead, the quarter century from 1910 became the heyday of guided climbing at the Hermitage (Table 9.1).

Hunting and tramping were activities that attracted much wider participation than climbing. The Tararua Tramping Club (TTC) was established in 1919: it was convened by W.H. Field, Member of Parliament for Otaki, who had been active for many years in scenery preservation and track development in the area, and F.W. Vosseler, a Wellington businessman. The generational tensions between the club leaders – many of whom were city professional people and believers in a 'safety first' policy in the hills – and their younger members echoed the tensions of class between the alpine climbers in the 1890s. Field and Vosseler saw the TTC's mission as introducing more people to the Tararuas 'so they could visit them and enjoy the great beauty and healthy recreation of these areas without risk'. Guy Mannering of the NZAC had written in similar vein in 1892 that 'the very presence of danger draws out of a man all the caution he possesses, and brings his most admirable qualities into play'.[34]

New Zealand tramping and climbing were, however, by no means male-only preserves: 'The presence of women was one of the most conspicuous features of club tramping.'[35] The first woman to walk the Milford Track did so in 1890. In the summer of 1912/13, her fifth season in the Southern Alps, Freda du Faur of Sydney made five guided ascents from the Hermitage, causing 'quite a sensation amongst those interested in alpine work' with the first traverse of the three peaks of Mt Cook, and the first traverse of Mt Sefton. She had earlier become the first woman to climb Mt Cook. As Table 9.1 shows, quite a high proportion of high climbs at the Hermitage involved mixed or women-only parties. However, social convention decreed that female participation was no simple matter: both dress and company were problematic. Miss du Faur records how

Table 9.1 High ascents from the Hermitage, 1914/15 to 1938/39

Season	High ascents parties	All-male parties	Mixed parties*	All-female parties†	Unguided
1914–15	42	24	5	13 (10)	1 M, 1 M/F, 1 F
1915–16	48	23	4	21 (19)	–
1916–17	32	11	8	13 (3)	2 M, 1 M/F
1917–18	42	15	17	10 (5)	5 M, 3 M/F
1918–19	17	6	5	6 (5)	1 M
1919–20	33	7	15	11 (11)	–
1920–21	26	16	9	1 (1)	–
1921–22	6	1	3	2 (2)	–
1922–23	7	3	4		
1923–24	17	12	4	1 (1)	2 M
1924–25	35	21	3	11 (11)	8 M
1925–26	10	5	2	3 (3)	1 M
1926–27	31	17	9	5 (1)	11 M, 5 M/F
1927–28	28	19	7	2	4 M
1928–29	37	20	7	10 (9)	7 M, 1 M/F
1929–30	24	17	6	1 (1)	1 M/F
1930–31	42	26	2	12 (12)	5 M
1931–32	20	12	1	7 (6)	3 M
1932–33	45	30	2	13 (13)	15 M
1933–34	22	11	2	9	9 M, 2 M/F, 3 F
1934–35	no list				
1935–36	50	42	1	7 (6)	26 M
1936–37	20	11	1	8 (5)	–
1937–38	no list				
1938–39	5	2		3 (3)	1 M

*Parties consisting of women climbers only, usually with male guides. Figures in brackets indicate number of parties consisting of a single woman (with guide).
† M= all male parties, M/F = mixed, F = all women parties. Numbers represent number of parties.
Source: Annual Reports of the Department of Tourist and Health Resorts, *AJHR*, H2

older women at the Hermitage, fearing for her reputation, implored her not to climb alone with a male guide.[36] Tramping clubs like the TTC resolved the issue by using chaperones and 'discouraging pairing off, known as "twosing"'.[37]

The TTC's initial tramps were not difficult, and it was January 1922 before it arranged one across the Tararuas from Otaki to the Wairarapa (the 'Southern Crossing', which Field had completed in 1912). Yet on this first strenuous outing, one member died from hypothermia, as did another tramper five months later, one of two travelling the same route independently. These incidents brought home the need for well-located huts for shelter, and better clothing. The first waterproof garments did not appear until the mid-1920s. Despite this, tramping flourished in the Tararuas. The TTC had over 300 members by 1923, and the attitude of its leaders was always that of safety first. One interpretation of this was to travel in large parties: John Pascoe, the well-known alpine writer and climber, who moved to Wellington from Canterbury in 1937, referred to the TTC's 'alarming caravanserai'.[38]

Students who preferred a less organised attitude created the Victoria College Tramping Club in 1922. Manual workers in the Hutt Valley – inviting Vosseler to speak at a meeting to discuss formation of another group – founded the Hutt Valley Tramping Club in 1923. In the same year, having written to the TTC for advice, the Otago Tramping and Mountaineering Club was formed: university students were prominent among a membership of 157 after 12 months. Two years later the Christchurch Tramping and Mountaineering Club came into being. It was an alpine specialists' group; like the Wellington and Otago clubs, it put a lot of effort into building huts in the backcountry. In 1931, 20 clubs around the country, including the NZAC, joined forces to establish Federated Mountain Clubs (FMC). Of the 20, eight were tramping clubs, five were skiing clubs and one was a climbing club, while six promoted a combination of activities. The New Zealand Deerstalkers Association, representing the interests of hunters, was set up in 1938.[39]

The first president of the FMC had been a founding member of the TTC, so safety issues were to the fore. In 1932, a subcommittee was set up to investigate mountain accidents to see what could be learned. Older club members expressed concern about the adventurousness of younger colleagues. Table 9.1 shows the growth in the number of unguided high climbs from the Hermitage. Climbers were also getting away from the ridges and snow slopes and targeting harder rock climbs and peak traverses. Partly this reflected the more widespread use of crampons. Anxieties within the NZAC were echoed at the time within the American Alpine Club. Training camps were organised – an idea borrowed from the Alpine Club of Canada. The FMC published *Safety in the Mountains* in the mid-1930s, and facilitated a national search and rescue system under the aegis of the police, based on one developed by the TTC.[40]

The growing popularity of mountain recreation, and these changing attitudes towards challenge and danger, were reflected in the growing fatality record in the hills (Figure 9.4). Those involved were usually hunters, climbers and trampers; the commonest causes of death were gunshot wounds, falls and hypothermia, followed by drowning.[41] Social attitudes, however, were moving towards those that had been expressed – ahead of his time – by Leonard Cockayne in 1900. The Labour government's Minister of Internal Affairs, promoting the policy of physical welfare, asserted that 'good physical and mental health should be the foundation of a good life, besides making the individual profitable to the nation'. The hills provided opportunities for individuals to test themselves in the wilderness in a way the farming frontier, now all but closed, had once done. And in 1953, the year of the successful ascent of Everest by a New Zealander, an MP urged that

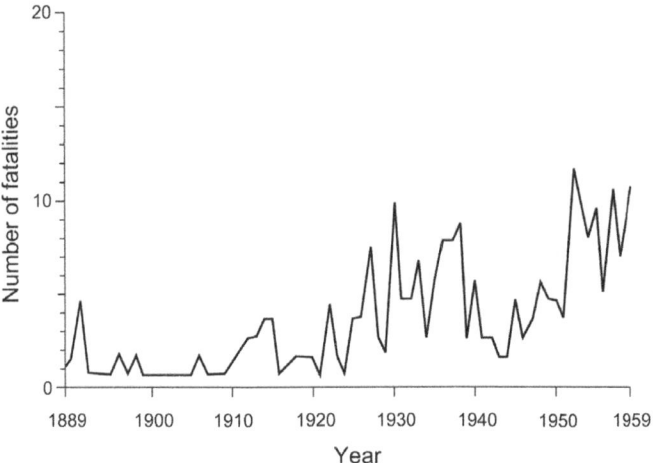

Figure 9.4 Graph of alpine fatalities between 1889 and 1959. Source: adapted from Margaret Johnston and Eric Pawson, 'Challenge and Danger in the Development of Mountain Recreation in New Zealand, 1890–1940', *Journal of Historical Geography*, vol. 20, no. 2, 1994, p. 179

nothing should be done 'to discourage young people from going into the mountains. [They] are one of the best training-grounds ... some of our best soldiers overseas ... have done mountaineering.'[42] Most of these training grounds were national parks.

Mountains and national parks

The pattern of establishment of national parks was a function of the relationship between mountains and the frontier of settlement, as well as of the popularity of particular places with tourists and recreationists. The concept of the national park was introduced into New Zealand at Tongariro in the 1890s (as discussed in the preceding chapter), but not until 1952 was there coordinating legislation to deal with each park in the same way. The underlying rationale for national parks was, however, clear. The Department of Lands asserted that 'the vast and magnificent national parks of Mount Egmont, Mount Cook, Tongariro and Ruapehu, and the West Coast Sounds, embrace country that can never be suitable for close settlement, but is of world-wide fame as the home of scenery'. Consistent with this, it stated that 'it has never been the practice of the Department to unduly withhold from settlement areas of rich soil and well adapted to pastoral or agricultural pursuits merely because they are also suited for scenery preservation. The needs of settlement are imperative ...'[43]

The distinction between mountainous national parks and settled rural areas therefore reflected the sharp dualism of 'people' and 'nature' so characteristic of settler societies. This was reinforced in the 1952 National Parks Act, which declared: 'Native plants and animal life should as far as possible be preserved, and introduced plant and animal life should as far as possible be exterminated.' National parks represented an ideal of 'naturalness', something more than merely the bits left over that could not be made to fit into the pastoral vision of Britain's southern farm.[44] But naturalness assumed particular expressions in particular places. Both Tongariro and Egmont national parks had their origins in a fear of the effect of the frontier on particular values bestowed on the wild: in the former case, revered peaks that Ngāti Tuwharetoa did not want violated by farms and mills; in the latter, to preserve forests for purposes of catchment control, so guarding against the risks of flooding new farmlands. Egmont and the other parks were also in the areas in which alpine tourism was most apparent at the turn of the century: Fiordland and the lakes, Mt Cook, and Arthur's Pass. A Canterbury MP, urging action, with pressure from Cockayne and the newly formed Christchurch Beautifying Association, claimed that a national park incorporating the Otira Gorge 'would prove ... a gold-mine in the future in the way of inducing tourists to visit that part of the colony'.[45]

Mountain areas that were less accessible or spectacular drew little attention from tourists or recreationists, and were less likely to become national parks. A good example is the glacier district of South Westland. South of Fox Glacier in 1906, Maud Moreland had found herself 'in a world untouched by man'. The famous Fox Glacier Hotel did not open until 1928, and only in the 1930s were bridging and roading put in place. The *New Zealand Motorists' Road Guide* of 1938 referred to Fox Glacier as 'the centre of the proposed National Park', but this status was not attained until 1960. An earlier request from the MP for Timaru that a national park be created in the accessible Rangitata Valley of South Canterbury, because of its scenery and potential for ice skating, came to nothing.[46]

There were other mountain areas that did attract plenty of human use but did not become parks. The Tararua Ranges, north of Wellington, is the best example. In 1936 a proposal emerged from within the TTC to have the Tararuas set aside as a national park. The idea was to construct a network of new huts and tracks with enhanced roading access as Wellington's centennial memorial, marking 100 years since the signing of the Treaty of Waitangi. Younger members were opposed, but eventually it gained sufficient support for official consideration. It was turned down as being too expensive and without wide enough appeal. Representations to the Minister of Internal Affairs were also unsuccessful; he may have reasoned that national park status would be incompatible with the potential for hydropower development on several Tararua rivers.[47]

The Waitakere Ranges are another interesting case. To the mountaineer Pascoe, they 'looked like a hedge on part of Auckland's low horizon', pleasant enough for 'hedge-hopping'. The Auckland City Council had begun buying land there for water-catchment purposes in 1899, as well as for scenic reserves. In 1924 the Auckland Blue Mountains Society called for creation of a park. The following year the Waitakere Association was founded to attain national park status for the area. In the late 1930s 'the Waitakere National Park' was the goal of those who wished it to become the chief centennial memorial for Auckland, as 'a perpetual source of health and enjoyment'. An Act of Parliament in 1941 created the 'Centennial Memorial Park', but as the country's first regional, rather than national, park.[48]

Accessibility was therefore not a sufficient condition for the creation of national parks, although it was often a necessary one. Nonetheless, parks needed to be beyond the frontier, and with little development potential for farming, forestry or power generation. Equally important was that their landscapes conform to the ideals of spectacular scenery that nineteenth-century artists and, increasingly, photographers had created as representing 'New Zealand'. On this score, neither the low-level Waitakeres nor the rugged Tararuas passed muster. A degree of exclusivity was required – although the in-club opposition to the Tararua proposal revealed the tension between this and access: the challenge of the hills was lost for some if they became too 'easy'. When parts of existing parks became very popular in the 1950s, this tension was partially resolved by the creation of a small number of wilderness areas within them, using the provisions of the 1952 Act. These were remote places without huts and tracks. The first designated wilderness area was Otehake in the Arthur's Pass National Park.

Conclusion

To a considerable extent the mountains of New Zealand have become known through the icons of the national park system, just as New Zealand itself is often known through its national parks. Gully's representations of alpine peaks may have been the picture that 'Home' readers of illustrated newspapers received of the colony in the 1870s, but posters and photographs played the same role for New Zealanders in the 1920s and 1930s. And just as mountains had been prominent in the 1898 postage stamp issue, so too did they feature in the next pictorial set, in 1935, with Mt Cook and Mt Egmont gracing, respectively, the cheapest and most expensive denominations. When the 'Peace' stamps were issued in 1946, the halfpenny showed the Southern Alps reflected in the still waters of Westland's Lake Matheson, while the ninepenny – one of the most popular of New Zealand's stamps – portrayed the Franz Josef Glacier through the altar window of the local church.

These examples underline a point that has been central to this chapter: that mountains are not 'natural landscapes' but features of the environment on which have been inscribed specific cultural and social values.[49] They are charged with meanings, even if to many Pākehā these meanings have, historically, tended to be more detached than for tribal Māori. In colonial times, mountains were often viewed as wastelands, unless commodified for purposes of tourism or, later, for character-building recreational pursuits. As in North America, significant alpine areas were then reserved as national parks, provided they were good for nothing else. After the passage of the 1952 Act, which brought into being a body, the National Parks Authority, responsible for extending that system, an increasingly large proportion of the national land area became part of the national conservation estate. It remained overwhelmingly mountainous in character, with not much in the way of high-country tussock, lowland native forest or wetlands.[50]

In the last 30 years, this has become less the case. National park status has been given to areas such as the river landscapes of the Whanganui, the karstlands of Paparoa and the Kahurangi wilds of northwest Nelson; and parts of the high-country tussocklands east of the Southern Alps have been designated conservation areas as the result of tenure review of Crown leasehold land.[51] This raises other issues, including whether there are the resources to care for such an extensive conservation estate. The landscape values that the tenure review policy seeks to preserve are the product of more than a century of grazing by merino, and the cessation of this practice may undermine the conservation objective. Nor – extending a now famous argument that William Cronon developed about 'the trouble with wilderness' in the United States – is it reasonable to assume that an ever burgeoning conservation estate somehow absolves New Zealanders from developing a more deeply embedded national ethic of living with nature in town and country.[52]

Arguably, however, this is exactly what the presence of such prominent mountain landscapes has prevented. As order was wrought across the land – leaving the ranges and mountains aside because they alone were invulnerable to conversion – it was these wild areas that assumed the role of representing New Zealand not only to itself but on an international stage as well. This is despite a history in which relatively few New Zealanders, most of them urban dwellers, have gained much personal intimacy with the mountains. Nonetheless, we have needed them. For some people, it has been the mountains, visible on city horizons, that have kept alive dreams of challenge and pioneering spirit. For others, they have meant the appeal of peace and contemplation, a welcome counterpoint to the drudgery of suburban routine. For most of us, however, they have been landscapes of renown, taken-for-granted national 'possessions' that do service as New Zealand's face to the world: rendering distinct, and distinguished, a homeland otherwise too often marked by little but its ordinariness.

Further reading

There is a range of essays relevant to cultural attitudes to mountains in Mick Abbott & Richard Reeve (eds), *Wild Heart: The Possibility of Wilderness in Aotearoa New Zealand* (Otago University Press, Dunedin, 2011). On mountain recreation, see Lee Davidson, 'Publicising Peaks: Early Promotion of Mountain Tourism' (*New Zealand Geographic*, no. 117, 2012, pp. 58–72); Kate Hunter, *Hunting: A New Zealand History* (Random House, Auckland, 2009); and Kirstie Ross, *Going Bush: New Zealanders and Nature in the Twentieth Century* (Auckland University Press, Auckland, 2008). Place-specific studies include Roberta McIntyre, *Whose High Country? A History of the South Island High Country of New Zealand* (Penguin, Auckland, 2008), with a comparative perspective in Eric Pawson & Hans-Rudolf Egli, 'History and (Re)discovery of the European and New Zealand Alps until 1900' (*Mountain Research and Development*, vol. 21, no. 4, 2001, 350–58).

10.

'Swamps which might doubtless Easily be drained': swamp drainage and its impact on the indigenous

Geoff Park

… Swamps which might doubtless Easily be drained, and sufficiently evinced the richness of their soils by the great size of the plants that grew on them … indeed in every respect the properest place we have yet seen for establishing a Colony.

—Joseph Banks, 1769[1]

The cultivation of a new country materially improves its climate. Damp and dripping forests, exhaling pestilent vapours from rank and rotten vegetation, fall before the axe; and light and air get in, and sunshine ripening goodly plants. Fen and march and swamp, the bittern's dank domain, fertile only in miasma, are drained; and the plough converts them into wholesome plains of fruit, and grain, and grass.

—Charles Hursthouse, 1857[2]

ONE OF THE TRULY GREAT landscape changes in modern times, according to the eminent historical geographer Clarence Glacken, has been marsh, bog and swamp drainage.[3] If the magnitude of the change from swamp to farm is any gauge, the flat lowlands of New Zealand must be among the world's most modern landscapes. Drainage has reduced the 670,000 hectares or more of freshwater wetlands that existed at the commencement of European settlement to some 100,000 hectares. Twenty per cent of New Zealand's indigenous birds use wetland environments as their primary habitat.[4] Evolutionarily, New Zealand's swamps contain some of its most ancient life forms. But lowland New Zealand today is dramatically altered from the swampy country in which its unique, indigenous ecosystems evolved – and in which the Treaty of Waitangi was signed.

The 85 per cent decline in New Zealand's wetlands since European settlement is one of the most dramatic known anywhere in the world – far higher than the countries in which modern agriculture began large-scale draining of swamps and marshes. The Netherlands and Britain are believed to have lost around 60 per cent of their original wetland areas; France only 10 per cent. The New Zealand figure is also considerably greater than in most countries that received what Alfred Crosby has termed 'the biological expansion of Europe'.[5] In the continental United States, for example, about 53 per cent of the original area of wetlands has been lost.[6] In many parts of lowland New Zealand, such as the Bay of Plenty, where less than 1 per cent of the natural wetland area remains, the change has been so comprehensive that the full extent of swamp and other wetland environments prior to European settlement can only be inferred from soil evidence.[7]

Imperial landscapes

The vanishing of New Zealand's swamps was a consequence of what has been called 'ecological imperialism'.[8] Their cardinal environments – the lowland plains where the country's most productive lands lie, where the prime export produce is grown, and a great many of us live – are like our laws of justice. They appear and function as they do to us today because of England's eighteenth- to twentieth-century extension into New Zealand – a cultural spreading that carried with it a deep antipathy to leaving country to the grip of the swamp.

Joseph Banks, the explorer–natural scientist, and Charles Hursthouse, colonist–farmer, believed certain things would have to be done to New Zealand were it truly to become the country that, to Hursthouse, deserved to be called 'South Britain'. No haunts of ancient peace to them, its swamps would have to go. Depicting swamps as 'wilderness' and 'waste' and forcing their native fluidity into the straight, criss-crossing channels of man-made drains was fundamental to the Western cultural construction of nature of which Banks and Hursthouse were prime agents and advocates (Figure 10.1).

Banks and Hursthouse were Englishmen, Lincolnshire lads both, native to the low, watery East Anglian country of marsh and fens; a vast, damp flatness broken only by

Figure 10.1 Excavating a drainage canal, Hauraki Plains, 1910. Source: Drainage Operations in the Hauraki Plains, *AJHR*, C8, 1910

the Gothic spires of its market towns' cathedrals. Inheritor of some two hundred of its farms by the time he visited New Zealand with Cook on the *Endeavour*, Banks was one of those men of empire, of inordinate power, who shaped the world as it came to be for the people of Australia, New Zealand and the South Pacific. No less than his enthusiastic discernment of the Pacific as 'an arcadia of which we were going to be kings',[9] and his influence on British colonisation, trade and imperial science, was his passion for creating pasture and cropland from nature's swamps. In an 1813 portrait, Banks, at the height of his power as President of the Royal Society, holds not an explorer's chart or botanical specimen, but the map of a fen drainage scheme.[10]

That drainage of the Lincolnshire Fens was concluding when Hursthouse, who emigrated to Taranaki with a New Zealand Company scheme, was a child. 'A few years since,' he recollected, 'the ague was the scourge of my native swamps in Lincolnshire; and fen infants, like myself, were only preserved by copious cups of bark and wine. But now, reed and rush and snake and buzzard, rat and eel have vanished before the plough; the "reek o' the rotten fen" is gone, and the ague is a tradition of the past.' Settlement and cultivation in New Zealand, he said, with more than a hint of Sanitary Movement speak,[11] would produce the same effects.[12]

The only great swamp that the *Endeavour* people encountered at close quarters was what is today the Hauraki Plains. But Joseph Banks knew what he had seen there, and what could now happen. To someone with such an eye for colonising opportunity, New Zealand's swamps had three values: their watery acres of tall trees were so vast he could 'see no bounds' to them; their rich soil would make 'ample returns of any European Vegetables sown in it';[13] and, other than those in the raised swamp village or fishing camp who welcomed the *Endeavour* party ashore 'in the most friendly manner immagineable',[14] they appeared surprisingly devoid of people.

The marshes of Banks's Lincolnshire estate had been busy with people since before Saxon settlement in about AD 700. Through medieval times, former tidal marshes around the Lincolnshire fen edges had been reclaimed piecemeal, divided by drainage ditches and surrounded by sea banks, behind which farmers became wealthy from the valuable new pastures.[15] Nonetheless, until the 1760s most of the fens still lay in their natural state. Their diversity of oak and alder woodlands, sedge, rush and reed swamps, open water and tide-swept saltmarsh sustained a way of life that was luxurious compared with that of similar people elsewhere. The poorest people had common rights access to fish, wildfowl, timber, peat, thatch reeds and hay. But they faced ruin when the fens and marshes were enclosed and drained by estate owners such as Joseph Banks.[16]

One of the reasons that the Pākehā settler culture has never held swamps in esteem as picturesque – compared with the wooded edges of rivers and lakes, for example – is that the confines of closed, wooded wetlands posed a problem for an aesthetics of sight accustomed to more open space and invulnerability in watery environments. When Englishmen raised and habituated to the flat openness of the Fens confronted the densely wooded swamps and wetlands of their far antipodean discoveries they were in for a culture shock of the most profound magnitude – and with durable after-effects.[17] To such a sense of swamp country, the towering, vine-tangled kahikatea swamp forests that the *Endeavour* shore party encountered in the Hauraki must have been dizzying. We know from the traditions of Hauraki people that, unknown to the English, the boats' passage was shadowed across the swamp country.[18] But their invisibility and the way that, beyond the fishing settlement, such palpably fertile a plain was 'completely cloathed' with heavily timbered forest would have given Joseph Banks a whole new quickened sense of what wilderness meant.

Twentieth-century archaeology would reveal otherwise. The ecological diversity of the Hauraki swamp made it about as rich in food resources and raw materials as anything in Aotearoa. Furthermore, the pattern of Māori settlement it sustained, reflected it. There were many more villages than Cook and Banks saw. Archaeologists estimate that swamp drainage and flood-control works have destroyed 70 per cent of the sites of dozens of kāinga up the labyrinth of myriad hidden waterways that English eyes were unable to discern.[19] But the seed that drove Edward Gibbon Wakefield's colonisation schemes of the 1840s – the impression, as Charles Hursthouse would recount, of 'immense districts of teeming fertility, literally without an inhabitant, [that] did not produce a mouthful of food for a single human being'[20] – was sown.

The history of swamp drainage is substantially the linear logic of modern book law meeting and beating the chaos of nature. As the draining of swamps produces the flat and relatively dry surface of the earth on which the straight, equilateral lines of the settlers' towns and cities – and the agricultural space to sustain them – can be laid out, it purposefully obliterates nature's organic flow lines (Figure 10.2). Nothing that meets the eye on a New Zealand coastal plain that has been the subject of a swamp-drainage scheme is yet a century old. No plant or animal it sees, other than the odd raupō plant or eel in a farm drain, is indigenous. Yet the recognisable combination of trees, pasture, and human structures makes it seem as if perhaps that they are all that was ever there. In countries like New Zealand and Australia, what is termed 'spatial history'[21] tends to interpret such a landscape as one of imperialism.

A landscape of swamp drainage like today's Hauraki Plains is undoubtedly an imperial landscape, a place of amnesia and erasure in whose shaping political passions have been mobilised; 'a potent, ideological representation that serves to naturalise power relations and erase history and legibility'.[22] But ridding country of its low-lying, wet, fecund places is an imperialism not confined to the former European colonies. Nor is it restricted to the colonisation phase from the second half of the nineteenth century, or its European antecedent, the agricultural revolution from the late seventeenth century in England's Fens and the polders across the Channel. Some would give swamp drainage and wetland reclamation an even wider cultural framework than a millennium or more in the Low Countries:

> Postglacial Europe, even the periglacial Mediterranean shores, was a swampy place. The most fertile soils were often those, from an agricultural and medical point of view, most in need of drainage. The reclamation of wetlands played an enormous role in the civilized history of the continent and an equal role, I believe, in the psychohistory of those for whom the symbols of a new duality were being shaped … the fearful potential of organic fructification and the muckiness of wet decay, as opposed to the order of the farm or the city itself.[23]

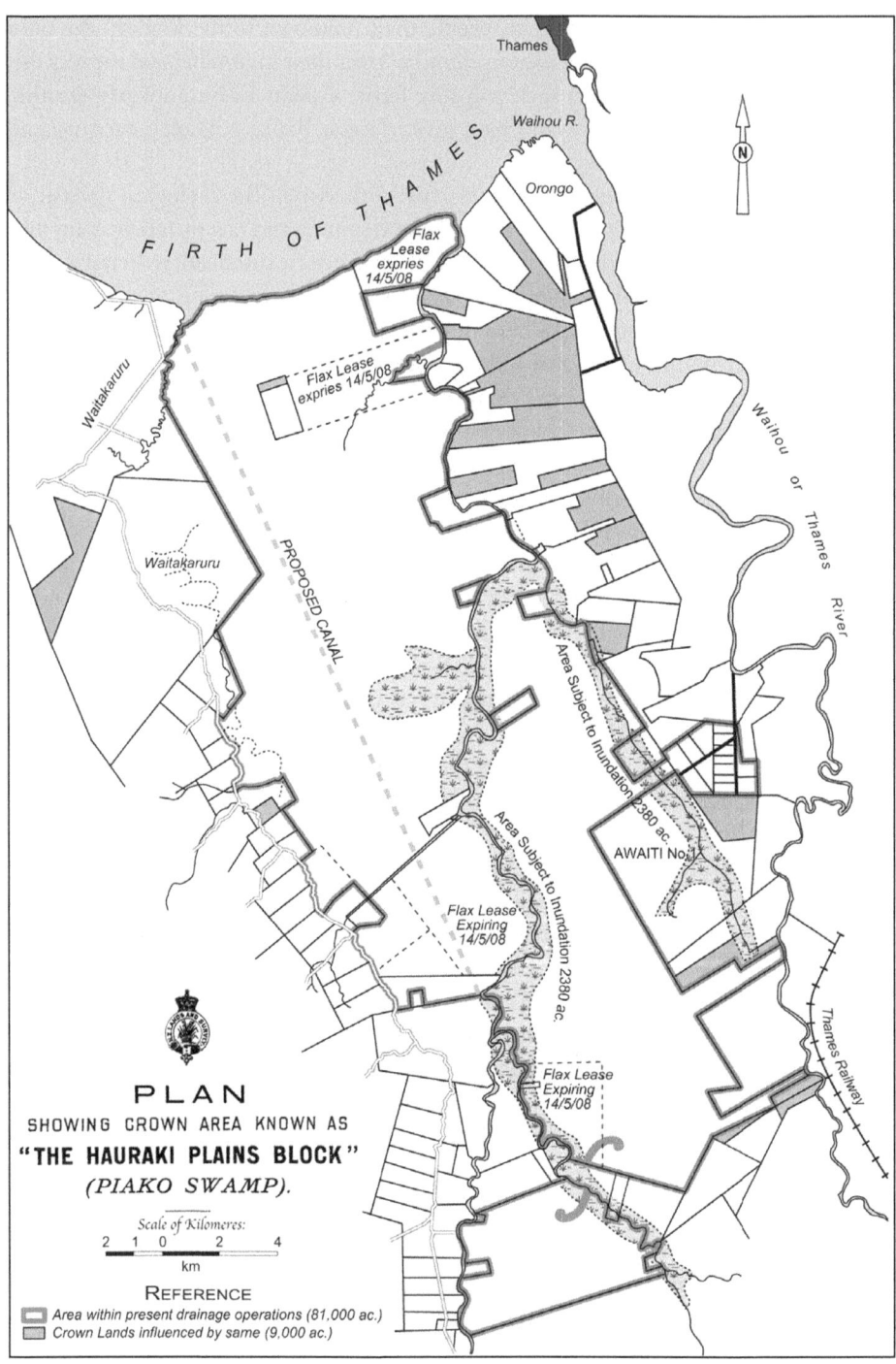

Figure 10.2 A plan of swamp drainage, Hauraki Plains, 1908. Source: Annual Report of the Department of Lands, *AJHR*, C1, 1908

Despite the postmodern trend of wetland restoration, the same dualism is everywhere today, as indeed swamp drainage and wetland reclamation continue to be wherever nature's wet, uncultivated places are in close proximity to culture.

Taming the wet wild

The significance to New Zealand historiography of fen drainage in eastern England between the 1760s and 1840s, and the agricultural revolution therein, is in part to do with the timing of the colonial project here, led as it was from that particular corner of the world. Its greater significance perhaps has been its demonstration of the massive scale and almost military-style organisation and control of the landscape (both water and land) that was necessary, which in New Zealand only the Crown could undertake – as settlers attracted to the great Rangitaiki Swamp in the Bay of Plenty were to discover.

In 1890 a commercial traveller proposed draining the Rangitaiki Swamp and converting it to farmland. The government surveyed the swamp into 500-acre sections, and the first settlers arrived to take up their land in 1891. Their piecemeal approach of each draining their own section failed because of the lack of adequate outfalls. So in the early 1900s the settlers formed a drainage board, it having become obvious by then that drainage of the swamp would have to be under a comprehensive scheme to be successful. The scale of the operation proved well beyond the board's resources. It lacked what the historian of Rangitaiki Plains Drainage Scheme called 'a vital prerequisite for the successful prosecution of any drainage proposals – control over Crown Lands and Native Land'.[24] Thus, by way of the Rangitaiki Land Drainage Act 1910, the government took over.

The following year work began on the long drains and cutting a direct outlet to the sea for the Rangitaiki River. By 1924 both the Rangitaiki and Tarawera rivers had been diverted out to sea, their floodplains drained, and their courses straightened and stopbanked. By 1925 the drainage of the Rangitaiki Swamp was substantially complete, except for a major drain outfall to the sea near Whakatane, for which Māori kāinga land was taken under the Public Works Act. Like other drainage schemes, it was dependent on Māori swampland, which, if Māori did not want to sell, they were forced to part with compulsorily. Of the land affected by the time of the scheme's completion, only 8 per cent was freehold land, 35 per cent was Māori Land, 42 per cent was land leased from the Crown, and 11 per cent unoccupied Crown land.[25]

The consequence a little north in the landscape of the Hauraki Plains Act 1908 was little different (Figure 10.3). That same year the government commissioned a survey of what the local county's centennial historian would describe as 'a desolate wasteland of swamp and alluvial flats'.[26] The following year government-facilitated works began clearing much of the Piako and Waitoa rivers and constructing stopbanks along both the foreshore and rivers to prevent tidal and flood overflow. Behind them they began criss-crossing the plain with miles of canals and drains. The Hauraki Plains Act also gave the Crown the power to take Māori swampland compulsorily. In the three years following its passage, almost 2000 acres of Māori land was acquired to facilitate what the Act termed 'the more effective carrying out of drainage works'. By 1911, 5200 acres of

Figure 10.3 Hauraki Plains swamp drainage, 1911–12. Source: Annual Report of the Department of Lands, *AJHR*, C8, 1911; C8, 1912.

the plain was drained and burnt ready for balloting to eager waiting settlers. By 1918, newspaper columnists were describing the 'splendid results' of making 'a wilderness carry a prosperous population and produce ever increasing wealth, of changing 'useless swamp into rich farm land'. Not too different at all from the conclusion Arthur Young had come to in England in 1799: 'fens of water, mud, wild fowl, frogs and agues, have been converted to rich pasture and arable'.[27]

One of Britain's most regarded historical geographers, H.C. Darby, has described the drainage of England's fens as giving 'man and his conquest of nature' its 'full meaning'. In a context such as the colonisation of New Zealand, with the imperative (in a promoted-as-flat but actually very unflat land) to obtain for agriculture any fertile, level ground, a large swamp like the Hauraki, Rangitaiki or Manawatu was readily constructed as an easily vanquished foe.[28] The Australian cultural theorist who described swamp drainage as 'a colonial device for subduing an ostensibly recalcitrant, even rebellious, indigenous population and wetland environment'[29] had Vietnam's Mekong Delta and Plain of Reeds more in mind than the Rangitaiki or Hauraki swamps in New Zealand. But his portrayal of drainage as a threat and obstacle to progress that the colonial project would have to subdue is not an inaccurate appellation of the New Zealand situation.

New Zealand was settled on the promise of soil: of 'level ground … space so ample … soil so fertile',[30] a North Island 'full two-thirds rich alluvial',[31] a space unknown, inland from the coast, but believed so low and level that James Cook's River Thames (the Waihou) flowed navigable all the way from Hauraki to the strait named after him. It was also settled on the premise that the native landscape of that low, sprawling country would be of no use as it was, and that large-scale conversion of it to agricultural production would be vital to the colony's success. It is this, the British settler perception of what Māori considered among their most productive resource environments (and thus refrained from clearing and cultivating) as land still in 'the primeval dress of nature' – 'wilderness' and 'waste', in other words – that is the root cause of swamp drainage having so comprehensively eliminated the indigenous.

The swamp became an obstacle to progress from the moment the first settlement scheme's colonists walked up the beach to discover that New Zealand's plains were far from the grassy, alder and oak woodland version. There was relief, in Charles Heaphy's description, that the Wellington shore of the New Zealand Company's intent at Petone at least had *something* of level ground, but shock at its being 'covered with high forest to within a mile and a half of the beach, when swamps full of flax … intervened'.[32] The next New Zealand Company shipload, the surveyors, discovered that where they were meant to lay out Britannia, 'a great many of the streets would run through swamps and marshes, in some places six and ten feet deep'.[33]

After a day in the swamp with his chief surveyor, the Company's Principal Agent William Wakefield acknowledged that while 'perfectly level, and of considerable extent … the marshy nature of the soil, and its liability to inundation' made it 'at present unfit for the settler'. It would nevertheless 'with draining become valuable'. [34] Within weeks the whole place did inundate. The settlers, waiting in tents and makeshift huts, protested at the choice of site, but Wakefield declared 'the swamps would be thrown out'.[35] Days later, however, he was persuaded to move the colony across the harbour to the present-day site of Wellington.

As the Company's draughtsman, Charles Heaphy was required to prepare drawings of the chosen site for the colony for advertisements back in England. However he actually depicted the swampy kahikatea and flax country behind Petone Beach, it appeared in the Company's promotional lithographs as sprawling, well-drained, well-wooded open plains country. To have presented thickly timbered swamps as thickly timbered swamps would not have attracted the ships of settlers that were needed. Ruth France's poem of the colonising moment may have urged: 'There is no need to remember swamp-grass / Or how the first women (let the rain pass, They had prayed) wept …'[36] But although excised and omitted so as not to unsettle potential settlers, the obstacle to progress at Petone would not be forgotten.

Modernity takes all

> If you bid me of the features most remarkable to speak,
> I should say a spur, a gully, then a swamp, then a creek,
> A little distance further on there will recur
> First a swamp, and then a gully, then a creek, and then a spur.
>
> —William Pember Reeves, 1889.[37]

Swamps abounded in lowland New Zealand in the early phase of European settlement. Prior to the discovery of refrigerated shipping, which spurred settler demand for flat land, they were a common part of the landscape. We today need to be very careful about reconstructing how lowland New Zealand looked and how it functioned ecologically when it had its swamps. The documents of colonisation on which environmental historians draw are not exactly cluttered with descriptions of swamp country. To the patriarchal cultural tradition that brought the demands of Western modernity to New Zealand, swamps were places of disease and decay, melancholy and horror, absolutely inimical to modern ways.[38] They were places to dredge, drain, or fill in, but rarely places to write about. Few nineteenth-century writers saw much in swamps to be admired, though William Pember Reeves saw a native quality in the landscape to be admired in its vanishing:

> Small streams ran out of the swamp … and disappeared in the shingle of the beach. When not disturbed with draining work, their water was sweet and clear. The swamps had been covered with tall flax, toetoe, rushes and small bushes, green and beautiful in the sunlight, but as the drains did their work, the peat sank, cracked and dried, the surface was systematically burnt and became stretches of black, hideous ashes and mud, poached up by the hoofs of cattle.[39]

Swamp *drainage*, on the other hand, was described because it was settlement history. It had Acts of Parliament in its name and purpose. The swamps themselves, however, were merely empty wilderness to the eyes of the literate, waiting for their place in history. Swamp drainage as history is, to an ecologist, a process by which a thriving, powerful expansionist culture learns how to wring wealth from a certain kind of country by cleverness and industry; comes, inevitably, to need more of it to swell its success; goes searching for it; finds it lying, seemingly vacant and unused, in another culture's land; and, declaring it 'wilderness' and 'waste', claims it and transforms it into wealth.[40] In New Zealand the declaring and transforming was effected by laws – laws of the Crown.

A powerful principle of English land law, that prior to its arrival in New Zealand had facilitated the landscape revolution of the enclosures in the Lincolnshire Fens, was the dictum enunciated by the seventeenth-century philosopher John Locke. Land 'left wholly to Nature, that hath no improvement of Pasturage, Tillage, or Planting', wrote Locke in his *Two Treatises of Government*, 'is called, as indeed it is, *waste*.'[41] To the eyes of colonists from civilised England, the swampy flats lying inland of so many a New Zealand beach were 'waste' at its most profound. To those who believed their civilising and 'reclaiming' of land from the wilderness to be 'God's work',[42] the only possible purpose of swamp country, unpeopled and uncultivated as it seemed, was what they believed it was waiting for: providing the productivity that would sustain the cities they could imagine their culture building here. That is why Cook called the river that took his *Endeavour* party up into their New Zealand swamp the 'Thames'.

But as the New Zealand Company settler William Swainson would later write of Māori and their whenua: 'Land, apparently waste is highly valued by them.'[43] The great swamp of Hauraki, the vast wetland systems of the Piako and the Waihou Rivers, had a human-sustaining productivity of a very different kind to that desired by James Cook's culture. In 1840 George Clarke, Protector of Aborigines in New Zealand, visited what his Instructions termed 'by far the largest block of land in New Zealand that is, in its whole extent, fit for cultivation' – 'to treat with the natives for such portions of their land as they may be disposed to part with and can conveniently spare'.[44]

It was early summer, but Hauraki Māori told him how 'every winter the swamps from the entrance of Piako to the interior for about thirty miles is an inland sea in which nothing but water and the tops of a few kahikatea trees are to be seen with canoes sailing in all directions over the expanse of water. The only place secure from these inundations … in winter season becomes an island.'[45] One 'island' was Oruarangi; the longest, permanently inhabited settlement site known in New Zealand,[46] and the site of the 'Indian village' that Cook and Banks visited when they came inland from the *Endeavour* in 1769. Another 'island', well up into the kahikatea forests beyond where they ventured, was Kakaramea; a major swamp pā from which Ngāti Maru traded timber in the 1790s, but merely a rise on a dairy farm today, indistinguishable from the stopbanks.

Clarke was in no doubt about the fertility of the Hauraki flat country. But its wetness overwhelmed him. He could see 'no probability of redeeming a country lying so low and receiving such an immense body of water from the interior'.[47] Water was not the only drawback: the plain was only 'part clean'. Much of it was kahikatea forest. As such, the Hauraki swamp was one of the biological pumphouses of the New Zealand ecosystem: where both biodiversity and nutrient turnover reached peak levels; where birds of forest, scrubland, and water and fish of fresh water and sea bred and fed. Just a few miles away, over a few low, wooded hills, was another even greater example: the Waikato swamplands, where the river that had once formed the alluvial plain of the Hauraki swamp had flowed since the Taupo eruption in the second century AD. By the time the century was out, the 'redeeming' of both the Hauraki and the Waikato was being effected by what the government officially called 'land improvement' schemes.

From prodigious ecology to geometric settlement

But in a landscape that actual European settlement had revealed to contain only too few Haurakis and Waikatos, the premium on any level acres was substantial. In the Crown colony that New Zealand was, swamps and swamp drainage became the subjects of legislation.

Central to English land laws are the distinctions between waterways and dry land. Swamps, rivers and lakes are interconnected ecologically; that is the key to their biological productivity and resource value. Many cultures make neither distinction nor division between them. Rather, the ecological processes that link them, like fish spawning and eels running to sea, are what matters. As traditional Māori fishers knew, river and lake fisheries are productive in relation to the areas of swampland they water.[48] But for a culture accustomed to its rulers using legal boundary lines as means of asserting governance, swamp environments provided a real problem. Where does a swamp become river or lake? What are the delineating features? Is a swamp merely a shallow lake with plants growing through the water? Or is it is dry land only temporarily or ephemerally wet? To such a perspective, like English law, that needs classificatory order and that is predicated on a hard and fast distinction between land and water, swamps were anomalous. Untamed places where all was yet to be done, their indeterminate nature, in space and time both, does not readily lend them to such absolute linear logic.

Māori settlement in 1840 had a particular pattern to it. It was concentrated along the floodplains and mouths of rivers and at estuaries and lagoons – the environments in which the indigenous flora and fauna tended to be most diverse and rich.[49] While the presence of European crops for 50 years prior had expanded Māori agriculture, the primary resources of the Māori economy continued to be the indigenous forests and fisheries referred to in the Treaty of Waitangi. Some of the richest and most productive resources of all were the coastal swamps in which those forests and fisheries co-occurred in the same immediate environment.

The high regard for lowland swamps in the traditional Māori landscape derived from the often vast areas of country that they watered and gave access to, from the birds that were attracted to them for food and breeding sites, and the native fish that came to spawn.[50] Dominating the swamps were rushes, reeds, flax, and the kahikatea, or white pine. Mature fruiting kahikatea were a seasonal mecca for birds and people. Waikākā (spring eels, mudfish), a traditional delicacy for presentation at feasts, hibernated during summer drought beneath kahikatea roots. Myriad indigenous fish species such as inanga, kōaro and kōkopu migrated through the estuaries and lagoons that would become taken by Crown law for swamp drainage scheme spillways, through into the pools enclosed by flax and raupō in the gaps in the kahikatea swamp forests. It was such conditions that made tidal swamp-plain rivers like the Waihou, Waikato, and Manawatu great fisheries prior to British agricultural settlement.[51]

Ecologically, these were landscapes of interconnection and interaction, the antithesis of the boundary lines and the subdivision of the country into legally separated units desired by English land laws. It is in this sense that the American ecologist Daniel Janzen has commented how 'what escapes the eye is the most insidious kind of extinction – the extinction of interactions'.[52] Using the body-as-system analogy, the ecological effect of swamp drainage in New Zealand was as if a major organ like the heart had been ruptured and removed. When the native birds no longer had the premium breeding and feeding zones that the lowland swamps provided, the hill forests, too, went silent.

The small-scale geographic diversity of New Zealand meant that, in most districts, the birds of interior forest country were never far from the relatively food-rich swamps of the coastal lowlands. A bird like kererū, the New Zealand pigeon, for example, would flock across country in huge numbers when a swamp food species such as kōwhai was in season. That, no less than the colonial evisceration of forest habitat, is why native bird numbers began plummeting dramatically when swamp drainage facilitated by government schemes really got going comprehensively in the 1900s.

The cumulative impact of that period's agricultural settlement on the customary Māori swamp resource was reported to the Waitangi Tribunal by the South Island tribe Ngāi Tahu:

> European settlement inevitably began to impinge on Ngāi Tahu mahinga kai resources … By the late nineteenth century, most sources of mahinga kai in the Otakou Block had been destroyed or enclosed by settler occupation … Streams, drains and lagoons were much more important to us than the main rivers. These smaller water bodies are the first to disappear when farmers or catchment boards start land drainage or river management works.[53]

The advent of refrigerated shipping in the 1880s opened up huge European markets for meat and dairy produce and made great demands for land on which to grow it. The concomitant draining of swamps and alteration of lakes and rivers to bring any level land into agricultural production were deemed in the 'national interest'. As the Waitangi Tribunal historian Ben White has observed, the country's future lay in sheep and cattle, not eels and kōura.[54] Swamps and the people who valued them as mahinga kai were caught between the settler–colonist imperative to drain and reclaim any low-lying land that could become fertile pastures and the Crown imperative to control waterways. Customary Māori rights to swamps and waterways, those with which Article the Second of the Treaty of Waitangi terms 'taonga' – 'fisheries', 'forests' and 'properties' – are concerned, were primarily about use rights, from which ownership derives (Chapter 17). Under English law the reverse applied; the right to use and manage a resource flowed from ownership of it. The way Māori were forced to reconceptualise their customary resource rights and laws in order that they be cognisable under English common law is one of the key issues of the colonial encounter in New Zealand.[55]

As White has said, while many colonists regarded the notion that Māori had rights in waterways as far-fetched, they considered idiotic any contention that Māori had rights in swamps. They believed tacitly that any rights to swamps on the land to which they held title were exclusively theirs – no less perhaps than Māori who sold lands containing swamps believed, from *their* customary standpoint, that they would continue to enjoy usufructuary rights in them. Little did they realise that not only would settler title to their swamps preclude their future access to them, but also that it would totally transform their mahinga kai – by removing them.[56]

A succession of Crown laws effected this, with increasingly deleterious repercussions for an economy with any reliance on freshwater resources. The earliest legislation pertaining to swamps vested power to drain land in provincial councils and other local authorities. With the 1876 Public Works Act the Crown formally empowered itself to control waterways. Any natural watercourse in which fluvial action occurred – thus any swamp – could be declared 'a public drain'. Councils could be empowered to undertake

Figure 10.4 A swamp plough at work on Mr A. Lusk's farm, Oaonui, Taranaki, photograph by Feaver Studios, c. 1914–15. Source: PHO2013-0002, Puke Ariki, New Plymouth

large-scale drainage operations, and could take any land required for drainage purposes. Ratepayer boards, which enabled farmers with a vested interest in drainage and the prevention of flooding to dominate decision-making and override other community interests, were set up by the 1884 Rivers Board Act, and in 1893, the Land Drainage Act. White has recounted how such boards acted to the detriment of Māori, of whom only the trustees of landowner trusts could claim ratepayer status.[57]

By 1913 the government was being recommended to eliminate swamps from the national landscape altogether (Figure 10.4). If there is a single tree species that characterises New Zealand's swamp environments, it is kahikatea, or white pine as settlers knew it. While Europe inched toward war, the Royal Commission on Forestry was deciding the future of land 'still under standing forest'. The government's policy of 'close settlement' and the 'pressing demand' for New Zealand's 'limited area of first class land' meant one kind of indigenous forest had to go:

> As is well known the soil of the white-pine swamps, when drained and the trees removed, forms one of the richest of agricultural land, which when grassed, is extremely useful for dairy farms ... Since no land is more suitable for occupation than that of the

white-pine swamps, when drained ... their value in this regard is a strong plea in favour of the removal of the trees forthwith.[58]

The Swamp Drainage Act 1915, extending government powers to undertake large-scale drainage operations, was enacted with a view to making more agricultural land available for settlement. Māori settlement land could be taken under the Act if, as in the case of the outfalls of the Rangitaiki Drainage Scheme in the Bay of Plenty, the Crown considered it essential for the completion of a drainage operation.

Wetlands rediscovered

'Wetlands,' said a recent Asian and Pacific UN Environment Programme, 'are vital life-support systems for many communities throughout the world. The traditional [i.e. Western] view of wetlands as wastelands is being overturned and there has been a rise in awareness of their importance.'[59] From the beginning of the twentieth century most elements of New Zealand's indigenous life that British colonists had considered 'wasteland' became protected in some way, as ecosystems as well as species (Chapter 8). But not swamps. As early as the 1930s, ornithologists had tied the national decline in indigenous birds to agriculture's transformation of lowland swamps. In the 1950s the impact of lowland swamp clearance on indigenous fauna became a conservation issue as, in district after district, once-common wetland birds like matuku (bittern) became rare. But it was not until the end of the 1960s that there was any acknowledgement by the state that swamps had intrinsic, indigenous values, over and above their potential productivity as farmland.

The recognition was initiated by the Wildlife Service undertaking a series of wetland surveys in the Bay of Plenty.[60] Such surveys led to the Service, by opposing local drainage schemes and proposing protection, becoming a counter-force to the Crown's traditional policy role of draining and developing wetlands. But it was not until the 1980s that New Zealand government sufficiently accepted the principle to state, as policy, that swamps have not been adequately represented in the reserve system: that 'if samples of the whole range of natural wetlands are to be retained as part of the landscape, many existing wetlands must be reserved'.[61] Just as wetland drainage quintessentially characterises modernity, wetland restoration, perhaps, characterises postmodernity.

Conclusion

It is no coincidence that the Crown's firm tightening of its animal protection laws as the nineteenth century ended and through the first decade of the twentieth, prohibiting hunting of native birds like kererū, straddles the same passage of time in which the Crown undertook its great swamp-drainage schemes. They were years of change and crisis in the Māori landscape, as the Native Land Court worked its way through tribes' estates, and as the natural life-support systems that had traditionally sustained them vanished (Chapter 17).

The Crown's concern was with what 'would be appreciated by many settlers'.[62] In turn-of-the-century parliamentary debate, swamps were seen as country whose settlers faced particular difficulties in helping the colony realise prosperity. William Massey described the situation in 1903 that led to the Crown becoming the agent in swamp drainage it soon was:

> A settler goes onto swamp land. The first thing he has to do is to get in his drains – not only his main drains but also his side drains. Then, if the swamp has too much water thereon, after the settler has got his drains in and got the water off he has to allow the scrub to grow on it before he can get a burn. That may take two or three years. After the scrub has grown sufficiently he has to wait for a dry summer. Then, when he gets it burnt off, he has to wait another year before he gets the grass to grow, and probably two to three years after that before it is sufficiently consolidated to carry heavy stock. That is the situation of many of the swamps in the North Island.[63]

Ecological landscape change takes time even when a colonising culture is in landscape transformation mode. Yet by the late 1890s, anyone who had anything to do with the land and nature was aware of the dramatic decline in bird numbers occurring. But the MP for Northern Māori, Hone Heke Rankin, objected in Parliament to 'any legislation that tends to take away any of the native rights to shoot or kill birds' as the closed season clauses in the Animals Protection Act 1895 were doing.[64] It would, he said, have a massive impact on his Tai Tokerau people. Heke repeated an argument that Māori had made consistently since the 1879 Māori Parliament – that the government should not be able to dictate bird protection or game management on Māori land.[65] His opposition was on the basis of rights. Castigating the new legislation for doing nothing to halt the loss of forest habitat that European settlement was causing, and for putting the burden on Māori who had kept their forests, he called for flexibility in the timing of closed seasons to allow for phased fruiting of the birds' berry trees.[66] 'The birds suffered,' he said, 'because settlers cut down the bush, not because of hunting.'

Te Tai Tokerau was one of New Zealand's swampiest regions, wet climatically and with countless low-lying valley flats winding inland from its many harbours. Much of the bush that settlers were downing was swamp forest, in order to get to the rich soil of the flats – then to drain, burn, fence, grass, and stock it. As a factor that collapsed the indigenous resource of Hone Heke Rankin's concern, the succession of changes to Te Tai Tokerau's swamps would have paralleled in magnitude the massive settler clearance of hill forests that was occurring.

It was in Te Tai Tokerau, too, that a coastal swamp resource had first initiated Māori into the Western trade economy (Chapters 3 and 4). The exchange of flax fibre for muskets began the massive readjustment of tribal power that preceded the Treaty of Waitangi. And in the concomitant change in Māori lifestyle and settlement pattern it produced lay the root cause, some colonists believed, of Māori becoming the vulnerable, vanishing race they appeared to be when swamp drainage was hitting its straps. Consumption, 'the chief disease that kills the Maoris', according to Alfred Newman in 1881, was widely considered the consequence of the shift from the 'old hill-forts' of their 'former wild' habitat to 'live at the edge of a bush or a swamp, almost always on low-lying, damp, ill-drained spots'.[67]

Swamps and Māori alike had little place in Newman's New Zealand. Both had threatened, but none of the country would be allowed to remain as Charles Hursthouse's 'bittern's dark domain, fertile only in miasma'. No New Zealand environment has seen greater loss of its indigenous life than its swamps. None has so little future prospect for native flora and fauna as the 'portion of the wilderness' for which Hursthouse foresaw the plough converting into his 'wholesome plains of fruit, and grain and grass'. 'The sanatorium of a large body of … wealthy invalid-emigrants', he called it: 'cultivating a few acres for amusement near the towns and profiting by the fine scenery and the really good society of the young Land'.[68]

Further reading

This essay has been left unchanged out of respect to its author, who died in 2009. A collection of his work was published as Geoff Park, *Theatre Country: Essays on Landscape and Whenua* (Victoria University Press, Wellington, 2006). Some later writings appeared in *Forest and Bird*, nos 316 to 339 (May 2005 to February 2009), under the byline 'Itinerant ecologist'. Other works on wetlands include Margaret Forster, 'Recovering Our Ancestral Landscapes: A Wetland's Story', in Rachael Selby, Pātaka Moore & Malcolm Mulholland (eds), *Māori and the Environment: Kaitiaki* (Huia Publishers, Wellington, 2010, pp. 199–218), Matthew Hatvany, 'Environmental Failure, Success and Sustainable Development: the Hauraki Plains Wetlands through Four Generations of New Zealanders (*Environment and History*, vol. 14, no. 4, 2008, pp. 469–95), Janet Hunt, *Wetlands of New Zealand: A Bitter-sweet Story* (Random House, Auckland, 2007) and, in an urban context, Pamela Wood, *Dirt: Filth and Decay in a New World Arcadia* (Auckland University Press, Auckland, 2005).

IV
Modernising

11.
The grasslands revolution reconsidered

Tom Brooking and Vaughan Wood

P.W. SMALLFIELD, retired Director-General of Agriculture, published *The Grasslands Revolution in New Zealand* in 1970. In his account of the substantial improvement of New Zealand pasturelands since the 1890s, he recalled the words of A.H. (Alfred) Cockayne, the well-known agricultural scientist and later Director-General of Agriculture. In the 1920s Cockayne had likened New Zealand's farming land to 'a magic hat which, while still retaining its original size, allowed the conjurer to draw a seemingly never ending stream of objects from its interior'. It appeared to Smallfield that spectacular production increases on but a slowly increasing land area between 1920 and 1966, and especially between 1949 and the 1960s, 'more than vindicated Cockayne's simile'. Smallfield attributed the threefold increase in production from the 1920s to the 1960s not to 'the conjurer's magic wand', as Charles Darwin earlier had on seeing the English-style farm established by the Anglican missionaries at Waimate, but to 'the combined and sustained efforts of farmers, research and advisory services, marketing boards, merchants and successive governments'.[1]

This chapter outlines the 'revolution' as it was understood by its major architects and promoters, identifies differences between them, and analyses New Zealanders' obsession with the development of grasslands at the expense of other land development strategies. It examines some of the ways in which this orthodoxy was challenged, as the grasslands revolution was questioned for its less desirable environmental impacts. It concludes by questioning whether 'productivist' approaches to primary production have nonetheless persisted in New Zealand.

What was 'the grasslands revolution'?

Smallfield, along with Sir Bruce Levy (Director of the Grasslands Division of the Department of Scientific and Industrial Research, 1937–51), the geographer Kenneth Cumberland and soil conservators Lance McCaskill and Douglas Campbell coined the term 'grasslands revolution' to celebrate the huge increase in farming output (from both the sheep and dairy sectors) that occurred after the acquisition of phosphate-rich Nauru Island in 1919, and especially after the introduction of aerial topdressing in 1949.

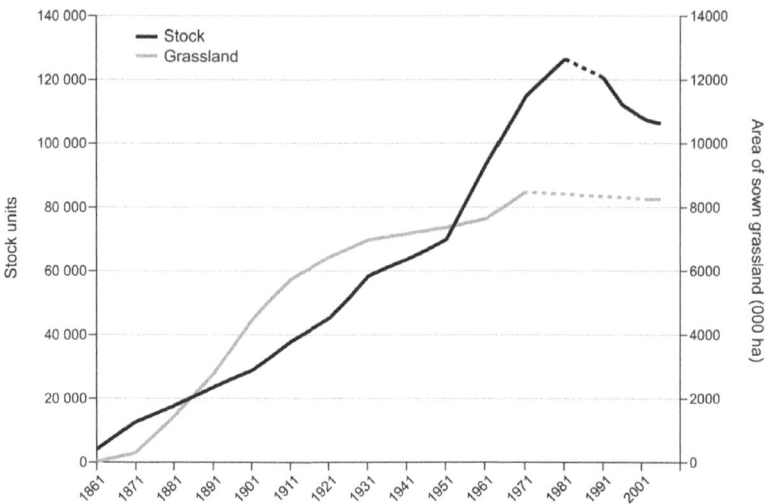

Figure 11.1 Graph of changes in the area of sown grassland (grey line) and the numbers of stock units (bold line), 1861–2005. Source: Tom Brooking & Eric Pawson, *Seeds of Empire: The Environmental Transformation of New Zealand*, I.B. Tauris, London, 2011, p. 10

The application of superphosphate and an increasing amount of variants (particularly serpentine) peaked at over 3 million tonnes in 1985, or about 2 per cent of the world's total. Sheep numbers had stagnated at around 20 million in the early 1900s, but then increased between the wars to around 30 million before rocketing from 32 million in 1949 to a peak of 70 million in 1982. This occurred with only a small increase in the area of cultivated land (Figure 11.1). Tonnages of fertiliser and sheep numbers fell dramatically from 1985, when the fourth Labour government removed subsidies and recast traditional agriculture as a 'sunset' industry.[2] Until then New Zealand had prospered on the basis of export agriculture, maintaining a standard of living from the 1920s to the 1970s that ranked between the fifth and third highest in the world.[3] Little wonder, therefore, that agricultural historian B.L. Evans and historian of New Zealand science F.R. Callaghan confirmed the notion of a revolution in grasslands farming by treating the men who directed it as heroic champions of the application of science to agriculture.[4]

The story told by both historians and agricultural scientists praised nineteenth-century farmers for converting bush to pasture because New Zealand's temperate and moist climate supposedly made it the country most suited in the British Empire, and even the world, to growing grass and carrying stock year-round. Rarely was this simple environmental determinism tempered by acknowledgment of the country's relative paucity of fertile soils. Instead, such authors argued that a combination of planting exotic pastures, using what were invariably described as 'English grasses', and application of artificial fertilisers easily compensated for any such natural deficit. They therefore praised enlightened pioneer farmers who had used fertilisers from the 1890s, especially blood and bone from the number of freezing works that had burgeoned with the growth of the frozen meat export trade.

But with the advantage of hindsight, these modernists also condemned the ignorance of the pioneers and hill-country farmers for pushing New Zealand farming past its ecological limits by 1920. This had posed problems for soldiers returning from World War I who been balloted to settle remote and marginal areas of land.[5] Difficult economic times compounded such apparently intractable problems as regenerating bush and soil erosion when, according to Cumberland, nature exacted its revenge.[6] The huge effort in laying down some 5 million hectares of English pastures by 1911 seemed threatened as productivity increased but at a more moderate rate once the initial virgin fertility had been worked out.[7] Failure forced many of the soldier-settlers to leave the land. Production, which climbed in the early 1920s in response to increasing mechanisation and the application of superphosphate made at fertiliser plants by mixing sulphuric acid with phosphate from Nauru, continued to increase but at a slower rate. Failure heaped up when commodity prices collapsed during the 1930s depression. The 'Bridge to Nowhere', built for returned soldier settlers in the Whanganui backcountry, became a potent symbol of the overconfident expectations that had been placed on New Zealand agriculture. War compounded the situation by interrupting supplies of phosphate and discouraging farmers from adopting more innovative approaches. Unbeknown to pessimists, however, war also held the key to recovery, through developing greatly improved aircraft. The Royal New Zealand Air Force and Douglas Campbell worked together to adapt military planes for the difficult task of dropping superphosphate from the air.[8]

Improvements in the understanding of soil nutrients (particularly the role of such crucial trace elements as cobalt and molybdenum), and advances in pasture ecology in the 1930s, also enabled farmers to develop the old gum lands of the Far North and the pumice lands of the North Island's Volcanic Plateau after World War II. The development of more vigorous strains of exotic pastures, especially clovers and perennial ryegrasses, along with Bruce Levy's notion of heavy stocking to create 'urine and dung showers', led to increases in production from the 1930s that became more rapid from 1949 onwards. Bulldozers apparently minced secondary growth into submission, while applied science and increased aerial topdressing raised soil fertility and increased carrying capacity of formerly marginal land. Research and development and government assistance propelled New Zealand to the world leadership of grassland farming and enabled it to maximise supposed 'natural advantages'.

The end result was the conversion of 51 per cent of New Zealand's surface area to grasslands by the 1970s (Figure 11.1). This far exceeded the world average of 37 per cent.[9] Overseas scientists marvelled at New Zealand's super-efficient grassland farming. Little wonder that agricultural scientists celebrated this conversion from a giant rainforest to smiling pastures apparently by then free from 'Old World' soil exhaustion and the curse of regenerating bush.

This glorification of the benefits of agricultural science, downplaying any negative impacts, survived largely unchallenged from within mainstream agricultural thinking until the 1980s, despite queries raised by some microbiologists and ecologists about impacts on water quality and Cumberland's work on soil erosion carried out in the 1940s.[10] Despite these earlier misgivings, however, Cumberland's 1981 television series and book *Landmarks* later popularised the notion that agricultural science had saved New Zealand from economic and environmental disaster.

Differing perspectives on the revolution

Agricultural scientists (especially agrostologists, that is, grassland scientists) were generally more enthusiastic than soil scientists. Levy, in particular, became less critical as the years passed. As a young instructor in the Department of Agriculture, he had acknowledged that the forest had a key role to play in New Zealand's future, and had adopted an ecological perspective that placed him in the vanguard of contemporary scientific thought in the first edition of *The Grasslands of New Zealand* (1923). By the time of the 1951 edition, he had become obsessed with grass and his sense of ecology did not extend beyond the paddock fence. In his 1973 final version, the retired scientist and international eminence wanted every possible hectare of New Zealand covered in high-producing pasture so that it could carry 111 million sheep.

Few authors could see beauty or worth in scrub or gorse, but Levy differed from others in not lauding the fact that the actual increases had occurred without much expansion of the farming area. On the contrary, he seemed annoyed by the reluctance of government and farmers to ensure that all marginal lands were producing to their full potential. Even the more critical Kenneth Cumberland, like P.W. Smallfield, expressed pride in so much more being produced from a virtually static area of land. He muttered darkly that the threatened removal of the herbicide 2,4,5-T (finally banned in 1987) would give gorse the 'half chance' it needed to flourish. Smallfield expressed hopes of a massive increase in sheep numbers and thanked science for making available to farmers some 243 chemical fertilisers and herbicides rather than the 'handful' present in the 1920s.[11]

Lance McCaskill, like Cumberland with whom he worked, was always more guarded because he realised that Levy's beloved 'dense sward' could not hold the land by itself. Even so, like Campbell, McCaskill praised aerial topdressing as a useful ally in the fight against soil erosion. McCaskill also portrayed the restoration of Molesworth station in Marlborough as a miracle of soil conservation, the lesson of which was that production could be maintained by resting pasture from grazing, and by grazing cattle as opposed to sheep.[12] Although, like Levy and other agricultural scientists, McCaskill and Campbell had considerable state support, they struggled to change farmer attitudes towards erosion control, compared with the success of grassland scientists in persuading them to intensify production through use of aerial topdressing and a ryegrass/clover regime (see also Chapter 12).

Levy, whose belief in heavy stocking and high fertiliser inputs is still shared by farmers in some parts of Southland, Canterbury, Taranaki and Waikato, acted as a radical with his 'unbridled pasture' approach. McCaskill held a more conservative position. Some mainstream grassland scientists such as Kevin O'Connor, successor to McCaskill at the Tussock Grasslands and Mountain Lands Institute at Lincoln, shared the soil conservators' reaction against Levy's excess, but still accepted that applied science had achieved much and would continue to do so for the foreseeable future. Another group, represented by P.D. Sears (Director of the Grasslands Division of the DSIR, 1956–63) and J.K. Syers (Professor of Soil Science at Massey University, 1972–85), positioned themselves somewhere between Levy and McCaskill. Sears, for example, advocated more extensive sowing of white clover in country suited to sheep farming, but also recommended that pasture be retired in steep areas and turned over to trees.

The 'grasslands revolutionaries' such as Levy, as well as more cautious moderates, remained oblivious to the extraordinary environmental damage to Nauru, and rarely questioned the energy costs involved in shipping phosphate to New Zealand and then distributing it by fuel-guzzling aeroplanes. In their view, sustainability was not a serious concern so long as New Zealand's grassland farmers remained far more efficient users of energy than the cropping farmers of North America and Europe.[13] This insistence on efficiency provided moral force to their argument that solutions to farming problems would always involve the use of artificial fertilisers, herbicides and pesticides. Like the New Zealand Institute of Agricultural Science (founded in 1953 by Campbell), they agreed that, while mistakes had been made, repair had succeeded and would continue to succeed; to them the gains far outweighed any problems and losses.

An essential part of Levy's approach was the use of a simple regime of ryegrass and clover on all New Zealand environments. He rejected the older, more complex pasture mixes (described in Chapter 5). He disdained earlier experiments with grasses like Chewings fescue in Southland and paspalum in Northland, even though these species seemed to suit the chilly climate of the south and the subtropical warmth of the north, respectively, rather better than ryegrass.[14] Cooperation between farmers and the Department of Agriculture in the form of trials on working farms also declined rapidly under Levy's leadership, in favour of experimental plots in the grounds of agricultural colleges and research stations. It seems that Levy and his supporters wanted to impose on the entire country a simple, one-pattern system that overrode environmental variation. Some of the more cautious supporters of applying science to farming worried that such reliance on expert knowledge might upset delicate local ecologies and so induce losses in productivity.[15]

Most discussions of the 'grasslands revolution' tend, however, to hide the fact that the changes occurred in stages rather than in one long, uninterrupted progression, and that the 'revolution' occurred at very different rates in the North and South Islands, and between different districts within them.[16] Initially, superphosphate brought most benefit to the dairy farmers of the North Island, who wanted to end time-consuming and costly reliance on supplementary crops. Growing a single crop of grass proved much more attractive. South Island farmers, in contrast, other than the grain growers of Canterbury and Southland dependent on such fodder crops as turnips, did not use superphosphate on any scale until the advent of aerial topdressing. Consequently, 2.1 million of the 2.4 million acres topdressed with artificial fertilisers in 1934 were in the North Island. Southern farmers preferred to retain short rotations of pasture and crops, and failed to increase production much between 1900 and 1950.

South Island sheep numbers fell behind those of the North Island from 1899 and increased only by 38 per cent from 11.3 million in 1910 to 14.8 million in 1950. In contrast, North Island sheep numbers increased by 47 per cent over the same period, from 12.9 million to 19 million. Indeed, A.H. Cockayne's initial concern with grasslands related to the South Island's relative underperformance. From the establishment of the Department of Agriculture's Grasslands Research Station at Palmerston North in 1926 (transferred to the DSIR in 1936), attention shifted to bringing pumice land, gum land and land that had reverted to secondary growth into production, and to repairing the damage caused by erosion on the east coast of the North Island. McCaskill was left to cure the South Island's ills. The drier basins and plains of the South Island also required irrigation before farming could push past nineteenth-century production levels. In combination, irrigation and aerial topdressing closed the gap somewhat between 1950 and 1980 (from 14.8 million to 32.8 million sheep, as against the North Island's 19 million to 36.9 million).

A rather different pattern emerges with both dairy and beef cattle, in that the bulk became increasingly concentrated in the North Island up to the 1980s. In 1920 there were 3.1 million cattle, including 893,000 dairy cattle. About 77 per cent of all cattle were domiciled in the North Island, and 72 per cent of dairy cattle. By 1949 dairy cattle numbers had increased to 1.7 million and total cattle to 4.7 million. About 83 per cent of cattle were domiciled in the North Island, and 87.7 per cent of dairy cattle. Numbers increased up to 1980, when there were about 5.5 million cattle in total and 2.9 million dairy cattle. The South Island did better in terms of total cattle, with about 23 per cent of the beef herd, but only 8 per cent of the dairy herd. So the South Island's recovery as a carrier of stock related much more to sheep than cattle, despite rapid growth in stock numbers across the period of the grasslands revolution.[17]

The upsurge in fertiliser usage of the 1950s gathered momentum through the 1960s, supported by both subsidies and the advocacy of the Department of Agriculture and the Grasslands Division. It began to slow only from the mid-1970s, when the rising costs of superphosphate and aviation fuel forced farmers to reconsider the rate of application (Figure 11.2). The amount of superphosphate and additives dropped over the green fields of New Zealand peaked at over 3 million tonnes in 1985. This then fell dramatically when, rather than phasing out subsidies, the fourth Labour government ended them abruptly.

Changes in infrastructure, advances in related technological areas and shifts in government policy are always necessary before new technologies can surge ahead. The remarkable feature of the grasslands revolution is that it advanced as rapidly as it did. Despite its efforts, the Department of Agriculture did not always persuade farmers to give up the practice of sowing several varieties of grass, or of using cocksfoot as a complement to ryegrass that struggled to provide year-round feed, especially when those older practices had worked relatively well in the past. The Great Depression of the 1930s also made it difficult for debt-ridden and cash-strapped farmers to change their operations quickly. New Zealand's underdeveloped roading system before the 1950s made reaching farmers in the field difficult for Department of Agriculture advisors. The serious erosion

Figure 11.2 Advertisement: 'Fertiliser helped us grow …' A romantic portrayal of the technology that delivered the 'grasslands revolution'. Source: *New Zealand Journal of Agriculture,* June 1977

and regular flooding that afflicted the eastern North Island between 1938 and the early 1950s further delayed wholesale adoption of the Levy system. Nevertheless, the great majority of farmers followed Levy's advice from the late 1950s.[18]

Why did it occur in such extreme form?

Why did the agricultural establishment become quite so obsessed with establishing grasslands as a kind of monoculture? The obvious answer is economics, because fat lambing, wool growing and dairying won New Zealand a considerable degree of postwar prosperity. Only worsening terms of trade, exacerbated by Britain's manoeuvring to join the European Economic Community from the 1960s, the dramatic collapse of wool as a commodity with a global move to synthetics in the same decade, and the oil shocks of the 1970s, threatened the continuance of that prosperity and forced a major rethink on how the country earned its living.

Concern to reduce escalating labour costs was influential in persuading agricultural scientists to advocate greater use of artificial fertilisers and chemicals. Smallfield admitted that cropping could provide improved feed but lamented that it 'needed too much labour for production and feeding out'. He praised chemical weedkillers for promoting 'a revolution in weed control from the days when we used to bend hour after hour'. A more organic approach seemed laborious, slow and expensive in comparison.[19] Attempts at controlling weeds before World War II showed that chemicals were much more effective than the raft of punitive legislation from 1900 aimed at coercing farmers into controlling weeds better.[20] The topdressing plane, especially once the specialist Fletcher had been developed from 1957, also liberated hill-country farmers from drudgery.

Yet much more than economic imperatives and fascination with high-tech innovation was involved in explaining the pursuit of a grasslands 'paradise'. Scientific conceptualisation and practice is never innocent and objective. It is rather a product of its time, especially of the climate of opinion, intellectual fashions, and changing cultural norms. During the period from the 1920s to the 1960s, faith ran high in the limitless possibilities of science and technology to resolve humanity's problems. Cynicism concerning the claims of mainstream science lay well in the future. Two world wars raised serious doubts about humanity's moral capacity but reinforced this blind faith by accelerating technological and medical advances. Gains in areas such as aviation were so spectacular that this optimism became 'scientism', or a kind of substitute religion.[21] Agricultural scientists, who held considerable power and influence in New Zealand over this period, shared this optimistic faith in 'man's' capacity to tame and improve on nature.

Noble intentions also played a vital role in driving the grasslands revolution, especially the determination to ensure the success of rehabilitation. Levy, Smallfield, Cumberland and McCaskill all shared the first Labour government's determination to avoid the pitfalls of the 1920s and to build a stronger economy and a better society.[22] So too did the report of the Sheep Industry Commission in 1949 and editorials in the *Journal of Agriculture*.[23] The high wool prices induced by the Korean War seemed to confirm this confident prognosis. The commissioners also talked incessantly about feeding a hungry world, as did Federated Farmers, which metamorphosed out of the older Farmers' Union between 1944 and 1946. New Zealand's future as a food producer seemed guaranteed by the global population explosion. Until postwar idealism waned in the 1960s, numerous conferences also echoed this optimistic reading (Figure 11.3).[24] It also reflected a much older and deeper belief in 'ruralism' – the notion that the country way of life is morally superior and socially preferable to city living. Levy, along with the *Journal of Agriculture* and *Straight Furrow*, portrayed the city as an unproductive and vice-ridden problem (Figure 11.4).[25]

Some other factors also influenced the narrow concentration on improving grassland farming. One was the direct impetus coming out of English agronomy from the 1920s as it attempted to catch up on the gains made in plant breeding in continental Europe. George Stapledon, champion of pasture improvement at the Royal Agricultural College at Cirencester (and later the Welsh Plant Breeding Station at Aberystwyth and the Grassland

Figure 11.3 Cover of the *New Zealand Journal of Agriculture*, July 1946, showing the Four Horsemen of the Apocalypse. The message is New Zealand's role as a bread basket feeding a hungry world.

Improvement Station at Drayton), visited Levy in 1926 and encouraged New Zealand experimentation. No one seemed to note the irony involved in the rejection by New Zealand agrostologists of Stapledon's suggestion that evolution equips indigenous grasses best to cope with environmental particularities. Instead they focused on his interest in grass and grass mixes and went ahead with improving imported English grasses.

Levy later worked with one of Stapledon's star students, William Davies, to develop even more vigorous strains of rye grass and clovers. Stapledon also moved agrostology away from a broader ecological perspective by fixating on the improvement of a narrow range of easily controlled grasses, especially ryegrass. To do so, he sought out and bred from grass strains, including those from New Zealand, that best encapsulated and had preserved the traits of the plants growing in old English pastures. He hoped that this invigoration of grass seed stock would expand agricultural production within Britain

Figure 11.4 Cover of the *New Zealand Journal of Agriculture*, September 1946. A portrayal of the virtues of rurality and New Zealand's ongoing role as Britain's outlying farm, with St Paul's representing the temple of Empire.

as well as within the Empire.[26] Lord Bledisloe, New Zealand's Governor-General at this time and a successful and enthusiastic farmer himself, articulated a similar agenda. In his 1934 address to the Grasslands Association Conference he praised grassland scientists as 'progressive far sighted patriots' and argued that grasslands constituted 'the true wealth' of New Zealand.[27]

Lingering imperialist ties and ongoing commitment to the project of colonisation underpinned such influences. Cumberland and Levy responded to the imperial rallying cry by suggesting that Māori land currently covered in weeds and gorse needed 'vigorous occupancy' if it was to be converted into the best pasture. Men of Levy's generation continued to envisage New Zealand as Britain's farm and could not tolerate the idea of places within it that did not operate at the highest levels of productivity. They ignored the irony that their revolution could never have been implemented without the land removed from Māori ownership by dubious purchase and confiscation, as demonstrated in Chapter 3.[28]

Certainly the agrostologists' obsessions mesh with the argument of Libby Robin, Australian environmental historian, that ecology became 'a science of Empire'. She points out that ecology grew out of the earlier 'Systematics', which had been developed to help administrators better understand the new lands of their ever-expanding domains.[29] Ecology gained official recognition in 1913 with the establishment of the British Ecological Society. This development lagged behind the United States, where land-grant colleges had linked the 'science of development' with agriculture in the 1890s. New Zealand followed over a generation later with the establishment of Massey Agricultural College and the Department of Scientific and Industrial Research (DSIR) in 1926, and the modernisation of Lincoln (founded in 1878) to provide the South Island with an up-to-date, American-style agricultural college. Developments in Australia also followed the American model but paralleled New Zealand chronologically. The key agency, the Commonwealth Scientific and Industrial Research Organisation, was also set up in 1926. Like New Zealand's DSIR, it represented a local institutional response to the urgings of politicians and administrators to increase the productivity and wealth of the Empire.[30]

As in New Zealand, most Australian 'applied science' soon focused on agriculture rather than manufacturing, reflecting the colonial division of labour and reinforcing dependence of the Australasian periphery on the British core (Figure 11.4). Australian agricultural scientists turned their attention to pasture improvement, pest control and breeding more productive sheep and cattle, producing what Robin labels 'an unnatural economy'. Severe soil erosion problems, however, forced both them and ecologists to carry out much more research beyond the farm and paddock. Yet even today Robin argues that the essential 'Western nature' of ecological science means that its practitioners still concentrate their labours on improving economically valuable 'imported' floras and faunas. Huge increases in artificial pastures also occurred in Australia after World War II, but the percentage of its area covered in introduced grasses never came to equal that of New Zealand.[31]

New Zealand's grasslands trajectory might, however, have been different if another year like 1938, with its disastrous floods and soil erosion on the North Island's east coast, had occurred. Instead landholders had only to accept the flood-control initiatives of catchment board engineers (see Chapter 12), rather than rethinking how bush clearance and overgrazing damaged the land. Many farmers consequently ignored the warnings of soil conservators regarding flooding. Indeed, the Sheep Industry Commission set up during World War II to discover how this key engine of the economy could operate more efficiently, when it finally reported in 1949, argued that farming had nothing to do with erosion. Despite the work of Cumberland and McCaskill, this view remained largely unchallenged within orthodox farming circles until Cyclone Bola in 1988 caused substantial damage again on the east coast of the North Island, where most bush cover had been removed from hilltops.[32]

From the 1960s the shift to town became relentless, as sons gave up the rigours of farming for more lucrative careers in the urban professions and joined the cohort

Figure 11.5 Advertisement for Tordon Brushkiller, 1999. The imagery of war against nature continues down to the present. Source: public handbill distributed at the time

of deserting daughters. Worse, political power, held for so long in the strong hands of farmers, slipped away faster than an East Cape hillside. The old order increasingly appeared to constitute a golden age, as suburb, rather than farm or small town, became the abode of most New Zealanders, as revealed in Chapter 13.[33] Increasing productivity through the application of fertilisers, herbicides and pesticides via aerial topdressing represented a tangible way of retaining prosperity, power and lifestyle. It also maintained the pugnacious, masculinist attitudes of landholders against nature, a perspective that persists among some to the present day (Figure 11.5).[34]

Contesting the orthodoxy

If the great majority of farmers believed in the heroic version of the grasslands revolution, a few, including the noted naturalist and sheep farmer Herbert Guthrie-Smith, had doubted the efficacy of so-called scientific farming since the troubled 1930s. Guthrie-

Smith farmed at Tutira in the ranges of northern Hawke's Bay. The massive floods and landslips of 1938 in this area confirmed his worst fears. The preface to the third edition of his book *Tutira*, for example, reveals considerable anxiety about orthodox farming and its wholesale replacement of trees by pasture. His essay published in 1940 in the illustrated school series *The Making of New Zealand* lamented that the soil had deteriorated so badly that grasses that 'used to flourish with enthusiasm grow now with philosophic calm, almost with the senility of age'. He concluded gloomily that grass could not hold the land, and that erosion had become an 'inevitable, uncontrollable process', without a massive attempt at reforestation. Yet his voice was unusual.[35]

Despite the almost complete dominance of mainstream grasslands agriculture, there was nonetheless a persistent, if low profile, interest in organic farming in New Zealand. Artificial fertilisers became more popular with grassland and arable farmers from the early 1900s as their cost fell.[36] In contrast, market gardeners – especially those of Chinese extraction discussed in Chapter 14 – and orchardists continued to use whatever natural fertilisers they could as these specialisations developed from the 1870s. An alternative approach, developed in Europe and Britain, began to have a somewhat bigger influence from the 1920s, especially on suburban gardeners. Yet organics as a self-conscious movement in New Zealand had only a very limited impact until the 1980s, when international consumers, especially in Europe, Britain and the United States, became much more interested in the safety and purity of their food. Outbreaks of diseases such as BSE (bovine spongiform encephalopathy or mad cow disease) made European consumers, in particular, concerned about contamination of the food production chain.

Questions were being asked about chemical farming rather earlier, and references to Rachel Carson's disturbing critique *Silent Spring* (1962) appeared in the *Journal of Agriculture* from 1965. But they did not become sustained until the 1980s, and topdressing planes dropped the insecticide DDT in a mix with superphosphate from the 1950s to the 1970s. Carson had also condemned the overconfidence and carelessness associated with 'scientism', but few seemed to notice in rural New Zealand. Dichlorophenoxyacetic acid, or 2,4-D, which Carson highlighted as particularly toxic,[37] remained the most common herbicide until 1997, when it began to be phased out and replaced with a less volatile ester formulation that is still in use.[38]

The *Journal of Agriculture*, Federated Farmers, and most agricultural scientists welcomed 2,4,5-T as an effective herbicide. Farmers and the *Journal* initially sided with 2,4,5-T manufacturer Ivon Watkins Dow against criticism of 'Agent Orange'. Many also argued that DDT residues did not constitute a long-term problem, even though some soils in Southland and Canterbury still require careful treatment because of excessive use of this insecticide.[39] The *Journal* condemned the failure to clean up thousands of toxic dumps often associated with dipping from 1975, and blamed the problem on an 'out of sight, out of mind' approach. In 1989, farmers and government finally drew up an Agrichemical Users' Code of Practice and established a New Zealand Agricultural Chemical Education Trust to promote wise use of agricultural chemicals, even though New Zealand had persisted in manufacturing 2,4,5-T long after anywhere else.[40]

Women seemed to be concerned earlier than men: the Women's Division of Federated Farmers called for greater attention to environmental matters from the 1970s.[41] The realisation that agriculture rather than manufacturing was the greatest national polluter did not dawn on the broader rural community until the late 1970s, when scientists calculated that New Zealand sheep and cattle produced as much organic waste as 150 million people.[42] Concern about effluents appeared from the late 1960s, as microbiologists confirmed that dumping of dairy waste into creeks and streams, along with fertiliser run-off, could enrich, or eutrophy, waters to such an extent that most animals and plants within the affected streams or lakes died. Alternatively, weed growth stimulated by the excess of nitrogen and phosphorus strangled all other forms of life and gave off noxious odours.

The *Journal* responded by suggesting several alternative ways of disposing of this waste. The 1967 Water and Soil Conservation Act tried to regulate such practices and established a National Water and Soil Conservation Authority to monitor this work, with mixed results. *Straight Furrow*, the official organ of Federated Farmers, began to advocate lower stocking rates from the early 1980s. Tree planting was endorsed by the 1949 Royal Commission on the Sheep Industry, despite its denial of any link between farming and erosion; and the *Journal* chipped in with lots of advice on how and what to plant from the 1950s. The Farm Forestry Association, founded in 1959, enthusiastically promoted planting of 'useful' species. Government, meanwhile, established the Queen Elizabeth II National Trust in 1977 to protect open spaces, including wetlands, by voluntary covenant. Such moves towards more cautious husbandry won the support of many in the mainstream farming community by the 1990s. Federated Farmers, working with Landcare Research, also revitalised the older farmer support groups of catchment board days in the form of landcare groups in the South Island high country from the late 1980s. These groups spread into the erosion-ravaged east coast of the North Island in the 1990s.[43]

Such questioning of the old orthodoxy had practical spin-offs as Federated Farmers and the Ministry of Agriculture and Fisheries began serious investigations of organics. Two MAF surveys resulted, in 1983 and 1987, the second of which gave guarded approval to organic farming. When, in 1983, Dr Bert Quin argued that New Zealand's pastoral farming had always been essentially organic because the great majority of farmers raised stock rather than crops, did not inject stock with hormones and kept chemical, herbicide and pesticide use to a minimum, Bob Crowder, lecturer in horticulture at Lincoln, and practising biodynamic farmers such as Ian Stephenson disputed this comforting interpretation. Some agricultural scientists, such as R.J. Haynes of the Agricultural Research Division at Lincoln and R.S. Swift, head of Soil Science at Lincoln, initially ridiculed such protestations as bad science.[44] Within a year, however, editor, journalists and readers accepted Crowder's enthusiastic review of Miguel A. Altieri's *Agroceology: The Scientific Basis of an Alternative Agriculture* without adverse comment. Mainstream farmers began to implement integrated pest-management schemes employing a whole range of controls instead of relying on highly toxic pesticides. Leading grassland scientists such as Kevin O'Connor joined the growing chorus of opinion in favour of making farming 'sustainable' from the mid-1980s.[45] Research by social anthropologist Hugh Campbell also suggested that 'new' horticulturists and viticulturists, largely from urban backgrounds, were more amenable to following alternative strategies, including organic, that paid more attention to traceability, animal welfare and quality standards than traditional grassland farming.[46]

Conclusion

Unsurprisingly, studies of high-country farming cited in the 1997 report on *The State of the Environment* suggest that struggling farmers in difficult environments were still responding more to cost signals than concern for sustainability.[47] Nevertheless, farmers, agricultural scientists, ecologists and regional councils are slowly coming to realise that the old paradigm of relentless scientific and technological progress does not provide an adequate understanding of the complex and special environment in which we live. One solid proof of this shift is *Pastures: Their Ecology and Management* (a completely new edition of the 1972 *Pastures and Pasture Plants*), edited by R.H.M. Langer, Emeritus Professor of Plant Science at Lincoln, and published in 1990. Unlike Bruce Levy's original work, this volume acknowledged the problem of environmental degradation and the worth of indigenous grasses, admitted the limits of intensive farming based on high technological inputs and called for the preservation of indigenous landscapes. Even that old demon, gorse, was reassessed, because its nitrogen-fixing ability enabled it to act as nursery for regenerating bush.[48]

Even so, the rapid acceleration of intensive dairy farming since 1990 confirms that New Zealand farming across the twentieth century has been driven by a 'productivist' impulse rather than concern for sustainability. Increasing productivity remains the concern of the majority of farmers and the institutional support they receive, whether from the Ministry of Agriculture and Fisheries (now the Ministry for Primary Industries) or Massey and Lincoln universities. Sector groups such as Federated Farmers continue to espouse essentially productivist objectives. Only about 65 of New Zealand's 10,500 or so dairy farmers contracted by the giant cooperative Fonterra claim to be 'organic'.[49] Nevertheless, there seem to be emerging understandings that New Zealand's future well-being lies with production systems that are more openly framed by consumer priorities that include welfare for land and animals.[50] But the twentieth-century history of the majority of grassland farmers, and of the advisors, bureaucrats, scientists, bankers, stock and station agencies and fertiliser companies who supported their endeavours, indicates that they had not truly settled but rather still tried to remake the land into something different.[51]

Further reading

The argument behind this chapter is developed at length in Tom Brooking & Eric Pawson, *Seeds of Empire: The Transformation of the New Zealand Environment* (I.B. Tauris, London, 2011), and approached at a different scale by Gordon Winder, 'Grasslands Revolutions in New Zealand: Disaggregating a

National Story' (*New Zealand Geographer*, vol. 65, no. 3, 2009, 187–200). The focus of Tom Brooking & Eric Pawson, 'Silences of Grass: Retrieving the Role of Pasture Plants in the Development of New Zealand and the British Empire' (*Journal of Imperial and Commonwealth History*, vol. 31, no. 3, 2007, pp. 417–35), is self-explanatory. Two rather different challenges to the grasslands orthodoxy are to be found in Bruce Wildblood-Crawford, 'Grassland Utopia and *Silent Spring*: Rereading the Agri-chemical Revolution in New Zealand' (*New Zealand Geographer*, vol. 62, no. 1, 2006, pp. 65–72) and Hugh Campbell et al., 'From Agricultural Science to "Biological Economies"?' (*New Zealand Journal of Agricultural Research*, vol. 52, no. 1, pp. 91–97).

12.

An interventionist state: 'wise use' forestry and soil conservation

Michael Roche

BY 1920 NEW ZEALAND's biophysical environment had undergone widespread, rapid and often irreversible change. The role of the state in this as an agent of land development has been well documented in mainstream historical accounts, although its involvement in environmental protection and restoration is less readily understood.[1] This chapter examines state-led forest and soil conservation from the 1920s to about 1960, years during which a belief in superabundant resources gave way to concerns about the prospect of resource depletion. It is an era that remains comparatively neglected by environmental historians, the period up to 1914 having been more fully explored in a search for conservation origins, while the 1970s and beyond have attracted attention for the emergence of environmentalism. Yet in New Zealand, as elsewhere, there were some noteworthy initiatives and controversies between the 1920s and the 1950s.

The development of state conservation from the 1920s was overseen by a small number of technical experts within the bureaucracy. The state initially focused its attention on Crown land rather than regulating use of private land. Tentative steps in the latter direction awaited passage of the Soil Conservation and Rivers Control Act 1941. Until then management of state forest assets was its major conservation concern. Tensions existed, however, between the experts, dedicated to what they saw as 'wise use' of these resources, and public preservationist groups. Both labelled themselves as 'conservationists'.

Having benefitted from the transformation of forest to pasture in the name of progress and development, by the 1920s the population at large expected the state to solve the suite of environmental problems that had emerged from the rapid expansion of the farming frontier since the 1890s. An earlier aesthetic appeal to preserve forest remnants for future generations (Chapter 8) had given rise to the Scenery Preservation Act 1903, as well as the creation of other parks and reserves. A more concerning fear was that of 'timber famine': resources seen formerly as limitless were hastily reappraised as only too finite. These issues were matched by those in Australia and the United States. Soil erosion was also becoming perceived as a regional problem, seen as hampering effective settlement, reducing national agricultural productivity and harming the prospects of individual landholders.

The imperial setting

While insights from American environmental history can inform local investigations, in one respect they provide little guidance.[2] This is due to New Zealand's past as part of the British Empire. As a consequence of this, official responses to timber famine in New Zealand were filtered through an imperial lens, especially for scientific state forestry. It was less so in the case of soil conservation proposals, where the exchange of information with Australia was important, and American developments were more influential. This is not to deny the effort and ingenuity of some local forestry and soil conservation solutions, but rather to emphasise the significance of the wider circulation of knowledge.

Such a concern with flows of people, ideas and materials sits comfortably with networked ideas of empire articulated by the geographer Alan Lester. He stresses that nodes could be offset from the metropolitan core of empire, which is especially true for empire forestry; that mutual connections between core and periphery were important; and that flows of ideas, people and material around peripheries were also significant.[3] The New Zealand experience, however, was also distinctive. In a unitary state, there was scope for innovation on the part of individuals whose ideas could underpin national policies. Local adaptation also took place, noticeably so in the case of models of imperial forestry practice.

During the nineteenth century, colonial governments throughout the British Empire set aside land as reserves for various purposes, extending to include 'wise use' as well as preservation purposes by the 1890s. By 1919, in New Zealand there were 512 scenic reserves totalling 306,888 acres, national parks of 1,654,234 acres (Figure 8.1), and state forests amounting to 1,654,214 acres. Such efforts palled somewhat in comparison to the 1.5 million acres of forest cleared off Crown land opened for settlement from 1896 to 1914. The relatively pragmatic expansion of state functions meant that different branches of a single department such as Lands and Survey were engaged simultaneously in both land settlement and forest conservation, albeit with the emphasis in favour of the former. The 'best use' of land was for agriculture, but it was recognised that some land was uneconomic for farming and that it was better to set this aside as forest reserves for the timber industry, for protection forestry purposes (i.e. for soil or water management), or for scenery preservation. The boundaries between such uses and, ultimately, administrative responsibility for such lands were, however, contested into the 1960s and beyond.

Forest conservation

In 1900 forest covered approximately 25 per cent of the country and the main agents of ongoing deforestation were small farmers (see Chapter 7). A symbiotic relationship between settlers and sawmillers existed in districts where forest areas were accessible to road and rail links. In more isolated areas forest was largely felled and burnt under the guise of 'improvement'. The timber industry was a major agent of deforestation in regions such as Northland and Westland. From early in the twentieth century, official concerns about a coming timber famine intensified, manifest in a series of official reports. In 1913, the Royal Commission on Forestry estimated exhaustion of merchantable supplies at 30 years hence, or 1943.[4]

The most recent state response to timber anxieties had been in 1897 with the creation of a Forestry Branch with the Department of Lands and Survey, though its brief was limited to tree planting. Private plantation forestry, dating back to the 1860s with the introduction of *Pinus radiata* from its native California, revealed rapid growth rates for this and other exotic timbers. By 1921 the Branch had created 38,461 acres in 13 plantations in both islands. Already Whakarewarewa, Waiotapu and Kaingaroa plantations amounted to nearly 25,000 acres, or 64 per cent of the total area of state plantings (Figure 12.1). Government Department of Agriculture biologist A.H. Cockayne meanwhile boldly predicted in 1914 that *Pinus radiata* would become the timber tree of the future in New Zealand, although it was yet to become an important component of state plantations.[5]

The state was also seeking to 'mine' indigenous forests more efficiently, at the same time as establishing an exotic plantation forest estate for future use. This policy was criticised by leading empire forester Sir William Schlich, Professor of Forestry at Oxford University, who considered supplying all future timber needs from exotic plantations 'a very bold measure'.[6] Colonial forester David Hutchins was more damning, likening the idea of replacing felled indigenous forest with exotic plantations to a 'belief in witchcraft'.[7] Both were assessing this from the perspective of empire forestry practice, focusing on the management of both indigenous forest and, to a lesser extent, plantation forest resources, as in India.

An allied state response was to seek efficient utilisation of remaining resources by imposing regulations for harvesting forests on Crown land. Millers cutting off freehold land remained comparatively unregulated. This eking out of forest resources culminated in a series of export controls imposed from 1918 to 1922 to prevent the sale of kahikatea, kauri, rimu, matai and tōtara. The slogan 'New Zealand timber for the New Zealanders' defined such efforts.[8] The controls reduced the quantity of timber exported to Australia but also sought to protect the local timber industry from North American timber imports.

In 1915 Hutchins, then visiting Australia, was invited to report on the state of New Zealand's forests.[9] The government intended that he should principally review the afforestation programme but Hutchins successfully redefined forestry to officials as management of natural forests in perpetuity. He was very slow in completing his reports and Prime Minister Massey's interest waned. But not that of Sir Francis Bell, Massey's senior political ally, who by 1919 had established a Forests Department separate from the Lands and Survey Department, divided the portfolios and taken that of forestry himself, in order to commence the process of administrative reform.

The task of appointing a professionally qualified director then began. It was done within an imperial framework, the interviews being conducted in London by Lord Lovat and R.L. Robinson, Chair and Chief Technical Officer respectively of the British Forestry Commission. Two men were shortlisted: A. Dunbar Brander, a Deputy Conservator of Forests in the Indian Forest Service, and L.M. Ellis, a Toronto graduate with prewar forestry experience with Canadian Pacific Railways, then employed as a Forestry Officer by the Board of Agriculture in Scotland. Ellis was appointed. Politicians and officials had agreed that the Forests Department needed to be headed by a professionally trained

Figure 12.1 Map of state indigenous and exotic forests, 1930. Source: State Forest Service, Annual Report of the Director of Forestry, *AJHR*, C3, 1930

forester. Ellis met the criteria but, with his Toronto degree and Canadian work history, he was not in the image of the colonial forester with an Oxford Diploma of Forestry and experience in India.[10] In any case, shrewd advocates such as Edward Phillips Turner, the head of the Forests Department, realised that North America offered a better analogue for New Zealand conditions.[11]

Ellis arrived in early 1920 to prepare a report on local conditions. He recommended a departmental model proposed earlier by Schlich and used elsewhere in the Empire and mirroring that of British Columbia. This contrasted with the Forestry Commission approach that proved more controversial in Australia. (Such commissions were intended to be semi-autonomous and free from direct political and commercial pressure.) Only 33 years of age, Ellis was energetic, totally committed to forestry and frequently at odds with other officials. He was emphatic that state forestry could make a much greater contribution to state revenues. His report shaped the Forests Act 1921–22. This legislation charged the State Forest Service (SFS; it replaced the Forests Department) with six main functions: the control and management of forest policy; control and management of permanent and provisional state forests; the planting and maintenance of nurseries; enforcement of leases, licenses, and permits; the collection of rents, fees, and royalties; and administration of the Forests Act. The Act also provided for Māori owners of forest areas to have them designated as Māori Forest and managed by the SFS in order to provide a 'permanent and perpetual periodic revenue stream', although the realities did not necessarily meet the aspirations.[12]

Ellis quickly set about regulating the timber industry, by now largely dependent on Crown forests, by putting in place a set of forest management strategies as well as developing a research and education programme. The first step was to concentrate responsibility for all forests on Crown land in the SFS; this involved regaining some control from Lands and Survey and the mining wardens. Once forest was under SFS jurisdiction, the onus was on the other departments to justify its release. Ellis was a natural resource manager who saw himself as a technical expert acting in the public interest. In 1926 he replaced the wasteful royalty payments on sawn output with tendering for standing-block sales in order to produce more efficient felling and sawing and to prevent monopolising of cutting rights.

Another priority was the National Forest Inventory (1921 to 1923). This yielded information on the quantity and quality of forest growth, the extent of fire damage, the effects of grazing, and regional distributions of forest types. It was far superior to earlier estimates of forest area. The 'rational control of forest exploitation to secure constant renewal and perpetuation of the resources' was a cornerstone of Ellis's forest policy.[13] Following European and American forestry principles, he sought to understand indigenous forest regeneration and growth rates in order to implement sustained yield forestry. On his return from the 1923 Empire Forestry Conference, Ellis positioned New Zealand in laudatory tones: 'It was evident that this Dominion unquestionably leads the Empire, with the exception of India, in forest-tree nursery technique; and in plantation practice.' But he also acknowledged two serious deficits, in forest education and 'in the application of sustained yield principles to our indigenous forests'.[14]

Ellis recognised the importance of building up a trained staff and favoured local university-based forestry training, though a model for this devised in 1925 had collapsed within a decade.[15] Short-term relief was provided, however, by the return of a cohort of demobilised New Zealand soldiers who had taken forestry degrees at Edinburgh, who were later augmented by local graduates. But while the SFS as a whole became more 'professionalised', this was not the case at the highest levels where, after Ellis's resignation in 1928, he was succeeded by internal appointees Phillips Turner (1928–31) and A.D. McGavock (1931–38). The latter was a long-time public servant with decidedly sceptical views about the claims of professional foresters. In 1939, when Alex Entrican became Director of Forests, a technical expert was once more in charge, albeit with a background in engineering not forestry. Not until 1961 would a qualified forester, A.L. Poole, again lead the Forest Service.

Afforestation embraced

In 1920 Ellis proposed creating a further 20,000 acres of exotic plantation forest over the period 1920 to 1926. The 3400 acres planted by the SFS in 1921–22 was a 'record within the Empire'.[16] In 1925 he went much further and announced an unprecedented new planting target of 300,000 acres to be achieved over 10 years. Having worked so hard to convince politicians and the public that forestry was about the management of natural forests, this was a considerable change of emphasis. But the plan was consistent with Ellis's belief in himself as a man of action, and his view that 'unproductive land' ought to be given over to afforestation. Crucially it was also driven by his calculations that by 1965 annual timber demand would be much higher than previously thought at 675 million superficial feet a year, by which time merchantable indigenous forest would also be exhausted.[17] His confidence was further buoyed by improved nursery techniques and new planting practices that had reduced the cost of planting from over £26 per acre to less than £2 by 1925, and the speed of tree growth: all of which helped in overcoming orthodox forestry objections to large-scale afforestation.[18]

The programme was acceptable politically because it coincided with the long-standing popular view that tree planting was the solution to a timber famine, and because there were sizeable areas of Crown lands unwanted for pastoral agriculture available for afforestation, particularly the 'bush-sick' districts of the Volcanic Plateau. Besides which, as C.M. Smith, one of the New Zealand Edinburgh forestry graduates, remarked, 'the British Empire does on the whole apply itself more to exotic forestry than do most nations'.[19] The Empire Forestry Conference of 1928, held in Australia and New Zealand, recommended continuation in New Zealand of the process of reservation of indigenous state forests with a view to their sustained yield management. It affirmed the planting strategy, particularly in the context of perceived worldwide shortage of softwoods plantations, pointing to the problems posed by the narrow range of age classes, but expressed real concern about the lack of attention paid to thinning.[20]

Ellis even instigated technical investigations into the suitability of producing *Pinus radiata* wood pulp. After his resignation in 1928, the afforestation momentum was maintained and the SFS and private afforestation companies also continued to plant. SFS efforts were considerably scaled back after 1934, by which time Ellis's 300,000 acres 10-year target had been exceeded, assisted by Depression work relief schemes.[21] Further company afforestation also virtually ceased that same year, when new legislation curtailed the activities of those companies that had raised capital for tree planting rather problematically by bond selling and making overstated claims about forest growth rates and likely financial returns. By then the SFS had planted 406,179 acres, the bulk of it at Kaingaroa in the central North Island, and private companies had added 272,326 acres more (Figure 12.1). Much of this planting predated the years of the Depression, despite popular mythology that it was a response to that event.

Ellis was something of a mentor to Alex Entrican, who became Director of Forests in 1939. He had recruited Entrican to the SFS in 1921. Entrican was emphatic that forestry was 'one of the few activities which is recognised as essential a function of government as the maintenance of law and order'.[22] A full inventory of indigenous forest resources was undertaken (1946–55), which provided vital information about indigenous forests on lands of all tenures. It revealed that many of those remaining were on Māori lands.

Entrican's major contribution came, however, not in the area of forest management but in forest utilisation and industrial processing. Responsibility for protection forestry, in a move symptomatic of head-office tensions, was even passed to the Deputy Director, A.L. Poole, in the 1950s. Entrican had gathered information on pulp and paper making in the 1920s and this became central to his vision for the forest sector, with the state as tree grower as well as wood processor. Originally, he envisaged a large, fully integrated sawmill and pulp-mill plant that would also process trees from private company plantations. The companies distanced themselves from the proposal, but by 1946 the Labour government was committed to proceeding. Ad hoc imperial forestry oversight of New Zealand forestry policy undertaken via three-yearly Empire Forestry Conferences commencing in 1920 had also ceased by this time, even though New Zealand SFS personnel continued to attend.[23]

The National government elected in 1949 was more sympathetic to private enterprise but, with planning already well advanced, could only rework the integrated plant scheme to give more scope for this. The tender for the forests from Kaingaroa was let to Tasman Pulp and Paper Ltd, a public–private consortium that included Fletcher Holdings, a New Zealand construction firm (Figure 12.2). Entrican served on the first board of directors and was dismayed when James Fletcher, the chairperson, wrested control away from government representatives on the board.[24]

Kaingaroa and Tasman Pulp and Paper Ltd are illustrative of the extent to which the state in New Zealand by the 1950s was involved in the forest economy as both a regulatory agent and a participant.[25] Kaingaroa Forest grew to be one of the largest contiguous plantations in the world. Although wood wasp (*Sirex noctilio*) posed something of a threat in the 1940s, the plantation monocultures survived; but by the 1970s they became the subject of contentious debate between forestry professionals who generally down-played the risks, and environmentalists who tended to accentuate them.[26]

Figure 12.2 The arrival of the first trainload of logs at Kawerau, site of the Tasman Pulp and Paper mill in 1955. The Director of Forests, Alex Entrican, is shown on the right. He was a strong believer in integrated state forestry, from indigenous management to afforestation through to industrial processing. Source: AAQA 6395 M1195, Archives New Zealand Te Rua Mahara o te Kāwanatanga

Indigenous forest management revisited

After World War II the SFS, renamed the New Zealand Forest Service (NZFS) in 1949, renewed its efforts at wise use indigenous forestry. The intention was to reduce harvest levels to a point matched by rates of regeneration. The situation for the NZFS was summed up in 1952 by A.P. Thomson, a future director-general, in noting that sustained yield management of kauri was 'technically possible' but only on an 'insignificant scale' once the government had decided to protect Waipoua Forest in Northland as a sanctuary.[27] Kauri output had been severely compromised by 1914, and the much more widespread rimu, one of a number of podocarp species, had become the timber of choice for general construction purposes. Podocarps, an ancient southern hemisphere coniferous family, were widely distributed but the little regeneration evident was restricted to a few localities, while a lengthy rotation of 250 years was required. Beech, on the other hand, regenerated readily and was considered manageable on an 80- to 130-year rotation. Ensuing ecological investigation of podocarp and beech forests suggested that regeneration processes were more complex than previously realised, and that management would be difficult. A selective logging system rather than clear-felling was proposed, but anxieties remained. This is encapsulated in the NZFS annual report for 1956, which commented that 'Above all there must be faith in the final outcome to visualise the seedling Rimu full grown' three and four generations hence.[28] Forestry science was now uncomfortably close to a matter of faith, at least as far as indigenous forests were concerned.

The NZFS faced other pressures in the 1950s from the demands of postwar housing policy, which ensured that government wanted to keep harvest levels high and timber prices controlled. This was not the ideal political climate for advancing indigenous sustained yield management. By the late 1950s Entrican's frustration was evident when he observed that 'sustained yield silviculture of kahikatea would promise better than with another indigenous soft wood species, were it not for the prior claims of agriculture to the rich lowland soils'.[29] Similar comments were made for other timber species such as tōtara and mataī. The longer-term timber supply problem posed by the lack of podocarp regeneration also caused the NZFS to consider options such as interplanting of exotic trees in selected indigenous forests and replanting of areas of clear-felled indigenous forest with exotic plantations.

While making sense in timber supply terms, these sorts of strategies set a collision course between the NZFS and environmental groups over forest management in the 1970s. Entrican's parting rebuke to the politicians was that 'the pioneering instinct that trees are a liability and not an asset has survived for over a century … as a result forestry has had to accept mostly submarginal or marginal land'.[30] The professional foresters' perspective had been well put by C.M. Smith when he spoke both of creating 'an induced indigenous [forest] association' capable of meeting long-term timber requirements and of the particularities of the New Zealand situation where the dual management of indigenous forests and exotic plantations was one without 'precedent nor precept' from abroad.[31] This also signalled the limits of state conservation; efforts were largely confined to Crown lands and there was a political reluctance to expand forestry onto lands where agricultural development was possible.

The first major challenge to SFS management proposals occurred over plans to initiate sustained yield management of Waipoua kauri forest. Originally purchased by the Crown for land settlement but gazetted as a state forest of 22,650 acres in 1906, it was surveyed two years later by botanist Leonard Cockayne, who unsuccessfully recommended that it become a national park. Descendants of the original Māori owners maintained into the 1930s that there was a rāhui – a traditional concept embodying physical and spiritual preservation – placed on the forest, and that it should be protected in perpetuity.

In 1921 Ellis funded William McGregor from Auckland University College to investigate kauri regeneration. An initial round of research was completed but McGregor's working relationship with the SFS quickly disintegrated. The combative McGregor then emerged as one of the protagonists behind the Auckland Institute and Museum's moves in 1932 to have the whole of Waipoua preserved. That same year a committee was established to seek the 'absolute and permanent protection' of the kauri at Waipoua.[32] The Depression and war delayed resolution of this preservation versus wise use controversy.

A Waipoua Forest Preservation Committee was reformed in 1946 and renewed its efforts, calling for a reserve of at least 10,000 acres. With support from prominent zoologist Professor Valentine Chapman and McGregor, a 'Preserve Waipoua' campaign gathered momentum, even though professional foresters derided its 'irrational and unyielding depth of conviction'.[33] In what was to be a decisive move, a petition was launched in 1947 by the Whangarei Progressive Society and the Whangarei branch of the Forest and Bird Protection Society, calling for the preservation of the entire forest as a national park. A nationally circulated petition of 50,000 signatures was subsequently presented to Parliament.

Recognising that it was losing the political contest, the SFS devised its own proposal for a 7200-acre kauri preserve. Entrican, meanwhile, lambasted the petitioners as wanting to create a 'tree cemetery'. The presence of the Pacific Science Congress in Auckland in 1949 provided additional scientific support for the proposed national park. The balance continued to tip in favour of preservation until, in 1952, Parliament gazetted the whole of Waipoua Forest as a sanctuary under the Forests Act 1949, which left it under NZFS jurisdiction but set aside for preservation. Waipoua was a bitter defeat for wise-use forest management, particularly as few lowland sites were available for the sustained yield management of indigenous forests. Indicative of the wider NZFS frustration of the times was Entrican's earlier snipe at the work of the catchment boards and the Soil Conservation and Rivers Control Council (SCRCC) over catchment protection and 'well meaning but unrealistic enthusiasts' who in his view advocated a 'lock-up-use-not policy of forest reservation'. This, he asserted, would not lead to the perpetuation of forest cover: 'true conservation is preservation by wise use'.[34]

The Waipoua debate was an occasion when other scientists and members of the public challenged the government foresters' professional judgement. In so doing, the Waipoua Forest Preservation Committee implicitly rejected the notion of the apolitical expert natural resource manager and appealed directly to public and politicians for a policy change. The campaign used tactics that anticipated those of the environmental movement in its challenges to the NZFS in the 1970s.[35]

Soil and water conservation

Land settlement in the nineteenth century involved forest clearance on lowland and hill country in the North Island, and in the South Island was accompanied by systematic burning of the tussock grasslands. The environmental consequences were recognisable early in the twentieth century (see Chapter 5).[36] By the 1930s land degradation was evident throughout the East Cape and Hawke's Bay, with the most evocative contemporary statement being Guthrie-Smith's *Tutira: The Story of a New Zealand Sheep Station*, first published in 1921.[37] Land deterioration and reversion to scrub was occurring in the interior Whanganui and King Country (Figure 12.3). A series of severe floods on the east coast of the North Island awakened political interest initially in flood-control engineering, and subsequently to the wider problems of soil erosion.[38]

In the South Island, concern about land deterioration was expressed to various commissions examining land tenure. The Southern Pastoral Lands Committee of 1920, for instance, investigated the extent and causes of land depletion and deterioration. At issue was the use of fire, which runholders asserted was a necessary management tool.[39]

Figure 12.3 Sketches of 'Nature's revenge'. The Forest and Bird Protection Society, in articles and cartoons, regularly identified an erosion problem in the 1930s. Its message was that forest clearance for farming on steep lands would lead to flooding and accelerated erosion. The sketch of the half-buried hut at the bottom is taken from a photograph of the aftermath of heavy rain at Pukatea, Admiralty Bay, near French Pass, in 1938. Source: *Forest and Bird*, no. 51, 1939, p. 5

Official discussion of water conservation dated back even further but was unrelated to land degradation; it was primarily concerned with ensuring water availability for the goldmining industry in Otago (see Chapter 6).

The Department of Agriculture by the 1920s was committed to improving farming productivity by ushering in an era of more intensive agriculture. The department sought to improve production levels by matching farm practices with soil fertility and land capability. The limits of settlement under prevailing technology had been reached, but with prices for New Zealand commodities in Britain falling, sizeable increases in production were necessary to maintain export income. With these came the deterioration of the North Island hill country.

The Department of Scientific and Industrial Research (DSIR), established in 1926, was also involved in identifying the severity and extent of soil erosion in the 1930s. This work included Soil Bureau surveys in mid-Canterbury and Taranaki as well as land utilisation mapping, notably in Hawke's Bay. Particularly important work undertaken by Norman Taylor revealed the extent of accelerated soil erosion across the North Island. He tactfully called into question the efficacy of continuing to farm some of the more remote and erosion-prone lands. Equivalent but, at the time, less controversial research was also published on the South Island tussocklands.[40]

The Department of Agriculture only reluctantly agreed with these assessments, but a turning point came in 1939 when R.P. Connell acknowledged that land deterioration, including soil erosion, was now a serious threat to the national economy. For Connell, at least, the situation was serious enough to warrant state action on the grounds that individuals could not effectively counter the problem.[41] There was not, however, complete unanimity among government scientists as to what constituted effective action. The eminent grasslands scientist Bruce Levy, for example, tended to argue that a good sward of pasture alone was sufficient to counter soil erosion problems (see Chapter 11).[42]

The Esk Valley flood of 1938 hastened political action. Ernest Marsden, Secretary of the DSIR, was able to orchestrate a motion from the Royal Society of New Zealand, calling for a royal commission to investigate vegetation preservation and soil erosion. Norman Taylor was asked to chair the resultant committee of inquiry in 1939, which recommended statutory measures to institute a soil conservation programme.[43] The Esk Valley flood gave moment to the claims of soil erosion campaigners and provided a dramatic example that was too significant in terms of size and damage to be ignored.[44]

Meanwhile, the efforts of Christchurch Teachers' Training College lecturer (and later Lincoln College staff member) Lance McCaskill and geographer Kenneth Cumberland of Canterbury University College were also important in keeping soil erosion problems in front of politicians and the public alike. McCaskill had benefited from a Carnegie Grant to tour the United States in 1939, where he studied the methods of the US Soil Conservation Service. His lantern-slide show for parliamentarians was decisive in persuading them to combine river engineering for flood control with soil conservation into a single bill in 1941.[45]

The Soil Conservation and Rivers Control Act 1941 ushered in a new era of land management in New Zealand. It signalled a change of approach from that of Frederick Furkert, the head of the Public Works Department, who believed that the 'abandonment of flooded country' implied 'a retreat of civilisation before the forces of nature'.[46] It established the Soil Conservation and Rivers Control Council (SCRCC), composed of representatives of government departments, local authorities and agricultural and pastoral interests. This body first met in 1942 and was chaired by a senior government engineer, a sign that it would adopt an incremental approach to remedying the situation.

Simultaneously, new regional authorities termed 'catchment boards' were voluntarily established across much of the country between 1943 and 1948. The voluntarism was deliberate: the SCRCC emphasised education, promotion and demonstration of soil conservation techniques rather than more direct regulation of land use. Even so there was some farmer concern about the SCRCC and its activities. This was apparent in evidence denying soil erosion given to the Royal Commission on the Sheep Farming Industry in 1947 and 1948, which led to a recommendation in its final report that catchment boards be abolished. Even National Party MPs spoke out against this, but the episode did point to deep-seated farmer concerns about restrictions on land use imposed by catchment boards comprising some elected urban members deemed to be antagonistic to farming.[47]

The SCRCC was able to make more rapid progress after the war. It focused on demonstrating the utility of soil conservation techniques, working to an implicit assumption that these were in the national interest. Publicity efforts included a series of bulletins, with poignant titles such as *Down to the Sea in Slips,* and films shown widely throughout the country. New techniques were showcased on demonstration farms, and a research programme was initiated which, by 1963, included 17 soil conservation reserves, each with specific regional erosion problems. These were used for hydrological research and land capability surveys, with a view to devising farm-level soil conservation plans.[48]

Aerial topdressing was perhaps its most innovative area of investigation. This was to prove a real boon for soil conservation in New Zealand (Figure 12.4). It facilitated the effective stabilisation of at-risk slopes in the hill and high country, and was particularly associated with Doug Campbell, the Senior Soil Conservator, who orchestrated the original trials in 1948. Campbell's efforts in promoting the benefits of aerial topdressing extended beyond New Zealand. The salient feature of the topdressing programme was that not only did it allow successively grander soil conservation targets to be set, it also enabled farmers to continue farming without having to sacrifice overall production levels. By the time he retired in 1967, Campbell perhaps sensed that it would not be possible for these two goals of soil conservation and increasing production to coexist indefinitely.[49]

Soil conservation was able to be promoted very much in utilitarian terms in the 1950s, both as being in the public interest and as leading to a rational system of land utilisation that would enable the farm sector overall to lift production levels. It quickly became administratively complex, however, and subject to a contest, behind the scenes, as to where soil conservators should be based. Initially part of the Public Works Department, they were separated from the river engineers and transferred to the Department of Agriculture in 1956 after a five-year bureaucratic struggle. Here they remained until they were returned to Works in 1967. Having the staff supporting the SCRCC divided between two departments was a hindrance to achieving soil conservation targets. This separation arose from the competing views of the Department of Agriculture – that soil conservation was a legitimate part of its farm extension activities – and of the Ministry of Works – that soil conservation was part of its flood-control programme.

Locally trained soil conservators emerged from Lincoln College after 1946 and with degrees from 1959. Massey College also offered degree-level qualifications from 1962. (This was in contrast to higher forestry education, which lapsed in 1934 and was not reinstated until 1969, at Canterbury.) Education campaigns and demonstrations aside, what gave real impetus to soil conservation and rivers control in the 1950s and 1960s was the availability of government subsidies to undertake on-farm work and regional flood-control schemes. From 1943 to 1962, a total of £14.5 million was spent on a range of river works and soil erosion subsidies and grants.[50] Expenditure varied regionally across the country, reflecting the mix of soil erosion and flooding problems; for example, river channel clearance work predominated in Canterbury, while tree planting was a

Figure 12.4 'Giving wings to soil conservation', 1955. Aerial topdressing provided a means of stabilising at-risk slopes in higher country. Doug Campbell of the Soil Conservation and Rivers Control Council was instrumental in organising the original trials and promoting topdressing as a means of soil conservation. He wrote two SCRCC bulletins on the subject, including this one, and spoke at many meetings in New Zealand and overseas.

particular feature of the Wairarapa, Hawke's Bay and East Cape regions. It has been argued that Māori lands were targeted unreasonably for afforestation.[51]

Soil erosion never generated the same sort of popular concern as deforestation in New Zealand, although the use of films to publicise its effects heightened public awareness in the 1940s and 1950s. Even so, soil conservation tended to be regarded by the public as a problem for the expert. The caveat was that on-farm soil erosion was a major problem in some regions, and the onus was on remedial action at the farm level. To this end, soil conservators helped farmers to develop land utilisation plans (Figure 12.5). Unlike foresters, soil conservators had to persuade farmers to adopt soil conservation techniques.

An interventionist state 223

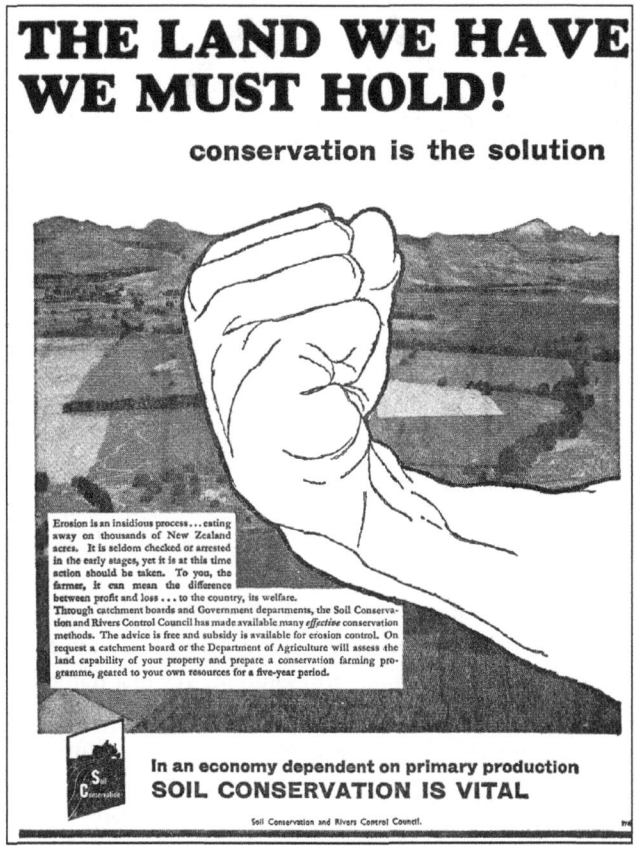

Figure 12.5 'The land we have we must hold!' The SCRCC made use of a range of masculinist images in urging farmers to adopt soil conservation techniques. Source: *New Zealand Farmer*, 17 August 1961

They were aided by the availability of subsidies through catchment boards for on-farm works.

River engineering remained, however, an important part of the SCRCC's work. This also tended to capture the bulk of the funding; flood control works, by their very nature were expensive. Less obviously, within the organisation the engineers tended to dominate senior staff positions. There was special recognition of their professional training within public service salary structures, which did not extend to the soil conservators. The legacy of the soil conservation programme and the major river engineering schemes in the landscape was significant but skewed in favour of the latter.

There was a loss of momentum in soil conservation in the late 1960s. The benefits of aerial topdressing had been realised and there were new and complex water quality and allocation issues facing the state and the community. This gave rise to new legislation in the form of the Water and Soil Conservation Act 1967, and the emergence of the

National Water and Soil Conservation Authority. Water, rather than soil, provided the most difficult matters in the 1970s and 1980s, including challenges to attempts to establish water classifications, granting of water rights for the Clutha hydro dam, and protests about hydro development threatening wild and scenic rivers (see Chapter 17). From the early 1980s, Māori increasingly challenged the monocultural nature of the legislation, in particular in three claims to the Waitangi Tribunal over the waste outfall at Motunui, piping of Rotorura's sewage into the Kaituna River, and pollution of the Manukau Harbour.[52]

Conclusion

State conservation in New Zealand from the 1920s to 1960s was informed by an amalgam of élite public and technical expert perspectives, with the former gradually giving way to the latter. Conservation, interpreted by such experts, was largely a matter of the 'wise use' of natural resources in the public interest, although it was extended in New Zealand as elsewhere to include aesthetic and ecological ends, in the form of limited preservation of flora and fauna.

Forestry provides a fascinating example of conflicting currents of thought, as popular notions of 'timber famine' and a belief in the inevitability of indigenous forest decline pointed to exotic afforestation as a long-term solution. The first trained foresters advising the government reinstated sustained yield management of indigenous forests, but in turn expanded their efforts towards large-scale exotic afforestation. Initially this was to win time to enable the mechanisms of indigenous forest regeneration to be understood, and later because of the possibilities conferred by industrial forestry and a local pulp-and-paper industry for broadening the export basis of the economy.

'Wise use' was a foil to the 'developmentalist' push within government and from within the agricultural sector. Increased state control over natural resources was one feature of modernising states until the mid-twentieth century. 'Wise use' secured the protection of some areas of Crown land, but ultimately it unravelled as a concept due to internal contradictions, particularly the long-standing tension between utilitarian conservation and the counter-position of conservation as inviolable preservation. This was especially so when the utilitarian position was espoused by state officials, and conservation advocated by members of the public took on a more environmentalist character. The opaque nature of expert decision-making, at least to the public, also made it vulnerable to later neoliberal calls for greater transparency.

Soil conservation was taken up by the state only in the 1940s, and there was a reluctance to prescribe land use on privately owned land: education and subsidised works were favoured strategies. Aerial topdressing as a new technology was critical to the success of soil conservation efforts on the steeper lands, but underpinning the whole initiative was the assumption, unchallenged until the neoliberal 'reforms' of the 1980s, that such work was in the national interest. The claim of the state and its experts to be the source of knowledge and power was soon to be successfully contested. From the late 1950s there were renewed moves by public preservationists to protect scenery, including forests, rivers and lakes, from development. The dominance of the technical expert model of natural resource management was under threat.

Donald Worster's popular model of environmental history does not adequately conceptualise much of a role for the state,[53] yet in New Zealand from the 1920s to the 1970s the state, nationally and locally, played an important role not only in terms of natural resource development and its regulation, but also in key aspects of environmental conservation. There is perhaps a tendency among scholars to seek shortcomings in the achievements of the state. Such shortcomings are apparent particularly in the failure to implement a sustained yield indigenous forest management, in the reluctance of successive governments to regulate agricultural land uses to the full extent provided for by existing legislation, and in the limitations of aerial topdressing as a long-term solution. The dominance of the expert, before public participation and environmental perspectives were officially recognised, to some extent perpetuated frequently antagonistic debates over indigenous forests. Looking back at the efforts of twentieth-century forest management, it might even be observed that 'wise use' of forests did not advance much beyond the eking out of resources, more efficient felling and milling, and recovery of larger sums of money from sawmillers for cutting rights.

What such a perspective overlooks is that the state was active in pursuing natural resource conservation from the 1920s, guided by some notion of 'the national interest' through the efforts of a few professionals. They had limited resources available to them, were typically running counter to the mainstream development ethos in government, and were often in advance of public opinion. On occasions, however, they developed bold and imaginative solutions to real environmental problems, as in the cases of afforestation and aerial topdressing. Some of these efforts were derivative, though afforestation techniques and aerial topdressing owed much to local innovations, but they still took place within larger networks. In the case of forestry, British imperial connections were influential from the 1920s to 1940s, while for soil conservation, US links and exemplars were more important in the 1930s.

Further reading

The imperial forestry context is discussed by Michael Roche in 'Colonial Forestry at its Limits: The Latter Day Career of Sir David Hutchins in New Zealand 1915–1920' (*Environment and History* vol. 16, no. 4, 2010, pp. 431–54) and in 'Forestry as Imperial Careering: New Zealand as the End and Edge of Empire in the 1920s–40s' (*New Zealand Geographer* vol. 68, no. 1, 2012, pp. 201–10). Astrid Baker, 'Governments, Firms, and National Wealth: A New Pulp and Paper Industry in Post-war New Zealand' (*Enterprise and Society* vol. 5, no. 4, 2004, pp. 669–90) examines the industry. On aspects of soil conservation, see Brad Coombes, 'The Historicity of Institutional Trust and the Alienation of Maori Land for Catchment Control at Mangatu, New Zealand' (*Environment and History*, vol. 9, no. 3, 2003, pp. 333–59) and Eric Pawson, 'Creating Public Spaces for Geography in New Zealand: Towards an Assessment of the Contributions of Kenneth Cumberland' (*New Zealand Geographer*, vol. 67, no. 3, 2011, pp. 102–15). James Beattie, 'Environmental Anxiety in New Zealand, 1840–1941: Climate Change, Soil Erosion, Sand Drift, Flooding and Forest Conservation' (*Environment and History*, vol. 9, no. 4, 2003, pp. 379–92) provides a wide-ranging review.

13.

On the edge: making urban places

Eric Pawson[1]

NEW ZEALAND SINCE European settlement has arguably always been an urban land. Notwithstanding the emphasis on primary production for the last 150 years, European colonisation has been urban-focused, the towns acting as 'beachheads' (Chapter 4). Wakefieldian ideals decreed that it should be so, to promote economic advancement, social stability and political control. New Zealand was systematically settled from planted towns, its important trades have always been coordinated through towns, and since the early twentieth century a majority of its people have lived in towns. In 1931 the guide to 'Auckland City and Province' was claiming that in terms of population, output and imports, it already accounted for one-third of the country's total. The everyday experience of most people for most of the twentieth century was therefore urban. In her poem 'Living Here', Cilla McQueen imagines that 'this place is just one big city' with its 'suburbs strung out in a long line' from north to south, between ocean horizon and mountain ranges.[2]

Until recently there has, however, been relatively little literature to assist in the development of a New Zealand focus for urban environmental history. For instance, only in the last decade have environmental historians in North America begun to respond to William Cronon's observation that 'cities in particular deserve much more work than they have received', and then generally at a city-by-city scale.[3] Stephen Dovers' plea for 'the glaring need for a consolidation of the environmental history of Australian urban areas' has not really been fulfilled. For New Zealand, Jock Phillips suggested 25 years ago in *A Man's Country* that there may be a historical reason for this. He argued that in the construction of male mythology, 'because it was the most distinctive aspect of Pakeha society, the frontier experience was universalised and the urban world was ignored'.[4] The choice of 'On the edge' as the title for this chapter is therefore a metaphor for marginalisation, a statement to the effect that the writing of environmental history has been heavily ruralised.[5]

Taking a broader perspective, however, there is an expanding foundation on which more robust approaches to urban environmental history can be built. First, urban histories of colonial settler societies provide a long tradition of valuable research that situates the town functionally and politically in relation to the country.[6] Second, the emergence of work in urban ecology generates an opportunity to think critically about the relations between urban activity and nature and the ways in which both are remade through these entanglements. This field of scholarship, with its deployment of the concepts of systems, feedbacks and hazards, has the potential to resolve the lack of clarity that some have seen as characterising earlier attempts at writing urban environmental history.[7] Drawing on these literatures, 'on the edge' is employed in this chapter to construct a framework with analytical purchase for urban environmental history. This framework consists of three themes that explore the roles that towns and 'the urban' have occupied relative to both 'the environment' and 'the rural' in New Zealand between about 1900 and the 1960s. The chapter concludes with the application of each of these themes in a case study of a South Island resort town.

A framework

The town was long promoted, and understood, as a vanguard of colonisation in 'new' lands. In this view, the making of urban places encapsulated and symbolised the taming of the 'howling wilderness', a biblical phrase in wide use in nineteenth-century Otago.[8] Colonial towns stood on the edge of 'civilisation', channelling the agents through which modernisation proceeded: capital, military and cultural power, information, infrastructure and onward connection to a wider world. As a functional perspective on urban history, this has been less readily recognised in New Zealand than in North America, where geographers such as James Vance and Jean Gottmann (the latter with his idea of the Seaboard hinge of settlements) long since conceptualised it, although it was picked up here 30 years ago by Neil Quigley. Such work is based on the thesis that it is the town that made the country, as William Cronon argued in his work on Chicago.[9]

The environmental discourse underlying this first theme was progressive: that culture was destined to triumph over nature. Indeed, wilderness was often represented as an empty stage, benignly awaiting the imprint of the European. As the French visitor André Siegfried said of the Canterbury Plains: 'No Maori came to trouble the peace of these new settlers, and the large and beautiful plain open before them needed nothing but cultivation.'[10] But the unproblematic nature of this viewpoint also belies a second sense in which the urban has been 'on the edge': the vulnerability of these new towns to the unpredictability of the 'howling wilderness'. A way of thinking that imposed false expectations on the environment was particularly open to hazards – whether native peoples contesting colonisation, or physical processes that were, and to an extent remain, little understood.

In many of the neo-Europes, the threat posed by indigenous competitors was accommodated, although often not without more prolonged and effective resistance than standard sources reveal. The inhabitants of towns such as New Plymouth and Wanganui in the mid-nineteenth century attested as much.[11] However, it was the openness of new urban settlements to environmental threats that became so important in the history of these places, the more so because urban growth greatly accentuated, rather than lessened, human exposure to natural variability. This is a matter of wider significance than the combating of air and water pollution in degraded cities that fills those accounts of urban environmental history that do exist. Rather, it involves the extent to which multiplying assets – and numbers of lives – were put at risk from flood, fire, storm and earthquake, because of the assumption that nature could be tamed. In fact nature was rendered more dangerous to people when towns were allowed to grow in harm's way – made worse by the accompanying historical amnesia about the frequency and regularity of hazardous events.[12]

The third sense in which the urban is 'on the edge' was of particular significance in the making of twentieth-century urban places. In New Zealand most people live and work on the fringes of cities and towns, rather than at their centre. It is a pattern prominent in other affluent countries, too, with which 'the broad economic, technological, and political forces associated with major [urban] environmental transformations – industrialization, automobility, suburbanization, and dependence on [government] initiatives' are shared.[13] At a cultural level, however, the meaning of suburbs has lain deeper. They can be represented as places that were believed to embody high levels of environmental amenity, or, put another way, as realms in which resolution was sought of the dangers of nature untamed and of the city degraded.

These three themes – of the progressive town built boldly on the edge of wilderness; of the vulnerable town placed edgily alongside unpredictable nature; and of suburbs in search of environmental amenity on the edge of town – form the basis for a more extended analysis of New Zealand's urban environmental histories.

The progressive town

A recurring motif of urban boosterism – as well as of local histories written in the first six or seven decades of the twentieth century – was the role of the town as a centre of progress.[14] This was celebrated by means of comparisons with other towns and other countries; by civic achievements in terms of cultural icons and the healthiness of the population; and by constant reference, implicitly or explicitly, to nature tamed. The emphasis was on how far and how fast urban development was understood to have come. Writing of Auckland in 1917, James Cowan described it as

> one of the wealthiest and most sightly of young colonial cities ... Though many parts of the province, forest covered and imperfectly roaded, are still in their rough pioneering days, Auckland City is well past that stage ... and the citizens have sufficient time and taste to cultivate the City Beautiful, and to add vastly to the graces of their nature favoured home.

For at least the first 30 years of the twentieth century, the motto of Auckland Corporation was 'Advance'.[15]

The 1931 guide to the city, *Auckland, The Gateway to New Zealand*, told 'intending settlers and those wishing to invest their wealth' that they could 'know that sure, certain and safe progress is inevitable'. 'To have achieved [so much] in so short a space of time ... reflects the magnificent future which assuredly must lie in front.'[16] The markers of this assurance were evident in the townscape. Art deco architecture and design, for instance, signified the technological optimism of the modern world, a universal style adopted from Europe and America. Today it is recognised as emblematic of the new Napier, rising phoenix-like from the devastating earthquake of 1931 (Chapter 15); but the style was widely adopted in urban New Zealand and Australia in the 1930s. It helped that its lightly decorated concrete-box structures were cheap and could be built quickly.[17] Two decades later another Auckland guide proclaimed: 'Where once were silent hills and valleys, to-day there stand tall spires and towers of stately churches; institutes of culture and of learning; business houses, modern apartments and fine homes of the people …'[18]

Such publications portrayed these points boldly on their covers and in their illustrations. In those about Auckland, Grafton Bridge was prominent: 'the World's largest single span concrete bridge', said the 1931 guide, along with pictures of the geometrically neat Albert Park, 'adorned with statuary, lawns, trees, and walks', the epitome of nature ordered and contained for human pleasure. However, the most striking photograph in this particular guide, entitled 'The Picture of Health', is that of the smiling boys of Auckland Grammar School informally posed in front of its Spanish Mission buildings (a style closely related to art deco). That they are boys, or that this is a select educational establishment, is not remarked upon. The caption reads, 'Auckland's good Anglo-Saxon stock, her marvellous climate, garden homes, recreational facilities, and happy democratic atmosphere, build sturdy citizens'. The environment was represented as both facilitating and being transformed in the pursuit of progress.[19]

The development of urban New Zealand was celebrated, as elsewhere, with major exhibitions. One of the largest was the Christchurch International Exhibition of 1906–07, which attracted nearly 2 million visits that summer. Subsequently, Cowan wrote about it, averring that 'in the essentials of modern progress the four cities are well advanced', and going on to praise Christchurch as 'architecturally … a very sightly city', to which, for him, its wide streets contributed.[20] Most towns, large and small, had been built on a generously proportioned grid. The desire for order was part of it, but more commonly explanations focused on its simplicity for the surveyor, and that it was evidence of mastery over nature, or at least the need for it.

This can perhaps explain why grids were persisted with in unlikely situations, such as New Plymouth (Figure 4.2), Wellington and Dunedin, and underlines the symbolic role of towns on the edge of the frontier. There is a more essential reason, however – although it reinforces this point. The grid was one of the 'civilising' influences of capitalist colonisation. It was the most straightforward way of delineating private property rights in land, of commodifying nature for those asserting a stake in it. It was the most efficient means of accommodating capital in space, providing a stage from which that capital could be used in the conversion of nature into resources. This after all was the driving force of modernisation.

The cover of the 1931 Auckland guide is particularly evocative of the modern (Figure 13.1). There are multistoreyed office buildings, surrounded by all the signs of growing traffic: motor vehicles and trains; a large ship, docks, and cranes as symbols of trade; and, across the harbour, generously spaced suburbs. The scene is framed with benign

230 Making a New Land

Figure 13.1 Evoking the modern. In 1934 George Bernard Shaw found Auckland 'bright, clean, sunny and happy' (quoted in David Eggleton, 'Introduction', *Here on Earth: The Landscape in New Zealand Literature*, Craig Potton Publishing, Nelson, 1999, p. 21). Source: the cover of the city guide, *Auckland, The Gateway to New Zealand*, Wilson & Horton, Auckland, 1931.

vegetation and bathed in sunshine. The sun was often used as a metonym for progress, hence the sunburst symbol of art deco. So too was the garden, which was seen as the antithesis of idle wilderness. Taranaki, for instance, called itself 'the Garden of New Zealand', with New Plymouth being promoted for 'Health, Sunshine, Happiness'. In the 1940s Gisborne ('a magnificently appointed modern town ... on which Nature, in her bounty, has seen fit to bestow her richest gifts') saw itself as both 'Gisborne the Golden' and 'The California of New Zealand'. This encapsulated, at one level, its orchards and vineyards, and 'Miles of splendid beaches'; and at a deeper level was meant to evoke, by association, prosperity and glamour.[21]

Many towns at this time had progress leagues, which went under various names. In Gisborne and Napier the '30,000 Clubs' were dedicated to raising their towns' populations to that level. The goals of the 'New Plymouth Tourist and Expansion League' were self-evident; in Christchurch the Canterbury Progress League aimed to complete the city's rail links to the West Coast and to Picton. When Rotorua attained city status in 1962, it claimed to be the 'HAPPY, FRIENDLY, PROGRESSIVE CITY'. Synonymous with this, at a time – during the post-World War II long boom – when there was unanimity over the extensive role of the state in society, was substantial government investment in urban fabrics. This took the form of new offices for its agencies, roading schemes and state housing. Wellington, as the centre of government, was a particular beneficiary, displaying many of the early high-rise buildings in New Zealand. It was said of the capital in 1960 that 'in the last ten years the erection of many handsome buildings have changed the skyline of the city, giving it, with the ever-increasing motor traffic, an atmosphere comparable with the great cities of Europe and America'.[22]

This was the period when, internationally, the city was being lauded as the crucible of modernism.[23] Hence the desire of New Zealand's main centres to be seen as equal to, or in advance of – to use the language of the time – those elsewhere. Commenting on the growing focus on the city for artists, Frank Whitford writes: 'Nature now seemed finally to have been mastered. The city, in which nature was confined in parks, tubs, and pots, seemed to be the symbol of that mastery.'[24] There is a sense, however, in which this statement is inaccurate, not least for New Zealand towns. They remained very much on the edge of nature in their vulnerability to hazards. In fact, as urban growth proceeded, the extent of this exposure escalated.

The vulnerable town

One of the revelations of recent hazard research is that the more that has been spent on control measures, the greater the loss of assets because of events such as flood, fire and quake. The resolution of this paradox lies in the realisation that people often interpret technical interventions in the environment as returning it to a 'safe' state, one that is again convenient for human endeavours. Damaging natural events are considered to have been banished, given the progressivist assumption of nature tamed. Urban growth has therefore proceeded apace in attractive but risky places. The most devastating critique of this view is Mike Davis's exposé of the growing vulnerability of twentieth-century Los Angeles to environmental hazard, based on what he calls 'disaster amnesia' and decisions to base development on short runs of optimistically interpreted data about natural variability.[25]

New Zealand is equally prone to these problems, with much of the country under threat from earthquake events, land instability and floods. The seismic risks to the city of Christchurch that became all too real in 2010 and 2011 were not unknown: a lifelines study carried out 15 years previously had warned of the likelihood of damage from ground liquefaction to built structures and underground services. The city sits on a hazardous site, as does Wellington. Yet there, too, the likelihood of earthquakes, revealed in the events of 1848 and 1855 (the second resulting in the emergence from the harbour of much of what is now the downtown area, as well as of the lip of land that carries the rail and road lines north to the Hutt Valley), seems to have been quickly forgotten.[26] Perhaps this was because the initial site for the settlement, at the Hutt River mouth in Petone where the grid was to be laid out, was fortuitously flooded – as described in Chapter 10 – not long after European settlers landed. Had the event not occurred at that point, a flood-prone site would probably have been chosen on the basis of not even a short run of environmental data. This is what happened with countless other New Zealand towns: hence the nineteenth-century epithets for places such as Blenheim ('Beavertown') and Westport ('the Venice of the South').

Between 1920 and 1953, 64 per cent of the 136 places with more than 1000 people listed in the 1956 Census experienced one or more floods. 'The most frequent flooding occurred in the metropolitan areas of Auckland (33), Hutt (36), Wellington (29), and Christchurch (26), as well as the small West Coast towns of Greymouth (33) and Hokitika (29).'[27] The Christchurch problem – because of its location at the deltaic edge of the Waimakariri River fan – was grasped by some in the early days of settlement. James Balfour, a Dunedin-based engineer, reported to the Canterbury Provincial Council in 1866 after Waimakariri floodwaters had several times poured down the Avon through the town. 'It must be remembered,' he wrote, 'that if the Government once commit themselves to the policy of confining the river absolutely to its present channel, they undertake a work which must be never ending and which must grow in magnitude from year to year.'[28] Politically and publicly, however, the message was barely understood.

Flood-control measures were taken to ensure that the Waimakariri remained, as best as could be managed, in its existing channel rather than switching – as is natural – to any number of other outlets. A report in 1928 put this propensity down to 'the continued rise of the bed of the lower course', as indeed is to be expected at the edge of a large gravel fan. The report drew another conclusion: 'It shews that the inability of the river to get rid of its shingle is due to a deficiency of nature which must be made good by the art of man.' Hence the philosophy of 'training of the river … [so as to make it] flow in an orderly manner'.[29] On this basis the Waimakariri was artificially canalised along the northern fringe of the fan, both to encourage it to carry more of its load into the sea, and to constrain its flow away from Christchurch and its suburbs (Figure 13.2). So far the scheme, modified over the years but essentially as conceived in the 1920s, has protected the city. But this has been at the price that Balfour foresaw – of ever-increasing investment in stopbanks, revetments and plantings – and with the effect of ever greater urban expansion over the flood plain. The flood hazard now posed by the Waimakariri is the largest in the country, quite apart from the now-realised risk posed by earthquakes from the liquefaction of the unconsolidated river sediments beneath the city.

Canalisation of streams and rivers was undertaken in many New Zealand urban areas in the first half of the twentieth century. It was as if they were being demonised in the same way as swamps (Chapter 10). In reality, control over nature was sought in order to develop flood plains rather than to allow them to remain sodden. Extensive works were built on the Water of Leith in Dunedin, which, because of its steep, small catchment, is very responsive to intense rainfall on the hills behind the city. The Otago Harbour Board concreted the walls of its lower course and provided a lined outlet to the sea – the Leith Canal – across reclaimed land just before World War I. The scheme had to be expensively revised in the 1930s after it was shown to be too constricting during extensive floods in 1923 and 1929, the latter washing away several bridges and inundating 500 houses.[30]

Hazards such as flood and earthquake have been the result of urban development carried out in the face of natural variability. Other hazards were the result of urban concentration. In the early years of urban settlement, what Pamela Wood has called 'crises of contamination', ranging from the filth and dirt of unmade, muddy streets to the pollution of water supplies from lax disposal of human and animal waste, 'threatened the

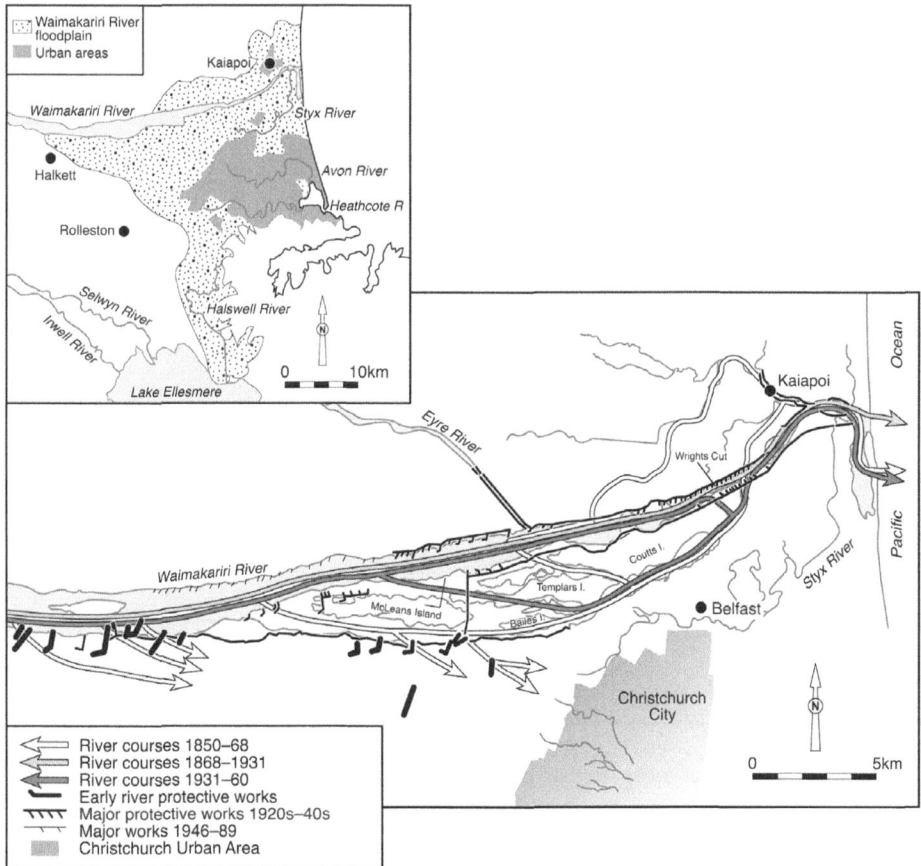

Figure 13.2 'Civilising' the Waimakariri River since the 1860s: its changing courses and flood-control works. The inset shows the relationship of Christchurch to the river's flood plain. Source: adapted from *Draft Waimakariri River Floodplain Management Plan*, Report R91 (9), Canterbury Regional Council, Christchurch, 1991

integrity of the settlers' vision of an ordered and healthy community'.[31] Human bodies, and the wells and streams on which they depended, were all too easily fouled. The advent of sewerage systems and reticulation of water supplies in the late nineteenth century usually resolved these problems at the local scale, displacing that of sewage discharge out of sight, for some years at least. Air pollution, however, remained very much on the nostrils and in view, although observers could interpret it in a positive light. For instance, one, standing on top of Christchurch's Cathedral tower in 1900, rejoiced in the 'intense activity that prevails' in the smoke-covered city.[32]

It seems to have been not until the early 1950s that the term 'smog' came into general use in Christchurch, although it had been introduced by the city's Sunshine League during its campaigns against the 'smoke nuisance' in the 1930s. Even when Mabel

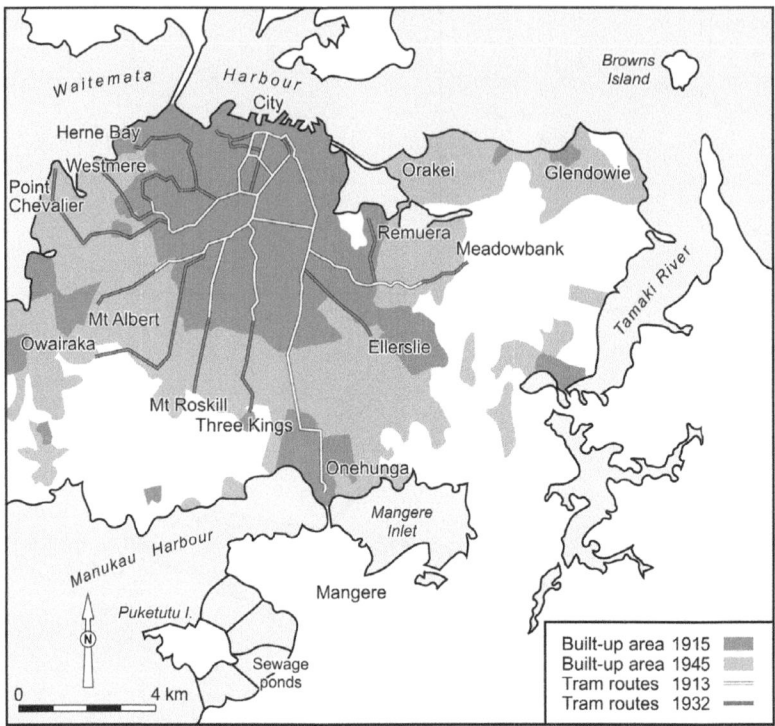

Figure 13.3 Map of Auckland suburban development, 1915–45. Source: adapted from Kenneth B. Cumberland, *Landmarks*, Readers Digest, Surry Hills, 1981, p. 247

Howard, as MP for the industrial constituency of Christchurch South, raised the matter in Parliament, the government view was that air pollution was essentially a South Auckland issue.[33] This was due to long-standing stenches from disposal of industrial and animal wastes into the Manukau Harbour, and from airborne pollutants from local factories (Figure 13.3). The district around Mangere Inlet was colloquially known as 'Lavender Flat'. Such was the problem that the government set up an official 'Fumes Inquiry' in 1955. Action thereafter was piecemeal: the Auckland City Council opened a new sewage treatment works in the area in 1961, but industrial pollution of the Manukau was still in evidence when the Waitangi Tribunal investigated this problem in the mid-1980s.[34]

Meanwhile in Christchurch, Howard's intervention drew attention to the role of domestic as well as industrial point sources in the generation of smog, both from chimneys and backyard incinerators – then the main means of disposal of household waste. She also highlighted the unequal way in which the costs of such pollution fell on the community: poorer people lived closest to smoke from factory chimneys and steam trains. However, the accentuation of the flood hazard by residential development situated alongside flood escape routes from the Waimakariri main channel potentially placed better-off residents at risk as well.[35]

The effects of hazards such as water and air pollution on the health of urban populations were factors used by rural interests from the late nineteenth century onwards to fuel animosity between town and country. This grew as New Zealand became more obviously urban. Miles Fairburn has argued that this animosity was based on an underlying desire for Arcadia – a safe orderly haven – that farming areas could no longer accommodate as the population grew, and that towns had demonstrated they were unfit to meet. Hence an underlying ambivalence about the town, in a land dependent on primary exports and sustained by frontier mythologies, was resolved in the twentieth century by the mass development of suburbs.[36] Urban history moved to its own edges.

The suburban town

The growth of suburbs, however, was well under way by the early 1900s. André Siegfried commented of his visit to Auckland in 1899 that

> The town, with its suburbs, contains 67,000 inhabitants … But, a common phenomenon at the Antipodes, the suburbs, to the number of a dozen, are almost as important as the city itself, having nearly 33,000 inhabitants … The suburbs of Auckland are renowned for their beauty. The town is confined into a narrow space; but on the hillsides the villas stretch at their ease, surrounded by luxuriant vegetation, and looking down on the magnificent roadstead …

To Cowan this beauty seemed to be entirely imposed: he wrote of the city that 'What she lacks in the seven lamps of architecture she more than makes up in the charms of public gardens … of streets which are leafy avenues, cool and fresh with their rows of spreading oaks and ash and elms and tall colonnades of whispering poplars'. Beatrice Webb, visiting from Britain in 1898, saw it differently, as have many later writers: 'unlovely bungalows … small and uninteresting' was her assessment.[37]

Home ownership reached high levels early, exceeding 50 per cent of households by 1916. Sales advertisements for suburb and section development are useful representations of the 'meaning system' underlying this drive for property. Such advertisements were selling a 'good', in both senses of the word, and were designed to enable consumers to envisage a suburban future.[38] Certain themes recur. First there are those relating to Arcadian environmental futures, identifying characteristics less readily available in the town. Sections in suburban Christchurch were frequently described in the 1920s and 1930s as receiving 'All the Sun that Shines!', as well as being well drained and possessing good gardening soil. In Dunedin promotion of new suburban subdivisions often focused overtly on health and the excellence of the view. Andersons Bay in 1910 was the 'Sunshine Township', and Calton Hill was promised in 1923 to be 'SUNNY! CHEERFUL!! HEALTHY!!!' Sun, of course, symbolised optimism as well.

A second group of themes emphasised access to urban infrastructures: on the edge but never isolated. Tramline extensions were a common selling point, as well as sewer, gas and electricity connections. In Dunedin proximity to schools and churches was highlighted. Suburban expansion was therefore promoted by drawing attention to the

Figure 13.4 'Salubrious St Clair'. Auction notice for suburban sections, 1913. Source: Hocken Collections, Uare Taoka o Hākena, University of Otago

ways in which it shared the advantages of both country and town. A third group of themes, usually interwoven with the others, centred on the benefits of investment. So, just as it was claimed in 1913 that 'THOUSANDS OF LIVES [are] made HAPPY and HEALTHFUL through residence in such a district as SALUBRIOUS ST CLAIR!', so prospective owners were assured: 'THOUSANDS OF POUNDS have been made by … far-seeing investors in ST CLAIR PROPERTY!' (Figure 13.4). Similarly, Mosgiel in 1927, given its 'Nearness to the City, is bound to grow, and is a delightful place in which to live. WATCH IT GROW!! REAP YOUR PROFITS!!'

The belief in the right to acquire and own property was institutionalised early in New Zealand. State lending support for home purchase extends back to 1894, and was greatly augmented from the 1920s.[39] In debates in Parliament, the role of property and good housing in integrating people into the social structure, promoting citizenship and diminishing threats of disorder featured frequently. The orderly suburbs in which family life would be underwritten were removed from the social as well as environmental dangers of both frontier and town. The first Labour government's state housing scheme opened similar opportunities to those unable otherwise to afford them, so 'democratis[ing] the middle class suburban life-style'.[40]

The lifestyle being sought (in the 'typical residential street' described by the 1931 guide to Auckland: 'garden-surrounded houses, tidy hedges, shady trees, broad footpaths, and well kept roads') was based on what the American environmental historian Samuel Hays has identified as 'beauty, health and permanence', or enjoyment of amenity, freedom from pollution, security and social stability.[41] Houses were built with balconies and sunrooms, with sections spacious enough for vegetables and play space as well as shrubs and flowers. The 1920s saw rapid suburban growth as the heyday of the Californian bungalow in Auckland (Figure 13.3). Auckland was said to be 'capable of almost unlimited expansion, as both unoccupied and partially developed lands are awaiting the necessary population and capital to bring them into full production'. Promoters of Penrose claimed in 1946 that 'in 25 years, this rural wilderness has been transformed into an intensely developed manufacturing area' by 'young and lusty industries'.[42]

Auckland adopted a regional transport plan in the mid-1950s, focused on motorways, and drawing from American design knowledge, a further sign of the international style of urban modernism popular in New Zealand. Despite some local enthusiasm for combining roads with local rail investment, central government was not prepared to fund rail – a position it long maintained.[43] The outline of the motorway network was in place by the mid-1960s, along with the Auckland Harbour Bridge. These capital projects furthered suburban growth that was also fuelled by large-scale state housing development in the South Auckland districts of Mangere and Otara. What was occurring there was also happening on the edge of other cities and towns. In part, state housing areas were used to accommodate Māori people who were moving from the countryside in great numbers in the 1950s and 1960s. For example, many Māori from rural East Cape and Hawke's Bay moved to the new forestry towns of the North Island interior and to Rotorua, as well as to Wellington and Christchurch.

It was the expansion of suburban Wellington into the Hutt Valley, where in the 1940s the Naenae state housing scheme caused controversy because of the loss of market gardens, that raised another matter: the environmental price of suburban development per se. In the early 1970s New Zealand's report to the first international environmental conference of government representatives in Stockholm described the national loss of good farming land to urbanisation as being 'most acute around Auckland',[44] turning round the progressive interpretation of suburban expansion of 40 years before. Such anxieties had already led in the 1950s to the implementation of town and country planning legislation, designed to encourage more efficient use of existing urban space and to give some protection to agricultural lands.

The regulatory role of the state, at national and local levels, thus became increasingly important in moulding urban development, although often its lack of investment in environmental care schemes left the contemporary suburban city with a range of problems, including traffic congestion in the Auckland metropolitan area in particular (no longer the marker of the modern that it was in 1960s Wellington); sewage disposal (the favoured long-term solution to Auckland's post-World War II water supply problems being a pipeline from the Waikato River into which towns such as Taupo and Hamilton still discharged their barely treated sewage); and growing exposure to hazards such as flooding. Analysis of two urban floods, at Invercargill and Paeroa in the early 1980s, showed that they were so expensive in damage terms because so much development had been permitted to occur in flood-prone areas by the respective local authorities since the 1950s.[45]

On the beach

Using the metaphor 'on the edge', this chapter has set out to construct a framework for analysis of New Zealand's urban environmental histories. Towns were created on the edge of wilderness, centres of civilisation for the reclamation of nature. Functionally and symbolically, they came to represent progress expressed as the conquest of nature. But nature has not been tamed, and towns remain on its edge, subject to environmental hazards of their own making. These hazards have contributed to New Zealanders' ambivalence towards the urban, which has in part been resolved by creating a landscape on the edge that is neither urban nor rural, but suburban. In conclusion, these three themes are drawn together in a short case study that focuses on a coastal town.

The town of Timaru was created in the 1850s to open up the district of South Canterbury. Its port exported wool from a landscape where previously (according to the progressive narrative of a 1926 guide) 'the ti-tree, tussock and flax clump rustled in the breeze, unheeded and unmolested'.[46] By the early years of the twentieth century the prosperity of the town was expressed in its public and commercial Edwardian architecture. Trade had been encouraged by construction of a breakwater, begun in 1878 to shelter shipping and much extended thereafter. Simultaneously, however, this breakwater interrupted the northward drift of sediment along the coast. The result has been the creation of both a hazard, as beach barriers to the north have been starved of gravel, thereby threatening low-lying coastal land, and a resource, as fine sand settled in the calm waters in the lee of the harbour to build a golden beach. Much of Timaru's twentieth-century history focused on this fortuitously produced beach in Caroline Bay.

Elsewhere 'the beach' has been described as a liminal zone, a place on the edge in yet a further sense, where behavioural norms of what is – and with whom it is – legitimate can be relaxed to provide pleasures for people otherwise bound by convention.[47] Yet this was not the image constructed for Caroline Bay. It was promoted as modern (for example, having in the 1920s 'the longest chain of [electric] lights on any waterfront in the Southern Hemisphere'),[48] as family-oriented and as 'sunny and safe' (Figure 13.5). And it was readily reached by rail. The interwar growth of family holidays was driven not so much by private motoring as by regular and excursion train services. New Zealand Railways (NZR) responded to the prospect of the car by encouraging demand for domestic holiday travel. Timaru was an ideal destination for jaded town dwellers from Christchurch and Dunedin.[49] Far from being risqué, its name was synonymous with that deemed healthy and non-threatening. Caroline Bay was dedicated to the reproduction, rather than the contest, of suburban values. And it was something of which to be immensely proud:

Figure 13.5 'Timaru by the Sea': promotional image from the 1930s. Source: South Canterbury Museum, Timaru

New York has Coney Island and Timaru has Caroline Bay. He who will not admit to the comparison is not a good citizen; for Timaru's Caroline Bay means just as much to Timaru as does Coney Island to New York … In very few towns in New Zealand are there such safe bathing places. Caroline Bay is safer than Takapuna at Auckland, than Island Bay at Wellington, than Brighton, than Sumner, than St Clair, than Lyall Bay, than Milford Beach.[50]

Continued accumulation of sand in front of the breakwater, however, has built the beach outwards to such an extent that the seaside facilities of the bay are now isolated a kilometre or so from the water's edge. Combined with changing holiday fashions and modes of transport, which have taken people away to adventure destinations by car as well as overseas by air, the town has been undermined both physically and economically. In a further irony, a replacement source of prosperity for Timaru – its industrial area of Washdyke to the north of the bay – is at risk of inundation as the barrier beach that protects it continues to shrink because of the loss of gravel trapped behind the breakwater. Timaru's environmental history has therefore been 'on the edge' in all three senses – progressive, suburban and vulnerable – that this chapter has employed to interpret the topic in New Zealand between the 1900s and the 1960s.

Further reading

Faces of the progressive town are represented in Dominic Alessio, 'Travel, Tourism and Booster Literature: New Zealand's Towns and Cities at the Turn of the Twentieth Century' (*Studies in Travel Writing*, vol. 14, no. 4, 2010, pp. 383–96) and Julia Gatley (ed.), *Long Live the Modern: New Zealand's New Architecture, 1904–84* (Auckland University Press, Auckland, 2008). The vulnerable town is discussed from different perspectives by Eric Pawson, 'Environmental Hazards and Natural Disasters' (*New Zealand Geographer*, vol. 67, no. 3, 2011, pp. 143–47) and Pamela Wood, *Dirt: Filth and Decay in a New World Arcadia* (Auckland University Press, Auckland, 2005). Work on the suburban town includes Ben Schrader, *We Call It Home: A History of New Zealand State Housing* (Reed Publishing, Auckland, 2005), while the Timaru case study is extended by Neill Atkinson in '"Call of the Beaches": Rail Travel and the Democratisation of Holidays in Interwar New Zealand' (*Journal of Transport History*, vol. 33, no. 1, 2012, pp. 1–20).

14.

The empire of the rhododendron: reorienting New Zealand garden history

James Beattie[1]

MANY OF THE PLANTS we appreciate and claim as our own, or think of as being from Europe, in fact originated in Asia. Rhododendrons, chrysanthemums, camellias, peonies, narcissi, and trees such as ginkgo or the many varieties of maples: all of them are Asian and many came to New Zealand in the nineteenth century. This chapter examines the complex pathways through which such Asian – and specifically Chinese – plants were introduced, using these as a window into both the history of gardens and garden-making in New Zealand, and the networks of exchange that underpinned such plant movements. Since most studies of botanical introductions to New Zealand have highlighted those from the Americas, Europe and Australia, focusing on Chinese plants presents new perspectives on colonial introductions and garden-making. The chapter begins by setting garden-making and plant collecting in New Zealand within its wider context, before presenting a series of case studies. It concludes by examining Chinese market gardens – and garden culture – as points of intersection among Chinese, European settlers and Māori since Chinese gardeners provided late-colonial settlements with the bulk of their fresh vegetables.

Concepts of plant distribution

In considering how plants reached New Zealand, scholars have emphasised how people transferred them from one place to another, often multiple times. Along the way, people also altered, or improved, them. Gardeners sought to create brighter and more attractive flowers by crossing them with other plants. Agronomists (agricultural improvers) attempted to increase the yield of edible varieties by lengthening their growing times or experimenting with new fertilisers. Tracing histories of plant exchange requires locating the networks of individuals and institutions as well as the technologies and markets that made such movement possible.[2]

In doing so, geographers and historians have studied the cultural meanings ascribed to plant species and varieties while at the same time recognising the role of biology and environment in shaping plant introductions. Jodi Frawley, for example, has explored the changing understandings of the mango after it reached Australia, revealing how cultural

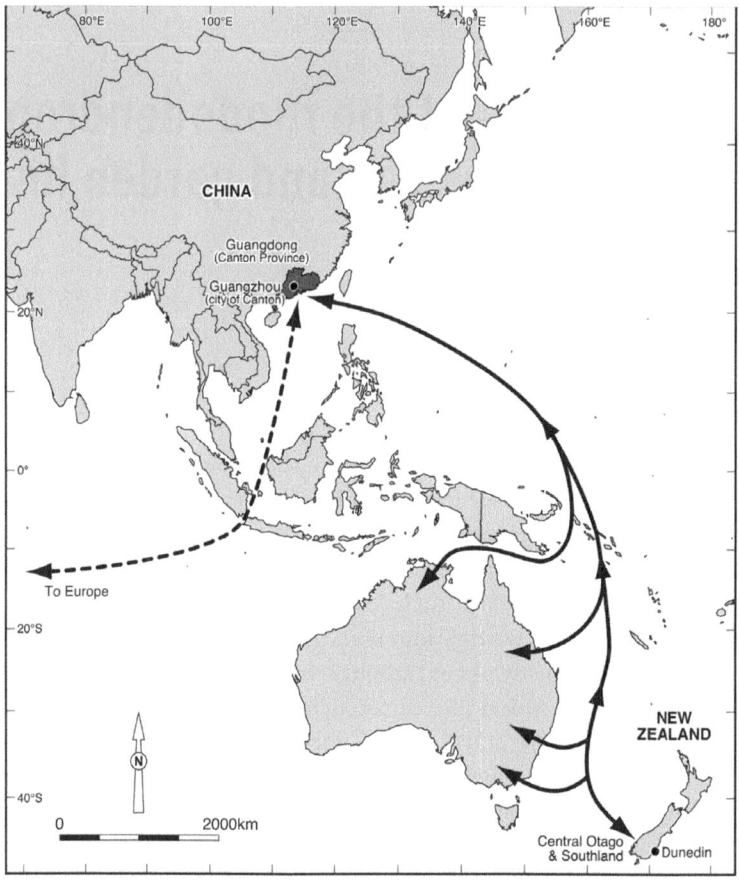

Figure 14.1 Map showing Cantonese networks of migration, information and plant exchange with New Zealand. Source: based on James Ng, *Windows on a Chinese Past: How the Cantonese Goldseekers and Their Heirs Settled in New Zealand*, Otago Heritage Books, Dunedin, 1993, p. 87

and even scientific descriptions of the plant changed in the process of its introduction. Through examination of patterns of plant dissemination, the role that different centres of exchange played at different times in facilitating introductions can be identified. Vaughan Wood and Eric Pawson have demonstrated how, between the 1880s and 1920s, an essential part the world cocksfoot-seed industry centred on Akaroa, Canterbury, thanks to its particular role in the development of a sought-after variety of the grass. Using the metaphor of a web to capture such patterns of plant distribution, Pawson has shown how examination of actors and institutions can help untangle the complex movement of plants from one place to another.[3]

This more recent scholarship is helping to modify the influential research of the historian Alfred Crosby. His groundbreaking book of 1986, *Ecological Imperialism: The Biological Expansion of Europe 900–1900*, forced a reconsideration of the process of colonisation by stressing its ecological dimensions. It made scholars consider how plants, pathogens and animals brought from Europe advanced imperialism by providing settlers with flora and fauna that, thanks to the conditions created by ecological change such as deforestation, often appeared to out-compete native species. Complementing the introduction of plants and animals, Crosby showed how new diseases severely weakened indigenous societies. Crosby's one-way model of plant transfers from Europe to New Zealand, which formed a central case study in the book, has been subjected to close scrutiny. Scholars are revealing the more complex pathways by which plants travelled the globe.[4]

This chapter's main focus on Chinese plant introductions into New Zealand and the transportation, financial and botanical networks that facilitated them complicates Crosby's thesis in several ways. First, it reveals that the same transportation networks bringing plants native to Europe and North America into New Zealand also carried numbers of other flora, including many from China. It was not simply a matter of European biota displacing those already present in New Zealand. Second, patterns of Chinese plant introductions took far more complex forms than Crosby envisaged, resembling perhaps more of a spaghetti junction of congested roadways than an orderly one-way street leading from Europe directly to New Zealand (Figure 14.1).

The chapter also considers the role of Cantonese (from the province of Canton – now known as Guangdong – in southern coastal China) as introducers of plants into New Zealand, problematising presentations of Europeans as the main agents of environmental transformation in the colony. Case studies of Chinese in New Zealand reveal individuals whose own networks were grafted onto, but were also sometimes separate from, those of the British Empire. Seen in the light of Chinese plant introductions, Crosby's memorable phrase that Britain's was the 'empire of the dandelion' might well be changed to 'empire of the camellia' or 'empire of the rhododendron'. Crosby's thesis about the role of plants in facilitating imperialism remains valid, but it needs to be expanded to encompass non-European flora and people as well.

Early New Zealand plant networks

On the eve of European colonisation of New Zealand, plants from the Americas, South Africa, the Antipodes and China were all the rage in Britain. '[N]ovelty as respects flowers is now a complete mania,' observed Joseph Paxton in the 1830s. This enthusiasm for new plants and for gardening, especially among Britain's middle classes, was cultivated through the publication of gardening journals, the soaring sales of commercial plant nurseries, the running of flower shows and the establishment of new gardening organisations such as the Royal Horticultural Society (1804). In an immensely competitive market for flowers, new introductions fetched high prices in Britain, but only for as long as supply remained low. For example, customers of England's Sunningdale Nursery paid 105 shillings per plant for the Chinese plant *Mahonia japonica* when it first appeared in 1854–55. In 1857, its price fell to 5 shillings and by 1862 the trade cost for *Mahonia* seedlings was '4s. per hundred'.[5]

Such introductions were testimony to Britain's growing commercial – and military – power. But in the case of China, British demand for plant material from the 'Celestial Empire' was severely limited. Under the terms of the Canton trading system (1757–1842), restrictions imposed on travel by all non-Chinese meant that British plant collectors, unable to go beyond the confines of Canton, had to focus on botanical material obtained in that city. This limited their collecting largely to plants from coastal and southern China. From 1842 Western powers began to carve out spheres of influence along China's coast, exacting trade concessions and payments, and obtaining extraterritoriality for their citizens. This process of 'informal imperialism' gave the British access to areas of China previously denied them, and thus access to plant material from northern and southwestern parts of the country.[6]

If New Zealand's participation in the global economy partially overlapped with the end of the first phase of British botanical interaction in China up to 1842, its colonisation neatly aligned with the second (see Chapter 4). Growing wealth and improved transportation networks benefited British and imperial collectors of Chinese plants and their hybrids. New Zealand's relatively late formal colonisation also meant that it received introductions of many Chinese plants and hybridised varieties via Britain and other places. The migration of Chinese to New Zealand from the mid-1860s, many of whom arrived with the goldrushes, further opened up new networks of plant transfer, new species and new gardening ideas.

One of the key early sources of Asian plants in Britain was the English East India Company (EIC). Drawing on its years of trading in the region, it began to cater to the demand for exotic species in Britain. It was encouraged to do so by people like Sir Joseph Banks (1743–1820), Director of the Royal Botanic Gardens, Kew. Kew Gardens cultivated and distributed many of the newly introduced species received from all around the world. Banks, along with commercial nurseries and private collectors, actively encouraged botanically inclined EIC employees to collect plants from all of the places reached by the company's vessels, from New Zealand and Australia, India and China, and the coasts of Africa. As a result, he developed an extensive network of European collectors around the world.[7]

Transporting a plant from one side of the world to the other was neither easy nor cheap. For example, the EIC surgeon and plant collector John Livingstone, writing in 1819, gloomily listed all that could (and very often did) go wrong in transporting live plants from China: from lack of adequate preparation to saltwater poisoning, neglect and the sinking of vessels. He estimated 'that one thousand plants have been lost, for one, which survived the voyage to England'; 'every plant [from China] now in England must have been introduced at the enormous expense of upwards of £300'.[8] By the time of New Zealand's colonisation, survival rates of live plants were increasing. Steamships reduced voyage time. The use of the Wardian case, effectively a mini-glasshouse, further aided live plant transportation, but it nonetheless remained a chancy affair.[9]

New Zealand's appreciation of botanical novelties from China (and elsewhere) relied on existing and emerging British and colonial botanical networks. Commercial nurseries, public botanical gardens and individual collectors in Sydney, Hobart and, later, Melbourne sent plants to New Zealand. Australian suppliers, in turn, obtained Chinese plants and hybrids from England, or through Calcutta Botanical Garden and sometimes directly from China. In the 1860s, reflecting the establishment of colonial towns and the rising wealth and security of its settlers, many more commercial nurseries opened in New Zealand to cater to evolving public gardening tastes. A greater variety of plants also came, thanks to quicker, more frequent and more reliable transportation networks.[10] The picture that emerges is of an increasingly complex series of botanical exchanges, centering on botanical institutions, private gardening societies, government agencies and, more often than not, commercial nurseries and individual collectors.

Gardening cultures: home, public and show gardens

Gardens in New Zealand, like those in Britain, fulfilled a variety of intersecting agendas, but there were also some important differences. In a new colony, gardening and acclimatisation appealed at once to biblical injunctions to improve and restock the earth with useful plants and animals – particularly important considerations in a place that was perceived to be lacking in useful plants. In the early years of settlement, the introduction of familiar edible plants also kept starvation at bay. At the same time as reminding settlers of home and of families left far away, economically valuable plants like wheat established the beginnings of an export economy in settler and Māori society; and, in the case of the potato, revolutionised Māori diet and lifestyles.

Botanical and acclimatisation gardens served as experimental sites for establishing the properties of native and introduced plants, providing everything from basic information about whether particular introduced plants would survive in New Zealand, down to their likely medicinal properties or usefulness for fencing and other purposes. Botanical gardens served also as centres of knowledge transfer and training – educative functions complemented, and sometimes compromised, by their role as sites of leisure. Public gardens, parks and domains catered to the expectations of settlers for freely accessible spaces in which to meet, read and exercise.

Compared to public gardens created by governments, councils and other organisations, private gardens made by settlers ranged from the very modest, like those established beside whalers' shacks, to the very grand, whose layout and rare plants symbolised the refinement and sophistication of their owners. Nineteenth-century beliefs in the environmental causes of disease also led colonists to regard the ecological functions of plants with approval. As green lungs, settlers believed certain plants drew from the air the unhealthy miasmas emanating from decaying animal and vegetable matter, in addition to helping to drain unsightly and unhealthy swamps.[11]

Garden-making performed important social and cultural functions. For Europeans, gardens were a marker distinguishing a civilised society from a barbaric one. Garden-making required high levels of societal organisation and, most importantly, the existence of a settled lifestyle. According to the European hierarchy of societal development at the time, just as these factors had laid the foundation for its own commercial development, so they would for other societies. Many colonists marvelled at the sophistication of Māori horticulture. Its absence, however, indicated to European eyes wasteland – land not being gardened – and invited purchase, confiscation and environmental transformation. Indeed, the metaphor of a garden – usually the Garden of Eden – provided a powerful motivation for colonisation, and for the ways Europeans sought to transform the landscape. Gardening also was a gendered, racialised and class-divided occupation. It offered an appropriate outdoors pursuit for wealthy women, but also, as for men, a personal realm of interaction with like-minded people. In Māori society, women traditionally tended gardens and harvests, and many Māori women worked as gardeners on Pākehā estates.[12]

Most gardeners grew edible and non-edible plants. In the first few years of 'pioneering', European gardeners generally favoured growing fruit and vegetables over ornamentals. In times of intermittent food supplies when landscape transformation through cutting down forest and laying down pasture was emphasised, only very wealthy gardeners could afford to grow ornamentals and rare flowers in large numbers. The few European ornamental gardens that did develop before the middle of the nineteenth century often served important symbolic, religious and political functions. Productive and ornamental missionary gardens provided means of educating Māori in European ways – or so the missionaries hoped (Chapter 3). Formal gardens and plant collections established by British officialdom signified civilisation, order and authority, even if their form did not translate into function.[13]

For example, the garden established between the 1830s and 1860s by the settler Thomas McDonnell (1788–1864) in the Far North at Horeke, in the Hokianga, spoke at once to the political and class pretensions of its owner and to the important symbolic functions of rare Asian plants that underlay them (Figure 14.2). McDonnell, a former opium trader, had strong connections with both India and China, regularly running drug and other cargoes between Calcutta and Canton. As a result of these networks, as well as exchanges with Kew Gardens and with visiting ships, he introduced several Asian plants, such as the hill cherry (*Prunus serrulata*) from China; it was one among several species 'probably in cultivation in New Zealand decades before its introduction into Europe'. Archival and botanical records reveal others, including the evergreen shrub *Cotoneaster franchetii*, quince (*Cydonia oblonga*), peach (*Prunus persica*) and loquat (*Eriobotrya japonica*).[14]

By the time McDonnell's garden was falling into ruin in the early 1860s, the trickle of European settlers into New Zealand was reaching a steady stream in some parts of the colony. From 1861 the goldrushes increased migration, bringing great wealth to the southern province of Otago, and growing numbers of people (Chapter 6). By the 1870s large-scale assisted migration greatly increased the European population. In 1840, New Zealand had about 100,000 Māori and 2000 non-Māori inhabitants. By 1851 the non-Māori population had reached over 26,000, rising to a quarter of a million by 1871 and three-quarters of a million by the turn of the century. While colonial towns mushroomed, land ownership and farming remained the real sources of power and influence throughout this period. Engaging in competitive consumption, wealthy landowners expressed their social pretensions through the construction of mansions and

Figure 14.2 Sketch of Horeke, 1840, pencil and ink. Note the extensive fenced gardens and buildings of McDonnell's property. Source: 'A view of the feast given by the Governor to the natives at the Huarake Hokianga Capt Macdonalds station Horeke [13 February 1840]'. Richard Taylor, 1805–1873: Sketchbook, 1835–1860, E-296-q-169-3, Alexander Turnbull Library collections, National Library of New Zealand Te Puna Mātauranga o Aotearoa

the establishment of extensive gardens stocked with rare plants, including many from China and India.[15]

In 1854, the former EIC judge Sir John Cracroft Wilson (1808–1881) arrived in New Zealand from Calcutta via Sydney. Onboard he had 'a menagerie for the public good', as a friend described it, along with many Indian seeds and plants.[16] These included 'Three Ward's cases of Scarlet Rhododendron', 'One Dozen cases of Bamboos', and 'Various kinds of seeds'. Dr Hugh Falconer (1808–1865), Director of the Calcutta Botanical Garden, had supplied Wilson with several different species of bamboo before his departure for New Zealand. The bamboos from Falconer reached Christchurch alive, but all were subsequently killed by frost despite being placed under shelter of the house veranda. Some of the other bamboos did not even make it that far. A fracture in the glass of the Wardian case killed '[t]he small hill bamboo (Nilgala)'.[17] These losses plunged Wilson into despair.

On his arrival he had given large amounts of the seed of the large hill bamboo to the well-known Christchurch gardener and nurseryman William Wilson (1819–1897). 'I kept them with the greatest care in my writing desk,' Wilson wrote dejectedly, but '[n]ot one seed germinated, and the loss of this plant, which would be perfectly invaluable in New Zealand, caused me more vexation than all my other losses in the horticultural line'. Contrasting his disappointment over the bamboo, the three young Himalayan rhododendrons he introduced were not only alive but 'flourishing under the care of' a friend of Wilson's, 'in a garden adjoining the house of Mr T.T. Brown in Christchurch'. Of the garden seeds and creepers that he distributed, several also fared well.[18]

In colonial New Zealand, Indian rhododendrons and azaleas proved immensely popular. Eagerly collected by private settlers, provincial botanical gardens and commercial nurseries, they were obtained from a variety of sources: from individuals, commercial nurseries and botanical gardens in Australia, Britain, North America and India. The Wellington settlers Thomas and Selina Drake, for example, received rhododendrons and azaleas directly from their uncles in Darjeeling, plants that somehow managed to survive one year in transit.[19]

Indian plants were also cultivated by commercial nurserymen and acclimatisation societies. Dunedin nurseryman William Martin (1823–1905) introduced the first rhododendrons and azaleas into Otago: by about 1880, he listed 19 varieties of Indian rhododendron. Of these, he was particularly proud of the yellow blooming *R. falconeri*, a bush spreading 'seven or eight feet high and fully as wide'. Every November when they flowered, Martin's 'Azalea Walk, a long curved path margined with azaleas and backed by the fine rhododendrons, eight feet or more tall', attracted as many as 500 visitors a day. Martin also earned himself a worldwide reputation for his hybridising of rhododendrons, the best example being 'Marquis of Lothian' (*R. falconeri spp. falconeri*), which in turn was sold overseas. As the *Otago Witness* observed in 1893, the appeal of rhododendrons and azaleas, as well as other plants, is readily explained when considering that 'India has perhaps a greater variety of plants than any other country in the world, having 15,000 native species, while the flora of the entire continent of Europe only embraces about 10,000'.[20]

Settlers also appreciated the colour and vibrancy of many of the newly introduced Chinese plants, often placing them alongside others from around the world. In 1841, Thomas (1818–1903) and Jane Mason (1818–1900) settled at Taita (Upper Hutt), north of Wellington; and they also leased or bought land elsewhere in the North Island. At Taita they developed a fine garden known as 'The Gums'. By the 1870s it contained about 8000 plants and 3 tons of bulbs, including a goodly number of harder-to-obtain Chinese species. In 1871, Thomas recorded 60 kinds 'of Camellias, besides Azaleas and Rhododendrons, all of which flourish here. Some of the Camellias are upwards of seven feet and have been in bloom for five months, and some are still to flower.' Camellias, first introduced into Britain in 1739, flourished in its cooler climate, as they would in parts of New Zealand. Newly discovered camellias, azaleas and rhododendrons continued to be introduced into Britain throughout the nineteenth century, many by Robert Fortune (1812–1880), who also brought in many varieties of Chinese tree peony. Mason seemed especially proud of his recent procurement of 'some 30 varieties of the Chinese Tree Peony (*Paeonia Moutan*)'. They 'seem quite at home [in The Gums] and are remarkable for their growth and beauty'. Without knowing their precise botanical nomenclature, it is impossible to pinpoint their origins. Botanical surveys of the site of Mason's garden, however, have revealed several other Asian species. These include *Sophora japonica* 'Pendula', the 'Chinese Scholar Tree', which, according to Winsome Shepherd, is 'the only old specimen known in New Zealand'.[21]

Figure 14.3 *Alfred Ludlam's House and Garden, 1850*, watercolour. It has a glasshouse and incorporates both native and introduced plants. Source: Francis Dillon Bell, 'The house of Alfred Ludlam Esq., River Hutt, New Zealand, 1850', accession 26,856, Hocken Collections, Uare Taoka o Hākena, University of Otago

Mason's plants came from friends, and from commercial nurseries in Britain and Australia – again demonstrating New Zealand's reliance on British networks with China. In 1841, Mason obtained 'a few Rose Tree seeds' from his uncle in York, England. 'Rose Trees' were rhododendrons; but which species and from where is unknown. In 1870 he obtained plants from Thomas Lang's famous Ballarat nursery in Victoria. As Lang himself specialised in collecting rare plants, especially from East Asia, this could well have provided a further source of Mason's Chinese plants. Mason also obtained tree peonies from the nursery of James Backhouse & Son, near York. In common with other gardens, the Masons laid out Chinese plants alongside other exotics from around the world.[22]

Mason's contemporaries Fanny (1822–1877) and Alfred Ludlam (1810–1877) developed a fine garden at 'Newry', Waiwhetū (Hutt Valley), further indicating the many plants available to settlers of wealth and discernment (Figure 14.3). Their garden preferences were outlined in Alfred's 'Essay on the Cultivation and Acclimatization of Trees and Plants', which sought to encourage 'the introduction and cultivation of trees and plants of the more rare and beautiful kinds' – activities which the Ludlams had spent several years pursuing. In particular, they promoted the growth of coniferae over other more commonly grown trees such as gums, wattles and poplars. East Asian varieties dominated Ludlam's list. For example, among the 86 plants listed under 'trees and shrubs', 47 per cent (40) came from East Asia. Overall, one quarter of all the plants listed by Ludlam were East Asian. Not all of these, he thought, would survive in New Zealand. For example, in contrast to the Chinese *Bambusa arundinacea*, 'The Black Bamboo, from the Himalaya Mountains is a much hardier variety', ideally suited 'for ornamental shelter'. Likewise, he praised *Wisteria sinensis* as 'beautiful' and 'very showy', covered 'in early spring … with bunches of Lilac colored, pea-shaped flowers'. 'Once established,' he observed, 'it grows with great vigour.' He also grew about 25 varieties of the hardy *Azalea indica* and about 50 varieties of *Camellia japonica*.[23]

Both these Hutt Valley gardens demonstrate the popularity of Chinese ornamentals among prosperous settlers. For the less wealthy living in the towns and rural areas of New Zealand, garden-making also represented civility and a respectable leisure pursuit. Reflective of these attitudes, horticultural and acclimatisation societies enjoyed immense popularity throughout the country.[24] Alongside their emphasis on useful knowledge, these societies encouraged introductions of new vegetables and flowers. Their activities and the plants offered for sale by nurseries give an insight into the plant preferences of settlers of more modest means.

Organised flower shows flaunted colonial development, not to mention the availability of new and existing introductions, and sometimes the networks underpinning them. Keith Stewart's history of roses, for example, demonstrates the rapid take-up of new varieties into the colony. Although roses are found over a wide geographical range, the new breeds from Asia became, he observes, 'the raw materials for a succession of new roses' developed in colonial New Zealand. Chinese varieties, such as Slater's Crimson China rose, when crossed with Damask rose, produced 'the large and influential Hybrid Perpetual family of roses, which was, after Centifolias, the second of the great modern rose families and particularly effective in the rose colonisation of New Zealand'. Stewart also speculates that the listing in 1854 by New Plymouth settler Benjamin Wells of 'Sanguinea Rose', some 33 years before its appearance in Britain, may have resulted from a direct introduction from China.[25]

Although Australian and British nurseries often supplied New Zealand settlers directly, and continued to do so, more nurseries began to open in New Zealand in the latter decades of the nineteenth century, offering ever greater varieties of plants. For example, the seven tea-scented roses listed by David Hay's Montpellier nursery in 1867 had reached 143 by 1891–92; the catalogue likewise ballooned from 16 pages in 1863 to 65 by 1867. Demand was high: 'nurseryman and seedsman must also do their utmost to keep pace with the times,' noted the *Otago Witness* in 1904. 'Were they laggard for even a year, their task to make up lost time would be a difficult one.'[26]

As new varieties flooded into New Zealand, so flower gardening enjoyed ever more popularity. In 1873, for example, the Canterbury Acclimatisation Society obtained from Archdeacon Davies via the Royal Society's Gardens in Hobart Town several Chinese plants as part of a plant exchange. The Canterbury Acclimatisation Society received Chinese bamboo, ginger, the Chinese palm (*Chamaerops fortunei*) and Chinese tree peony. Commercial nurseries sold large numbers of azaleas, camellias and hydrangeas, as well as other East Asian and foreign varieties. Horticultural and, later, chrysanthemum societies devoted special sections to Chinese and Japanese varieties, doing much to popularise them. As well as chrysanthemums, Chinese primulas were popular indoor plants from the 1880s.[27]

Most of these Chinese plants would have been obtained from sources in Britain, Australia and the United States. But some direct introductions from China also continued into the twentieth century through China–New Zealand missionary connections, botanical friendships and the migration of botanical-loving Europeans from the Chinese treaty ports. For example, E.H. Wilson (1876–1930), one of the next generation of plant hunters after Fortune, visited New Zealand in 1922, sending David Tannock (1873–1952), Dunedin's Superintendent of Reserves, 'two small plants of Taiwania [sic], the great conifer from Formosa [Taiwan]'. By this time, a vogue for rock gardens had taken hold among some gardeners, leading to introductions of East Asian alpines alongside New Zealand ones. From the 1900s to the 1930s, too, Japanese-inspired gardens, stocked with Japanese and New Zealand plants, were appearing in many places.[28]

Cantonese networks, plants and gardening techniques

Just as Asian plant varieties were changing New Zealand's landscape, so too were Chinese gardeners. As Mrs Williams' 1882 migrant guide noted:

> If you are near a town, possibly an itinerant Chinese gardener may come round to your door … in which case you would get vegetables more easily than you can rear them in the first one or two years of your inexperienced residence here … The Chinese surpass others in rearing vegetables cheaply and successfully for the market; and at the same time, set a good example to their neighbours by rising at 3 and 4 a.m.[29]

They came to work Otago's goldfields, but by the 1870s and 1880s many had turned to labouring, market gardening or laundering. In getting to New Zealand, the Cantonese relied on their own extensive migration and business networks. These networks provided financial, organisational and emotional support, facilitating everything from travel and accommodation to the carrying of letters and remittance money. Just as lineage and native place associations tied together family, clan and county in China, so these structures operated beyond China's territorial boundaries. Cantonese in New Zealand worked mining claims in clan groups, set up fruit stores along clan lines, and maintained family life despite being geographically dispersed. In environmental terms, these networks facilitated introductions into New Zealand of Chinese vegetables, flowers and agricultural techniques; the transfers were kept up by ongoing connections between Canton, Australia and New Zealand.

Most Chinese goldminers cultivated a small patch of ground to supplement their diet of tinned and dried foods (Figure 14.4). Private gardens supplied Cantonese tables rather than European ones. Transferring seeds along their own migrant networks, Cantonese introduced and grew species familiar to them. Alexander Don (1857–1934), Presbyterian missionary to New Zealand's Chinese, records eating '"white cabbage", grown of seed from China' in Bannockburn. His tour diaries – extending over decades – mention many

Figure 14.4 Unidentified Chinese man and Rev. Alexander Don outside a dwelling in Waikaia, c. 1900. Despite the state of the hut, the garden is neat and well tended. Source: McNeur Collection, 1/2-019146-F, Alexander Turnbull Library collections, National Library of New Zealand Te Puna Mātauranga o Aotearoa

Chinese introductions into New Zealand: bok choy, Chinese cabbage, and other popular edible plants such as spring onions, Chinese sugarpeas and kohlrabi and, less commonly, coriander, white radish and Chinese chives, as well as boxthorn and chrysanthemum greens. As James Ng notes, the Cantonese commonly grew 'Chinese cabbage or "bak toi" (literally "white stemmed vegetable"), and "guy toi" ("bamboo mustard cabbage")'.[30]

By the late 1870s, with easily accessible gold running out, market gardening assumed greater importance for the Chinese. In many late nineteenth-century New Zealand settlements, Chinese market gardens and vegetable hawkers were commonplace; as subsequently were Chinese greengrocers. In 1881, the *Otago Witness* complained of the near-monopoly of Chinese on Dunedin's vegetable trade; and a similar sentiment was echoed in Auckland in 1887. Wellingtonians, observed parliamentarian J. Duthie in 1896, 'are almost solely dependent on them for vegetables … but [for] these industrious Chinamen the people would generally go short'.[31]

The nature and size of Chinese market gardens and orchards varied over time and by region. New Zealand's Chinese population was initially concentrated in Otago and Westland (Figure 14.5). Most of the small towns in these regions had a market garden of one size or another; some had several. In 1889, Dunedin could claim 110 Chinese market gardeners, most (80) in South Dunedin. In addition to growing produce for Dunedin's residents, many Chinese market gardeners probably also supplied visiting ships. Towards

Figure 14.5 Well-known market gardeners Ah Sam and Joe Quin with vegetables, in Roxburgh, 1903. Source: Box 18, PAColl-7581-96, Alexander Turnbull Library collections, National Library of New Zealand Te Puna Mātauranga o Aotearoa

the turn of the century, market gardens were shifting to the main centres and into the North Island. J.F.E. Wright, farmer and former provincial councillor, admired how in only three and a half years, four Chinese gardeners had transformed 'a sparsely-grassed paddock' in Owhiro Valley (Wellington) into a 'beautiful market-garden, supplying fresh vegetables to the inhabitants of Wellington'. Some gardens were long-lived; others not so. For example, Chan Dah Chee (Chan Ah Chee, 1851–1930) established a succession of market gardens in and around Auckland from the 1880s until 1920. In contrast, Ah Sing lost his 3-acre Donnelly Creek (near Ross, West Coast) garden to a flood after less than three years.[32]

Chinese market gardeners grew vegetables for Europeans using Cantonese gardening techniques. Small-sized land holdings in South China, coupled with high population densities, had made them highly skilled at coaxing large harvests from small areas. They successfully adapted these techniques to the very different climates and soils of the goldmining areas of Australia, California and Otago. Typically, Chinese gardeners planted out vegetables in tight rows, fertilising them with manure and other organic waste. A series of articles published in the Otago newspapers in the 1870s held up Cantonese market gardening techniques in Australia as a model for emulation by Europeans in Otago.[33] Such techniques, coupled with their cheap labour and hard work, laid the basis for Cantonese commercial success.

Gardening success increased interactions between Chinese and Europeans – for good and ill. Some Chinese employed Europeans and Māori: for example, Owhiro Valley Chinese employed Europeans to dig, and to transport 'manure to their garden'. But success also drew criticism. Racially discriminatory legislation entered the statute books in 1881. Economic hardship, political expediency, nascent nationalism and scientific racism accelerated anti-Chinese sentiment. In 1887, Wellington's John Merbett attacked Chinese use of human manure and the practice of what he termed 'artificially' forcing vegetables 'through the ground'. (Ironically, contemporary New Zealand garden guides recommended the use of human manure in gardens.[34]) Many others cited the perceived unfairness of Chinese competition with Europeans through the former's more efficient cultivation methods and longer work hours. Several anti-Chinese leagues also sprang up.

Other critics labelled Chinese market gardens 'centres of fever and death-dealing miasma', as one correspondent melodramatically described those in Whanganui. Prosecutions against Chinese market gardeners followed on health grounds in several cases, but not in all. For example, amidst fears of Chinese gardens spreading typhoid, the *Auckland Star* 'affirmed that the Chinese cannot make the plants they raise become the agents of disease by using any ordinary or known manure'. Likewise, after inspecting an unnamed Chinese market garden, Wellington's Medical Officer of Health pronounced its yards to contain 'no offensive accumulations or deposits. The conditions found did not justify any action being taken under the city's health bylaws.' Concerns over racial health and national fitness also spilled over into fears of miscegenation between Māori women and Chinese (and Indian) men, culminating in a 1929 report prompted by worries over the employment of Māori women and children in Chinese-operated market gardens.[35]

Despite deep inequalities, neither the agency of the Chinese nor their colonial supporters should be overlooked. Colonial interactions with the Cantonese can perhaps best be considered as lying along a continuum: at one extreme, outright racism and violence; at the other, occasionally cordial relations; and, in between, cool curiosity. Although framed paternalistically, some colonists defended Chinese against racist outbursts, especially when sentiment against the Chinese intensified from the 1880s. One correspondent, for example, failed 'to see the force of the arguments' Merbett presented against Chinese gardeners as they 'are a law abiding and an industrious class'. The perceived virtues of the Chinese, just as much as their perceived vices, were understood through their garden-making. In a society valuing thrift, hard work and sobriety, few could deny the Chinese these virtues. Detractors consequently levelled other criticisms: that Chinese morally and physically polluted colonial values and people; or that Chinese work practices threatened the interests of the emerging labour movement.[36]

Just as gardening generated tensions among Chinese, European and Māori, so too could appreciation of plants offer avenues of mutual interest and interaction, as they had earlier in Canton between some Europeans and Cantonese. At the 1871 Royal Horticultural Society of Otago Show, Wong Koo received special mention for 'some specimens of Chinese Narcissus' and gained second prize for his 'exhibits of lilies and feather ornaments'. If the 'Chinese Narcissus' mentioned were *Narcissus tazetta* var.

chinensis (Chinese sacred lily or daffodil), then these could well have been introduced from China. Evidence suggests it was not otherwise commonly available in New Zealand until after 1891. As a bulb, the lily could be easily transported from China and stored for several months. Although not native to China, it had been in cultivation there for over 1000 years and was well integrated into its gardening culture. Several other Chinese gardeners entered colonial horticultural competitions. Tinwald's Chinese market gardeners scooped up five prizes in the vegetable section at Ashburton's 1905 Horticultural Society Show.[37]

In addition to market gardening, Chinese worked as agricultural labourers and gardeners on European-owned stations. For example, Otama Station, Riversdale (Southland) employed 'an experienced and industrious Chinese gardener' who received praise for the variety, appearance and abundance of the vegetables he raised.[38] A handful of individual Chinese also succeeded as entrepreneurs at the height of legislative and social racism. Ah Chee, mentioned earlier, initially earned a livelihood selling vegetables in 1870s Auckland. By 1882 he had established with Ah Sec the 3-acre 'Garden of Prosperity', near Parnell. The garden lived up to its name admirably. Profits from it supported the purchase or lease of further market gardens, and diversification into other enterprises: shops on Auckland's Queen Street, restaurants, ginger and banana plantations in Fiji, and sheep and poultry farms in New Zealand. In 1905, he purchased a 14-hectare market garden at Avondale; he also leased several other gardens.

Characteristic of many successful Chinese merchants, Ah Chee sponsored other Chinese workers, and relied on networks of kith and kin to run his businesses. For example, Ah Chee's nephew, Sai Louie, oversaw his uncle's Fijian business interests, while the English- and Chinese-language skills of his wife, Joong Chew Lee, proved equally indispensable to the enterprise. Business success translated into acceptance in European society. The Ah Chees had servants, and regularly visited China. Their market garden thus served as a conduit for ongoing networks of trade with China, and as a basis for material and social success in New Zealand and China.[39]

Another profitable sideline for Ah Chee was the collection and sale of edible tree fungus, *Auricularia polytricha*. This was pioneered by John Ah Tong, but Chew Chong (c. 1830–1920) developed the industry commercially. Having identified the fungus on his travels through New Zealand, Chew Chong started buying it in large quantities in 1868. At a time when colonists were struggling to make a living as they removed the North Island forest, the money to be earned collecting the edible fungus proved to be a lifeline. In 1870 Chew Chong opened a business in New Plymouth, with branches in Inglewood and Eltham. After buying the fungus from Māori and settlers, he would then send it to Dunedin for shipment, mostly to China but also to Sydney's Chinese community. Although Chinese authorities did not keep early excise records, those from 1872 to 1904 reveal imports of edible fungus valued at £375,000; from 1880 to 1920 New Zealand fungus exports totalled £401,551 (Figure 14.6). As a result of his success in this enterprise and in pioneering the dairy industry, Chew Chong became a member of the Taranaki Chamber of Commerce. He married Elizabeth Whatton in 1875. He was also well regarded in Māori society: in 1879 he was awarded the title of regional chief by

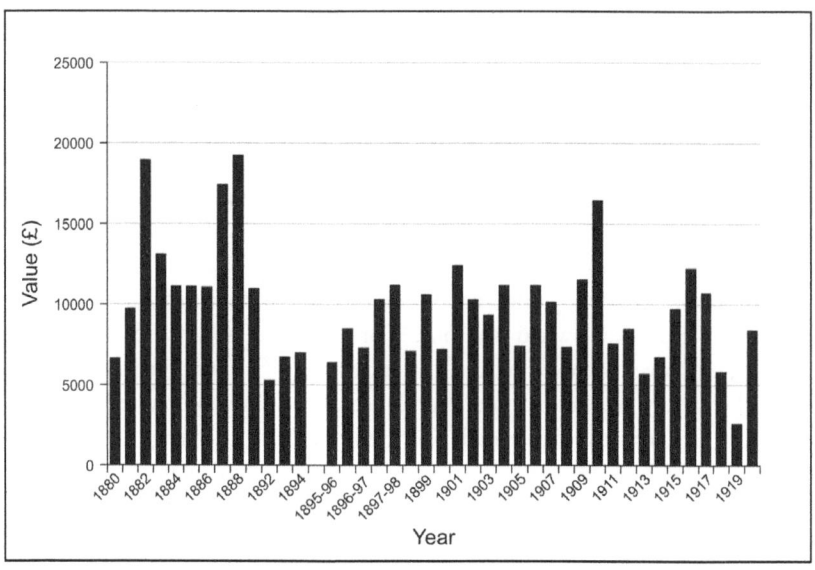

Figure 14.6 Value of New Zealand exports of fungus. Source: Import, Export, and Shipping Returns, AJHR, 1880–1920

the United Tribes of New Zealand. A contemporary newspaper noted he 'well merited the high esteem in which he was held by all classes because of his high principles and generous instincts'.[40]

Ah Chee and Chew Chong's successes might well have been exceptions to the norm, but they and other market gardeners nonetheless illustrate the ongoing importance of transnational Chinese networks in introducing garden practices and plants into New Zealand. In these networks, the imperial centre (Britain) was largely absent; more important were colonial ports (especially Hong Kong), and Cantonese entrepôts and transportation networks. The remittances carried by these networks also contributed funds towards Canton's own environmental transformation – a topic that, along with introductions of New Zealand plants into Canton by Chinese and settlers, merits further investigation.[41]

Conclusion

Whether brightening settler gardens or declaring colonists' social aspirations, whether providing the basis for new hybrids or advancing garden societies, whether growing the nursery trade or providing a leisure pursuit, Chinese ornamentals have played a crucial role in these islands' gardening culture, economy and social life. Examining their history complicates presentations of New Zealand's environmental transformation as having been undertaken solely with European and North American plants. Their introduction by settlers and nurseries also underlines the importance of non-governmental agents and networks in driving environmental change. And the role of Chinese as market gardeners, plant acclimatisers and agents of landscape change adds a new dimension to interpretations that assume the overwhelming significance of Europeans in these respects – findings that add to emerging research stressing Māori transformation of nature in this period.

What was the impact of Chinese market gardening in New Zealand? Chinese garden culture exposed New Zealand to new plant networks, new gardening techniques and new plants. Chinese market gardening opened up a variety of new experiences for colonists, Māori and Chinese alike. Garden-making offered an opportunity at once to judge and to criticise Chinese, but also to praise and provide employment, share garden cultures and buy produce. Examining these Chinese garden histories and their networks can contribute significantly to a reorientation of New Zealand's environmental and garden history, helping thereby to explain many of the origins of the plants growing in our gardens today and why several of our cities are graced by Chinese scholar gardens built to commemorate the Chinese market gardeners of the nineteenth century.[42]

Further reading

The themes in this chapter are explored in more detail in James Beattie, J.M. Heinzen & John P. Adam, 'Japanese Gardens and Plants in New Zealand, 1850–1950: Transculturation and Transmission' (*Studies in the History of Gardens & Designed Landscapes*, vol. 28, no. 2, 2008, pp. 219–36), Matt Morris, 'A History of Christchurch Home Gardening from Colonisation to the Queen's Visit of Christchurch' (PhD thesis, University of Canterbury, 2006) and several articles in a special issue of *Studies in the History of Gardens & Designed Landscapes* (vol. 31, no. 2, 2011), including James Beattie, 'Making Home, Making Identity: Asian Garden Making in New Zealand, 1850s–1930s' (pp. 139–59) and Duncan Campbell, 'Transplanted Gardens: Aspects of the Design of the Garden of Beneficence, Wellington, New Zealand' (pp. 160–66). The role of botanic gardens is discussed by Franklin Ginn, 'Colonial Transformations: Nature, Progress and Science in the Christchurch Botanic Gardens' (*New Zealand Geographer*, vol. 65, no. 1, 2009, pp. 35–47). Lily Lee & Ruth Lam, *Sons of the Soil: Chinese Market Gardeners in New Zealand* (Dominion Federation of NZ Chinese Commercial Growers, Pukekohe, 2012) is a detailed study of its topic.

V
Perspectives

15.

Postcolonial environments

Katie Pickles[1]

THIS CHAPTER CHARTS new terrain for the examination of postcolonialism and the New Zealand environment. Postcolonialism is a broad term with many – often contradictory – meanings, that is liberally applied to various regions of the world. Despite its now tropic use, postcolonialism is particularly important as a framework when considering New Zealand's environmental history. While internationally cultural approaches that focus on people often dominate discussion, in New Zealand a strong contemporary context of redress for Māori land loss, combined with the social, political, cultural and economic implications of treaty claims, means that the environment is necessarily central in postcolonial scholarship.[2] Adding a layer to this environmental emphasis is New Zealand's identity as a place where settlers from Europe occupy a position as both colonisers and the colonised, trying to make a 'Better Britain' out of the very ideologies that they seek to improve on. Yet, as the Hawke's Bay and Canterbury earthquakes have shown, geological uncertainty also renders the landscapes of colonialism vulnerable and compromised.

The chapter employs the cultural approach of sensory history to develop new ways of thinking about and researching the environmental colonial past. It also builds on cultural geography's tradition of reading landscapes as text. A critique of dominant colonial ways of seeing the environment has long existed. Much environmental history from the mid-twentieth century focused on re-evaluating colonial attitudes, and looked back on the colonial past through the framework of the earth being transformed for progress.[3] More recently Eric Pawson has explicitly referred to 'postcolonial' as 'a continuous engagement with the lasting effects of colonialism'.[4] This is a particularly apt observation for New Zealand, where the interaction between people and place is ever changing. Environmental historians have been well aware of the limits of colonial attitudes and the hazards that these have posed. Such understandings of environmental history have led to eerie predictions. As Tom Brooking wrote in 2004, 'Strict building codes and clever engineering initiatives have helped reduce the risks of earthquake damage, but a large event is inevitable, especially in Wellington, which like San Francisco is sited on a fault line'.[5]

The next section of this chapter offers an overview of the new area of sensory history, considering its uses for the examination of postcolonial environments. It is followed by two cases studies, which explore environmental events as sensed and felt experiences. The first discusses Canterbury's nor'west wind as a sensory and transcultural part of environmental history. The second turns to the Hawke's Bay and Canterbury earthquakes. The chapter concludes by suggesting that postcolonial moments can disrupt previously comfortable understandings of people's places and environments.

Sensory history

An interest in the role of the senses in history and society is gaining momentum. Such scholarship is replacing older paradigms of explanation and is challenging conventional theories of representation. New approaches enable recognition that all people have senses, and hence those previously excluded and marginalised from history, particularly along gender, race and class lines, become the subjects of study, and on new, 'embodied' terms. Mark M. Smith argues that sensory history is 'a habit of thinking about the past, an engrained way of exploring not just the role of sight in the past but the other senses, too'.[6] Interested in culture and the everyday, Elizabeth A. Foyster and Christopher A. Whatley argue that 'we neglect the senses to our peril, since the senses were a crucial part of the everyday'.[7] Their perspective is typical of the social and cultural emphasis that now dominates the historical literature.[8]

The new field of sensory history offers a path for advancing scholarship in postcolonial approaches to the environment. One major way is the interrogation and questioning of Western knowledge systems, as influenced by poststructural theorists. In particular, it draws on Michel Foucault's contribution of deconstructing the evolution of these knowledge systems. Foucault argued that in the epistemology of the late sixteenth-century West, 'resemblance' was how people acquired knowledge. Associating objects if one seemed like another 'organized the play of symbols, made possible knowledge of things visible and invisible, and controlled the art of representing them'.[9] Such understanding was replaced by the dominant principle that geographer Robert D. Sack has termed 'action by contact'. Suggesting that in a scientific age people lost touch with their senses, he argued that in social science action must occur by contact, noting that 'It is the way in which space and its properties can be seen to affect things.'[10] Sack argued that it is when we reject the principle of action by contact that the mythical-magical mode is re-entered. Sack agrees that now hidden ways of knowing once had enormous impact, not only in the conduct of social relations, but on the use of the land and the environment.[11]

Historian Keith Thomas has examined people's relationship with the environment, noting that people in the sixteenth and seventeenth centuries did not formulate a distinction between magic and medicine, let alone a distinction between magic and science.[12] In England at the time, magic and the supernatural aided those suffering disease and disaster. With the growing search for scientific explanation of the universe, magic declined. Thomas charts the rise of a rational age of science and progress where a belief in Mother Nature lost out to human control. Then, as now, there was a tension between ascendancy over nature versus partnership with it, a theme taken up in postcolonial scholarship.

Given their primary concern with the relationship between people and place, it is unsurprising that geographers have a longstanding interest in the senses. For example, in his 1990 monograph, *Landscapes of the Mind*, Douglas Porteous urged that the modern world not take the senses for granted.[13] Porteous's work considered the nonvisual senses in environmental perception. He broached the intangible relationships between humans and their environment (attachment, aesthetics, ethics and spirituality). By the turn of the century, approaches redolent of an earlier cultural geography were gathering momentum in other disciplines, while within geography there was a new interest in emotions and affect.[14] Internationally, there has been much work by geographers that questions the construction of 'nature' itself, as well as the place of humans and culture in relation to it.[15]

Sensory history, then, is poised to reconsider environmental beliefs and practices in the past, through its breaking down of the division between people and environment, and working at the very interface of that relationship. It also offers promise to move beyond Eurocentric mindsets and engage in partnership, both between people of different cultures and with the environment itself. This is a potentially rich area for development in New Zealand where Māori epistemologies are important. In seeking explanations for environmental disasters, and the sense of being at the mercy of the earth, colonial mindsets of progress and domination are fundamentally called into question. For example, explanations for earthquakes are necessarily diverse and transcultural, with the Māori god Rūaumoko, a Pākehā god and Mother Nature all variously apportioned responsibility.

In this respect, sensory history has also been particularly influenced by the feminist embodiment and ecofeminist literatures. Ecofeminism is a combined vision of feminist and ecological perspectives. As defined by Christine Dann, ecofeminism looks simultaneously at the exploitation of women and of the environment.[16] Carolyn Merchant has argued that 'women and nature have an age-old association' and that 'The female earth was central to the organic cosmology that was undermined by the Scientific Revolution and the rise of a market-oriented culture in early modern Europe.'[17] It was this latter tradition that settlers brought to New Zealand. In the American context, Annette Kolodny has critiqued settler discourses of possessing and taming 'frontier' societies.[18] These discourses also frequently consider the earth as female, constructing it as fertile, benign, accommodating and available to be colonised. In contrast, Figure 15.1 portrays a Māori Mother Nature as responsible for the Canterbury earthquakes. It captures the collision of modern secular beliefs with the perpetuation of nature as gendered female, although in this case nature, seen as acting against human interests, is depicted as malevolent.

Drawing on theories of embodiment, Joy Parr usefully argues in her book *Sensing Changes* that 'the body is a synthesizing instrument that defies the categorical and linear discipline of language and science. The environments and technologies with which we live, play and work lead us to develop specific modes of bodily interaction

Figure 15.1: 'Mother Nature', Al Nisbet cartoon. Source: adapted by the cartoonist from one that appeared in the *Press* on 12 July 2011, Fairfax Media/*Press*

and perception.'[19] Introducing Parr's work, Graeme Wynn sums up the importance of her sensory history, writing that she 'urges us to recognize that people make sense of the world through their sensing bodies and that human bodies are conditioned by the circumstances of time and place'.[20]

While the senses are infused throughout Māori knowledge, they are only very quietly present in Pākehā New Zealand scholarship regarding the environment. Yet colonialism involved engaging with the environment in profoundly sensory ways, and it is time to make that history explicit. There is plenty of ground to explore. For example, there is a paradox between the fear of the environment and the real dangers it posed (for example from 'the colonial death' of drowning) and the confidence of colonial development that positioned setters as triumphant, successful pioneers. Rollo Arnold is one author who has – briefly – examined the importance of the senses in the settler mindset. Underpinning his work was an implicit awareness of the importance of the elements, including fire and water, and how settlers had to learn to cope, and ultimately succeed in taming nature. On settlers' engagement with fire through bush burns in the mid-1880s Arnold wrote:

> Usually they filled their lungs with air redolent of the long leagues of the Tasman and the Pacific, but each year the reeking haze of the burning season set eyes smarting and nostrils tingling. By entering our story through the fires we have been able to put the bush where it belongs, as an important element in the colonists' consciousness, both as a major factor in their economy and as a feature unique to their islands, one of the aspects of New Zealand's individuality.[21]

This sensory sort of approach has exciting implications for how we understand the history of the environment – and its postcolonial implications – in New Zealand. These implications will now be explored in two more detailed case studies.

The nor'west wind

Canterbury's nor'west wind provides a case study of a new sensory approach to the environment. As one of the dominant winds of the region, it is a powerful sensory experience that has had considerable capacity to shape the collective personality of Canterbury and Cantabrians. It is a foehn wind that brings heavy rain and snow to the mountains on the West Coast of the South Island, and then blows hot, strong and gusty across the Canterbury plains, creating a visible 'nor'west arch' in the sky. The wind demonstrates the interplay of the physical and human landscape of the region through a sensory medium. It has had a transcultural effect, being important to both Māori and settler traditions. Winds can unite cultures and are an important part of life for all who have lived in Canterbury. While there are sensory interconnections about winds between Māori and Pākehā, the cultural knowledges that explain these can be quite different.

The longest standing knowledge about the nor'west wind is Māori. Here special significance to Māori epistemology must be observed. A view to ngā atua Māori (the Māori gods) and their place in the scheme of these sorts of environmental occurrences is a vital consideration, and Ngāi Tahu scholars have been quick to identify the important part played by the wind. Te Maire Tau writes that 'If we are to understand the Ngāi Tahu relationship with the local landscape, we must start with the elements that act as references to the gods.' Tau suggests that the connection between Ngāi Tahu and the Canterbury landscape 'starts with Raki's first wife, Pokoharua Te Pō – the source of all the winds, incantations and tapu. Thus the origins of the natural world commenced with the wind, or hau – the breath of life. To Māori, "hau" is the "vitality of man" and the "vital essence of the land".'[22]

From Raki's relationship with Pokoharua Te Pō came Uru Te Maha – a name that means 'the source of the westerly winds'. From this came Tāwhirimātea (manifestation of the wind) and eventually Te Mauru, the Ngāi Tahu name for the northwest wind. It was believed that the wind blew from the mountain that dominates the North Canterbury region: Maungatere (Mt Grey). The significance of the northwest wind to Ngāi Tahu was evident in its proverbial name, Te Hau Kai Tangata – 'the wind that devours humankind'. There was an association with unease and death. In a chant delivered before commencing battle, Moko, a Ngāi Tahu ancestor, 'emphasised his connection to the land by reference to Maungatere and told the opposing warrior that he, Moko, was the harbinger of death by reference to the nor'wester'.[23]

As New Zealand settlers made sense of their world through reference to familiar folklore and by inventing new traditions that mimicked old familiar reference points, it is unsurprising that just as the elements were a part of everyday Māori life, so too they became important to settlers.[24] Air was a vital element for settlers who had journeyed across the sea to New Zealand. They were familiar with tales and experiences of other great winds of the world, and made comparisons. Peter Holland's extensive work on settlers and the environment in southern New Zealand records colonial administrator Henry Sewell's description of the Canterbury plains as 'a howling wilderness'.[25] Tau writes that 'Pākehā know the impact that this wind has on the region: their paintings and poetry often illustrate its dramatic effect.'[26] Indeed, there are evocative writings of the effects of the nor'wester. Demonstrating the Pākehā understanding of the wind, at the turn of the twentieth century the *Cyclopedia of New Zealand* retold and transmitted knowledge about 'the most remarkable of Canterbury winds, the nor-wester':

> … no one can speak with authority about New Zealand winds who has not raced the nor-wester as it issues from the Rakaia Gorge or some other narrow outlet of the Dividing Range. Towns on the plains, which, like Ashburton, lie opposite the river gorges are exposed to its full fury … The temperature of the air is oppressively high, and the atmospheric conditions are strangely ELECTRICAL … But after blowing in furious gusts from this direction, the wind generally hauls round to the south-west and the sou'wester, accompanied by heavy rain, as a rule lasts as long as its forerunner – in each case about three days. The rapid changes of temperature and atmospheric conditions, and their mechanical regularity, are two of the most notable features of the Canterbury climate.[27]

It is clear that the nor'west wind made a big physical and cultural imprint on settlers. Writer and poet Jessie McKay (1864–1938) was born at the Rakaia Gorge. James Cowan wrote that 'she felt her land in its every mood'. Of the nor'wester she wrote:

> A tinder earth, a burning blue
> With eyes of Nemesis glaring through,
> Heavy as death and hot as hate!
> Windy brown to the mountain-gate
> Windy brown to meet the sky!
> All the sap of the earth is dry.

But relief comes in the evening, 'the hour between the lights', when the breeze of solace comes down from the Southern Alps.

> … the maidens of the cool
> Vast Eden of the after-glow
> Dream-heavy from the cooling snow,
> Their wings drop comfort as they glide,
> To cure the world at eventide:
> And more – they left the gate ajar
> Of Eden, where their dwellings are;
> For here, unsealing ear and eyes,
> Returns the Wind of Paradise![28]

Due to 'new' social and cultural histories taking a historiographical secular turn in the second half of the twentieth century, the spiritual side of white settler society was, at best, assumed.[29] But in the case of the history of the nor'west wind, sensory history continues in evidence through the work of writers and artists, such as that of Bill Sutton, whose work evokes the embodied experience (Figure 15.2).

As recognised by an early settler Mark Stoddart, the nor'wester is one of the great winds of the world. In his poem about Canterbury, Stoddart wrote:

Figure 15.2 William Sutton, *Hills and Plains, Waikari 1956*, oil on canvas. Source: 89/143, 1956, Christchurch Art Gallery Te Puna o Waiwhetu

I've witnessed all the winds that blow, from Land's End to Barbadoes –
Typhoons, pamperos, hurricanes eke terrible tornadoes.
All these but gentle zephyrs are, which pleasantly go by ye,
To howling, bellowing, horrid gusts which sweep down the Rakaia.[30]

The nor'wester has a deep psychological effect on many people subjected to its strong, hot, dry nature. The association of the nor'wester with unease and death was not confined to Māori. As one former resident of the Waimakariri Gorge put it, 'I absolutely hated them. They make you feel suicidal. They undress you. If your mouth is open the flies go in.'[31] The nor'wester is a furious wind that makes some people furious. In order to escape it, Canterbury farm women move inside, close all windows and draw curtains as the nor'wester approaches.[32] The wind has been statistically linked to increases in suicide and domestic violence. Research by the environmental scientist Neil Cherry indicated that the mechanism is a buildup of positive ions, strongest inland from the coast, and electric fields of this strength have been shown to upset the melatonin/seratonin balance in some people. He wrote that 'About 10 percent of people affected by the nor'wester feel elated and wonderful. But the rest feel depressed, irritable and lacking energy. People feel they can't cope with everyday things … There is irrational anxiety and a sense of foreboding.'[33] The most extreme behaviours that can be associated with the wind are violence and suicide. Juliet Peter's painting captures the dark and brooding elements associated with the nor'wester (Figure 15.3).

Figure 15.3 Juliet Peter, *Nor'west, 1939*, linocut. Source: 83/64, 1939, Christchurch Art Gallery Te Puna o Waiwhetu

The transcultural sense of foreboding that accompanies the nor'west wind is well founded. History reveals that the nor'wester can signal environmental disaster: Canterbury's strongest winds blow from the northwest, and a northwest gale on 1 August 1975 flattened tree plantations and damaged buildings. The 26–27 December 1957 storm that caused severe flooding in the mountains was also a nor'wester. Northwest rain in the mountains causes the Waiau, Hurunui, Waimakariri, Rakaia and Rangitata rivers to flood. Floods in colonial Christchurch were a result of the Waimakariri spilling over its banks.[34] Meanwhile, as the rate of evaporation is high during nor'westers, especially from shallow gravelly soils, droughts result from the winds. A nor'wester that followed the earthquakes in Canterbury after 22 February 2011 whipped up the drying dust from liquefaction and spread it around Christchurch.

The cultural identity arising from the sensory experience of the nor'west wind has been picked up for commercial use, signalling Canterbury characteristics that are connected to the environment. For example, the Nor'wester Café in Amberley calls for patrons to 'join us in the heartland of New Zealand, where we cook, serve, and care'. 'Following our name sake, the famous hot dry wind that has shaped Canterbury, the Nor' Wester has been blowing its patrons away since 1997.' The nor'wester also inspired an award-winning beer, the microbrewery Dux de Lux's 'big thirst quencher' Nor'wester ale: 'As dry winds parch the plains, try our big taste beer – Grand ... It's a STRONG ALE.'[35]

The material history and sensory experience of the nor'wester is writ large in the landscape. Illustrating the connection between culture, colonisation and land use are the shelterbelts that were planted across the Canterbury plains, oriented to protect against the effects of the nor'west wind.[36]

Earthquakes

Earthquakes are a major and often heightened sensory experience that cut to the heart of the relationship between people, culture and environment. Located astride the colliding Pacific and Australian tectonic plates, and part of the 'Pacific ring of fire', earthquakes are a common occurrence in New Zealand. While most such events have occurred in rural areas, the Hawke's Bay and Canterbury earthquakes were so damaging and life-threatening because they affected urban places. The Canterbury earthquakes have broad implications for how environmental history is viewed. In particular, they have necessitated a 'postcolonial moment' where the past is rethought through an acute sense of vulnerability, and a resurgence of the connection between people and their environment.

James Belich has suggested that for New Zealand history, disasters have been fixed in minds, becoming 'a point where private and public histories intersect'.[37] This can be extended to apply to sensory histories. The Hawke's Bay and Canterbury earthquakes are two disasters that loom large in New Zealand imaginations. Amid a global depression, on 3 February 1931 a magnitude 7.8 earthquake hit Hawke's Bay, killing 256 people. The Hawke's Bay region quickly attempted to move on to a new modern day. The contemporary art deco architecture that featured prominently in the reconstruction became an important and celebrated part of Napier's post-quake heritage. Not all buildings were quickly replaced: Napier did eventually rebuild its destroyed Anglican cathedral, which bore a strong Gothic resemblance to Christchurch's Anglican cathedral in the Square, even sharing Benjamin Mountfort as an architect. But it took until 1965 for the replacement Hawke's Bay cathedral to be mostly completed, and it was 2005 before the last three windows were installed. It was executed in a modernist style, and bore little resemblance to its predecessor, instead evoking a Spanish mission in South America.

Despite the different eras, there are similarities between the Hawke's Bay and Canterbury earthquake experiences – from the disasters themselves to the immediate reactions and the processes of rebuilding. The Hawke's Bay earthquake did produce many aftershocks; but less obviously so than the Canterbury earthquakes, which began on 4 September 2010 and continued as a prolonged series of events on separate faultlines over more than two years. Both the Hawke's Bay earthquake and the most damaging Canterbury event of 22 February 2011 occurred during the daytime and in the summer. In both places, people were going about their daily business, to be suddenly assaulted by the strong shaking of the earth. In Napier and Hastings in Hawke's Bay, chimneys fell down, churches crumbled and fires broke out in the centre of town.[38] Unlike Christchurch, Napier's town centre is close to the waterfront, and people fled to the beach. In the mammoth geological upheaval the tide went out on the Ahuriri Lagoon and never returned, stranding sea creatures and, ultimately, providing land for the future airport.[39]

Most of the Hawke's Bay fatalities were in urban areas, as were those in Christchurch 80 years later. As with the collapse of the CTV and PGG buildings in Christchurch, the Napier nurses' home and technical college, and Hastings' Roach's Department Store and Grand Hotel, became death traps. Also as in Christchurch, in downtown Napier and Hastings there were deaths from falling elaborate masonry, in many cases as people ran outside into the streets. In Hawke's Bay there was a rapid response from the military as HMS *Veronica* was in port and the Navy spread news of the disaster, before rushing to assist with rescue and recovery. Reinforcements including doctors, nurses, medical equipment and supplies, tents, shovels and picks were sent by sea from Auckland and arrived the following morning. Eighty years later in Christchurch, the army helped police and international rescue teams, and Australian police flew in to complement national services. As a point of difference, in a global age, professional urban rescue teams arrived from around the world.

The embodied experience was one of horror that became etched in people's minds. The cities shared a sombre experience of the search for survivors and the identification of bodies amid continuing aftershocks. In Hawke's Bay, a temporary morgue was set up in the Courts building; the protocol there was much less developed and culturally advanced than that at Burnham military camp after 22 February 2011. In both events, comparisons were made to both natural disaster and human-made warfare. Of the temporary morgue in the Napier Courthouse an eyewitness account recorded: 'One has seen heaps of mangled bodies after a minehead accident on the Rand and the sights of uncleared battlefields, but there was a pathos and horror of its own about this house of death.'[40] Making the connection between war and natural disasters, the September 2010 and February 2011 Red Cross earthquake appeals, which totalled over $128 million in donations, were referred to as 'the largest New Zealand Red Cross Appeal since the Second World War'.[41] In Hawke's Bay, with many survivors bewildered and in severe shock, a communal interment was held at the Park Island Cemetery on 5 February 1931.[42] Closure and commemoration for Canterbury victims at the Avonhead Park Cemetery interment site took considerably longer.

Electricity, sewerage and chimneys were all down in post-quake Hawke's Bay – as they were in Christchurch. Leaving Christchurch by air was an option for some. Residents initially fled Napier and Hastings and returned later. Most left in cars, but some desperate people fled on foot. Palmerston North was a popular destination: 'survivors from Hastings to Wairoa spent the night outside'. In Havelock North, Bill Ashcroft and his family 'didn't sleep, but sat watching the fires in Hastings'. Mary Hunter, camping with about 30 people on Cobden Road, found the 'dust and smoke from the fires in the town was very disagreeable and made us all filthy'.[43] In Napier, the army erected tents to house more than 2000 people in Nelson Park within two days of the earthquake. And more than 6000 people were evacuated in the following fortnight from the transfer station at Hastings Park, with another 2000 or so travelling independently. An estimated 70,000 people left Christchurch after 22 February, although, as in Hawke's Bay, many subsequently returned. Statistics New Zealand figures at the end of August 2011 showed that 4200 people had permanently left Christchurch since 22 February 2011, compared with 2500 people during the same time period the year before.[44]

The suspension of everyday life happened in both places after the earthquakes. Writer Fiona Farrell captures the physical and mental disjuncture that the Canterbury earthquakes caused in *The Broken Book*.[45] There were reports in the media of shocked and emotionally distraught people struggling to cope with the sequence of earthquakes. In Hawke's Bay, because it was summer and warm, terrified people could stay out of damaged homes and camp outside. Water was delivered daily by truck. Frequent aftershocks rocked the area in the weeks following the major earthquake, with approximately 525 aftershocks recorded in two weeks, the largest a magnitude 7.3 on 13 February. The sensory experience for many in Napier was 'severe shock experienced by most people during the disaster, which left many tearful, hysterical or with a lost, bewildered look'.[46]

Yet things moved quickly to a recovery phase, and in March 1931 the Hawke's Bay Earthquake Act was passed. At the time there was no Earthquake Commission (EQC) to call on. It took until 1944 for the Earthquake and War Damage Commission to be set up in response to the need for effective insurance following the 1942 Wairarapa earthquakes. In Hawke's Bay, in September 1931 Parliament noted a large number of hardship cases where people were unable to pay their taxes after the quakes.[47] The government provided some assistance, but most reconstruction was funded by the private sector and charitable donations. The Hawke's Bay Earthquake Act gave sweeping powers to two appointed commissioners, who were in office until May 1933: John Saxon Barton, an accountant, barrister and magistrate; and engineer Lachlan Bain Campbell.

They placed a moratorium on new building, and first sought to clear away the debris. Fletcher Construction Company created a temporary 'Tin Town' with areas for shops and commerce, while Barton and Campbell, with their committee, sought to work out ways to provide for secure urban reconstruction. The colonial architecture was abandoned; in its place, modern town-planning principles that involved 'perpendicular lines to create an illusion of height' were adopted. The building of a shopping centre at Napier's business heart was prioritised.[48] In January 1933 a week-long carnival celebrated the completion of the 'New Napier' (Figure 15.4).[49] The 'New Napier' poster captures a region looking to the future in a strong, safe, healthy and modern frame of mind.

In a parallel process, the Canterbury Earthquake Recovery Act was passed on 18 April 2011. Soon afterwards, two men were appointed to oversee the recovery effort: Canterbury Earthquake Recovery Minister Gerry Brownlee, and the Head of the Canterbury Earthquake Recovery Authority (CERA), engineer Roger Sutton. As in Napier, the government selected Fletcher Construction Company to lead the Canterbury rebuild. Christchurch's post-quake pop-up shopping mall 'Re:START', constructed of shipping containers, was a metallic temporary measure that evoked Napier's 'Tin Town'.

In both places, royal commissions were held after the earthquakes. There were attempts after the Hawke's Bay disaster to raise standards to prepare New Zealand for future earthquakes. The smaller size of the urban areas in the Napier and Hastings, and the fires that razed downtown Napier, assisted in the rapid rebuilding. And in the early 1930s the colonial era buildings in Hawke's Bay were in a style that was seen at the time

Figure 15.4 The 'New Napier': Napier Carnival poster, 1933. Source: AAOK W3241, Box 1 22, R18837553, Archives New Zealand Te Rua Mahara o te Kāwanatanga

as dated and old-fashioned, rather than – as was the case in Christchurch 80 years later – valued and nostalgic heritage. Ironically, it has taken a postcolonial turn of events to re-value the colonial built environment of Christchurch as never before.

Our postcolonial moment

Overall, the Canterbury earthquakes have necessitated a postcolonial moment for Canterbury, New Zealand and other parts of the world where settlers have transformed and created physical and built environments based on imperial mentalities. On 22 February the statue of the city's founding father, John Robert Godley, Director of the Canterbury Association, was shaken off its pedestal after 143 years at the centre of the city in Cathedral Square, in front of familiar landmark the Christ Church Cathedral (Figure 15.5). A statue of early provincial superintendent William Rolleston fell backwards off its

Figure 15.5 The fallen statue of John Robert Godley, Cathedral Square, Christchurch, 24 February 2011. Source: Fairfax Media/*Press*

plinth to lie beheaded by the forces of nature. British imperial adventurer Captain Robert Falcon Scott's likeness was also knocked over and needed repairing. These damaged statues are symbolic of the end of a colonial era in Christchurch. The city appeared to be breaking free from its colonial past. The 22nd of February was a postcolonial moment – most strongly for Christchurch; but it also set off warning bells at a national level about future-proofing the relationship with the environment. This necessitates questioning previous colonial practices and attitudes, the remnants of which abound in the landscape.

In the apparently formerly peaceful, charming garden city of Christchurch it has taken a major environmental event to clarify the move to a new stage where society is complex, plural and self-determined. A postcolonial moment has not come about through traditional warfare; and it is deeper than grassroots movements: it was instead led unstoppably from under the earth.

The nature of colonial settlement was to start with a blank slate, leaving out the indigenous past. In 1850, Captain Thomas and his assistant Edward Jollie had set about naming the streets with familiar Old World reference points. Firmly cementing Christchurch as a part of the British Empire, the names of streets were taken from bishoprics in England, Ireland, Wales and other colonies. Durham, Armagh, St Asaph, Montreal and Madras are streets whose names promoted colonial ties. Perhaps most symbolic of the all-encompassing imperial stamp was the Anglican cathedral in front of which Godley's statue stood. A long time in the planning, the main building was ready in 1881 and provided the iconic centrepoint for colonial Christchurch.

Surveyors were aware that the swampy nature of coastal Canterbury was unsuited to urban development. The site chosen was the first reasonably dry and slightly elevated area handy to the limits of boat navigation on the Heathcote and Avon rivers. Largely relying on experience of elsewhere, colonists were blind to the pitfalls of 'liquefaction' – the result of soil losing its strength and stiffness during earthquake-induced shaking – that has recently destroyed so much. They optimistically believed that nature could be conquered. And so Christchurch became 'the city on the swamp', with all its attendant problems of ill health, smog and difficulties with drainage.[50] As Eric Pawson has argued, subsequent city planners and engineers have continued rather than challenged previous colonial attitudes, most recently absolving human behaviour from responsibility for the effects of the Canterbury earthquakes and instead firmly advancing 'nature' as at fault.[51]

The Pākehā settlement of Christchurch is now so distant that most people today are probably unaware who launched the city on its colonial pathway. The four avenues that are the boundary of the inner-city disaster zone are named after the city's first four superintendents of the period of provincial government: FitzGerald, Moorhouse, Bealey and Rolleston. Their proud Halswell quarry-stone Provincial Council Chambers was also damaged in the Canterbury earthquakes. Indeed much of Christchurch's visible colonial heart was fractured and became a place associated with death and destruction. Despite nineteenth-century settler experience of earthquakes, the initial, most easily built wooden buildings had been replaced by more permanent, fire-resistant and progressive modern stone and brick structures. Typical of the colonial era, the style of buildings was largely copied from elsewhere. Permanent, proud, mainly stone Gothic Revival buildings became a visual part of placing Christchurch firmly within the British Empire. That architectural style united Christchurch with other British settlements of the same era in Australia, Canada, India and Britain. Its destruction in Christchurch in the earthquakes represents most evocatively the loss of the colonial past.

Very suddenly, the earthquakes have hurried along a process of decolonisation. Many of the features of colonial Christchurch had already been transformed. While Anglican was initially the largest Christian denomination, it was never the sole one. At the beginning of the twenty-first century only 51 per cent of Christchurch citizens identified as Christian, and 35 per cent had no religion.[52] The reality of spiritual belief was increasing diversity and a move away from Christianity. The Christ Church Cathedral existed for many as a hollow icon with good branding potential – a tropic image to be placed on a letterhead, or stamped on a wheelie bin. It called out 'Christchurch' in all its multiplicity of meanings, promising everything and nothing at once.

Economic restructuring had seen the artisan roots of the city, and its key role as an agricultural service town, transform and diversify. Before the Canterbury earthquakes the central city had become a tourism site, with the area within the four avenues a colonial theme park featuring the Gothic Revival architecture, statues, city plan and gardens. Meanwhile American-style suburban drift occurred during the second half of the twentieth century. Roads became increasingly clogged with cars, and shopping malls went from strength to strength. The sea and rail transportation so vital in the colonial era gave way to an extensive roading network, and the airport became a pivotal hub.

Culturally, as the city grew, it harboured a unique mixture of conservative, radical, global and local identities. Although most people still identified as European, the percentage declined, and more ethnic groups became part of Christchurch. Asian migration has dramatically changed the city, fostering a regional Asia–Pacific identity. Māori migration from the North Island to Christchurch in the post-World War II years, and Pacific Island migration population, have added much; and the current cry of 'Kia kaha' ('Be strong') is evidence of the hope of a bicultural society. The impact of the earthquakes on suburban Christchurch has made it clear that, akin to other postcolonial cities such as Vancouver and Melbourne, the city does have distinct ethnic suburban enclaves. A special issue of the *New Zealand Geographer* in 2009, edited by Julie Cupples and Kevin Glynn, explores some of the city's 'counter-cartographies'.[53]

Most symbolically, the felled Christchurch works of art recall images of statues of former leaders, such as Saddam Hussein and Josef Stalin, being eagerly and emotionally pulled down by societies erasing the old and advancing the new. Some former colonies of the British Empire had removed statues that evoke colonisation. In India, Queen Victoria statues are in storage; and in 1963, Quebec's Front de Libération du Québec (FLQ) blew the head off a Queen Victoria statue, making clear its desire to break with the past. With the head still detached, the statue has been displayed in museums as a political relic from the past. Following suit, in 2013 Godley and Scott were displayed as part of a 'Quake City' exhibition run by the Canterbury Museum in the centre of Christchurch.

Conclusion

If the Hawke's Bay earthquakes saw the advance of a modern age, indicators of a postcolonial moment have appeared in Canterbury in many embodied sensory reactions. Particularly interesting is how feelings and emotions have been 'written on the body', with people expressing their feelings directly by placing a tattoo of the Christ Church Cathedral on their body, or draping a memory bead of a cracked house with a broken chimney as a macabre lucky charm around their wrist. As in previous historical eras, resemblance is alive and well, with expressions borrowing from and combining religious, secular and high and low culture. Images from the landscape are often fused in a number of media: gold, silver, cotton, wool and wood. Items include socks that feature an earthquake line on them, and a tea towel featuring the resurrected Christ Church Cathedral on an Edmonds baking-powder tin with the slogan 'Sure to rise' and the words 'Christchurch, 22 February 2011, kia kaha and (munted)'. The days of repressing the senses and denying engagement with the local environment are gone. Rather, the senses are recognised as vital in recovering from the earthquakes.

Due to seismic events, the landscapes of colonialism in Canterbury are vulnerable and damaged. Ways of living in the environment, and the basic relationship between people and place, have been challenged. Just as the physical landscape and built environment of Christchurch have fundamentally changed, so too must the attitudes of people adjust to remake a place that is situated in its locale, rather than one that adheres to an imperial mindset. The challenge of building postcolonial environments has started.

Further reading

Mark M. Smith, *Sensory History* (Berg, Oxford and New York, 2007) is a key introductory work, and Joy Parr's *Sensing Changes: Technologies, Environments and the Everyday, 1953–2003* (UBC Press, Vancouver, 2009) uses a sensory approach to environmental history. The varied postcolonial historical approaches being taken in New Zealand are identified in Giselle Byrnes & Catharine Coleborne, 'Editorial Introduction: The Utility and Futility of "the Nation" in Histories of Aotearoa New Zealand' (*New Zealand Journal of History*, vol. 45, no. 1, 2011, pp. 1–14). James Beattie's 'Recent Themes in the Environmental History of the British Empire' (*History Compass*, vol. 10, no. 2, 2012, pp. 129–39) places New Zealand in the wider imperial context. A picture of colonial Christchurch waiting to be disrupted is portrayed by the editors and other contributors in Julie Cupples & Kevin Glynn (eds), 'Counter-cartographies: New (Zealand) Cultural Studies/Geographies and the City' (special issue of the *New Zealand Geographer*, vol. 65, no. 1, 2009, pp. 1–89).

16.
An updated history of New Zealand environmental law

Nicola Wheen[1]

NEW ZEALAND ENVIRONMENTAL LAW has always enabled resource development to promote social and economic growth.[2] Despite significant changes since the 1960s, the law remains very accommodating in this respect. Only in the conservation estate are environment and conservation interests legally paramount, appearing in the general body of environmental law in terms and contexts that allow them to be compromised and avoided. Although important recent legislation promotes the high, well-meaning principle of sustainability, this is then defined and applied so loosely that it loses impact and provides only a limited check against development and resource exploitation.

By the time English common law was introduced in the nineteenth century, Māori environmental management had assumed a strong underlying conservation ethic. Concepts like whānaungatanga, arohatanga, manaakitanga, mauri and utu directed human interactions with the environment. Straddling both the physical and metaphysical, this indigenous law was enforced by mechanisms such as tapu, mana and rāhui. The courts' misguided view that '[o]n the foundation of this colony, the Aborigines were found without ... any settled system of law'[3] allowed English common law to take over. Environmental law became one branch of the law of property and focused on defining entitlements to resources according to land ownership.

The legislative takeover

Fuelled by the pace of European settlement, legal change in the form of statutory encroachments into the common law came quickly. Initially these were fragmented and overlapping, reflecting local government patterns of the time, and the prevailing '... tinkering, and engineering approach to "resource" management'.[4] With time, many have been integrated into fewer, more comprehensive pieces of legislation. The aim of such encroachments was seldom to conserve, but rather to establish 'a legal and administrative order within which property rights could be allocated, transferred, and protected, [and] resources could be exploited'.[5]

Legislation enabled access to resources such as gold and other minerals (see Chapter 6), geothermal energy, fossil fuels and water. Usually it did so by assuming Crown ownership or control of resources, and then authorising allocations of new rights to users on application. Further, public bodies were empowered to use resources to attend to the needs of settlements for supply of water, provision of roads and bridges, sewage disposal, electricity and so on. Agricultural interests were protected by legislation authorising the construction of works to control floods and erosion, and restricting or preventing the introduction of plant and animal pests.

Virtually no recognition or protection was given to Māori customary rights over natural resources, or to Treaty of Waitangi promises that these rights were guaranteed.[6] When provision was made to protect the quality of natural resources, this was only where the health, safety or economic interests of the settlers were at stake. For example, the early Marine Boards, Marine, and Harbours Acts controlled and prohibited discharges of ballast into harbours, but until 1950 this was only to prevent damage to shipping; the Thermal Springs Districts Act 1881 was made not to protect the 'Hot Lakes District, in some respects one of the most wonderful regions in the world', but rather to prevent it 'from falling into the hands of private capitalists who might monopolise its advantages for selfish purposes, and at the same time throwing it open to European influence and energy';[7] and the steps taken in 1953 to protect groundwater quality were for domestic, farming and industrial uses. The primary object of these provisions was not nature conservation.

Where access to resources was already open under the common law, legislative intervention took the form of restrictions and controls to prevent overexploitation. For example, restrictions on the time, place and means of fishing, or on the size or weight of the fish, could be imposed to protect parts of the sea fisheries resource. Little was done, however, to protect forests. Even when, in 1949, the notion of 'balanced use' was introduced to direct state forest management, this was expressed as one principle among many, with the emphasis on production, water and soil conservation, scientific objectives, and recreation and amenity values.

Some steps towards nature conservation were taken in early legislation, but these were relatively small (see Chapter 8) and the emphasis was undeniably on recreational and scenic (that is, human-centred) values. Late nineteenth-century moves to set aside reserves, domains or national parks so as to provide scenic, historic, nature and recreation areas for public benefit and enjoyment were fragmented. They involved a series of Acts empowering various public authorities. Sometimes the law provided that the reserve, domain or park should be managed in order to protect the animals and plants living there, or the quality of water in rivers, streams or lakes flowing through or on the land.

Plants, animals and water were also protected, piecemeal, in other Acts. The Oil in Territorial Waters Act 1926 and the Oil in Navigable Waters Act 1965 purported specifically to control oil pollution in waters within New Zealand's territorial limits. The Waters Pollution Act 1953 had more general application but was practically ineffective. The animals and birds at first protected by legislation were the game species, but since 1953 all wildlife not specifically declared to be game or unprotected has been protected from hunting.[8] The Native Plants Protection Act 1934 made it an offence to take, without authorisation, protected native plants from public reserves, Crown land, roads or private land. The order that nominates protected plants may be invalid, so the Act has hardly been enforced and anyway its penalties are derisory.

Both the main development-oriented body of environmental law and the more fragmented collection of conservation legislation have been subjected to a great deal of change in the last 40 years or so. The law's provision for environment and conservation has expanded, at first on an ad hoc basis in response to particular problems, but later in more comprehensive and integrated ways. This reflects the growing politicisation of the environment, which was kickstarted by the Manapouri controversy.

The Manapouri controversy

After World War II, with state development priorities shifting towards industrialisation, the need for cheap power became more insistent.[9] The government embarked on a 30-year hydroelectricity programme, and dams were built at Tongariro, in the Mackenzie Basin and on the Waikato River. Lake Manapouri – arguably the most beautiful of New Zealand's southern lakes – and its neighbour, Te Anau – the South Island's largest lake – had first been earmarked for hydroelectric development in 1903. Then in September 1959 a proposal to develop these lakes' hydroelectric potential to supply power for aluminium production was announced. To that end the National government entered into an agreement with an Australian company, Consolidated Zinc Proprietary. Within a year, however, the company had cried too poor to build the power scheme.

Undeterred, the government secured the enactment of the Manapouri–Te Anau Development Act 1963; the water rights reverted to the Crown, enabling it to develop the hydro resource itself. An underground power station of a scale unprecedented in the southern hemisphere was constructed. Water from the lakes was diverted from its

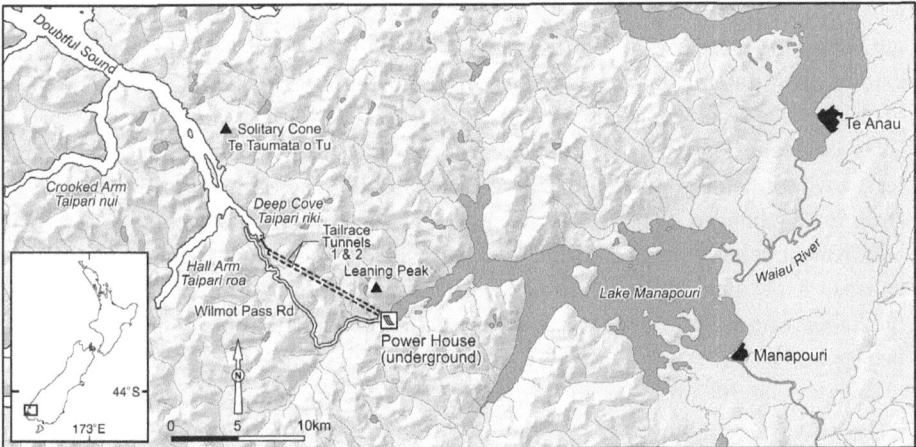

Figure 16.1 Map of the location of Lake Manapouri, showing the power scheme and tailrace tunnels. The second, parallel tunnel opened in 2002. Source: adapted from *Discover Manapouri*, Meridian Energy, n.d.

Figure 16.2 'The view stinks but it's worth millions!' Sid Scales cartoon, *Otago Daily Times*, originally published 22 October 1959; republished 26 February 1970. Source: Hocken Collections, Uare Taoka o Hākena, University of Otago

natural course into the Waiau River, down vertical penstocks onto turbines and then out through a tailrace tunnel to a new outfall at Deep Cove on the Fiordland coast (Figure 16.1). Power output would be maximised by raising lake levels, and a rise of 8–10 metres was proposed.

Manapouri's turbines first turned in 1969 – and only then was the need for some kind of environmental impact assessment realised. The issues focused on the likely effects of raising lake levels on wildlife habitats, on the Waiau River and, especially, on the lake's shoreline (Figure 16.2). Public feeling was mobilised in the country's first large-scale environmental campaign, to 'Save Manapouri'.[10] The campaign peaked when the government changed in 1972 and the new Minister of Energy instructed the Electricity Department to keep the lakes within their natural levels. The new Labour Prime Minister, Norman Kirk, then approved terms of reference for the appointment of Guardians for the Lakes, who were to devise and supervise the implementation of guidelines for lake management.

Six leading members of the Save Manapouri campaign were appointed as the first guardians, beginning a new relationship between conservation and development interests that endures today. These steps were formalised with the amendment of the Manapouri–Te Anau Development Act in 1981. The provisions for raising the lake levels were repealed and a new provision for guidelines to protect ecological stability and recreational values of the shorelines (at the same time as optimising energy production) was inserted. Guidelines were devised and officially published in the *New Zealand Gazette* on 3 December 1981. One of the guardians, botanist and conservationist Sir Alan Mark, argues that the campaign was a 'milestone in the transition from the pioneering era of resource exploitation to one aimed at integrating conservation with development, and associated with the sustainable management of our natural resources'.[11]

The road from Manapouri: environmental gains

The importance of Manapouri should not be underestimated but must be placed in context. The 1960s was a period of rapid growth in levels of public concern for environment and conservation throughout the Western world. The Manapouri campaign was a product, as much as anything, of its time. With the 1981 amendment, it clearly had a direct bearing on New Zealand environmental law. The government also responded to post-Manapouri public expectations by creating new environmental institutions, chief among which were the Nature Conservation Council, the Environmental Council and the Commission for the Environment. One role of the Commission was to implement the Environmental Protection and Enhancement Procedures adopted in 1973. Under these procedures, environmental impact assessments were required for all government works likely to have discernible environmental effects. Environmental impact assessment has been an integral part of all of New Zealand's environmental law since.

The politicisation of the environment that occurred with the Manapouri campaign also had flow-on effects for the wider body of New Zealand environmental law. In 1967 water management was completely overhauled. The Water and Soil Conservation Act put in place a single licensing system to control water use for industry, town supply, fisheries, wildlife habitats and recreation. The Act took a catchment-wide approach and so helped to focus local government by advancing regionalism. To be authorised, proposed uses of water had to be proven beneficial when weighed against anticipated losses. The environmental flaws inherent in this balancing approach were vividly illustrated by the 1978 decision approving the hydro development of the Rangitaiki and Wheao rivers in the Bay of Plenty. This was despite the anticipated loss of an outstanding trout fishery and the destruction of the habitat of two endangered duck species.[12] One critic, lawyer Tony Black, argued that the decision illustrated the law's 'decided bias towards resource development ... For in the process of balancing tangible economic benefits and tangible social benefits against intangible and generally misunderstood environmental benefits the tangible wins out every time.'[13]

Environmentalists urged that some way of conserving rivers and lakes of national significance was needed. In 1981 Parliament introduced a water conservation scheme to preserve the natural, wild, scenic, recreational, wildlife or scientific features of water bodies, and the Motu in the Bay of Plenty became the first river for which a water conservation order was issued.

Public pressure by conservationists in the post-Manapouri period also brought improvements to the law relating to the marine environment and forests. 'New Zealand's past is steeped in the blood of whales and seals ...';[14] indeed, the last whaling station did not cease operations until 1964. Then one of the most widely supported of all environmental campaigns – 'Save the Whales' – forced a turnaround in government policy and the law. The Marine Mammals Protection Act 1978 (which protects these animals in their natural habitat) was the legislative result, and followed hot on the heels of the Marine Reserves Act 1971 (under which reserves can be established to protect scenic areas, natural features or marine life in the national interest and for scientific study).

The 1971 publication of a government proposal for large-scale harvest of South Island beech forest, much of which is adjacent to alpine national parks, generated another major environmental campaign. Two new and powerful lobby groups – the Native Forests Action Council and the Friends of the Earth – were formed. In response to mounting public concern, the emphasis on natural, intrinsic and scientific values was increased in the reserves and national parks legislation when this was consolidated in 1977 and 1980. In 1976 'balanced use' became the foremost management goal for the Crown's own production forests, making water and soil management, land conservation, indigenous flora and fauna, recreation, education, and historic, cultural, scenic, aesthetic, amenity and scientific values all relevant concerns, along with production.

Finally, the 'growing concern for the protection of the environment' in the post-Manapouri period was behind changes to town and country planning legislation.[15] These recognised the need to preserve the natural character of the coast and of river and lake margins, and to conserve the physical environment. More specifically for urban environmental quality, the Clean Air Act 1972 made a small inroad into air pollution. Its general duty to adopt the best practicable means to minimise air pollution was only enforceable against industry, and although it authorised the creation of zones where domestic pollution could be controlled, only one limited zone – Christchurch City – was ever created. Despite all these improvements, however, the main body of environmental law continued to enable and facilitate resource exploitation.

The road from Manapouri: developmentalism resurgent

Manapouri and Te Anau were only two of the lakes targeted for hydroelectric development. Others, including Waikaremoana, Monowai and Hawea, were also affected, along with the Waikato, Tongariro, Waitaki and Clutha rivers. The controversy caused by the damming of the Clutha River, which drains around one-twelfth of the country's land surface and flows through some of its most dramatic scenery, constitutes the next thread in this history (Figure 16.3).

Serious investigations into the Clutha River's hydro potential began in 1962. In 1965 the investigations were officially announced and rumours – later to prove justified – that the historic township of Cromwell would be flooded and that the levels of Lake Wanaka would be raised began to circulate. The 1972 general election, which brought the Kirk-led Labour government to power, also saved Lake Wanaka. The Manapouri example was followed with the passage of the Lake Wanaka Preservation Act 1973, and guardians were appointed to protect its natural values.

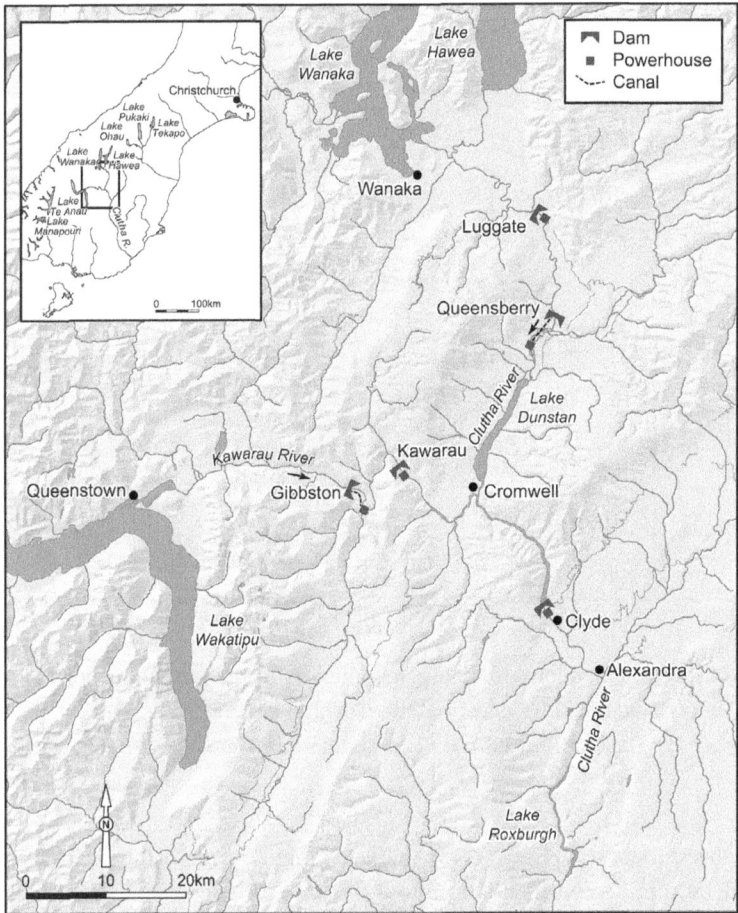

Figure 16.3 'Mastering the vast energy potential of the Clutha River'. The map shows Scheme F, approved in 1976. By 1983 the two dams on the Kawarau River had been abandoned because of geological problems. The two dams on the Upper Clutha towards Wanaka were later abandoned by the fourth Labour government. Source: adapted from *Clutha Power*, Ministry of Works and Development, revised December 1983

However, also in 1972 an official report put forward six alternative schemes for the hydro development of the Clutha. The option finally selected, by a National government in 1976, was the second most destructive, involving construction of a 195-metre-high dam at Clyde. Old Cromwell town and several of the pipfruit orchards in the Clutha Valley would be flooded. Public opposition mounted as it became apparent that part of the government's goal was to generate cheap power for a new aluminium smelter.[16] The smelter proposal was itself highly contentious: the intended site was at Aramoana, directly across Otago Harbour from Taiaroa Head, where the only mainland royal albatross colony in the world is situated. Both the dam and the smelter disputes moved into the courts (Figure 16.4).

Figure 16.4 Government ministers kneel before the graven image of Aluminium, as the Minister of Energy offers a jug of Clutha water into the sacrificial sink. Sid Scales, *Otago Daily Times*, 13 December 1980. Source: Hocken Collections, Uare Taoka o Hākena, University of Otago

The government's application for the water rights to dam the Clutha was initially approved, despite a recommendation from the local catchment board that it be declined in favour of a smaller scheme. Some protesters appealed, but their appeal was dismissed. When the Aramoana smelter proposal finally lapsed, because of the withdrawal of likely corporate investors, another appeal was taken to the High Court. In it the objectors claimed that the proposed end use of the power to be generated by damming the Clutha had not been taken into account in the original decision, and that this was contrary to law. The decision in *Gilmore v. National Water and Soil Conservation Authority* (1982), that the end use of the power was relevant, was the final straw. Irked by the protesters' success, the government decided to take legislative action. It secured the necessary support of two minor party members of Parliament just as the court decided that, once the end use of the power was considered as required by *Gilmore*, the objectors' appeals must succeed.

The Clutha Development (Clyde Dam Empowering) Act 1982 conferred the water rights to dam the river on the Minister of Energy. In so doing, it effectively reversed the court's decision, to the government's advantage. This gives the Act a constitutional notoriety unmatched in New Zealand's legal history. It breached an important constitutional convention and underscored the government's willingness and ability to manage and control parliamentary processes to its own ends. It illustrated a determination to see the dam built no matter what, and a derisory attitude towards the objectors and the courts.[17]

In 1979 the National government had also been responsible for enacting the National Development Act. This was the legislation under which the Aramoana aluminium smelter was initially considered. It provided a fast-track procedure to assess and authorise proposed works deemed to be of national importance, with final decisions resting with government ministers. The Act was an integral part of National's 'Think Big' policy, which promoted large-scale industry exploiting New Zealand's natural resources. Both the policy and the Act were very controversial. Ministerial decisions were vigorously challenged in the courts,[18] but schemes such as petrochemical developments in Taranaki, using the country's limited supplies of natural gas, proceeded. Although the Act was repealed in 1986, the idea that large-scale projects should be approved using special, high-level political processes, rather than the regular community-level decision-making subject to judicial control, has persisted.

The fourth Labour government and environmental reform

There is little doubt that 'the perceived indifference, if not open hostility' of the National government towards environmental interests was exploited by the Labour Party in the run-up to the 1984 election.[19] Labour offered a marked shift in environmental policy, along with institutional reform to provide a more effective voice for conservation.

These reforms would, however, take place 'within the libertarian ideology of the government. At the core of this philosophy is an almost doctrinaire belief in the market as the dominant mechanism in the economy.'[20] Policy generation within the fourth Labour government (1984–90) was dominated by Treasury and a core group of half a dozen 'free market' ministers. Environmentalists were not in principle opposed to this approach: 'for many ... the State had been the main culprit of environmental degradation'.[21] This made the ensuing reforms easier to implement. The extent of the programme was such that it was, in several respects, unfinished by the end of Labour's second term in office in 1990.

Labour began by disestablishing the existing 'mixed mandate' environmental agencies. These included the Ministry of Works and Development (which had been responsible for hydroelectric schemes, as well as for oversight of water conservation orders) and the Department of Lands and Survey (which was a land development agency as well as having a role in administration of the conservation estate).

Conservation assets and functions were allocated to a new 'proactive conservation oriented' Department of Conservation.[22] It would manage natural resources to preserve and protect them under the Conservation Act 1987, and its presence would enable a level of integration hitherto absent in conservation management in New Zealand. Today the Department enjoys high levels of public support but nevertheless has been constrained by consistent underfunding.[23] A new Ministry for the Environment, with a policy role, and the new office of Parliamentary Commissioner for the Environment were also established. The Commissioner has been described as an 'environmental ombudsman',[24] whose 'influence lies in the power to embarrass'. New Zealand's Parliamentary Commissioner has operated 'independently and effectively for over 25 years'.[25]

Commercial assets – land, coal, forests, fisheries – and functions would be transferred to new state-owned enterprises, which were instructed to operate as successful businesses. But as the government began the transfers, it confronted a major hurdle in the form of the Treaty of Waitangi. The legislation the government would use to make the transfers prevented it from breaching the principles of the Treaty of Waitangi, or affecting Māori rights. Hoist with its own petard, the government was forced into the courts and the Waitangi Tribunal to defend its plans.

The Māori renaissance that these events were part of has had a major effect on New Zealand environmental law over the last 30 years. Other key influences were Labour's decision to extend the powers of the Waitangi Tribunal, and government decisions since 1984 to include Treaty principles in most important environmental legislation. The Tribunal was established in 1975 to investigate Māori claims that the government had breached Treaty of Waitangi principles, including a duty on government to protect Māori rights and interests in natural resources. In 1985, this jurisdiction was extended back to claims dating from 1840, resulting in the Tribunal being able to investigate 'much wider claims concerning loss of ownership and control of resources'.[26] While including Treaty principles in environmental legislation has not succeeded in delivering 'appropriate levels of control, partnership and influence' to kaitiaki Māori,[27] it has given Māori a formal voice in environmental management and some significant victories.[28]

'After 1989, under pressure from Māori successes in the courts, the Crown began a "deliberate strategy" to regain political control over Treaty claims.' It developed its own Treaty Settlement Process, which was led from 1995 by the Office of Treaty Settlements. Many settlements have now been reached and have resulted in settlement legislation introducing a range of mechanisms safeguarding access to customary resources, enabling iwi participation in environmental management, transferring ownership over specific resources to claimant groups, and creating new relationships between Māori and environmental agencies.[29]

The local government and environmental planks in Labour's programme of reform took several years to develop, and fell to be completed by the National government of the early 1990s. First, the environmental functions of almost all residual specialised agencies (such as catchment boards) were devolved on to general-purpose territorial authorities and regional councils. This nationwide tier of regional administration, based to a considerable extent on catchment boundaries and introduced in 1989, was to provide the framework within which newly formulated resource management law could be administered.

The resource management law reform process proper was launched in January 1988, culminating in the passage of the Resource Management Act in October 1991. This Act introduced the concept of sustainability into New Zealand law, and also went some way towards integrating New Zealand's environmental law and making it more coherent. The Act replaced nearly 60 existing Acts, including the Town and Country Planning Act 1977 and the Water and Soil Conservation Act 1967. Created after unprecedented consultation, it covers the use, development and protection of air, land and water. It put in place a three-tiered (national, regional and local) management system and a scheme for allocating rights to use land, air and water. When the Resource Management Bill was introduced into Parliament in December 1989, Ministry for the Environment staff predicted:

> a new era in environmental management has begun. Our existing ... laws ... have evolved in a piecemeal fashion, resulting in a set of complex, overlapping, and sometimes conflicting rules. As a result, the outcomes for the environment are often inadequate ... The ... Bill proposes to change all this. In a world where concern for the environment grows stronger every day, New Zealand has recognised that the clean-up must begin at home.[30]

Sustainability and a libertarian ideology

The Resource Management Act seeks to facilitate sounder environmental decision-making by prescribing guiding principles and objectives. Before any proposed use of land, air or water is authorised, the actual and potential environmental effects on the environment must be assessed and a whole range of interests, including development, amenity and intrinsic values, and the principles of the Treaty of Waitangi, must be considered. Most importantly, decisions made under the Act must promote sustainable management. By adopting sustainability as its core purpose, the Act broke new ground. Sustainability has since become the 'golden thread' of resource management in New Zealand, finding its way into other legislation, including the Forests and Fisheries Acts. But the Resource Management Act also illustrates the libertarian ideology of the reforms begun by Labour in 1984. It is 'not prescriptive [but rather] focuses on the regulation of the effects of activities'.[31] Under its terms, limited markets in water permits have developed in some catchments.

The same features appear more strongly in the Fisheries Act 1996, under which commercial fisheries are managed. This Act aims to provide for the conservation, use, enhancement and development of fisheries resources while ensuring sustainability. The fisheries are managed using a quota management system (QMS), which was introduced in 1986 in the single most radical change ever made to commercial fisheries management. A total allowable catch (TAC), designed to maintain, or move fisheries stock towards, the maximum sustainable yield (MSY), is set for each nominated fishery. Within the TAC, a total allowable commercial catch (TACC) is determined. Rights to fish on individual tradable quota, representing a proportion of the TACC, are allocated to fishers. This is a wholly market-based system, and was introduced to promote industry efficiency and conserve the fishery. Furthermore, its introduction forced government to settle outstanding Treaty of Waitangi commercial fisheries grievances by propelling the claims into court as Māori resisted the effective privatisation of commercial fisheries.

Sustainability was introduced to privately owned indigenous production forests by the Forests Amendment Act 1993. Labour's institutional reform had split state forest management between the commercial and conservation estates, and put the bulk of the Crown's indigenous forests into the latter. Cutting rights to most state indigenous production forests were corporatised and, in some cases, subsequently privatised, though the Crown retained large tracts of indigenous production forest on the South Island's West Coast. These were brought into the conservation estate after 2000, after attempts to control logging under accords negotiated between conservation interests, the Crown and the timber industry were abandoned by the new Labour-led government.

Under the 1993 Act, indigenous timber cannot be felled or milled and indigenous timber product cannot be exported unless the timber has been harvested sustainably. 'Sustainable forest management' is defined as management that 'maintains the ability of the forest growing on that land to continue to produce a full range of products and amenities in perpetuity while retaining the forest's natural values'.

By the end of the millennium, sustainability stood as the management focus of commercial fisheries, indigenous production forests, and land, air and water. The interests of future generations had been added to environmental and conservation policy. In 2012, the natural resources of the exclusive economic zone and continental shelf were added to this list of resources to be sustainably managed. The climate system could, perhaps should, be next.

New Zealand's main response to its international duties to combat climate change has been to introduce an emissions trading scheme. This scheme is designed to be an all-encompassing price mechanism for reducing the country's net greenhouse gas emissions below business-as-usual levels. This mechanism fits nicely with the libertarian approach continued by governments since Labour's reforms in the 1980s, albeit this time with the backing of the international community of party states to the Framework Climate Change Convention and its Kyoto Protocol.

The adoption of sustainability in legislation has not, however, always led to an improvement in environmental outcomes. According to *Environment New Zealand*, the second national State of the Environment report, New Zealand is the world's twelfth highest per capita emitter of greenhouse gases, is experiencing worsening water quality and has overfished 15 per cent of its assessed commercial fishing stocks. Of our approximately 16,000 known marine species, 444 are threatened; and between 1997 and 2002 'native land cover decreased by an estimated 0.12 per cent ... including a decrease of 17,200 hectares of native vegetative cover'.[32]

This gap between environmental ideals and practice arguably exists because of the government's willingness to accommodate pressures for development. This willingness is demonstrated both when legislation is used to provide routes – or allowed to leave spaces – for development to proceed more easily, and when the environmental standards provided are too vague to be meaningful.

Developmentalism persists

Although the National Development Act was repealed in 1986, a call-in mechanism was introduced with the Resource Management Act. This circumvents the usual consent process, providing a more expeditious route to approval for projects of national significance. The power to call-in applications was at first used infrequently but amendments resulting in the involvement of a new Environmental Protection Authority in the process seem to have resulted in more applications being considered in this way.

In Canterbury, developmental pressure for access to freshwater resources has led both Labour and National-led governments to intervene in local resource management issues. Because it holds over 60 per cent of New Zealand's freshwater resource, Canterbury was seen as a region with 'huge economic growth potential and [as] a key driver in making New Zealand a more productive and competitive economy'. To the government, it was 'vital that a clear vision ... for economic growth and sustainable management of the natural resources in the Canterbury region' was developed.[33]

In 2003 Meridian Energy proposed Project Aqua, the first major hydroelectric scheme since the Clyde Dam. This proposal involved six dams for the lower Waitaki River, and the diversion of over 70 per cent of the river's flow (Figure 16.5). However Meridian soon abandoned the project, citing uncertainties in the process generated by the failure of the regional council, Environment Canterbury, to have an allocation plan in place for the river. Concerned about the effect of this on Meridian's (and other potential) projects, the government intervened with the Resource Management (Waitaki Catchment) Amendment Act 2004, establishing a board to make an allocation plan for the river, and allocate abstractive rights to its waters.

In 2010, under pressure from irrigators seeking to intensify the burgeoning dairying industry in the region, the government again became frustrated with Environment Canterbury. The Council's 'lack of a proper allocation plan, increasing problems with water quality and ... failure to progress opportunities for water storage'[34] were, it said, putting 'at risk the region's prospects for economic growth'.[35] Again government intervened and replaced the elected council with appointed commissioners under the Environment Canterbury (Temporary Commissioners and Improved Water Management) Act 2009.

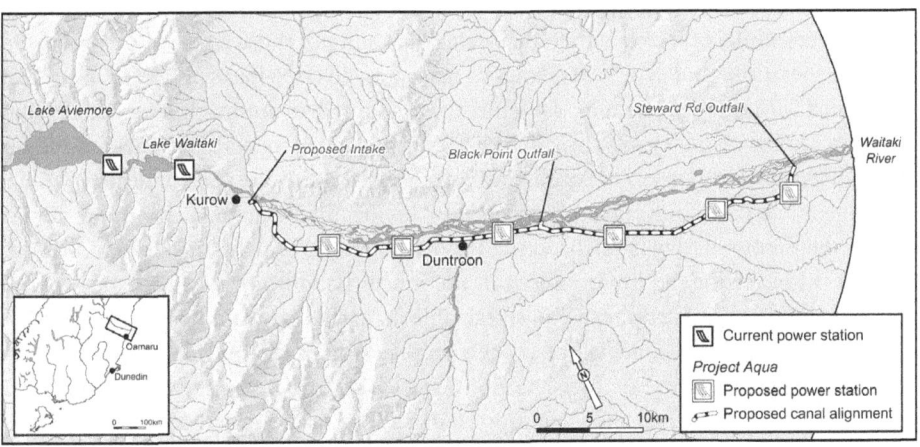

Figure 16.5 Map of the Project Aqua proposal, showing the location of the canal and six power stations, alongside the Waitaki River. Source: adapted from Meridian Energy press release, 2 July 2003

Governments, however, have been slower to intervene to curb development on behalf of the environment. Dairying has intensified greatly over the last 20 years, and this has put a great deal of pressure on freshwater quality in regions like the Manawatu, Canterbury and Southland. Today, 'increasing pollution from non-point-sources, such as diffuse run-off from pasture … poses the greatest challenge for water management in New Zealand'.[36] Regional councils have the main responsibility for managing freshwater under the Resource Management Act, but have failed to implement appropriate policies for responding to non-point and diffuse pollution, adequately to monitor policy once it is introduced, or to prosecute non-compliance adequately.[37] The pollution problem has been documented for many years, and in 2011 central government finally responded by making a national policy statement under existing resource management legislation.

The second reason why there is a gap between environmental ideals and environmental outcomes, even under legislation that promotes or provides for sustainability, is that sustainability is a very general concept, both inherently and as defined in the legislation, and this allows for environmental goals to be compromised.

High (but indeterminate) principles

Of all the principles adopted by international environmental law and policy worldwide, sustainability is both the most important and 'the most wonderfully wide'.[38] In New Zealand, it has been expressly incorporated into the Forests Act 1949, the Environment Act 1986, the Resource Management Act 1991, the Fisheries Act 1996, the Hazardous Substances and New Organisms Act 1996, the Energy Efficiency and Conservation Act 2000, and the Exclusive Economic Zone and Continental Shelf (Environmental Effects) Act 2012.[39]

For the purposes of these Acts, 'sustainable management' is defined in 'open' language. It means managing or providing for the use (and development and protection) of resources while, or in a way that, maintains the ability of those resources to provide for future generations and avoids, remedies or mitigates adverse effects on the environment.

Faced with these definitions, the courts have adopted an approach intended to 'allow the application of policy in a broad way'.[40] A very general 'overall judgment' approach that 'allows for comparison of conflicting considerations and the scale or degree of them, and their relative significance or proportion in the final outcome' has been applied under the Resource Management Act.[41] This dashes hopes – including those of Geoffrey Palmer, Minister for the Environment in the fourth Labour government, and architect of the Act – that the Act's purpose of sustainable management would 'overturn the whole philosophy of the decision-making process' in place under the previous legislation.[42] By defining and reading these crucial sections as creating balancing tests, Parliament and the courts have weakened the legislation. Balancing tests are inherently biased towards development, 'for in the process of balancing tangible economic … and … social benefits against intangible … environmental benefits the tangible wins out every time'.[43]

The same balancing is evident under the Fisheries Act, which aims to promote utilisation of fisheries while ensuring sustainability. According to the Supreme Court this purpose 'expresses ... two competing social policies' that must both be 'accommodated' in decisions made under the Act. However in recognising 'the inherent unlikelihood' that both policies can be fully accommodated in practice, the Act ultimately requires that 'utilisation must not ... jeopardise sustainability'.[44]

The balancing approach is ecologically derelict in its failure to provide uncompromising environmental standards, and the statutory definitions of sustainability are so wide they can support almost any decision on a given set of facts. On this approach, allowing residential development on land with food production potential would 'not sustain the resource represented by the soil for the reasonably foreseeable needs of future generations'. However 'an equally valid argument could be advanced' that blindly protecting soils of high value 'does not sustain the resource represented by the community ... for its foreseeable future needs if it deprives future generations of the ability to live in an expanded community'.[45] New Zealand Law Commissioner D.F. Dugdale postulates that some of the 'voluntary ambiguity' of recent legislation may be 'attributable to ... politic uncertainty, a desire for an appearance of agreement despite the existence of nettles left ungrasped'.[46] But by using high principles like sustainability in environmental law, and not defining and prioritising them tightly enough, Parliament has created a situation where the law's bark is far greener than its bite.

The saving grace of the most recent legislation, from an environmental perspective, could lie in its adoption of the precautionary principle. This principle is a policy response to scientific uncertainty; it 'requires greater weight to be given to environmental and public health protection in the all too common situation where there is insufficient scientific information available on which to base decisions'.[47] It is an 'information principle' in the Hazardous Substances and New Organisms Act (which controls imports and developments of new organisms, including genetically modified organisms), the Fisheries Act, and the Exclusive Economic Zone and Continental (Environmental Effects) Act. Precaution has also been read into the Resource Management Act.[48]

Conclusion

For the last century or so New Zealand environmental law has comprised an increasingly comprehensive body of legislation designed to manage (and promote) resource development. There have been, and still are, statutes that aim to conserve natural resources, but generally conservation and environment have been, and still are, relegated to the periphery. Today this relegation occurs mainly because of the vague standards and definitions used to guide environmental decision-making. Of these, the foremost is sustainability. This is a fine principle, to be sure, but

> the existence of a principle in itself never saved a tree or cost a logging company any money. Where the benefit and the pain come from are the laws implementing that policy, and the administrative decisions derived from it.[49]

The implementation of sustainability has been hampered by the commitment of successive governments to developmentalism, and a legislative preference in New Zealand for widely framed and flexible definitions. The aim is to be embracing and to leave plenty of discretion to decision-makers so that they can tailor outcomes to the facts of each case. The problem with discretion is that it leaves (too much) room for environmental bottom lines to be compromised and so tends to result in developments proceeding, so long as the environmental impacts are somehow reduced. As Bruce Pardy argues, 'significant long-term environmental changes can be caused by the accumulation of small impacts. Compromise allows environmental death from a thousand inconsequential cuts.'[50] The underlying value preference in New Zealand favours development and the management of environmental effects. Taking a precautionary approach to uncertainty surrounding these effects will help but without either very clear and uncompromising ecological bottom lines, or a transition to a greener value paradigm, this bias will continue to manifest itself in environmental decision-making.

Further reading

Klaus Bosselmann, David Grinlinton & Prue Taylor (eds), *Environmental Law for a Sustainable Society*, 2nd edn (Monograph Series, vol. 1, New Zealand Centre for Environmental Law, Auckland, 2013) covers the Resource Management Act and some of other issues discussed in this chapter. Nicola R. Wheen & Jacinta Ruru, 'The Environmental Reports', in Janine Hayward & Nicola R. Wheen (eds), *The Waitangi Tribunal/Te Roopu Whakamana i te Tiriti o Waitangi* (Bridget Williams Books, Wellington, 2004, pp. 97–112) reviews Tribunal reports on environmental topics and Nicola R. Wheen & Janine Hayward (eds), *Treaty of Waitangi Settlements* (Bridget Williams Books, Wellington, 2012) includes several chapters on settlements about land, forests, fisheries, freshwater and pounamu. On ambiguity in environmental law and the lack of ecological bottom lines, see Bruce Pardy, 'In Search of the Holy Grail of Environmental Law: A Rule to Solve the Problem' (*International Journal of Sustainable Development Law & Policy*, vol. 1, 2005, pp. 29–57).

17.

Ngāi Tahu and the 'nature' of Māori modernity

Michael J. Stevens

Ngāi Tahu say they want to be a future people but then they start behaving like Pākehās, or like the general capitalist model ... Māori are generally ... not thinking fundamentally differently.

—Sir Tipene O'Regan[1]

WHEN EAST POLYNESIANS settled the islands that now constitute New Zealand they encountered an environment that was larger and colder than any they had previously known. In response to this, and left in isolation, Māori culture developed. The term Māori, however, used to collectively describe the descendants of these Polynesian settlers, did not exist until the sustained presence of Europeans made it necessary.[2] Strictly speaking, therefore, 'Māori' is often used anachronistically. This corporate term evolved out of the word māori, English translations for which include ordinary, natural and normal. With wai meaning water, waimāori simply means freshwater, and in contradistinction, waitai means saltwater.

A question commonly heard in Māori settings is 'Ko wai koe?' This is generally translated as 'Who are you?' but literally asks, 'What are your waters?' An older, rhetorical, version of this question recorded in Ngāi Tahu tradition is 'Unutai?' – Which tides do you come from?[3] Answers to these questions, delivered through the likes of mihimihi, pepeha and waiata, commonly reference water bodies from specific tribal areas. Water, then, is central to Māori self-definition and self-identification. This is true for Ngāi Tahu, the iwi that holds mana whenua over the majority of the South Island.[4]

Ngāi Tahu ki Te Wai Pounamu

Ngāi Tahu communities were radically reshaped by sustained European contact in the first half of the nineteenth century. Despite this, many mahinga kai continued to be valued.[5] Between the 1840s and the 1860s, however, most of these were lost to tribal members through the logic and practices of colonisation. In Tony Ballantyne's words, a Ngāi Tahu 'vision of plenty [was] displaced and paucity became a key condition of the tribe's existence'. Colonists, on the other hand, took possession of 'wastelands' hitherto controlled by Ngāi Tahu and 'enjoyed the fruits of their ability to swiftly actualise their vision of [the South Island] as a land of plenty'.[6] In short, 'Ngāi Tahu [were] deprived, with their land, of the capital base with which to participate in the new economy that it engulfed it.'[7] Tribal representatives communicated grievances to the colonial state as early as 1849 and lodged a comprehensive historical claim with the Waitangi Tribunal in August 1986, soon after it received retrospective powers of inquiry back to 1840. As a result, Ngāi Tahu was one of the first iwi to conclude a major Treaty of Waitangi-based constitutional property settlement with the New Zealand government.[8]

As with the Tainui (Raupatu) Claims Settlement Act 1995, the Ngāi Tahu Claims Settlement Act 1998 became a benchmark for subsequent Treaty settlements.[9] Similarly, the way in which Ngāi Tahu is structured administratively as a result of its Treaty settlement is commonly examined. On the whole, this framework is commended. Thus, as Ngāi Tahu lead negotiator and elder statesman Sir Tipene O'Regan puts it, many tribes have partly adopted the 'Ngāi Tahu script'.[10] An outline of this script is hence relevant beyond Ngāi Tahu and the South Island. The particular organisational design of Te Rūnanga o Ngāi Tahu (TRoNT), the tribe's mandated iwi authority created by private statute in 1996, is therefore a specific point of focus in this chapter.

In short, TRoNT keeps its wealth-generating activities – its commercial subsidiary entities – separate from its wealth-distribution – its social development initiatives (Figure 17.1). Through reference to the management and use of freshwater, an increasingly valuable but stressed resource in New Zealand, this chapter highlights some of the consequences of this dyadic approach. In so doing, it suggests that readers will come to appreciate the different layers and constituent parts of iwi authorities such as TRoNT. In other words, it illustrates how the tribal appellation 'Ngāi Tahu' is legitimately attached to a range of bodies, and how and why these can pursue different, sometimes contradictory, things.

Ngāi Tahu was denied a collective personality and largely 'Left Out of New Zealand' for 150 years. A reversal of these things in the late twentieth century has led to a number of positive developments for Ngāi Tahu whānui. Having said that, many of the old challenges persist,[11] while being brought into New Zealand has created a new set of

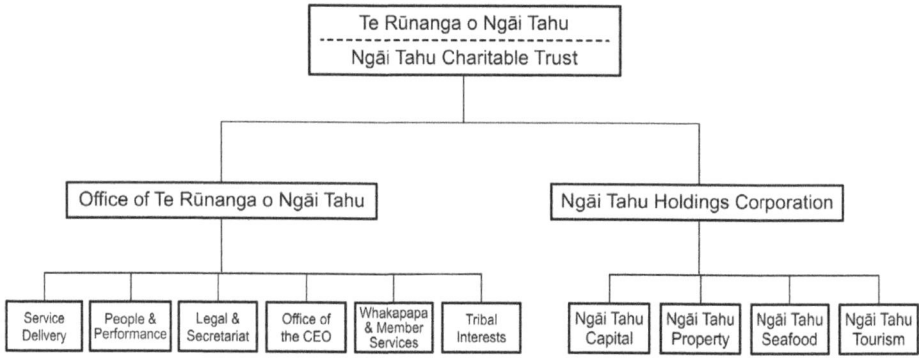

Figure 17.1 The organisational chart of Te Rūnanga o Ngāi Tahu. Source: TRoNT

social, economic and environmental challenges for Ngāi Tahu, as evidenced by a focus on waimāori.

While the management and use of freshwater is a national-level issue, it is regionally uneven. It is critical in Canterbury, where in May 2010 the government replaced the elected councillors of Environment Canterbury (ECan), the Canterbury Regional Council, with seven commissioners. This followed a damning report over ECan's failure to successfully manage Canterbury's freshwater demands over the preceding two decades. The commissioners have been tasked, among other things, with the completion of a resource management plan for water in Canterbury – the Canterbury Water Management Strategy (CWMS).[12]

In undertaking this work the commissioners are directed by nine fundamental principles. These are outlined in a schedule of their empowering legislation. The third of these principles is 'kaitiakitanga'. This holds that '[t]he exercise of kaitiakitanga by Ngai Tahu applies to all water and lakes, rivers, hapua, waterways and wetlands, and

Figure 17.2 TRoNT's 18 regional papatipu rūnanga. Source: TRoNT

shall be carried out in accordance with tikanga Maori'.[13] This focus on the interests and aspirations of Ngāi Tahu is also reflected in the choice of one of the seven commissioners: Donald Couch, of Rapaki. A former papatipu rūnanga representative and Deputy Kaiwhakahaere of TRoNT, Couch has experience in resource management and local government in both New Zealand and Canada.

This turn of events led to the development of a co-governance agreement between ECan and Ngāi Tahu (via the Te Waihora Management Board) in relation to Te Waihora/Lake Ellesmere in 2012. This was bolstered in February 2013 with the launch of 'Tuia', a relationship agreement between ECan and the papatipu rūnanga within its local government boundaries. TRoNT consists of 18 regional papatipu rūnanga, and 10 of these fall within ECan's area of responsibility (Figure 17.2). The Waihora agreement is the first of its type not created by a Treaty of Waitangi-based settlement. This co-management agreement and the Tuia framework have been rightly described by Mark Solomon, the Kaiwhakahaere of TRoNT, as 'a new approach to management of natural resources in Canterbury'. The ECan Chair of Commissioners, Dame Margaret Bazley, agrees and describes the developments as significant for both Canterbury and New Zealand as a whole. As she says, 'We are forging a way in which iwi and regional government can work together for common goals.' In her view this is particularly important to the success of the collaborative CWMS.[14]

The CWMS divides Canterbury into 10 geographic zone committees, each of which is tasked with producing an implementation plan for their catchment in a short space of time. Ngāi Tahu has representation on each of these committees and likewise a regional one that deals with issues common to all 10 areas. In recent years rūnanga representatives from across Canterbury have therefore spent long hours in meetings with farmers, environmentalists, recreationists and officials trying to figure out a new approach to freshwater management. One of these, Raewyn Solomon, who has represented Te Rūnanga o Kaikōura on the Hurunui–Waiau Zone Committee, is grateful for the opportunity. She is critical of Resource Management Act processes (see Chapter 16) and the adversarial approaches they give rise to. She instead feels that, 'It takes conversations, education and a receptiveness to other people's values to manage water so that everybody and everything can benefit in an equitable way.' And this, according to her, is what the CWMS approach has enabled. However, because decisions are made by consensus, she points out that she does not agree with everything in the zone implementation plan she helped create. Recalling a now dried-up water body that her father used for eeling, Solomon ties her measure of success to mahinga kai. In her words, 'I don't feel like I've succeeded in my profession if there's no water in rivers. That's really our measure.'[15]

Much of the foregoing suggests that Ngāi Tahu is distinguishable from a number of other iwi in a number of ways – which it is. However, it is ultimately representative: the challenges it faces in relation to waimāori are shared by most iwi. As Mark Solomon commented in late 2012, 'the very same issues are raised and the same questions are asked all around the country – the only difference is the name of the iwi or hapu and the name of the water body'. This chapter on Ngāi Tahu therefore continues the tale of 'the ordinary struggles of an ordinary tribe'.[16]

Ngāi Tahu and waimāori

The Ngai Tahu Claims Settlement Act 1998 gave effect to the Deed of Settlement agreed to by Ngāi Tahu and Crown negotiators in 1997 and enabled the Crown to transfer a total of $170 million in cash to TRoNT.[17] Combined with various 'bolt-ons' (for example, cultural redress such as the implementation of dual placenames), Ngāi Tahu negotiators believed that the Crown's total settlement package was both the 'best that could be achieved in present circumstances' and 'sufficient redress to re-establish an economic base for the tribe'.[18] Key to the sufficiency of the settlement's cash component was that it could be used to purchase, at market value, certain Crown-owned assets including commercial properties, farms and forestry blocks. This 're-capitalisation of an iwi',[19] as lead claim negotiator Sir Tipene O'Regan describes the Treaty settlement, is the basis on which TRoNT funds social and cultural development within Ngāi Tahu whānui.

Mahinga kai, which is primarily understood among Ngāi Tahu communities as the harvesting of uncultivated foodstuffs, was a central pillar of the Ngāi Tahu Claim. Indeed it was the final of the tribe's 'Nine Tall Trees', an umbrella of grievances laid before the Waitangi Tribunal in 1987. This ninth tree claimed that Ngāi Tahu had been denied access to and protection of mahinga kai since 1840. According to the Waitangi Tribunal, mahinga kai was 'one of the most emotionally charged elements of the Ngai Tahu claim' in part because it is was central to the 'whole social fabric of tribal and intertribal life'.[20] The ongoing importance of mahinga kai to Ngāi Tahu communities, and the threats faced by surviving sites and species that underpin it, helps to explain why TRoNT issued a Freshwater Policy Statement in 1999.

This policy statement outlined Ngāi Tahu associations with freshwater resources, ways in which Ngāi Tahu sought to participate in freshwater management, and the environmental outcomes it desired. The document did not discuss issues relating to the ownership of water other than to say it believed that this question remained unresolved. The document was underpinned by six kaupapa that TRoNT wanted to 'govern the formulation of water policies and plans within the rohe of Ngāi Tahu'. The very first of these, in a clear nod to mahinga kai, stated that water plays a unique role in the traditional economy and culture of Ngāi Tahu. Another stressed the need for a shift in 'behaviour from one that prioritises consumptive and … inefficient use towards one that recognises and provides for cultural and ecological values as priorities'.[21]

In 2000, TRoNT presented a plan for the environmental, social, cultural and economic development of Ngāi Tahu whānui over the following 25 years. This plan, *Ngāi Tahu 2025*, self-described as 'a living document', was approved by TRoNT in March 2001 and launched with much fanfare. The aspirations it outlined related to matters from resource management and regional development to cultural revitalisation and investment planning. With respect to the latter, TRoNT set a benchmark of becoming the 'dominant economic force in Te Waipounamu'. In terms of the environment, *Ngāi Tahu 2025* expressed a desire to increase the rangatiratanga and kaitiakitanga of Ngāi Tahu over 'wāhi tapu, mahinga kai and other taonga tuku iho' – essentially, to increase the management and care of important sites, natural (mainly food) resources, and other valued things handed down from the past. These were described as being 'of paramount importance' and 'the cornerstone of the spiritual, historical, cultural, social and economic well-being of Ngāi Tahu'. Pollution, habitat destruction, species extinction, the degradation of water quality and quantity, and intensified and changing land use were all identified as being key Ngāi Tahu concerns.[22]

TRoNT expressed a desire to protect, restore and manage or co-manage key sites and waterways according to 'cultural standards' over the next 25 years. It also set itself the goal of managing its own assets 'in a manner consistent with Ngāi Tahu environmental practices and policies'. Central in all of this was the development of 'State of the Takiwā' environmental reporting framework. This is underpinned by a holistic 'Ki uta, ki tai' – 'from the mountains to the sea' – approach to resource management and a catchment-by-catchment focus.[23]

Ki uta

With this background in mind, TRoNT's January 2010 response to proposals for wide-scale irrigation and intensive dairying in the Mackenzie Basin and Upper Waitaki is unsurprising (Figure 17.3). The iwi authority was alarmed by the scale and intensity of the applications to take millions of cubic metres of water covering more than 27,000 hectares of land. TRoNT's legal advisor argued that the science presented in support of the proposals was patchy and the effects on cultural values were unclear. He contended that the applicants wished for a 'suck-it-and-see' approach to be adopted, which Ngāi Tahu considered intolerable. Suggesting that the Upper Waitaki is still an important source of mahinga kai, one Ngāi Tahu submitter, Mandy Waaka-Home, stated that Ngāi Tahu should not 'have to suffer the indignity of gathering and eating food from an environment that is knowingly polluted'.[24]

During the resource consent hearing for these large-scale water-take applications, TRoNT's legal advisor broke from written evidence to deplore dairy farming's negative effects in Canterbury: 'Since large-scale dairy farm conversions, Ngai Tahu has witnessed the degradation of many of the rivers on the east coast of the South Island.' The response of the Mid-Canterbury president of Federated Farmers was that water quality and quantity problems should not be blamed on one source alone. Even so, a 2009 Environment Court report found that one in five of the region's dairy farms appreciably breached compliance rules, and more than half were not fully compliant.[25]

The concomitant consent applications to ECan from three companies to intensively farm 17,850 dairy cows on just over 8500 hectares in the Mackenzie Basin were to be heard in March 2010. As well as its scale, the nature of this proposal aroused considerable public interest – and opposition. The cows were to be housed in 16 large cubicle sheds 24 hours a day for 8 months of the year, and 12 hours per day for the remainder of the year. The amount of effluent produced would be equivalent to a city of 250,000 people – approximately 1.8 million litres per day. The Parliamentary Commissioner for the Environment, Jan Wright, therefore argued that the matter was of national significance and urged the Minister for the Environment, Nick Smith, to 'call in' the resource applications. This has the effect of bypassing local authority and Environment Court hearings. Two of Smith's Cabinet colleagues similarly expressed concerns after a site visit.[26]

In late January 2010, Smith did indeed call in the applications and established a specialist Board of Inquiry to consider the application. Chaired by an Environment Court judge, this five-member panel included a water engineer and lake ecologist as well as Edward Ellison, who was simply described as Ngāi Tahu.[27] This label masks more than it reveals. Ellison is a former Deputy-Kaiwhakahaere of TRoNT (during the development of both the Freshwater Policy Statement and *Ngāi Tahu 2025*) and a former member of the New Zealand Conservation Authority. He is however primarily engaged in sheep farming on whānau land at Otakou, on the Otago Peninsula. He thus has valuable knowledge of the nature and value of both mahinga kai and commercial agriculture, and their sometimes competing interests.

In November 2011 the companies' resource applications for water extraction were declined. This was despite their putting forward 'a more traditional farming alternative' and proposing a model of 'adaptive management'. The latter centred on changing management plans if and when water quality was negatively affected. The board saw merit in this approach but ultimately feared that there could be irreversible 'serious adverse consequences' for groundwater, streams and rivers before plans could be adapted. Landscape considerations as well as Ngāi Tahu 'cultural and spiritual values' were also weighed up against the economic benefits of irrigation.[28]

Te Waihora

In August 2011, a few months before the board released this decision, the Minister for the Environment announced an $11.6 million package to clean up Te Waihora/Lake Ellesmere, a 20,000-hectare lake southeast of Christchurch (Figure 17.3). This was launched on the shores of Te Waihora, on Ngāti Moki marae, the focal point of Te Taumutu Rūnanga (Figure 17.2). Formerly a proverbial food basket for Ngāi Tahu, well into the twentieth century, the lake was especially noted for its eels, flounders and waterfowl. It was Te Kete Ika o Rākaihautū – the Fish Basket of Rākaihautū, 'a lake that sustained life'. However, it is now considered unsafe for people to wade into. In fact by 2011 Te Waihora was the most polluted of New Zealand's 140 lakes. Given that at nearly 200 km² it is also the nation's fifth-largest lake, this distinction is particularly depressing.[29]

The story of Te Waihora forms part of a larger narrative. In late 2012 the Ministry for the Environment reported that only 20 per cent of New Zealand's monitored recreational river sites are safe for swimming. Sickness and infection, especially in children, are highly likely from the other 80 per cent of sampled areas. A Ministry spokesperson stressed that these sites were chosen 'for their susceptibility to risk factors that might make swimmers sick, as well as their popularity for recreation' and therefore suggested that they 'were not representative of the overall water quality in the country's swimming spots'.[30] Even so, for a nation that generates a substantial amount of its foreign exchange earnings

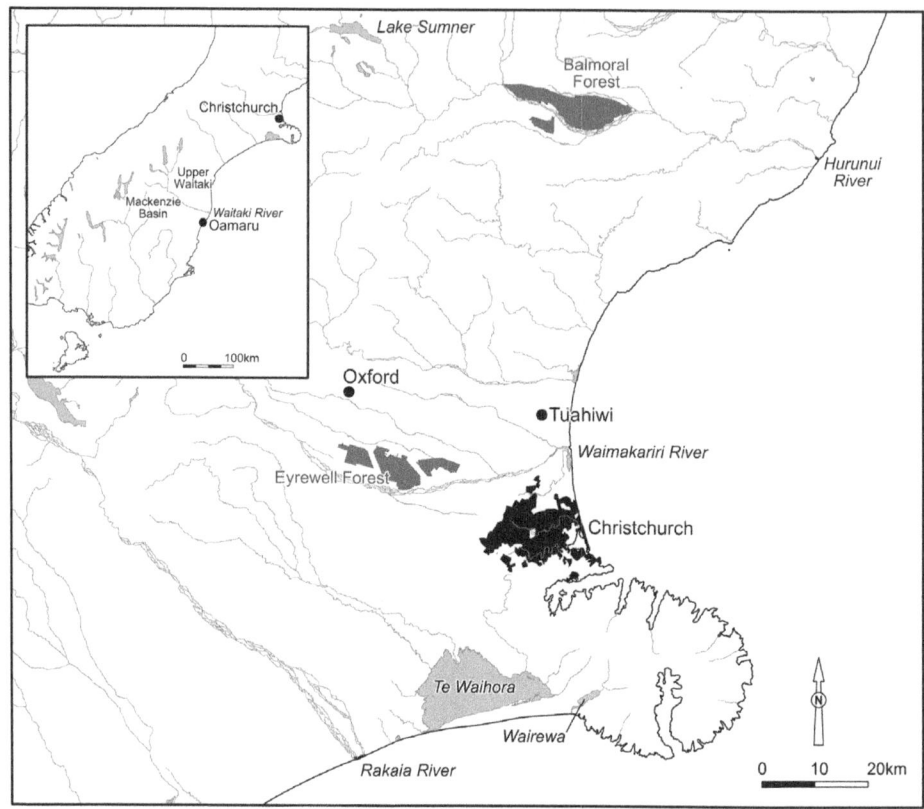

Figure 17.3 Map of places in and around Canterbury mentioned in the text.

through tourism, particularly ecotourism, and substantially bases its marketing on its environment being '100% Pure', this is not good news.

For Māori, particularly those who continue to reside in tribal heartland communities, this state of affairs is heartbreaking rather than newsbreaking. Māori kin groups have observed and voiced their dissatisfaction with environmental transformations, including those that have negatively affected waterways, since the earliest phases of colonial settlement. Herries Beattie, for instance, recorded an incident from the early twentieth century when a Ngāi Tahu elder visited a pond where he had earlier gathered harakeke. This had been cut and cleared and the pond drained. The elder said he felt like weeping.[31] Freshwater degradation is bad for all people; however, it is arguably disproportionately worse for 'heartland' Māori. This is because many of these 'flaxroots' people have an ongoing commitment to aspects of traditional lifeways – the harvesting and preservation of eels, for instance. While these activities have various levels of legal recognition, including through Treaty of Waitangi-based constitutional property settlements, they require healthy ecosystems to have real meaning. Māori underrepresentation in the farming sector, which is a major contributor to declining freshwater standards, arguably compounds the tragedy: many whānau bear the brunt of environmental decline but very few enjoy direct economic benefits of its causation.

Te Waihora, a case in point, was a key focus of the mahinga kai aspect of the Ngāi Tahu Claim. The Waitangi Tribunal was therefore presented with evidence to show that large parts of the lake had been drained for farming and settlement, and its vestiges increasingly polluted. Even so, flounder and eels persisted, only to be worked hard by commercial fishers. While less than 1 tonne of flounder was caught in Te Waihora in 1970, by 1979 that had increased to 272 tonnes, which represented 15 per cent of New Zealand's total flounder catch. As for eels, the commercial fishery also developed in the late 1960s and early 1970s. In 1975–76, 646 tonnes were landed from the lake. This represented 30 per cent of New Zealand's total eel take. In 1979–80, despite increased fishing effort, a much lower figure of 348 tonnes was landed. A 300-tonne annual quota was set around 1980 and by 1986 this had been reduced to 136.5 tonnes.[32] This severely reduced the ability of local Ngāi Tahu to continue to source mahinga kai from Te Waihora.

In 1991, the same year that the Waitangi Tribunal issued its first of three reports into the Ngāi Tahu Claim,[33] Te Waihora Management Board was formed. This body advised the then Ngāi Tahu Māori Trust Board (succeeded by TRoNT) on matters pertaining to the lake in its dealings with local and central government. In terms of the Ngāi Tahu Claims Settlement Act 1998, legal ownership of the lakebed was vested in TRoNT. Since that time the Management Board has worked with the Department of Conservation (DOC) on a joint management plan for the lakebed and adjacent DOC-administered lands. TRoNT, for its part, is estimated to have spent more than $3 million to help clean up the lake and enhance its mahinga kai values over the same time.[34] Ngāi Tahu involvement with the clean-up effort has been described as evidence of a maturing relationship between the government and the iwi. Nick Smith was said to be constantly surprised at the depth of interaction between the two entities. He suggested, approvingly, that it goes beyond what was envisaged in the tribe's Treaty settlement with the Crown. Regardless, Te Waihora remains less a food source than 'a sink for farming run-off and the direct discharges of sewage'.[35]

According to Professor Ken Hughey of Lincoln University, which is based near Te Waihora, the 2011 restoration package is woefully inadequate. It is for two years, whereas he argued that a 25-year plan was required. In other words, if a clean-up of Te Waihora is to be successful then tens of millions of dollars are required. Tony Hawker of Fish & Game New Zealand echoed Hughey's sentiments and described the initial funding as a 'drop in the bucket'.[36] Hawker also raised the spectre of increased nutrient leaching by way of the Central Plains Water irrigation scheme. This proposed scheme, which its backers describe as one of the largest construction projects ever undertaken in the South Island, comprises a water storage reservoir, a headrace canal and a network of irrigation channels to irrigate 60–80,000 ha of land situated between the Rakaia and Waimakariri rivers, which bound Te Waihora to the north and south respectively.[37]

As it stands, 70 per cent of New Zealand's irrigated lands are already found in Canterbury. But the economic value of the region's agricultural operations is immense: farmers spend around $750 million every year on goods and services provided by Christchurch businesses, which accounts for 60–70 per cent of the city's economic activity.[38] These figures were calculated before the devastating earthquake that struck the city on 22 February 2011. The economic stimulus generated by farming is now particularly important. There is accordingly a real tension between people's desires for healthy freshwater systems on the one hand, and the economic benefits delivered by way of intensive agriculture on the other hand. TRoNT is not immune from this cognitive dissonance.

Ngāi Tahu Property dairy farming proposal

Prior to the 2011 launch of the Te Waihora clean-up campaign, Ngāi Tahu Property (a subsidiary entity of Ngāi Tahu Holdings Group), announced that it was about to establish three dairy farms. These were to have 1000 cows each and be based on 1200 hectares of ex-plantation forestry land at Eyrewell, near Oxford, in North Canterbury. This was announced as merely the first stage in a project that might ultimately lead to the development of 16,000 hectares of dairy operations. As one ECan commissioner has observed, 'within Ngāi Tahu there is exactly the same debate going on about the balance between economic outcomes and cultural and environmental outcomes'.[39]

How have these divergent responses to dairying within Ngāi Tahu come about, and can they be reconciled? The answer to the first part of the question is found in the division between the commercial and social development arms of TRoNT (Figure 17.1). Another part of the answer is found in differences between the central administrative core, TRoNT, and its constituent regional papatipu rūnanga (Figure 17.2). This can be illustrated through the activities of Ngāi Tahu Property, relating to concerns about irrigation held by flaxroots Canterbury Ngāi Tahu, which came to a head in 2009 in relation to the proposed Hurunui Water Project (HWP).

The HWP scheme sought consent to create large-scale water storage and irrigation in North Canterbury by raising the level of Lake Sumner through a weir and building a 75-metre-high dam on the south branch of the Hurunui River (Figure 17.3). This would make irrigation available for 42,000 hectares of land in the Hurunui and Upper Waipara catchments, and potentially supplement an existing irrigation scheme in the Balmoral area. Ngāi Tahu Property is a significant landowner in the Balmoral area and had a seat on the board driving the HWP. However, two local papatipu rūnanga, Ngāi Tuahuriri and Ngāti Kuri, opposed this scheme. Along with the Office of TRoNT they therefore supported an application for a Water Conservation Order for the Hurunui River. 'This seemed somewhat paradoxical,' notes one scholar, 'as Ngāi Tahu as a whole were supporting two opposing groups.' This 'discrepancy of two arms of the tribe' is largely due to the fact that the 'commercial domain in the tribal organisation is somewhat independent'. In this instance of disagreement between Ngāi Tahu-owned commercial entities and flaxroots Ngāi Tahu, the governors of TRoNT 'forced the property company to step aside', as Mark Solomon put it.[40]

In late 2012, Ngāi Tahu Property's 'pilot' dairy farms at Eyrewell became operational. Again, this seems to directly conflict with the 'Ngāi Tahu' view that opposed dairy expansion in the Mackenzie Basin. A number of people, both Ngāi Tahu and non-Ngāi Tahu, have commented on this. One online rural news source described Ngāi Tahu as 'vocal objectors to dairy proposals … [who] appear to have different ideas with their own dairy ambitions'.[41] Its author did not distinguish between papatipu rūnanga and TRoNT, or between the latter's commercial and non-commercial functions: it was all simply 'Ngai Tahu', a unitary (and duplicitous) entity.

Subsequently, and in another context, Mark Solomon explained that, '"Ngāi Tahu" is a term that can be used to mean a number of different things, from the corporate structure, to an individual marae-based community (Papatipu Rūnanga) through to individuals of Ngāi Tahu whakapapa (genealogy).'[42]

Commenting on the Eyrewell pilot farms in early 2013, Mike Sang, the chief financial officer of Ngāi Tahu Holdings (formerly chief financial officer with agricultural supply company PGG Wrightson) was reported as saying that these were being assessed in economic and farming terms. If they are judged successful on these measures, Ngāi Tahu will apparently 'plough ahead' with further conversion of former forestry lands. The only direct reference to ecological enhancement was the comment that shelterbelts would include 'native plantings'. It was also mentioned that over 9000 hectares of the land earmarked for further expansion was located near the Hurunui River where 'the tribe had invested in an irrigation project around the Hurunui which was going to the consenting phase' and was 'also involved in a project to source water from the river via an irrigation scheme'.[43]

Following the publication of this interview, a correspondent to the *Press* noted that 'there appears to be a cultural conflict of interest when comparing this [commercial] stance to those of local hapu and runanga, where the protection of their land and water is a legacy for future generations'. The *National Business Review* similarly noted in 2009, in the context of Ngāi Tahu Property's withdrawal from the HWP, that it was 'walking a tightrope to assuage cultural concerns of grassroots members … while seeking to advance redevelopment of its Balmoral forestry for dairy conversion'.[44]

Can TRoNT's seemingly divergent aims in relation to dairy farming be reconciled? Ngāi Tahu Property chief executive, Tony Sewell, projects a belief that they can. Following the announcement of Ngāi Tahu Property's dairy plans, Sewell stated that the proposal gave Ngāi Tahu the opportunity to 'take ownership positions and leadership positions' vis-à-vis dairying and water management. The then General Manager of Tribal Interests (which oversees Toitū Te Whenua, the tribe's environmental advisory team) supported this view and argued that it enabled Ngāi Tahu to create and then meet standards that meet iwi expectations. Having thus raised the bar he felt that it would give the tribe 'greater leverage to … expect the rest of the community to step up'. Sewell described this transition from reactive opponent of dairy farming, to active but measured proponent, as a 'philosophical shift' on the part of TRoNT. He also sought to assure Ngāi Tahu readers that the whole enterprise would be based on 'a lot of science and … modern thinking'.[45]

Members of Ngāi Tuahuriri, an important Ngāi Tahu hapū, have witnessed declining freshwater quality in recent years. This includes the loss of good drinking water at their Tuahiwi settlement north of Christchurch and the presence of toxic algal blooms in nearby 'playground' rivers. This is why this papatipu rūnanga had opposed local dairying practices. Even so, rūnanga spokesperson Clare Williams felt that dairying could be conducted sustainably and that Ngāi Tahu Property could achieve this. In support, she cited Tumara Park, a subdivision in Burwood, a Christchurch suburb, which Tuahuriri originally opposed. However, the company built a stormwater management system in a way that protected an adjacent wetland, thus easing rūnanga concerns.[46] Even so, the subsidiary entities of Ngāi Tahu Holdings are not renowned for acting in accordance with the traditions and values of flaxroots Ngāi Tahu. Ngāi Tahu Seafood is a case in point.

Ki tai

One of the 'three pillars' of TRoNT's commercial base, Ngāi Tahu Seafood is a substantial owner of commercial pāua quota in southern New Zealand, much of it purchased on the open market. A deeply valued food in Māori families and communities past and present, pāua has also become a valuable export commodity, mostly in Asia. Though harvested by hand via freediving, rather than with underwater breathing apparatus, pāua stocks have come under considerable pressure – to the point of virtual extinction in some places. This has spurred Ngāi Tahu groupings, including papatipu rūnanga, to apply for mātaitai, usually with assistance from Toitu Te Whenua in the Christchurch-

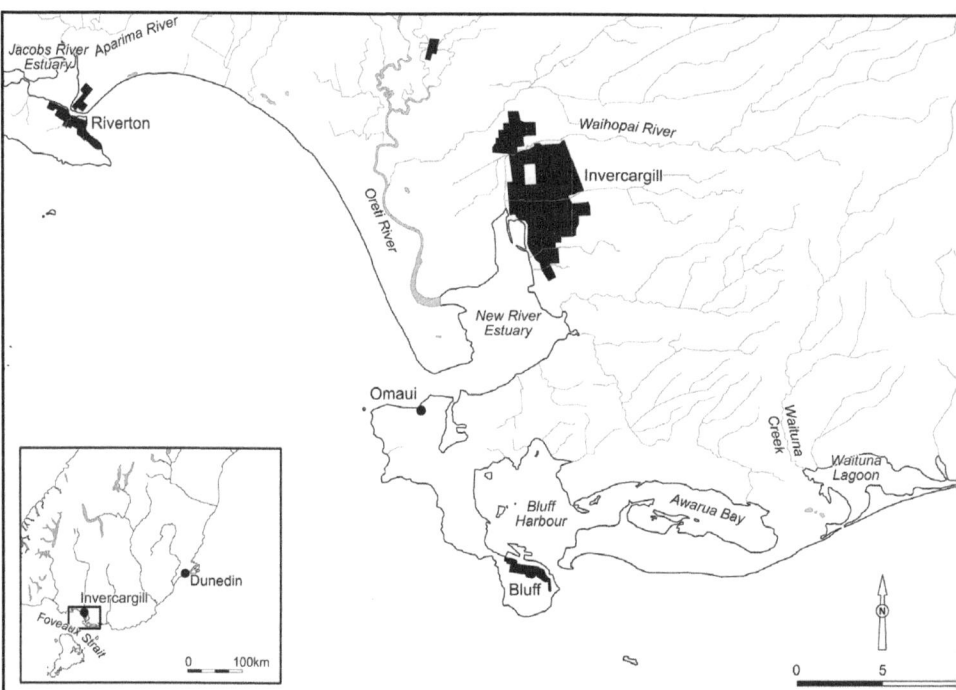

Figure 17.4 Map of places in Southland mentioned in the text.

based office of TRoNT. Mātaitai are a form of reserve designed to give effect to legally recognised non-commercial Māori fishing rights. They reduce the impact of commercial fishing, potentially allowing overfished areas to recover, and enable local community management of local resources.[47] The default position of commercial pāua fishers, therefore, is to oppose mātaitai applications or at least have them reduced in size. In the mid to late 2000s, Ngāi Tahu Seafood, as a quota owner, apparently did just this.[48] Once again, members of the public, if they noticed, saw 'Ngāi Tahu' applying for mātaitai and 'Ngāi Tahu' opposing mātaitai.

In early 2013, a proposal to open up areas of coastline in Otago and Southland currently closed to commercial pāua divers was vigorously opposed by a diverse segment of the community, including Ngāi Tahu individuals and families, many of whom were supported by Toitū Te Whenua which also submitted against the proposal on behalf of TRoNT. However, TRoNT did not publicly announce a formal position. Many felt that its silence spoke volumes. One of the areas proposed for opening up is on the western coast of Omaui, which is located near the mouths of the Oreti and Waihōpai rivers downstream from Invercargill, and near the industrial port town of Bluff, the base of Awarua Rūnaka, another papatipu rūnanga of TRoNT (Figure 17.4).

Omaui was an important village in Ngāi Tahu history, both before and during the colonial encounter; it was one of five parcels of land in Southland reserved for Ngāi Tahu under the Murihiku Purchase in 1853.[49] For most of the twentieth century, Ngāi Tahu families (including my own) engaged in the seasonal tītī/'muttonbird' harvest, sourcing rimurapa and other types of mahinga kai from its coastline. The rimurapa was used to construct pōhā – vessels that hold preserved food – in this case juvenile tītī.[50] However, after a wool scour was built on the edge of the New River Estuary in the mid-1970s the kelp became polluted and completely died. The closure of the wool scour in recent years,

Figure 17.5 Kura-mātakitaki Stevens and his great-grandfather, Tiny Metzger, prepare pōhā in Bluff, using rimurapa from Omaui in advance of the 2012 tītī harvest. Source: author

perhaps assisted by upgrades to its neighbour – Invercargill's Wastewater Treatment Plant – has however led to a remarkable recovery in these kelp beds. So much so, that our whānau was able to harvest rimurapa for pōhā from there in 2012 for the first time in nearly 40 years (Figure 17.5). This was fortuitous. Our replacement source of rimurapa during those years, Kaka Point, near the mouth of the Clutha River, was itself seriously polluted in 2011.[51]

Whether or not this positive development at Omaui will persist, however, remains to be seen. In November 2012 it was reported that the eutrophic area of the New River Estuary, which empties into Foveaux Strait via Omaui, had increased dramatically since the year 2000. Mud, nitrates and phosphorus were being washed into the estuary at such a rate that between then and 2012, the eutrophic area of the Waihopai Arm had increased from 3 hectares to 123 hectares. Sediment was 200 mm higher in 2011 than in 2007, and seagrass that was abundant as late as 2007 is now almost nonexistent.[52] Levels of noxious hydrogen sulphide – which creates a rotten-egg smell when matter breaks down without oxygen – have increased. A similar picture is unfolding in the Jacobs River Estuary near Riverton, an important natural feature within the catchment of Oraka-Aparima Rūnaka, another papatipu rūnanga of TRoNT.[53] Both the New River and Jacob's River estuaries thus face the same fate as the internationally significant Waituna Lagoon and Awarua wetland system northeast of Bluff.

Waituna

Although it is a registered Ramsar site, intensive dairy farming takes place on the margins of 1350-hectare Waituna Lagoon. Moreover, consented dairy cow numbers have more than doubled within its 20,000-hectare catchment since the year 2000. There are now 40 dairy effluent discharge consents within the catchment, up from 28 in 2000.[54] The lagoon has undergone increasing eutrophication, and in mid-2011 it was announced that it is on the verge of 'flipping' into a 'toxic and turbid algal soup'.[55] At the official Ngāi Tahu commemoration of Waitangi Day 2012, which was held on Te Rau Aroha marae, Bluff, the Upoko of the Awarua Rūnaka hosts, Sir Tipene O'Regan, addressed those assembled on the topic of Waituna Lagoon.

O'Regan argued that the ecological health of the lagoon, and other wetlands similarly, was linked to the core of Ngāi Tahu tribal health. Without them 'we wouldn't be able to put food on the table'. It was not simply about the preservation and restoration of Mother Nature for the sake of it, but about maintaining the practices and rituals of mahinga kai: 'a central component of who we are as a people'.[56] Ngāi Tahu historian Te Maire Tau describes the elder recorded by Herries Beattie, referred to above, as a victim both of colonisation and change, and also of his own mind and belief system. In erasing his pond, an object with which he strongly identified, the elder's 'image and therefore self had also been erased'.[57] O'Regan's comments indicate that sentiments like this persist among Ngāi Tahu. The whole episode also illustrates that a constitutional property settlement, even one that responds directly to customary rights in the area of mahinga kai, is no protector of the ecologies necessary to give them meaning.

Commenting on Southland's degraded estuaries, the *Southland Times* wrote that the 'finger of suspicion points at farming ... [but] it doesn't just point – it jabs ... rudely in the ribs'.[58] Dairy farming in particular has come in for heavy criticism off the back of its massive expansion in the region. In 1994 the dairy cow population in Southland was 114,378. By 2009 that number had grown to 589,184 (just over 10 per cent of New Zealand's total).[59] However, the newspaper editorial also pointed out that general land use, stormwater and sewage are also contributing factors to water degradation, which is correct. In February 2012, the Awarua plant of South Pacific Meats, located on the edge of the New River Estuary, discharged a large volume of meat processing effluent onto land, some of which made its way to coastal waters. This was found to be the result of both design and management faults. The Environment Court judge who sentenced the company (part of the Talley family empire, one of the largest privately owned companies in the South Island) commented that it had shown no remorse but instead a reluctance to admit its ongoing failures.[60]

Conclusion

The current chief financial officer of TRoNT, Arihia Bennett, appointed in August 2012, lives at her traditional kāinga, Tuahiwi. More than that, she and her family designed and built a multigenerational household. She is thus rightly described as 'a person who lives by the values that are espoused by many [Māori], but not necessarily followed'.[61] Notwithstanding, she was recently told that she 'did not know much about mahinga kai'. The led to her to reflect on childhood days:

> Dad bringing home his beloved kina ... Mum and her magnificent ways of presenting crayfish. The muttonbird tins that would arrive at a certain time of the year.
> Whitebaiting adventures starting at absurdly early hours. The neighbours always had eels hanging in their garage before they popped them into their homemade smokehouse.

She concluded that she knew how to benefit from mahinga kai as a child, but her appreciation 'of the cultural and social values that lay beneath was somewhat lacking'. Her new role, however, has catapulted her into the world of mahinga kai, 'and more specifically the worry of polluted waterways, eroding ecological habitats and land use intensification'. As she says, 'I now have my eyes wide open when I travel up and down Te Waipounamu watching the kilometres of irrigation and wondering about all that water.' Not only are 'the blinkers off', she has undertaken to be 'guided by our whānau to ensure I strategically advocate in our best interests to bring about the restoration and preservation of our precious environment'.[62] Bennett's sentiments dovetail with Mark Solomon's comment in late 2011 that iwi members were sceptical of the Eyrewell dairying proposal being led by Ngāi Tahu Property. At that time he stated that unless the farms are 'fully sustainable with no negative effect on waterways' they would be 'sure to stop the project'.[63]

It remains to be seen whether or not this dairying project is environmentally sustainable, and if it is not, whether TRoNT will abandon the scheme. There will be significant implications either way. One interesting factor to reflect on in closing are those Ngāi Tahu individuals who are registered with the iwi authority but who do not live in the tribe's heartland communities, or even within its takiwā: more Ngāi Tahu actually live outside the South Island than within it.[64] However, all registered members have the opportunity to share in direct distribution benefits such as Whai Rawa – the tribe's matched savings scheme; the Ngāi Tahu Fund – a dedicated fund for promoting and reviving cultural traditions; kaumātua grants; and tertiary education scholarships. TRoNT funds all of these things out of profits generated by the subsidiary entities of Ngāi Tahu Holdings Group. But one wonders if heartland Ngāi Tahu communities might seek to place limits on activities of these subsidiary entities if or where they negatively impact on the lifeways of these communities, and if they might be successful in doing so. Given that this would reduce funding available for direct distribution, how would the majority of tribal members – those removed from heartland communities – respond? These are the types of challenges that TRoNT and other iwi authorities might be increasingly forced to grapple with in post-Treaty settlement environments.

In 1987, the same year that the Waitangi Tribunal began to hear the Ngāi Tahu Claim,[65] a collective of Ngāi Tahu families in Kaikoura established a whale-watching ecotourism operation. This was one response to macroeconomic reforms and resultant high unemployment that devastated a number of Ngāi Tahu communities, including Kaikoura, in the 1980s. The founders of Whale Watch Kaikoura made enormous sacrifices, including mortgaging their own homes, and overcame considerable Pākehā antagonism, to develop what has become one of the country's leading tourism experiences.[66] The whole town of Kaikoura has benefited immensely from this. The flow of assets and capital to TRoNT following the Ngāi Tahu Claims Settlement Act 1998 thus held out hope to many heartland Ngāi Tahu people that similar regional ventures might subsequently unfold. One reason in particular why heartland Ngāi Tahu families and communities held Whale Watch in high esteem was because of its three core principles: the company undertook to pursue things that were culturally acceptable, economically viable, and environmentally sensitive.[67] Time will tell whether or not a TRoNT-owned dairying operation, or some of its commercial fishing activities, can or will be held to these values.

Te Maire Tau describes the way in which Ngāi Tahu functioned after World War II 'as small rural family collectives engaged in farming, fishing and traditional food-gathering'. He notes, though, that this variant of the tribe 'is disappearing into the mists'. 'Explaining this transition,' he adds, 'will be the task of future historians.'[68] In doing so, these historians, in keeping with established explanations of colonialism in New Zealand, will point to continued environmental degradation done *to* Ngāi Tahu interests by non-Ngāi Tahu interests – colonialism was always unfinished business,[69] but it is an ongoing business too. However, these same historians might also point to environmental degradation done to Ngāi Tahu interests *by* Ngāi Tahu interests. For while TRoNT has growing political power in the South Island landscape, which can be deployed to achieve better environmental outcomes according to Ngāi Tahu values, much of this power rests on its commercial base, the operations of which can potentially impinge on the environmental values and traditional lifeways of heartland Ngāi Tahu families and communities. Unutai?

Further reading

For more insight into the freshwater issues facing Ngāi Tahu, and its varied responses, the quarterly magazine issued by TRoNT, *Te Karaka*, contains a wealth of information, especially issues 33, 34, 38 and 49 to 54 inclusive. For an outline of freshwater components of Treaty of Waitangi-based settlements, see Linda Te Aho, 'Ngā Whakataunga Waimāori: Freshwater Settlements', in Nicola R. Wheen & Janine Hayward (eds), *Treaty of Waitangi Settlements* (Bridget Williams Books, Wellington, 2012, pp. 102–13). For an overview of the Māori legal position vis-à-vis freshwater, see Jacinta Ruru, 'Property Rights and Māori: A Right To Own a River?' in Klaus Bosselmann & Vernon Tava (eds), *Water Rights and Sustainability* (New Zealand Centre for Environmental Law, Auckland, 2011, pp. 51–75); 'The Right to Water as the Right to Identity: Legal Struggles of Indigenous Peoples of Aotearoa New Zealand' in Farhana Sultana & Alex Loftus (eds), *The Right to Water: Politics, Governance and Social Struggles*, Routledge, Abingdon, 2011, pp. 110–22; 'Undefined and Unresolved: Exploring Indigenous Rights in Aotearoa New Zealand's Freshwater Legal Regime', *Journal of Water Law*, vol. 20 no. 5–6, 2009, pp. 236–42.

18.

Mastering the land: mapping and metrologies in Aotearoa New Zealand

Andreas Aagaard Christensen

THE ENVIRONMENTAL HISTORY of New Zealand is one of the clearest and most recent examples of the way humans make a home for themselves in newly explored territory. New Zealand was the last major land area in the world to be colonised by people and, given its extraordinary natural history, the first settlers could hardly have been more surprised when they arrived in the thirteenth century. At the time of this first Polynesian settlement, New Zealand was a land not only without humans, but without any terrestrial mammals except for a few species of bats. In their absence the avifauna had proliferated, and in ecosystems developed with birds as the only large grazers, the flora had developed in ways not seen anywhere else, leaving only limited plant foods available for humans. This must have made New Zealand not only a challenging but also an initially incomprehensible land for newly arrived Polynesians as well as Europeans. This fact makes their success in forging cultural landscapes from the new land all the more interesting for students of environmental history.

As an example of such processes, New Zealand illustrates the way human newcomers learn to master an environment, change the land and its resources, and in the process change themselves. From the 'fragile plenty' of the first Māori to the cultural landscapes in which they lived at the time of the first European discovery, to the settler economy and the modern society of today, New Zealand is an example of the way a society develops on the basis of natural resources which change as the society itself changes. Newcomers to any environment meet it with a set of technologies and a culture which they bring with them and which changes continuously, as it aligns with experience gathered in that environment. The environmental histories told from a multiplicity of viewpoints in this volume are contributions to our understanding of this central dialectical relationship, which over time led to the creation of the landscapes and ecosystems of contemporary New Zealand.

This chapter picks up on a theme which has been touched on in most of the preceding chapters, but which has not been fully unfolded. It argues that while conditions and events changed the relationship between society and environments repeatedly, the history of New Zealand was always a history of spaces and of the ability of its inhabitants to control space and resources cognitively, socially and physically. With this perspective in mind the chapter outlines the history of production of spatial knowledge about the environments of New Zealand. This is not only to provide an overview of understandings of the environment, but also to investigate and illustrate the close ties between knowledge and practice: between understanding the environment and changing it.

From the narratives employed by Māori settlers to codify and recall important aspects of the environment, to the early maps of European explorers, and from the survey maps of the nineteenth century which commodified the land, to the environmental management and monitoring efforts of the postwar period, New Zealand environments were continuously charted, measured, monitored and mapped. It was through such measuring practices that the resources of New Zealand were discovered, appropriated and organised socially. And the practice of measuring was regulated through systems of measurement called 'metrologies': socially acceptable ways of codifying and describing aspects of the environment, which were brought to New Zealand and adapted to the conditions there by both Maori and European colonisers. The single most important and contested resource to be measured was the land itself, making maps and other map-like metrologies central to understanding the way spatial resources were organised and changed over time.[1] The two types of encounters evident in the history of New Zealand – the encounter between cultures and ecosystems and between two cultures – have one critical characteristic in common: they happened in contested spaces and were centred on the legitimacy and power to take possession of spatially distributed resources for sustenance and wealth.

Māori spaces: narratives and performance

When the first Māori colonists arrived in New Zealand from East Polynesia, their use of the available resources made a drastic imprint on the environment. Starting from an initial state of settlement based on fishing and the hunting of seal and moa, the Māori were able to sustain a rapid population growth, which slowed when the most easily accessible protein food sources became exhausted. In the aftermath of the extinction of large avifauna and with the decimation of seal colonies on the coast, Māori society evolved into a stable state characterised by the establishment of cultural landscapes supporting horticulture alongside hunting and fishing activities. In this period most of the great forest areas of both islands, which contained few other sources for human sustenance than the moa, were removed with fire and replaced by successional fern and tussock cover that could survive continuous burning.[2] This allowed for a diet of starch-rich fernroot to develop and opened up large areas for easy inland travel. It also allowed for the country to be examined more easily, and the improved spatial resolution of the landscape is likely to have influenced the development of the institutions of territory and tribal customary land rights which took place in this period.[3]

As Atholl Anderson has shown in the second chapter of this volume, the organisation of Māori society went through a series of changes in parallel with the changes it wrought on the environment. Horticultural settlements of the type still prevalent at the time of the first European encounters in the eighteenth century represented a different use of space from the early more transient societies that had less investment in specific landscape resources. The development of horticulture as the primary provider of sustenance necessitated a number of new social institutions and led to the gradual establishment of territories and institutions for resource use allocation between whānau, hapū and iwi.[4] Continuous improvement of the resource base in cultural landscapes of this type further advanced the need for such social institutions. The settlement types that were most widespread at the time when European settlers arrived in the nineteenth century were the result of such processes of social adaptation to environmental change and changes in land use over a prolonged period of time.

While knowledge of the processes that led to the creation of nineteenth-century Māori land use systems is limited, the tenure system that was the eventual result is well understood. As Evelyn Stokes demonstrates in Chapter 3, the Māori society which European settlers were faced with when they arrived was structured around a system of tenure based on specific types of customary land use rights. It was a relationship between land and people centred on use value, which represented a striking contrast to the Judeo-Christian parallel conception of resources and land as something wholly governed by proprietary rights. It was also a conception equally estranged from ideas of capital gain and investment in measurable resources – the kind of systematic transfer of value between social contexts that is characteristic of most Western societies.

Māori tenure and spatial knowledge infrastructure was usufructuary. It was based on a distribution of land use rights among agents, not a distribution of the land or the resources themselves.[5] Traditions and histories of land use (including land use conflicts) relating to specific areas constituted the main point of reference for the establishment of rights to further land use, and as such the tenure system was based indirectly on occupation. Histories of occupation and land use thus formed the basis for claims made regarding legitimate access to resources. This meant that resource use rights overlapped geographically and spatially, with people from different social groups potentially using the same areas at the same time for different purposes, or for the same purposes in a way agreed on or enforced socially. This was a system which did not embrace an institution of private ownership or a concept of alienation in the way known in the west, but which nonetheless was governed through politics of kinship and enforced by military power when needed.[6]

These usufructuary and multifunctional rights to resources in the landscape were communicated and reiterated by way of whakapapa (narratives) recounting mana whenua: genealogies of historical occupation and resource use.[7] This is a system found in many societies that developed without a written language, where narratives were remembered with direct reference to landscape features used as mnemonic cues when retelling the history of the environment.[8] Because Māori topographical knowledge was stored and structured primarily using a narrative technique of this kind, the retelling and transfer of such knowledge about the environment was often performed in the landscape setting with direct reference to the features mentioned. Drawings made on the ground or on other temporary media were also used, but primarily in situations where direct reference to the landscape was impossible or impractical, and as a minor part of the practice of remembering through spoken narratives.⁹ In short, it may be argued that the topographical knowledge infrastructure of Māori society was presentational. It was centred on presenting the environment to a listener through the use of narrative, rather than representing it in an object, a symbol or a text.¹⁰

When the British explorers and colonisers – the Pākehā – arrived, they brought with them a completely different way of conceptualising space, a different way of understanding the relations between resources and people, and a different political order: all incongruent with those of Māori.

European encounters: placing New Zealand on the globe

Documented European experiences with New Zealand stretch as far back as 1642 with the arrival of Abel Tasman, who charted the western coastline of New Zealand and returned to Europe with the first evidence of a land long mythologised in European scholarship.¹¹ Europeans may have visited before and many European maps at the time charted a coast at the approximate location of New Zealand, said to have been 'first discovered by a Spanish ship severed from her fleet and driven here along in the southern sea'.¹² Such ideas, whether based on experiences or tales, abounded in Europe at the time and had done since the sixteenth century, forming an intricate mesh of geographical imaginations of the great southern sea. Tasman was, however, the first to bring back documentary evidence, and evidence was important beyond any measure in European culture at this time.¹³

The evidence that Tasman brought back was in the form of maps: in essence a special type of drawing developed since antiquity, which was used to store and communicate experience with surface features of the planet.¹⁴ Maps were regarded as documents of great authority in Europe at this time (as modern maps are today) and were used both as practical spatial tools for orientation and taxation purposes, and as devices to claim territory and demonstrate power and majesty.¹⁵ Because of this authority, maps were used not only for communication, but also as evidence: as replacement experiences. Maps provided a way to effectively transport aspects of a transient experience from the Pacific to Europe or any other destination – and from surveyor to decision-maker and landholder.

The first explorer of New Zealand to tap effectively into the power of representation inherent in the map medium was Lieutenant James Cook, who arrived in 1769 and charted the littorals of New Zealand as the second European observer. His maps of the coast and coastal waters were of such lasting importance to later European and New Zealand navigators and colonisers, that parts of them were still in active use until 1997.¹⁶ With his maps, New Zealand became a possible destination for Europeans, a place among other places on a world map, illustrated in terms of distance and direction from other known territories. This, in essence, was the nature of early European depictions of New Zealand. Through their measurement practices Cook and the navigators who followed him prepared a vital piece of infrastructure for later colonial endeavours, while at the same time adding images of antipodean environments to the appetite for land which was developing in Europe at the time.

Whalers, sealers, lumbermen and merchants in flax and fish were the next to arrive in New Zealand, using their own experiences in combination with the spatial infrastructure provided by Cook and his successors to navigate there. By trading with the Māori those early colonists, the first Pākehā, were able to secure access to the resources and labour they needed to extract valuable commodities for sale on the world market, which was coming into existence at the time as a consequence of such activities. The most important terrestrial commodities were kauri spars for shipbuilding, flax for the weaving of canvas and rope, ship supplies and the range of marine animal products still available. The growing interest in inland resources led to a need for knowledge about the interior of New Zealand, and many of the early maps made to serve that purpose were done with the help of Māori. Willingly and unwillingly, they shared their acquired knowledge of the environment with Europeans, who captured fragments of the information on maps and in written language in order to transmit it to others.[17]

In such instances, where knowledge of the environment was transferred between cultures, the difference between Pākehā and Māori modes of storing and shaping spatial knowledge became clearly visible. There are numerous examples of such exchanges, and in most cases the setting was that of a knowledgeable Māori person recounting a narrative while illustrating it for his European audience with lines and surfaces marked in sand, with chalk on a ship's deck, or on another such surface.[19] The Māori who were employed or forced to service British map makers were *performing* spatial knowledge – shared by way of oral spatial mnemonics – while to the British it was the map as an objective instrument of knowledge storage and retrieval which was thought to perform.

When compared to the Māori tradition of reciting whakapapa to establish mana whenua, the objective and legitimising character of European maps illustrates something quite extraordinary in the makeup of European culture – namely the particular significance of the idea and practice of objectivity itself. For while objectivity was often held high as an ideal for knowledge production in Western societies (and still is today), it was also the privileged counterpart to a host of alternative attributes of knowledge. The ideal of objectivity thus structured an important dichotomy between value and truth, which enabled a seemingly well-defined division line to be drawn in civic life, between authoritative knowledge production on one side, and political process on the other. As the French philosopher Bruno Latour has pointed out, the culture of objectivity that enforces this distinction between autonomous knowledge (objectivity) and independent morality (subjectivity) forms one of the constitutions of Western systems of authority.[19] And maps, which were most often designed with a specific interest in mind, but which were also considered to be authoritative evidence, were a tool able to transgress the partition of civic life and be both scientifically objective and politically biased at the same time. In this way maps were able to provide potentially unquestionable authority for political claims that would otherwise lack such authority, thus '[rendering] ordinary political life impotent through the threat of an incontestable nature'.[20] In comparison with whakapapa whenua, the map was a powerful tool in the colonial setting. For while the European map could be construed as a piece of objective evidence (able to replace fieldwork and direct observation), the whakapapa whenua narrated aspects of the political, social and spiritual history and reality of the landscape – a completely different purpose, which was of much less significance to the Pākehā officials charged with subdividing New Zealand.[21]

As many of the previous chapters have highlighted, these key differences led to continuous conflicts and bitter grievances, because Māori and European conceptions of authority, legitimacy, land and tenure expressed in terms of the spatial cultures described here, were often mutually exclusive. As a consequence, a large part of the recent history of New Zealand has had to do with differences stemming from disputed spaces. As Evelyn Stokes and Danny Keenan have shown, the conflicts reached their highest point with the work of the native land courts from the 1860s, when British ways of conceiving of tenure and spatial evidence was openly forced on the Māori both officially and in practice.[22] Tenure existed in the eyes of the native land court only if substantiated with evidence in the form of maps or similar documents or statements able to relate claims to the land in question. There are numerous accounts of Māori customary land users singing and narrating their claims to usufructuary rights over prolonged periods in the court, but to no avail. In effect, land could only be held by those who could conform with the spatial technologies endorsed by the court. This position forced many Maori chiefs to react violently, bringing the country into a prolonged series of civil wars which eventually ended in the 1870s when the Crown subdued Māori resistance.

Figure 18.1 Surveyors cutting down bush to make survey pegs, 1903. Pegs were used as proof of tenure inscribed on the landscape and their use was often contested. In 1843, both Māori and Pākehā lives were lost in one such altercation, known as the Wairau Affray, in the northern part of the South Island. Source: 1/2-111812F, National Library of New Zealand, Te Puna Mātauranga o Aotearoa

From then on, because of the success of European colonisers in repressing open resistance, the spatial organisation of New Zealand environments became modelled mainly on European traditions. The rural New Zealand we see today, with its well-delineated grid of property boundaries and its clearcut lines of division between conservation estates and productive land, is partly the result of that heritage of British cultural practices and environmental knowledge technologies. The whole territory is filled out by properties defined by way of cadastral maps and marked by hedges, roads, land use borders and fencelines in the physical landscape. This structure emerged during the latter half of the nineteenth century, when initially there were no such points of reference for marking boundaries and when the standard way to show property boundaries was by using wooden pegs hammered into the ground (Figure 18.1). In this way possessions could be defined even in landscapes where there were no visible or otherwise evident relationships between landscape features and proprietary rights: a practice which attains significance and meaning only in the context of a culture which embraces institutions of private property and documentary spatial evidence.

Making space for settlement: surveying, services and security

When European settlement gained pace in the 1840s and the New Zealand population began to grow rapidly, spatial infrastructure was systematised for the first time. Colonists of many trades were arriving from Europe and were in need of land. And in order to buy it from the Crown that land needed to be well defined on paper. Thousands of surveys ensued to fill the need, and the great majority of maps made of New Zealand land areas in the first part of the nineteenth century were survey maps. A lands and deeds registry was set up in 1841 to care for the storage of land records and maps, and in 1852 six provincial survey offices were established, to take care of the subdivision of land under provincial government.

At this time New Zealand was part of a Pacific frontier in the process of being included into the world commodity market – making it possible for the first time to build export economies.[23] Investments were flowing into New Zealand and maps were needed in order to secure the capital being invested. But as the blank spaces between the isolated survey maps were slowly filled out, these maps alone were no longer enough to secure the safety of capital, the cornerstone of the colonial economy. What was needed was a cadastre: a system of survey maps organised and embedded within a larger scale map of cadastral blocks, which would ensure that all claims were coordinated and never overlapped. The creation of such a system was undertaken in 1877, initiated following criticism from British observers and experts visiting New Zealand at the time. They included a prominent surveyor, Mr Henry Spencer Palmer, representing the Ordnance Survey of Great Britain. He arrived in New Zealand in 1875 and in his report to the government on the state of the surveys, he said that in his opinion the current survey efforts were of little use compared to 'a cadastral map on the correctness of which all men may agree, and which will give safety and value to Crown grants, and protect individuals from litigation, and Government from the risk involved in the issue of land titles under the Land Transfer Act'.[24] The cadastral system that was planned based on these recommendations was not finished until the 1960s, but remained a project of high priority to governments in the prolonged period until it was complete.

The earliest survey maps took account of only just enough context for title boundaries to be discernible in the field. The property boundary pattern with its straight rectangular outlines was the central motif, and its pure geometrical form and economic implication was imposed on a barely legible underlying landscape. The excerpt in Figure 18.2a is an early example. It maps out land along the Whanganui River into a blank region of unknown territory beyond the capital frontier. The rationality of the map is about access along the riverbanks and on the river itself, where the availability of water for use by settlers on their new properties made land attractive. Land is the central resource which, combined with water and access, attains use value. This is then translated into exchange value through the social performance of the map. Seen from an anthropological perspective, the map can be seen to reflect a culture of individuality and a social order governed by the institution of private ownership, without which the production of such a map would have been inconceivable. From the era of early settlement onwards, larger and larger areas were taken over first by the Crown through barter with Māori chiefs and then by private land holders through the sale of land from Crown holdings using maps of this kind. In this way territory was commodified and usufructuary rights alienated, transforming land from a simple resource into capital, which could be sold or rented out from owner to labourer.

In the concluding decades of the nineteenth century, mapping efforts progressed to encompass larger and larger tracts of land, and surveyors began to produce regional scale maps that divided the country into coordinated blocks for further subdivision (Figure 18.2b). Such maps provided a much needed context for land sales and the development of infrastructure across larger areas, and this became standard practice in the 1880s and onwards. Like the early survey maps, the information contained in the regional surveys was limited to features of critical importance for land sale and development, with special care given to the mapping of wetlands and rivers that could block and allow for access. The maps of this period illustrate how surveyors were attempting to provide more comprehensive outlines of territory on the ground, while trying to overcome the limited perspective available when drawing maps of a landscape portrayed from above. Ridges were often marked only with shadings designed to indicate where difficult terrain had stopped the map maker from exploring further, and road plans given on the maps were like snakes of detailed measurements in territory which was otherwise unmeasured. Elevation information – which was critical to most road and land use planning – was only made available locally if needed, and even though such maps gave the impression of a bird's eye view of the land, they reflect a restricted perspective on a still largely unexplored but rapidly appropriated land mass.

During the first half of the twentieth century, map making in New Zealand expanded from its initial concern with resources and properties, to include a widening selection of physical landscape features. The introduction of aerial surveys in the 1920s offered a new perspective for map makers, and the perceived need for maps to be used by the armed

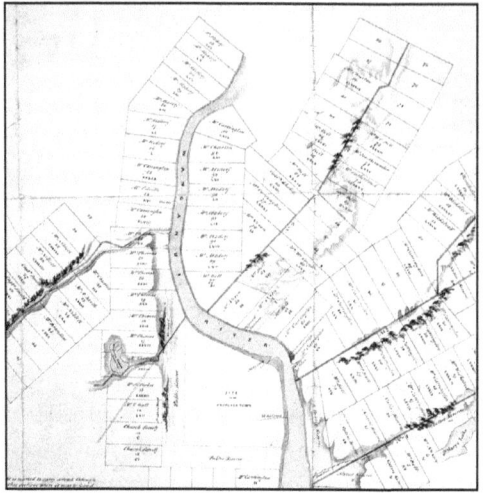

A. 1841, A cadastral map with topographical information

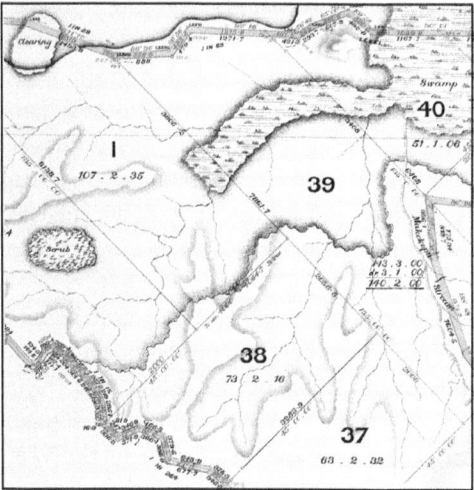

B. 1886, A topographic map with cadastral information

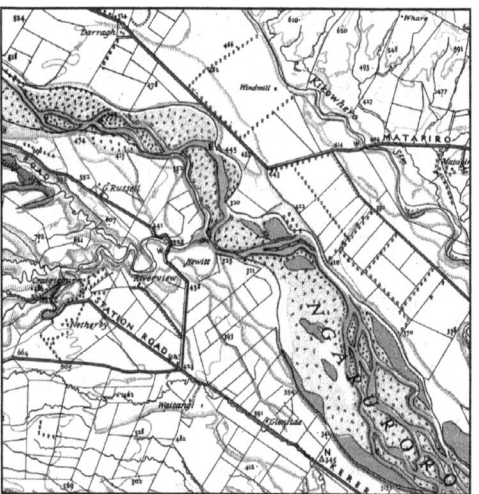

C. 1939, A military-topographic 'landscape' map

Figure 18.2 A hundred years of New Zealand map making. From an initial interest in commodification of land, it progressed to an interest in providing infrastructure and order, and climaxed with military maps to defend what had been gained. *Map A*: a survey of land titles detailing country sections laid out along the Whanganui River, by Kenneth Webster. *Map B*: an instructional example of how to design topographical maps by the Department of Lands and Survey. *Map C*: detail from the first sheet of the NZMS1 series. Sources: Map A: excerpt from 'Plan of the country sections laid out on the Wanganui', 1841, MapColl-832.41gbbd/1841/Acc.15388, National Library of New Zealand Te Puna Mātauranga o Aotearoa; Map B: excerpt from 'A specimen plan of a portion of a section survey of a block', in *Regulations and Instructions of the Survey Department of New Zealand*, 3rd edn, Government Printer, Wellington, 1886; Map C: excerpt from 'Napier and Hastings', New Zealand Map Series 1, sheet N134, Department of Lands and Survey, Wellington, 1939

services in preparation for war spurred a growing interest in detailed topographical mapping. Plans were made for a comprehensive topographical survey of the entire country, and when the first sheet of the New Zealand Map Series (NZMS) was published in 1939 it marked the establishment of a map-making tradition which is still practised today (Figure 18.2c). The new maps provided spot heights and contour lines, depicting the surface of the earth as a continuous surface for the first time. They also contained information on the location and extent of all the groups of phenomena we would expect to find in a modern map: trees, shrubs, field divides, roads, buildings and water features, all draped across a well-defined surface. This was due to the fact that the topographic maps were made primarily for military purposes. They were designed in the context of pre-World War II fears of a Japanese invasion and were intended to supply the information needed for officers to direct regiments and artillery fire in combat situations.[25] As a consequence, they were also well suited for other planning and management purposes.

The first topographic map series was completed in 1975, when all of New Zealand had eventually been photographed from the air and stereoplotted to obtain contour line information. Since then the maps have been updated regularly in a number of subsequent topographical mapping schemes, which have changed only slightly since their introduction in the 1930s. They have come to represent a contemporary baseline understanding of the environment, which would have been alien to many surveyors in the nineteenth century, but which now seems rather ordinary. This is because the maps depict the landscape in ways that correspond with the spatial culture of contemporary society. As Jan Kelly has expressed it, the maps which form an active part of our spatial culture today are characterised by being 'ordinary maps, so explicable that the transition to "real" landscape becomes seamless in our minds and the symbols are as easily read as is a written language, or are seen as the living landscape itself'.[26] This cultural condition was developed over a prolonged period of time and has left its clear mark on the landscapes of New Zealand from the 1840s until today, given that many of them have not only been mapped but also moulded into shapes relevant to social understandings of the environment. The early maps of New Zealand, and the environments they depict, reflect a social order which was implemented spatially through European settlement and which was based on concepts and practices of capitalist production, individual freedom and private property rights.

Socioenvironmental expansion: from settlement to society

The combined subject matter of New Zealand maps has expanded over time and today map making is a tradition with as varied a content as most other art forms and sciences. New Zealand society has also changed dramatically and map-making efforts have reflected those changes, as well as changes in the environment. From the early navigational maps and surveys for settlement, to the topographic–military map making in the period between the wars and into the postwar period, when environmental management became a topic of interest to map makers, maps have reflected shifts in the ways in which society has understood its environment. These shifts can be represented

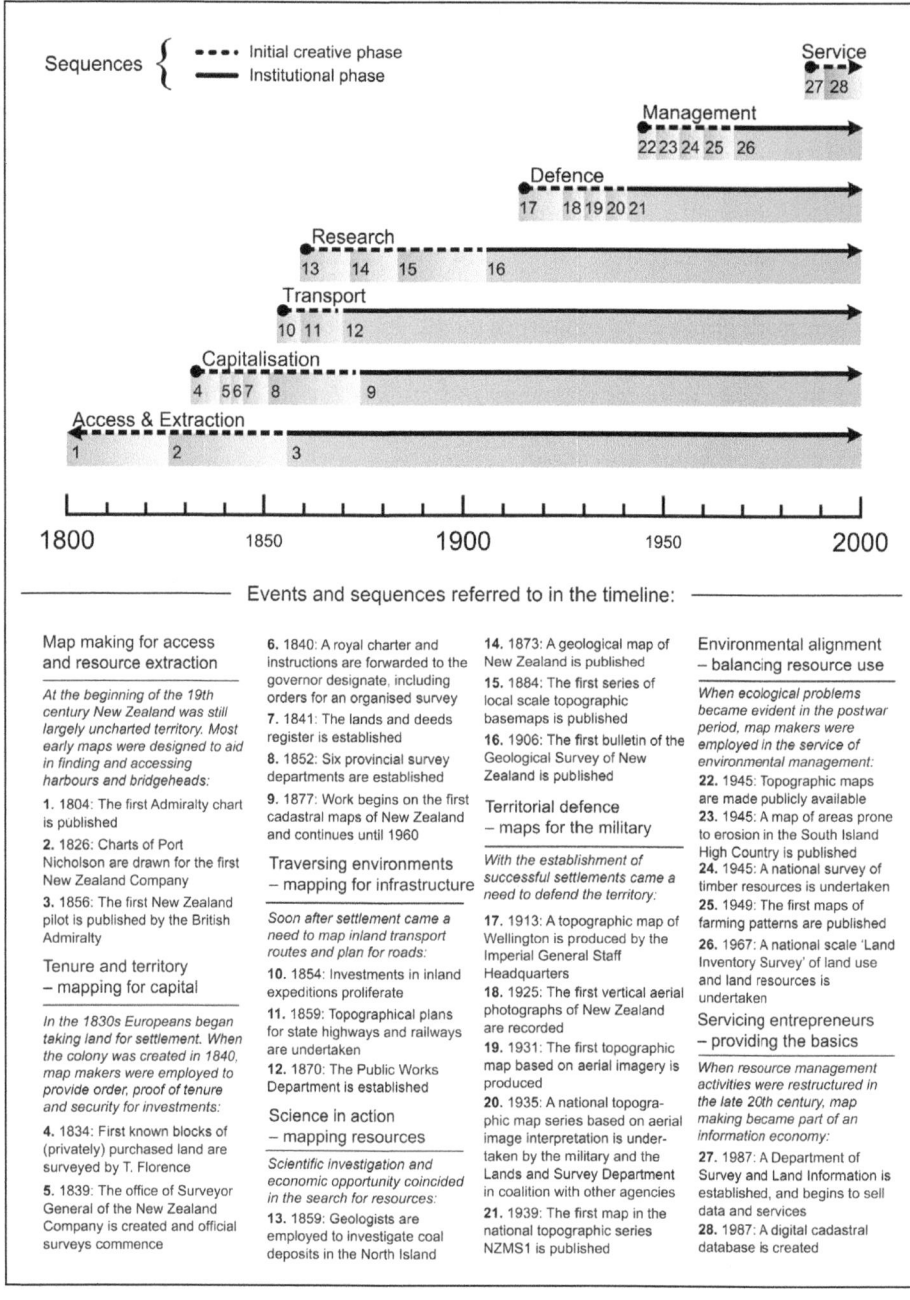

Figure 18.3 Key sequences in the production of geographical knowledge of New Zealand environments, 1800–2000. The reasons for mapping and measurement have changed over time, reflecting shifting societal needs. Only the seminal events in each novel map-making practice have been included. Source: based on information in Brian Marshall, 'From Sextants to Satellites: A Cartographic Timeline for New Zealand', *New Zealand Map Society Journal*, no. 18, 2005, pp. 1–110

as key sequences in the production of geographical knowledge (Figure 18.3). The major topics of interest for mapping and measuring have been: (1) access, (2) capitalisation, (3) transport, (4) research, (5) defence, (6) management and (7) service. This sequence reflects the way New Zealand society has evolved over time, but is also testimony to a history of constantly expanding investment in environmental knowledge production, where phenomena included in map-making practices are never dropped or considered obsolete. This history is parallel in many ways to the sequence of ever-expanding claims placed on New Zealand environments by society, which allowed map-making practices to evolve from being a field-based local or regional endeavour to becoming a large-scale knowledge industry that exploded in the postwar period with the introduction of cheap aerial imagery.

Recording the environment in the service of the state

In 1923 the New Zealand government bought its first airplane fitted with a camera, and in 1936 the private company New Zealand Aerial Mapping (NZAM) was formed to supply government departments with spatial information. From then on aerial photography became an important part of the way New Zealand was mapped and understood, and NZAM became the major producer of aerial imagery in the country, both in respect of civil and military recordings.

In the years following its creation NZAM worked primarily for the Public Works Department, the Department of Scientific and Industrial Research and for individual farmers in need of imagery when conducting land improvements. Many of the early recordings were consequently of wetlands, coalfields and mineral deposits in peripheral areas of rural New Zealand. And in the same way the surveys and topographic maps of the preceding decades had led to changes in the environment, most of the environmental knowledge gained through aerial photography was used directly to change the environment through drainage projects, infrastructure development and resource extraction.

When war broke out in September 1939 such work was interrupted, and all aerial map-making efforts were directed towards the provision of maps for the army, which desperately needed new ones because the only available topographic maps were 'devoid of any but the main cultural features, being designed mainly for land boundary title purposes' (Figure 18.4).[27] The initial priority was to map the more densely populated North Island extensively, to map the important 'fortress areas' around the major cities which constituted key defensive terrain for the army, and to map aerodromes and training areas. A total of 177 map sheets at a scale 1:25.000 were completed on this contract, establishing NZAM as a close ally to the government departments involved in map-making activities.[28]

The peace of 1945 led to the demise of military investment in map making and the opening up of new markets consisting mainly of smaller clients interested in local or regional photomosaic surveys. As the director of NZAM put it, the postwar period was about '[producing] aerial survey maps in quantity, making them available at a rate which even the smaller local body will find reasonable'.[29] The transition to civil aerial

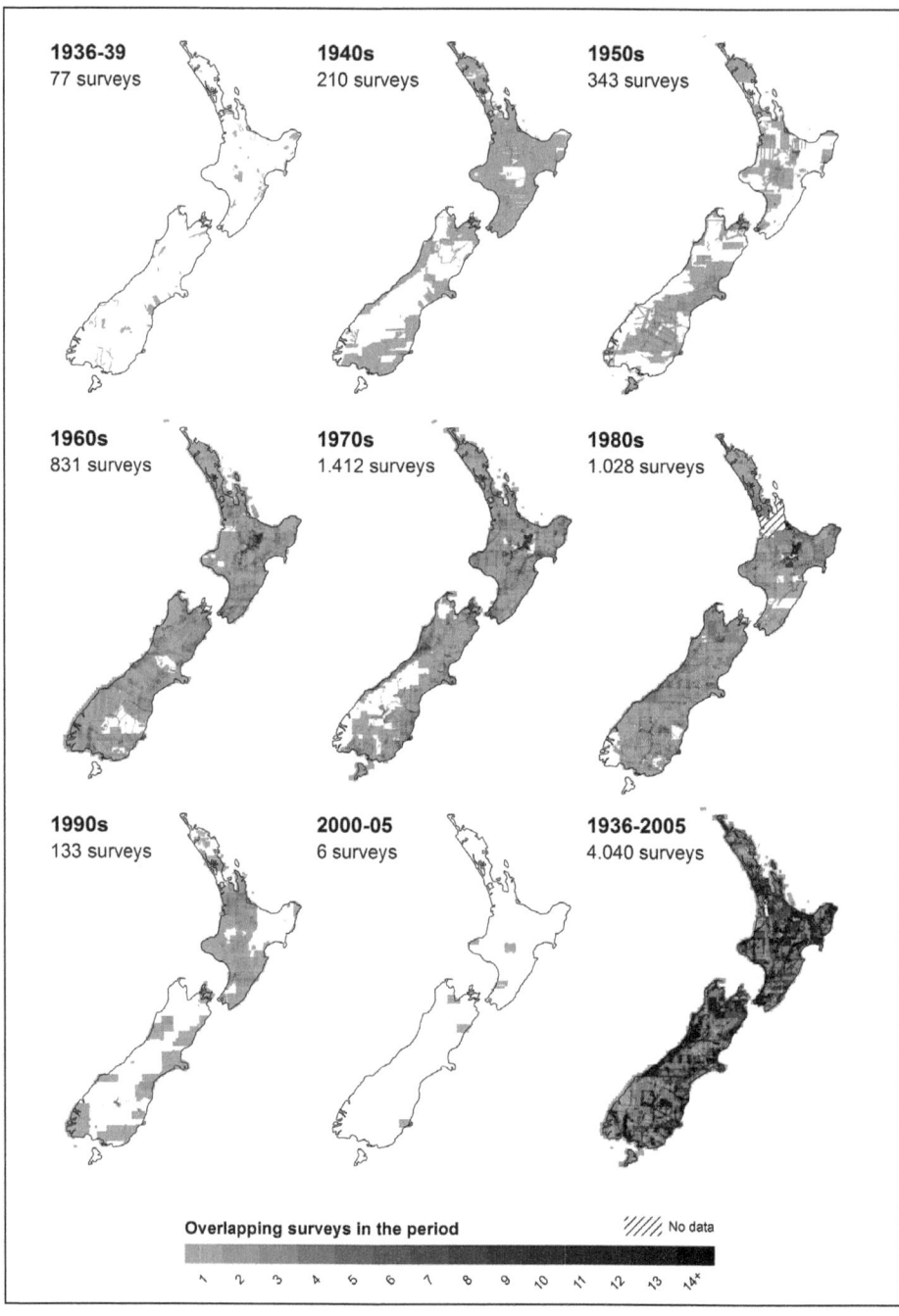

Figure 18.4 Recording New Zealand, 1930–2010. The geographic distribution and magnitude of aerial photography recorded for the New Zealand government for each decade from 1930 to 2010. Source: based on records of photography conducted by New Zealand Aerial Mapping for the New Zealand government, available from Land Information New Zealand (LINZ) at www.koordinates.com

Mastering the land 323

Figure 18.5 Production systems under siege in the Wairarapa district, 1959. A countryside battle plan is in the making at a conference in Wellington in an attempt to stop the spread of hazardous organisms through the use of centralised systems of spatial intelligence.
Source: EP/1959/1566-F, National Library of New Zealand Te Puna Mātauranga o Aotearoa

photography turned out to be unproblematic, with numerous local government agencies interested in new coverage. Additional revenue was gained through the sale of archival images. These images became an asset from the late 1940s because it was possible for the first time to construct datasets containing time series of the same area, allowing for processes of environmental change to be mapped.

Local and national government interests in the environment had changed after the war and were now inclined towards a managerial approach to environmental knowledge production. Monitoring became increasingly important, while surveying and improvement efforts plateaued. This trend continued well into the 1980s with revenue figures for archival imagery sales 'edging ahead of income from flying commissions'.[30] Mapping efforts had reached a threshold. Most of the country had been explored and settled at great initial cost, with the creation of extensive agricultural, pastoral and silvicultural land use systems as a result. Spatial knowledge production efforts were beginning to shift focus accordingly, to bring human-induced processes of environmental change into the crosshairs of map-making efforts. Such interests – spurred on by the declassification of many important data sources in 1945 – led to a continuous production of studies of processes of environmental degradation and hazards in the postwar period. Maps were increasingly used as management devices, for example as the basis for efforts to safeguard the biological economies of the countryside and monitor resource use (Figure 18.5). This continued for 40 years until the reform period of the late 1980s, when shifts in government emphasis completely changed how the environment was recorded and data made accessible.

Reform, privatisation and the information economy

In the late 1980s the government institutions responsible for producing and preserving spatial knowledge were restructured as part of a broad-ranging government reform that affected most departments, including those involved with environmental management. With the Survey Act of 1986 which came into effect on 1 April 1987, the mapping functions of the former Lands and Survey Department were moved into a new Department of Survey and Land Information (DOSLI): almost the same name as its predecessor, but with the significant addition of the word 'information'.[31] After a few years, DOSLI itself was replaced by another agency, Land Information New Zealand (LINZ). The word information in this context signified a new self-reflective stance in the vision for government map-making activities. It highlighted the transient and relational aspects of working with information rather than authoritative and objective knowledge per se, representing a new way of understanding knowledge.

Something similar occurred in most Western societies at about this time and had to do with cultural and technological developments that had grown in importance since the 1950s. The invention and spread of specialist computing systems after World War II and the development of information science as a discipline in the 1950s had seen the concept of information take on an increasingly important role both in academic contexts and in the wider public domain. Information in this sense was not understood to be synonymous with knowledge, or even with knowledge applied in the form of surveys and maps: information was knowledge communicated, denoting a process in which data reached a receiver who would then decipher the data into information. As Capurro and Hjørland put it 'information is not a pure observable, but a theoretical construct. It is "interpreted data".'[32] In this critical respect, regarding the difference between data, information and knowledge, the map-making paradigms of old were in conflict with the new culture and technology which were beginning to define society.

Until the late twentieth century the map had largely been considered an objective depiction of reality, its truthfulness thought to depend on the precision of its design and indirectly on the authority of its author.[33] With the 'information concept' of map making, map users in a way have been allowed into the backstage area of cartography. Access to raw spatial data in the form of computer-readable files became available to professionals and later to the wider public. Maps were not finished images any more, but pieces of (objective) information stitched and collaged together by subjectively motivated map designers to meet their own specific requirements. This amounted to nothing less than a revolutionary liberalisation and decentralisation of shared topographic knowledge in New Zealand. One of the most significant events in this respect was the establishment of a digital cadastral database in 1987, which made it possible to access, edit, view and sell spatial economic data and data derivatives in a hitherto unseen flexible and decentralised way. The creation of a similar topographic database in 2002 led to comparable effects in the realm of topographical knowledge.

The neoliberal imperatives inherent in the reforms of the 1980s and 1990s went hand in hand with ideas of synergy, dynamic effects and growth, which were understood to be stimulated through competitive, objectively informed individual entrepreneurship, an ideal closely resembling the metaphor of the egotistically benevolent 'invisible hand' described by Adam Smith in his seminal work two centuries earlier.[34] In its 'Statement of Intent' published in 2012, LINZ made this plain when among its priorities the government mapping agency stated an intention to 'accelerate growth in the use of location-based information technologies by government and business as a key tool for growth and decision-making'.[35] With the application of ideas of this kind set in motion by the government reform of the 1980s, an information market had been set in place. These major changes, affecting most if not all public knowledge production regarding the environment, constituted a turning point in the history of New Zealand cartography and metrology and heralded a return to the liberal ideals of public management of the environment through support of private entrepreneurship, prevalent during the first European discovery of the country. It remains to be seen what the long-term effects of such policies will be.

Conclusion

Since the arrival of humans in the thirteenth century, the environmental history of New Zealand – with its myriad facets of human–environment interactions – has been closely intertwined with a parallel history of spatial knowledge production. The production of such knowledge can be construed as integral to the way human societies have attained power over the environments of New Zealand, in their continuous effort to attain sustenance and wealth. This was the case with Māori society, which developed from an initial state of transient resource consumption to settlement based on combinations of hunting, fishing and horticulture. This occurred through the development of usufructuary tenure institutions that were needed to stabilise a society based on contested spatially distributed resources. It was also the case when European settlers arrived approximately six hundred years later. They brought their own tenure institution and a set of technologies for spatial knowledge production with them, which led to drastic changes in both understandings and use of the environments of New Zealand.

Due to the eventual success of the European colonisers in subduing Māori resistance, the structure and disposition of the terrestrial environments of contemporary New Zealand is largely a result of interactions with the environments based on European

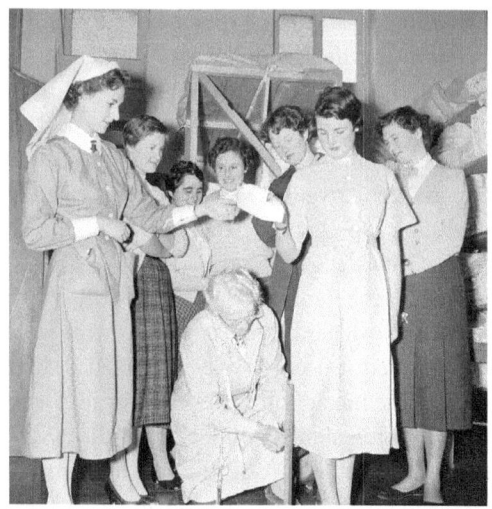

Figure 18.6

Top: Metrologies for decency, anno 1956. Even these young nursing students had to conform with objective measures for correct conduct, in this case embodied by the skirt-length measurement stick. Source: P/1956/1455/F, Alexander Turnbull Library collections, National Library of New Zealand Te Puna Mātauranga o Aotearoa

Bottom: Normalising the produce: wool classers at work with New Zealand wool, Hornchurch, England, during World War I. British quality paradigms, product standards and measuring technologies were widely applied in New Zealand to meet the expectations of export markets.
Source: 1/2–013990-G, National Library of New Zealand Te Puna Mātauranga o Aotearoa

spatial culture and technology. The principal facet of such interactions was the particular idea of objectivity that was introduced through the application of devices and practices for measuring resources, products and properties. Metrics have since become such an integral part of modern society as to be all but invisible in terms of everyday practice (Figure 18.6). Historically their application was expressed most clearly in the way the settler economy appropriated and alienated space through surveying, defended it through mapping, and developed it through monitoring efforts. But as many of the previous chapters in this volume have illustrated, land was merely one among a wide range of resources to be measured, appropriated and ordered by way of European metrologies. For example, Jim McAloon has shown that the hunt for valuable kauri spars for shipbuilding – which led to the destruction of vast tracts of forest in the beginning of the nineteenth century – was motivated by a practice of measuring where 'the requirements were precise as to length, diameter, and taper, with one in ten thousand good enough for a mainmast'.[36] The same later became true for other commodities like New Zealand wool and meat, and indeed most other export products. Today these are ever more carefully measured to determine their value and secure access to export markets.[37] Such practices clearly illustrate how New Zealand resources and environments have been adapted historically to outside needs and specifications by way of metrological applications. The environmental history of New Zealand is in part a history of the application of such knowledge production practices, the culture and social order that supported it, and the environmental transformation that ensued.

Further reading

Little has been written about maps and metrologies in a New Zealand context. Russell Kirkpatrick's introduction to his *Bateman Contemporary Atlas of New Zealand: The Shape of Our Nation* (David Bateman, Auckland, 2nd edn, 2005) reflects on the history of map making, and Brian Marshall, 'From Sextants to Satellites: A Cartographic Timeline for New Zealand' (*New Zealand Map Society Journal*, no. 18, 2005, pp. 1–110) provides a lot of background information. *Te Ara – the Encyclopedia of New Zealand* contains a number of articles about surveying and map making, and Giselle Byrnes' book *Boundary Markers: Land Surveying and the Colonisation of New Zealand* (Bridget Williams Books, Wellington, 2001) provides a postcolonial critique of settler culture. In an international comparative context two publications should be mentioned: Roger J.P. Kain & Elizabeth Baigent, *The Cadastral Map in the Service of the State: A History of Property Mapping* (University of Chicago Press, Chicago, 1992), which contains a chapter on New Zealand; and Phillip Lionel Barton's chapter on 'Maori Cartography and the European Encounter' in David Woodward & G. Malcolm Lewis (eds), *The History of Cartography*, vol. 2, book 3 (University of Chicago Press, Chicago, 1998, pp. 493–532).

19.

Epilogue

Eric Pawson and Tom Brooking[1]

IN CLOSING THIS COLLECTION of essays about New Zealand's environmental histories, it is appropriate to pose some wider questions. These questions may have arisen in the course of reading the chapters; they might arise in discussions about the practice of environmental history. First, to what extent is environmental history an interdisciplinary pursuit? Is it more than history with nature added in? Second, what can be learned from environmental history about the future? Will an understanding of the past better inform the ways in which we approach our place on earth? Third, to what extent do the concerns of environmental historians mirror those of environmental scientists and managers? Do they have things to say to each other?

These questions are not readily answered, but are worth posing in the spirit of encouraging moments of opening and new possibility. The humanities and the sciences have often talked past each other, some would say no less than in the late 1950s when the British scientist and novelist C.P. Snow identified what he described as 'the two cultures'.[2] It is not enough to assume, for example, that just because we are better informed about the history of human environmental engagement, this will somehow contribute to resolving the big issues of our times, like human-induced climate change. The International Panel on Climate Change (IPCC) may have moved beyond the strict science focus of its early work, and the fifth Assessment Report in 2014 will provide more analysis of human behaviours. But there is a long way to go. In this regard, the Australian environmental historian Tom Griffiths has argued eloquently that the human costs of bushfires in his country are paid over and over, with little learning from the past. Likewise, in New Zealand, longstanding knowledge of the risks of ground liquefaction in seismic events did nothing to prevent the development of hazardous land in the 20 years before the Christchurch earthquakes.[3]

In this context, this Epilogue sets out to do two things. The first is to identify a short series of issues around which there is, or could be, common cause between environmental historians on the one hand and environmental scientists and managers on the other.[4] In both cases, we are using these terms broadly to denote those interested in the past, and those concerned with effective understandings of the present and future. These issues arise from a selection of themes identified in, and between, the contributions to this book. Second, we sketch some ideas about 'doing environmental history' in both academic and everyday ways.

Themes for further research

There are many ways to approach what Sörlin and Warde call 'the problem of the problem', in other words how to identify 'a readily articulated set of issues at stake in environmental history'.[5] Here we focus on three themes that we see as central to the future of both the environment and environmental history in New Zealand. These are: water, land, and environmental instability. New Zealand may not be subject to the meteorological and climatic variations that typify Australia, for example, but it is still a very weather-dependent country. This comes in part from its ongoing reliance on primary sector exports, and in part from its rather fragile landscapes. For example, the long, dry summer of 2012–13 was good for many, but meant water shortage for others. Farmers through much of the country faced prolonged drought; town dwellers in Wellington and some other places were subject to strict water use restrictions. In other years, water shortages have rendered power supplies from hydro lakes insecure.[6]

This sort of empirical approach is one way of expressing the nature of environmental relationships. But beyond it lie more complex social and environmental issues. These concern the ownership of particular resources, who has the right to use them, and the conditions under which use is permitted. The ways in which apparent abuses are regulated, given ecological, economic and social interdependencies, raises questions about whether there are better ways to live within the confines of available land and water. It is debatable whether environmental histories show us that people are becoming 'truly settled',[7] or are still engaged in a sequence of resource frontier activities that could be said to have characterised the last century and a half. The interaction of human behaviour with changing climatic or ecological conditions may also expose critical environmental thresholds, such as soil erosion during storms on deforested land, or flooding of coastal subdivisions as the result of removal of the natural buffer that sand dunes provide.

Understandings and uses of water have been central to many of the chapters in this book. Exploitation of resources began at the coast (Chapters 2, 3 and 4). Inland, wetlands were rapidly all but obliterated (Chapter 10) and rivers turned into mining-waste discharge channels (Chapter 6). These chapters also show that the management of water and land go hand in hand, as illustrated by the emerging challenges of soil erosion (Chapter 12), or of irrigation and hydro development (Chapters 11, 16 and 17). Today it is intensive milk production based on large herds in areas formerly considered too dry for dairying, especially Canterbury, that prompts public concern about water use and nutrient run-off into streams and lakes. Politically this drives the search for governance models through initiatives like those proposed by the Land and Water Forum to facilitate environmentally responsible development.[8]

Water availability and quality issues affect both Māori and Pākehā, although Māori have always had particular interests in water, as Chapter 17 shows. Specific water bodies have regularly been subject to Treaty claims before the Waitangi Tribunal; its careful environmental histories point to the nature of the issue, and often to means of resolution. Early examples are the Motunui–Waitara claim of 1983, in which offshore discharges from new industrial developments in North Taranaki were proved likely to worsen existing pollution of rich seafood-bearing coastal reefs; and the Kaituna claim of 1984 that prompted steps to clean up Lake Rotorua after years of fouling from urban sewage discharge.[9] In 2012, the Ministry for the Environment could still say that 'there are many freshwater swimming spots that should be avoided', in both lakes and rivers.[10] How this has come about can be the subject of case-by-case investigations in environmental history. A complementary approach is, by aggregating the cases, to highlight shared or specific factors. These might include identification of patterns of human activities and pollution flows within catchments, and the nature of the threshold conditions beyond which contamination impinges on interests in clean water, such as recreation and fishing.

Some of these matters were clear to an earlier generation of writers and scholars such as Herbert Guthrie-Smith and Kenneth Cumberland, as Chapter 1 reveals. The latter is a good example of a researcher whose work had real impact on water and land management through the creation of the mid-century pattern of catchment boards.[11] The nature of land use practices has been central to those chapters that have focused on the rapid and extreme nature of environmental transformation in New Zealand (Chapters 5, 7, 10, 11, 12, 13, 17, 18). One of the signatures of this transformation has been the division of the land into spheres of production and conservation (Chapters 7, 8, 9). This reflects the nature/culture divide inherent in the Pākehā world view, something produced in 'endless practices' in 'endlessly consequential' ways.[12] One such consequence has been the emergence of a form of econationalism that proclaims the virtues of 'clean and green', and '100% Pure New Zealand', but draws on an uneasy tension between the histories of caring 'for' nature and exploiting it (Figure 19.1). 'The trouble with wilderness', as William Cronon observed, is that to have so much of it might seem to absolve people from respectful environmental relationships elsewhere.[13]

Some have asked if it is possible to frame more optimistic stories than those of environmental degradation that seem to exemplify much writing in environmental history.[14] The patterns of environmental learning revealed in Chapters 2, 5, 8, 11, 12 and 13 show various forms of accommodation with the land, and Chapters 15, 16, 17 and 18 reveal how in quite different ways we have sought to inhabit it. There is potential for a great deal of applied environmental history in this regard, at both micro and macro scales. This could include examination of marae-based environmental work; the work of farmers who covenant land to protect key features, or who win regional farm environmental awards based on learning from experience; or initiatives that seek to declare or reclaim urban 'commons'.[15] These behaviours actively question the precepts of 'improvement' (Chapter 1) and express ways of being in a 'more-than-human' world. Documenting examples of such engagements, and exploring how these are occurring in particular places and the conditions under which they are occurring, are useful steps. To take it further, and to assemble or frame the examples in order to discern their commonalities, and to reveal why change is more likely to occur in some places, on some types of property, or under some conditions rather than others, is invaluable.[16]

Figure 19.1 'The height of happiness', Railways Studios, 1927, colour lithograph.
Source: S12–188a, Hocken Collections, Uare Taoka o Hākena, University of Otago

An overriding feature of New Zealand's environmental histories has been the use of land and water for successive waves of resource exploitation, for which reason this remains a country heavily dependent on a narrow range of primary exports. In the nineteenth century, wool, wheat and timber booms followed sealing and goldmining (Chapters 3, 4, 5, 6 and 7); in the twentieth century, the preoccupation was with the extraction of maximum value from grass-based commodities (Chapter 11). Such resource 'quarrying', of living 'off' rather than 'with' the land, has been framed through systems of law, surveying, cartography and metrology (Chapters 16 and 18). Yet almost every chapter has also revealed both how fragile and unstable (in human terms) are the land and waters so measured and appropriated, as much in urban as in rural places (Chapters 13, 15 and 17). Chapter 15 also argues that the acquisition of intimate environmental knowledges through the senses by Māori has not always been replicated among Pākehā. The assumption of the colonial project of improvement has been that of an essentially benign, if not fixed or knowable, environmental stage, which is why the project has in its turn been portrayed as 'fraught and vulnerable'.[17]

Some familiarity with environmental history readily undermines complacent assumptions. Its narratives might serve to place a more effective awareness of environmental instabilities in wider consciousness, although this is by no means an assured outcome. An account of the great Wellington earthquake of 1855, for instance, reveals how poorly prepared were many structures for a shake, despite the experience of one only a few years before in 1848: people had forgotten, and then forgot again that which they rationalised as aberrant.[18] The destruction of New Zealand's second-largest city in the earthquakes of 2010 and 2011 might dent such rationalisations, but only if it opens up an ongoing national conversation about how such risks are to be apprehended and managed. In the same way, the likely consequences of human-induced climate change for New Zealand's grass-based farm economy and water-based power generation system are anything but trivial. Not only are some of the most important agricultural districts, the dry eastern parts of both islands, becoming drier still, but diminished accumulation of snow and ice in the mountains will affect spring melt. Both have implications for irrigation, on which the country's largest export industry – dairying – depends.[19] Yet people have adapted to environmental variability in the past: for example, Māori developing rāhui to regulate harvesting of natural resources (Chapter 2), Pākehā farmers matching plants and animals to local conditions (Chapter 5), and modifying land use to conserve soils (Chapters 11 and 12). To reveal such adaptations is one point of intersection between environmental histories and futures.

Doing environmental history

It has been said that 'breakthroughs often occur when people trained in different disciplines come together'.[20] Certainly those who call themselves environmental historians often make a claim to interdisciplinarity. As a process, interdisciplinarity can take place in a number of ways, none of which is easy. No one discipline has moral high ground or can claim pre-eminence. The art of learning each other's approaches, concepts and suppositions requires effort. As Donald Worster has pointed out, this means that the 'environmental historian must learn to speak some new languages as well as ask some new questions'.[21] For those trained in the discipline of history, an important part of their work is to master narrative and context; for the geographer the focus is more on place and spatial relations. The ecologist is concerned with the interactions among organisms and between them and their environments. A willingness to cross these sorts of borders can bring special insights. An example is David Livingstone's discussion of the reception of Darwin's theory of evolution being contingent on the interests of people in the places in which it was heard, in this case Charleston, St Petersburg and Wellington. This goes some way to explaining why ideas of 'ecological imperialism' and a belief in the superiority of northern hemisphere plant species have had such lasting hold on the popular imagination in New Zealand.[22]

Ecological imperialism is a social not a scientific idea. Invasion biology, which has a tradition of historically informed writing, seeks to explain why and how these imported species have been so successful in colonising new niches.[23] It gives an entry point into an understanding of ecological processes being in continuous change, rather than working towards some end-state of 'equilibrium' in which all is harmony and balance. Equilibrium in this sense is also more of a cultural construct than a scientific explanation; for example, there has been no triumph of the invaders, at least not in any readily predictable sense; while indigenous plant species have proved remarkably resilient. So too has the native bird population where habitat destruction has been contained or reversed. But biodiversity loss has been identified as 'New Zealand's most pervasive environmental issue',[24] and the ways in which both Māori and Pākehā colonisers undermined the spatial ecology of native biodiversity is starkly laid bare in many parts of this book, particularly Chapters 2, 5, 7, 10, 11 and 14.

A benefit of attempting to practise interdisciplinarity is that different disciplines are methodologically diverse. Not everyone proceeds by way of locally focused research, although it can be very useful to do so in order to answer particular questions or to gain insight into specific issues. The danger of 'the broad outlines' of environmental history being buried 'under the weight of ever more case studies'[25] was referred to at the end of the first chapter, although equally it is necessary to be aware of 'easy generalizing'.[26] There are many ways in which research can proceed, using analytical or thematic frames of reference, as do most chapters here, or with more specifically conceptualised or theorised approaches (of which Chapters 13, 14, 15 and 18 are examples).[27] Analyses can be framed from syntheses of the latest research in a field, or through examining the nature or validity of a popular or academic notion by assembling the evidence (e.g. Chapters 2, 11 and 16), or by telling the story from a particular, insider perspective (Chapters 5, 17).

An approach that is growing in popularity is to think in terms of spatial scales and seek to uncover the global or imperial setting of a specific problem, as is explicit in Chapters 4, 12, 14 and 18. The long reach of international processes and forces into the New Zealand landscape means that a contextualised analysis of this sort is a valuable alternative, adjunct or precursor to the case study. By interrogating the sources, places where change was catalysed or retarded are highlighted, and these differences can reveal more about the broader roles of various agents of change. For example, records of imports during the colonial period showed that farmers in some regions adopted artificial fertilisers before others, and that once here, their use took hold in some places, but was only ephemeral elsewhere.[28] Equally the aggregation of micro-scale observations (as demonstrated for charcoal deposits in Chapter 2 or pasture species mixtures in Chapter 5) is often necessary to provide a convincing picture.

If as an environmental historian it is beneficial to think beyond a specific discipline, and beyond the local case study, it is also necessary to think beyond the traditional historical confines of primary archival and secondary printed sources. Donald Hughes' description of environmental history quoted in Chapter 1: 'a kind of history that seeks understanding of human beings as they have lived, worked and thought in relationship to the rest of nature',[29] points indirectly to the value of seeking to understand these lives in real context. Christof Mauch writes about how environmental history is different 'because it forced me to pack my hiking boots along with my laptop … it helped me to

Figure 19.2 Blasting tree stumps in Taranaki, c. 1900. The photograph shows, in the middle distance, a new stand of pine trees to the left, and the burnt trunks of indigenous forest to the right. Source: PHO2013–0001, Puke Ariki, New Plymouth

realize that the relationship between nature and culture is defined not just by boxes full of documents, but also by explorations of nature's own archives'.[30]

Nature's own archives show how unrepresentative paper archives can be. For example, it is unusual to find historical photographs of commonplace land transformation processes, such as burning the bush or broadcasting grass seed in the ashes of the fires. Although the tree stumps left behind were often used as emblems of change in colonial art, it is rare to find images of the removal of these stumps in newly delineated paddocks. One example of stumping appears in Figure 19.2. These processes were taken for granted at the time, and often there was no photographer on hand anyway. They are the sorts of things about which talking to farmers, for example, can reveal more. Donald Worster suggests that to appreciate nature's forces, 'we must now and then get out … and ramble into fields, woods and the open air'. But more than rambling, it is discussion with environmental actors – such as urban gardeners, rural land holders, mountain recreationists, environmental scientists and managers – to uncover new material and discover points of common interest that matters. In this way we may get better purchase on what is perhaps the greatest challenge for environmental history: how to learn from the past and to act accordingly in the future.

Conclusion

Environmental history can and should be more than history with nature added in. Otherwise it may rest as a history of environmental ideas, or of wilderness, which is indeed its origin in the United States.[31] The discipline of geography gave it a far broader base in New Zealand as long ago as the 1940s; today historians, geographers, Māori and legal scholars, archaeologists and landscape ecologists are working productively together as we intend this book to demonstrate. But to develop an edge environmental history needs to frame its stories in ways that engage not only those interested in the past, but also those whose role it is to analyse and to manage environmental behaviour now and in the future. Half a century ago, C.P. Snow wrote of the arts and sciences that 'when the two senses have grown apart, then no society is going to be able to think with wisdom'.[32] The same could still be said if what we learn about the past is detached from how we act today and tomorrow.

Further reading

Two essential resources for the study of environmental history in New Zealand are Malcolm McKinnon (ed.), *New Zealand Historical Atlas* (David Bateman, Auckland with the Department of Internal Affairs, Wellington, 1997) and *Te Ara – the Encyclopedia of New Zealand* at www.TeAra.govt.nz. The Australia and New Zealand Environmental History Network is at http://environmentalhistory-au-nz.org, a site that includes the freely accessible journal *ENNZ: Environment and Nature in New Zealand*. The *New Zealand Geographer* also frequently carries relevant articles.

There is a range of accessible international resources that engage with important ideas in environmental history. The Rachel Carson Center for Environment and Society at LMU Munich, http://www.carsoncenter.uni-muenchen.de/publications/index.html, has publications authored by prominent environmental historians which are available online at no cost. The key overseas academic journals are *Environmental History*, which tends to a North American focus, and *Environment and History* and the *Journal of Historical Geography*, which are both broadly international in coverage. See also Shepard Krech III, J.R. McNeill & Carolyn Merchant (eds), *Encyclopedia of World Environmental History* (3 vols, Routledge, New York and London, 2004).

Notes

1. Introduction

1. Tom Brooking, Eric Pawson & Hamish G. Rennie, 'New Zealand', in Sara G. Beavis, Michael L. Dougherty & Tirso Gonzales (eds), *The Berkshire Encyclopedia of Sustainability*, vol. 8. *The Americas and Oceania: Assessing Sustainability*, Great Barrington, MA: Berkshire Publishing, 2012, pp. 200–05. Aotearoa is the Māori name for New Zealand.
2. The 'large land, uplifted high' is from the journal of Abel Janszoon Tasman, commander of the first known European voyage to sight New Zealand's shores. It refers to the mountains of the northwest of the South Island: M.F. Vigeveno (trans.), 'Tasman's Journal or Description', in *Abel Janszoon Tasman and the Discovery of New Zealand*, Department of Internal Affairs, Wellington, 1942, p. 45. For the deep, geological history of New Zealand before the arrival of humans, see Hamish Campbell & Gerard Hutching, *In Search of Ancient New Zealand*, Penguin Books, Auckland, 2007, and George Gibbs, *Ghosts of Gondwana: The History of Life in New Zealand*, Craig Potton Publishing, Nelson, 2006.
3. Tim Flannery, *The Future Eaters: An Ecological History of the Australasian Lands and People*, Reed Books Australia, Chatswood, NSW, 1994, p. 55. This is not to say that New Zealand has an especially large number of bird species. It does, however, have an enormous number of invertebrates and non-vascular plants. Flannery was referring to the prevalence of birds.
4. Kenneth B. Cumberland, 'A Century's Change: Natural to Cultural Vegetation in New Zealand', *Geographical Review*, vol. 31, no. 4, 1941, pp. 529–54; Peter Holland, 'Cultural Landscapes as Biogeographical Experiments: a New Zealand Perspective', *Journal of Biogeography*, vol. 27, no. 1, 2000, pp. 39–43.
5. J.M. Powell, 'Historical Geography and Environmental History: An Australian Interface', *Journal of Historical Geography*, vol. 22, no. 3, 1996, p. 293.
6. Jacinta Ruru, Janet Stephenson & Mick Abbott (eds), *Beyond the Scene: Landscape and Identity in Aotearoa New Zealand*, Otago University Press, Dunedin, 2010; Jacinta Ruru, Janet Stephenson & Mick Abbott (eds), *Making Our Place. Exploring Land-use Tensions in Aotearoa New Zealand*, University of Otago Press, Dunedin, 2011; Rachael Selby, Pātaka Moore and Malcolm Mulholland (eds), *Māori and The Environment: Kaitiaki*, Huia Publishers, Wellington, 2010, especially the essay by Merata Kawharu, 'Environment as Marae Locale', pp. 221–37.
7. J. Donald Hughes, *What Is Environmental History?*, Polity Press, Cambridge, 2006, p. 1.
8. Sverker Sörlin & Paul Warde, 'Making the Environment Historical – An Introduction', in Sverker Sörlin & Paul Warde (eds), *Nature's End. History and the Environment*, Palgrave Macmillan, Basingstoke, 2009, pp. 1–19.
9. William Main, *Bragge's Wellington and the Wairarapa: Images of a City and its Provinces from the 1860s to the 1890s as Recorded by Photographer James Bragge 1833–1908*, Millwood Press, Wellington, 1974; and 'Bragge, James', from *The Dictionary of New Zealand Biography. Te Ara – the Encyclopedia of New Zealand*, updated 30-Oct-2012: www.TeAra.govt.nz/en/biographies/2b37/bragge-james. In the 1870s, Wellington interests were working hard to extend the city's hinterland: David Hamer, 'Wellington on the Urban Frontier', in David Hamer & Roberta Nicholls (eds), *The Making of Wellington, 1800–1914*, Victoria University Press, Wellington, 1990, pp. 227–54.
10. Joan M. Schwartz, 'The Geography Lesson: Photographs and Construction of Imaginative Geographies', *Journal of Historical Geography*, vol. 22, no. 1, 1996, pp. 16–45; James R. Ryan, 'Photos and Frames: Towards an Historical Geography of Photography', *Journal of Historical Geography*, vol. 26, no. 1, 2000, pp. 119–24; J.B. Harley, 'Maps, Knowledge and Power', in Denis Cosgrove & Stephen Daniels (eds), *The Iconography of Landscape. Essays on the Symbolic Representation, Design, and Use of Past Environments*, Cambridge University Press, Cambridge, 1988, pp. 277–312.
11. Quoted in I.R. Matheson (ed.), 'The Birth of Palmerston North', supplement to the *Evening Standard*, 13 March 1871, p. 53.
12. Howard Morphy & Kate Flint, 'Introduction', in Kate Flint & Howard Morphy (eds), *Culture, Landscape and the Environment. The Linacre Lectures 1997*, Oxford University Press, Oxford, p. 1.
13. Paul Slack, 'Introduction', in Paul Slack (ed.), *Environments and Historical Change. The Linacre Lectures 1998*, Oxford University Press, Oxford, 1999, pp. 1–9.
14. Rachael Selby, Pataka Moore & Malcolm Mulholland, 'Introduction', in Selby, Moore & Mulholland, *Māori and the Environment*, p. 1.
15. This is not to deny that Pākehā relations with nature have not had multiple expressions; it is to say that they

are generally characterised by this nature/culture divide: but see James Beattie and John Stenhouse, 'Empire, Environment and Religion: God and the Natural World in Nineteenth-Century New Zealand', *Environment and History*, vol. 13, no. 4, 2007, pp. 413–46.
16 Edward Gibbon Wakefield, *View of the Art of Colonization with Present Reference to the British Empire: in Letters Between a Statesman and a Colonist*, Parker, London, 1849, p. 118.
17 On improvement, see Eric Pawson & Tom Brooking, 'Introduction', in Tom Brooking & Eric Pawson, *Seeds of Empire: The Environmental Transformation of New Zealand*, I.B. Tauris, London, 2011, pp. 1–12; W.D. McIntyre (ed.), *The Journal of Henry Sewell*, vol. 1, February 1853 to May 1854, Whitcoulls, Christchurch, 1980, p. 427.
18 H.C. Jacobson, *Tales of Banks Peninsula*, 3rd ed., Akaroa Mail, Akaroa, 1917, p. 296.
19 Cumberland, 'A Century's Change', p. 529; Eric Pawson, 'Creating Public Spaces for Geography in New Zealand: Towards an Assessment of the Contributions of Kenneth Cumberland', *New Zealand Geographer*, vol. 67, no. 2, 2011, pp. 102–15.
20 A.H. Clark, *The Invasion of New Zealand by People, Plants and Animals: The South Island*, Rutgers University Press, New Brunswick, 1949, p. v.
21 Carl O. Sauer, 'Theme of Plant and Animal Destruction in Economic History', *Journal of Farm Economics*, vol. 20, no. 4, 1938, p. 765.
22 Kenneth B. Cumberland, *Soil Erosion in New Zealand: A Geographic Reconnaissance*, Soil Conservation and Rivers Control Council, Wellington, 1944. L.W. McCaskill, *Hold This Land: A History of Soil Conservation in New Zealand*, A.H. & A.W. Reed, Wellington, 1973.
23 Alfred W. Crosby, *Ecological Imperialism: The Biological Expansion of Europe, 900–1900*, Cambridge University Press, Cambridge, 1986.
24 William Cronon, 'Foreword. A Passion for Small Things', in Herbert Guthrie-Smith, *Tutira. The Story of a New Zealand Sheep Station*, reissue, Godwit, Auckland, 1999, p. xi; and Guthrie-Smith, *Tutira*, pp. 195 and xxiii.
25 Crosby, *Ecological Imperialism*; William Cronon, 'Foreword'; Thomas R. Dunlap, *Nature and the English Diaspora. Environment and History in the United States, Canada, Australia, and New Zealand*, Cambridge University Press, Cambridge, 1999.
26 Malcolm McKinnon (ed.), *The New Zealand Historical Atlas*, David Bateman, Auckland & the Department of Internal Affairs, Wellington, 1997; Eric Pawson, 'The New Zealand Historical Atlas', *Journal of Historical Geography*, vol. 23, no. 4, 1997, pp. 496–99.
27 James Belich, *Making Peoples: A History of the New Zealanders from Polynesian Settlement to the End of the Nineteenth Century*, Penguin Books, Auckland, 1996; James Belich, *Paradise Reforged: A History of the New Zealanders from the 1880s to 2000*, Penguin Books, Auckland, 2001; Michael King, *The Penguin History of New Zealand*, Penguin Books, Auckland, 2003.
28 Paul Star, 'New Zealand's Changing Natural History. Evidence from Dunedin, 1868–1875', *New Zealand Journal of History*, vol. 32, no. 1, 1998, p. 60.
29 William Cronon, 'A Place for Stories: Nature, History and Narratives', *Journal of American History*, vol. 78, no. 4, 1992, pp. 1349–76.
30 William Cronon, 'Introduction: in Search of Nature', in William Cronon (ed.), *Uncommon Ground: Toward Reinventing Nature*, W.W. Norton & Co., New York, 1995, p. 25.
31 Vera Norwood, 'Nature' in Shepherd Krech III, John McNeill and Carolyn Merchant (eds), *World Environmental History*, vol. II, Routledge, New York and London, 2004, pp. 875–83.
32 Cronon, 'Introduction', p. 24.
33 Ruru et al, *Making Our Place*.
34 Karl Marx, *Capital. A Critique of Political Economy*, vol. 1, with an introduction by Ernest Mandel, Penguin, London, 1990, pp. 931–40. Recent work referred to includes Brooking & Pawson, *Seeds of Empire*, and Tony Ballantyne, *Webs of Empire: Locating New Zealand's Colonial Past*, Bridget Williams Books, Wellington, 2012.
35 Rollo Arnold, *New Zealand's Burning; Settler Kaponga 1881–1914. A Fragment of the Western World*, Victoria University Press, Wellington, 1997; Felicity Barnes, *New Zealand's London: A Colony and Its Metropolis*, Auckland University Press, Auckland, 2012, especially pp. 123–88.
36 James Beattie, *Empire and Environmental Anxiety. Health, Science, Art and Conservation in South Asia and Australasia, 1800–1920*, Palgrave Macmillan, Basingstoke, 2011; Eric Pawson, 'Plants, Mobilities, Landscapes: Environmental Histories of Botanical Exchange', *Geography Compass*, vol. 2, no. 5, 2008, pp. 1464–77; Ian Tyrrell, *True Gardens of the Gods. Californian-Australian Environmental Reform, 1860–1930*, University of California Press, Berkeley and Los Angeles, 1999;
37 See for example, Robert Darnton, *The Great Cat Massacre and Other Episodes in French Cultural History*,

Vintage Books, New York, 1985; Miles Fairburn, *Almost Out of Heart and Hope: The Puzzle of a Colonial Labourer's Dairy*, Auckland University Press, Auckland, 1995; Charles W.J. Withers, 'Constructing "The Geographical Archive"', *Area*, vol. 34, no. 3, 2002, pp. 303–11.

38 Richard Boast, *Buying the Land, Selling the Land: Governments and Māori Land in the North Island 1865–1921*, Victoria University Press/Victoria University of Wellington Law Review, Wellington, 2008.

39 Giselle Byrnes, 'Surveying Space: Constructing the Colonial Landscape', in Bronwyn Dalley & Bronwyn Labrum (eds), *Fragments: New Zealand Social and Cultural History*, Auckland University Press, Auckland, 2000, pp. 54–70; Julie King, 'Facing up to Fox: The Colonial Watercolours of William Fox', *Art New Zealand*, no. 95, 2000, pp. 84–87. Franklin Ginn, 'Colonial Transformations: Nature, Progress and Science in the Christchurch Botanic Gardens', *New Zealand Geographer*, vol. 65, no. 1, 2009, p. 35, seeks to show 'the colonial project ... as fraught and vulnerable, rather than as the extension of panoptic European power'.

40 Frank Uekoetter, *Consigning Environmentalism to History? Remarks on the Place of the Environmental Movement in Modern History*, Rachel Carson Center Perspectives 2011/7, Ludwig Maximilians Universität München/Deutsches Museum, Munich, 2011, p. 4. Alan Grey implicitly made this point about New Zealand when he set out to write a book drawing together many years of unpublished student thesis research: *Aotearoa and New Zealand: A Historical Geography*, Canterbury University Press, Christchurch, 1994.

41 Richard White, 'Foreword', in Eric Pawson & Tom Brooking (eds), *Environmental Histories of New Zealand*, Oxford University Press, Melbourne, 2002, p. iii-iv.

PART 1 ENCOUNTERS
2. A fragile plenty: pre-European Māori and the New Zealand environment

1 Richard Taylor had visited Te Rangatapu in 1843 and was similarly equivocal about whether the moa bones represented hunted birds: Atholl Anderson, *Prodigious Birds: Moas and Moa-hunting in Prehistoric New Zealand*, Cambridge University Press, Cambridge, 1989, p. 99. Except in relation to Figure 1, all dates mentioned in this chapter are calendrical ages.

2 John T. Holloway, 'Forest and Climates in the South Island of New Zealand', *Transactions of the Royal Society of New Zealand*, vol. 82, no. 2, 1954, pp. 329–410; Kenneth B. Cumberland, 'Man's Role in Modifying Island Environments in the Southwest Pacific, with Special Reference to New Zealand', in F.R. Fosberg (ed.) *Man's Place in the Island Ecosystem: A Symposium*, Bishop Museum Press, Honolulu, 1963, pp. 187–206, quote p. 193. The role of climatic change remains debated, e.g. Patrick J. Grant, 'Effects on New Zealand Vegetation of Late Holocene Erosion and Alluvial Sedimentation', *New Zealand Journal of Ecology*, vol. 12 (supplement), 1989, pp. 131–44; and opposing views in M.S. McGlone, 'The Polynesian Settlement of New Zealand in Relation to Environmental and Biotic Changes', *New Zealand Journal of Ecology*, vol. 12 (supplement), 1989, pp. 115–29; Janet M. Wilmshurst, 'The Impact of Human Settlement on Vegetation and Soil Stability in Hawke's Bay, New Zealand', *New Zealand Journal of Botany*, vol. 35, no. 1, 1997, pp. 97–111. The supposed pre-Māori population was often called 'Maruiwi' or 'Moriori', the latter term being, in fact, the correct name for the original population of the Chatham Islands.

3 Barry Brailsford, *Song of Waitaha: The Histories of a Nation*, Ngatapuwae Trust, Christchurch, 1994.

4 Atholl Anderson, 'Prehistoric Polynesian Impact on the New Zealand Environment: Te Whenua Hou', in Patrick V. Kirch & Terry L. Hunt (eds), *Historical Ecology in the Pacific Islands: Prehistoric Environmental and Landscape Change*, Yale University Press, New Haven, 1997, pp. 271–83.

5 Douglas G. Sutton, 'A Paradigmatic Shift in Polynesian Prehistory: Implications for New Zealand', *New Zealand Journal of Archaeology*, vol. 9, 1987, pp. 135–55; Patrick V. Kirch, 'Review of Geoffrey Irwin, *The Prehistoric Exploration and Colonisation of the Pacific*', *Pacific Studies*, vol. 20, no. 2, 1997, pp. 119–29.

6 Atholl Anderson, 'Research Report: Was *Rattus exulans* in New Zealand 2000 Years Ago? AMS Radiocarbon Ages from Shag River Mouth', *Archaeology in Oceania*, vol. 31, no. 3, 1996, pp. 178–84; R.N. Holdaway, 'The Arrival of Rats in New Zealand', *Nature*, vol. 384, 1996, pp. 225–26; Richard N. Holdaway, 'A Spatio-Temporal Model for the Invasion of the New Zealand Archipelago by the Pacific Rat *Rattus exulans*', *Journal of the Royal Society of New Zealand*, vol. 29, no. 2, 1999, pp. 91–105; Atholl Anderson, 'Differential Reliability of 14C AMS Ages of *Rattus exulans* Bone Gelatin in South Pacific Prehistory', *Journal of the Royal Society of New Zealand*, vol. 30, no. 3, 2000, pp. 243–61. Also, radiocarbon dating of land snails exhibiting rat-gnawing from various localities in Northland, found none dated earlier than about AD 1200. Conclusive results on dating rat introduction summarised in: Janet M. Wilmshurst, Atholl J. Anderson, Thomas F.G. Higham & Trevor H. Worthy, 'Dating the late prehistoric dispersal of Polynesians to New Zealand using

the commensal Pacific rat', *Proceedings of the National Academy of Sciences, USA*, vol. 105, no. 22, 2008, pp. 7676–80.
7   M.S. McGlone, 'The Polynesian Settlement'; Sutton, 'A Paradigmatic Shift'; Matt S. McGlone & Janet M. Wilmshurst, 'Dating Initial Māori Environmental Impact in New Zealand', *Quaternary International*, vol. 59, no. 1, 1999, pp. 5–16; R.M. Newnham, D.J. Lowe, M.S. McGlone, J.M. Wilmshurst, & T.G.F. Higham, 'The Kaharoa Tephra as a Critical Datum for Earliest Human Impact in Northern New Zealand', *Journal of Archaeological Science*, vol. 25 no. 6, 1998, pp. 533–44.
8   Atholl Anderson, 'The Chronology of Colonization in New Zealand', *Antiquity*, vol. 65, no. 249, 1991, pp. 767–95; T.G.F. Higham & A.G. Hogg, 'Evidence for Late Polynesian Colonisation of New Zealand: University of Waikato Radiocarbon Measurements', *Radiocarbon*, vol. 39, 1997, pp. 149–92; B.G. McFadgen, F.B. Knox & T.R.L. Cole, 'Radiocarbon Calibration Curve Variations and their Implications for the Interpretation of New Zealand Prehistory', *Radiocarbon*, vol. 36, no. 2, 1994, pp. 229–32.
9   Rosalind P. Murray-McIntosh, Brian J. Scrimshaw, Peter J. Hatfield & David Penny, 'Testing Migration Patterns and Estimating Founding Population Size in Polynesia by using Human mtDNA sequences', *Proceedings of the National Academy of Sciences, USA*, vol. 95, no. 15, 1998, pp. 9047–52.
10 Data forthcoming in Atholl Anderson, Judith Binney & Aroha Harris, *Tangata Whenua: An Illustrated History*. Bridget Williams Books, Wellington, 2014.
11 Janet M. Wilmshurst, Terry L. Hunt, Carl P. Lipo & Atholl J. Anderson, 'High-precision Radiocarbon Dating Shows Recent and Rapid Initial Human Colonization of East Polynesia', *Proceedings of the National Academy of Sciences, USA*, vol. 108, no. 5, 2011, pp. 1815–20.
12 Quote from D.J. Lowe, R.M. Newnham, T.G.F. Higham, J.M. Wilmshurst, M.S. McGlone & A.G. Hogg, 'Dating Earliest Human Impact and Settlement in New Zealand', in Douglas G. Sutton (ed.) *The Origins of the First New Zealanders*, 2nd edn, Auckland University Press, in press.
13 A.K. Pawley & M.D. Ross, *Austronesian Terminologies: Continuity and Change*, Pacific Linguistics, Series C–127, Australian National University, Canberra, 1994.
14 Atholl Anderson & Matt McGlone, 'Living on the Edge: Prehistoric Land and People in New Zealand', in John Dodson (ed.), *The Naïve Lands*, Longman Cheshire, Melbourne, 1992, pp. 199–241; Helen Leach, *1000 Years of Gardening in New Zealand*, Reed, Wellington, 1984; Stanton E. Tuller, 'Summer and Winter Patterns of Human Climate in New Zealand', *New Zealand Geographer*, vol. 33, no. 1, 1977, pp. 4–14.
15 T.H. Worthy & R.P. Scofield, 'Twenty-first Century Advances in Knowledge of the Biology of Moa (Aves: Dinornithiformes): A New Morphological Analysis and Moa Diagnoses Revised', *New Zealand Journal of Zoology*, vol. 39, no. 2, 2012, pp. 87–153.
16 See Anderson, 'Prehistoric Polynesian Impact', on the windward–leeward model.
17 R.N. Holdaway, 'Introduced Predators and Avifaunal Extinction in New Zealand', in Ross D.E. MacPhee (ed.), *Extinctions in Near Time: Causes, Contexts and Consequences*, Kluwer Academic/Plenum, New York, 1999, pp. 189–238, quote p. 216; T.H. Worthy, 'What Was on the Menu? Avian Extinction in New Zealand', *New Zealand Journal of Archaeology*, vol. 19, 1997, pp. 125–60.
18 R.N. Holdaway, 'New Zealand's Pre-human Avifauna and its Vulnerability', *New Zealand Journal of Ecology*, vol. 12 (supplement), 1989, pp. 11–25, quote p. 15.
19 R.N. Holdaway & C. Jacomb, 'Rapid Extinction of the Moas (Aves: Dinornithiformes): Model, Test and Implications' *Science*, vol. 287, pp. 2250–54; Anderson & McGlone, 'Living on the Edge'; Anderson, 'Prehistoric Polynesian Impact'; Worthy, 'What was on the Menu?'; F.J. Petchey, 'New Zealand Bone Dating Revisited: A Radiocarbon Discard Protocol for Bone', *New Zealand Journal of Archaeology*, vol. 19, 1997, 81–124.
20 C.L. Batcheler, 'Moa Browsing and Vegetation Formations, with Particular Reference to Deciduous and Poisonous Plants', *New Zealand Journal of Ecology*, vol. 12 (supplement), 1989, pp. 57–65; G. Caughley, 'New Zealand Plant-Herbivore Ecosystems Past and Present', *New Zealand Journal of Ecology*, vol. 12 (supplement), 1989, pp. 3–10; Ian W.G. Smith, 'Māori Impact on the Marine Megafauna: Pre-European Distributions of New Zealand Sea Mammals', in Douglas G. Sutton (ed.), *Saying So Doesn't Make it So*, New Zealand Archaeological Association Monograph 17, Dunedin, 1989, pp. 76–108.
21 Anderson & McGlone, 'Living on the Edge'; John Ogden, Les Basher & Matt McGlone, 'Fire, Forest Regeneration and Links with Early Human Habitation: Evidence from New Zealand', *Annals of Botany*, vol. 81, no. 6, 1998, pp. 687–96; M.S. McGlone, 'Polynesian Deforestation of New Zealand: A Preliminary Synthesis', *Archaeology in Oceania*, vol. 18, no. 1, 1983, p. 23.
22 Caughley, 'New Zealand Plant-Herbivore Ecosystems', p. 7; M.N. Clout & J.R. Hay, 'The Importance of Birds as Browsers, Pollinators and Seed Dispersers in New Zealand Forests', *New Zealand Journal of Ecology*, vol. 12 (supplement), 1989, pp. 27–33; M.J. McSaveney & I.E. Whitehouse, 'Anthropic Erosion of Mountain Land

in Canterbury', *New Zealand Journal of Ecology*, vol. 12 (supplement), 1989, pp. 151–63; M.S. McGlone, A.J. Anderson & R. Holdaway, 'An Ecological Approach to the Polynesian Settlement of New Zealand', in Douglas G. Sutton (ed.), *The Origins of the First New Zealanders*, 1st edn, Auckland University Press, Auckland, 1994, pp. 136–63.
23 Anthony Lorrey, Paul Williams, Jim Salinger, Tim Martin, Jonathan Palmer, Anthony Fowler, Jian-xin Zhao & Helen Neil, 'Speleothem Stable Isotope Records Interpreted with a Multi-proxy Framework and Implications for New Zealand Palaeoclimate Reconstruction', *Quaternary International*, vol. 187, no. 1, 2008, pp. 52–75.
24 Anderson, 'Prehistoric Polynesian Impact'; Atholl Anderson, *The Welcome of Strangers: An Ethnohistory of Southern Maori, A.D. 1650–1850*, Otago University Press, Dunedin, 1998.

3. Contesting resources: Māori, Pakeha and a tenurial revolution

1 Waitangi Tribunal, *Muriwhenua Land Report*, GP Publications, Wellington, 1997, pp. 23–24.
2 Robert McNab, *Historical Records of New Zealand*, Government Printer, Wellington, 1914, vol. 2, p. 550.
3 John Savage, *Some Account of New Zealand, Particularly the Bay of Islands and Surrounding Country*, John Murray, London, 1807; Waitangi Tribunal, *Muriwhenua Fishing Report*.
4 John Liddiard Nicholas, *Narrative of a Voyage to New Zealand*, James Black, London, 1817, vol. 2, pp. 209–10.
5 Judith Binney, 'Tuki's Universe', in Keith Sinclair (ed.), *Tasman Relations: New Zealand and Australia*, Auckland University Press, 1988, p. 19.
6 Rhys Richards & Jocelyn Chisholm, *Bay of Islands Shipping Arrivals and Departures 1803–1840*, Paremata Press, Wellington, 1992; Harry Morton, *The Whale's Wake*, John McIndoe, Dunedin, 1982.
7 Richard A. Cruise, *Journal of a Ten Months' Residence in New Zealand*, Longman, London, 1824, p. 157.
8 James Buller, *Forty Years in New Zealand, Including a Personal Narrative, An Account of Maoridom, and of the Christianization and Colonisation of the Country*, Hodder & Stoughton, London, 1878, pp. 27–28.
9 R.P. Wigglesworth, 'The New Zealand Timber and Flax Trade 1769–1840', PhD thesis, Massey University, 1981, pp. 135–36.
10 Evelyn Stokes, 'Kauri and White Pine: A Comparison of New Zealand and American Lumbering', *Annals of the Association of American Geographers*, vol. 56, no. 3, 1966, p. 446.
11 Ernst Dieffenbach, *Travels in New Zealand*, John Murray, London, 1843, vol. 1, pp. 227–28.
12 Binney, 'Tuki's Universe', p. 24; see also Morton, *Whale's Wake*, pp. 166–69, 216.
13 Quoted by Wigglesworth, New Zealand Timber and Flax Trade, p. 193.
14 Edward Markham, *New Zealand, or Recollections of It*, Alexander Turnbull Library, Monograph no. 1, Wellington, 1963, p. 40; see also Wigglesworth, New Zealand Timber and Flax Trade.
15 Jeffrey Sissons, Wiremu Wi Hongi & Pat Hohepa, *The Puriri Trees Are Laughing: A Political History of Nga Puhi in the Inland Bay of Islands*, Polynesian Society, Auckland, 1987.
16 *Muriwhenua Land Report*, p. 392; see also Evelyn Stokes, *A Review of the Evidence in the Muriwhenua Land Claims*, Waitangi Tribunal Review Series No. 1, Wellington, 1997.
17 See Waitangi Tribunal, *The Ngai Tahu Report*, Brooker & Friend, Wellington, 1991.
18 See Waitangi Tribunal, *The Taranaki Report: Kaupapa Tuatahi*, GP Publications, Wellington, 1996, and *The Ngati Awa Raupatu Report*, Legislation Direct, Wellington, 1999.
19 Alan Ward, *A Show of Justice: Racial Amalgamation in Nineteenth Century New Zealand*, Auckland and Oxford University Press, 1973, p. 185.
20 I.H. Kawharu, *Maori Land Tenure. Studies of a Changing Institution*, Clarendon Press, Oxford, 1977, p. 17.
21 Ward, *A Show of Justice*, pp. 185–86.
22 David V. Williams, *Te Kooti Tango Whenua: The Native Land Court 1864–1909*, Huia Publishers, Wellington, 1999, pp. 44–46.
23 *Appendices of the Journals of the House of Representatives*, Session II, G1, 1891, p. 145.
24 Waitangi Tribunal, *The Pouakani Report*, Brooker & Friend, Wellington, 1993, pp. 307–08.

4. Resource frontiers, environment and settler capitalism, 1769–1860

1 Chinese and Indian connections are emphasised by Tony Ballantyne, 'Sealers, Whalers and the Entanglements of Empire' in his *Webs of Empire: Locating New Zealand's Colonial Past*, Bridget Williams Books, Wellington, 2012, pp. 124–36. On the Pacific Frontier, see J.M.R. Young, 'Australia's Pacific Frontier', *Historical Studies, Australia and New Zealand*, vol. 12, no. 47, 1966, pp. 373–88 and D.R. Hainsworth, 'Exploiting the Pacific Frontier: The New South Wales Sealing Industry 1800–1921', *Journal of Pacific History*, vol. 2, 1967, pp.

59–75. Of other works dealing with this period, Anne Salmond's *Two Worlds: First Meetings Between Maori and Europeans, 1642–1772*, Viking, Auckland, 1991, and *Between Worlds: Early Exchanges Between Maori and Europeans, 1773–1815*, Viking, Auckland, 1997, are the most recent detailed studies to 1820, although more concerned with culture than environment. Alan Gray, *Aotearoa and New Zealand: A Historical Geography*, University of Canterbury Press, Christchurch, 1994, addresses many themes of the present chapter.

2 The observations on revolutionary capitalism are from Karl Marx & Friedrich Engels, *The Communist Manifesto*, Penguin, Harmondsworth, 1967, pp. 81–84. Ecological revolution is defined by Carolyn Merchant, *Ecological Revolutions: Nature, Gender and Science in New England*, University of North Carolina Press, Chapel Hill, 1989, pp. 2–3.

3 See, generally, Keith Thomas, *Man and the Natural World, Changing Attitudes in England 1500–1800*, Allen Lane, London, 1983, esp. pp. 281–85. Cook's instructions are quoted in Salmond, *Two Worlds*, p. 99.

4 Cook on trade is in J.C. Beaglehole (ed.), *The Journals of Captain James Cook*, Cambridge University Press, Cambridge, 1968, vol. 1, pp. 186, 194–95. Flax and timber are described by Beaglehole (ed.), *The Endeavour Journal of Joseph Banks: 1768–1771*, Angus & Robertson, Sydney, 1962, vol. 2, p. 10. Other observations are Beaglehole, *Cook's Journals*, vol. 1, pp. 176, 269–70, 277–78; Beaglehole, *Banks's Journals*, vol. 2, p. 4.

5 Isabel Ollivier & Cheryl Hingley (trans.), *Extracts from Journals Relating to the Visit to New Zealand of the French Ship St Jean Baptiste*, Alexander Turnbull Library Endowment Trust, Wellington, 1987, pp. 136–38; Salmond, *Two Worlds*, pp. 331–38, 383, 406.

6 Beaglehole, *Cook's Journals*, vol. 2, pp. 126, 133–35; Salmond, *Between Worlds*, p. 52. Forster is quoted in Salmond, *Between Worlds*, p. 63.

7 Beaglehole, *Cook's Journals*, vol. 2, pp. 167–69, vol. 3, pt 1 p. 63; Michael E. Hoare (ed.), *The Resolution Journal of Johann Reinhold Forster, 1772–1775*, Hakluyt Society, London, 1982, vol. 2, pp. 287–98, who gives a longer list of vegetables than Cook; C.M.H. Clarke & R.M. Dzieciolowski, 'Feral Pigs in the Northern South Island, New Zealand: I. Origin, Distribution, and Density', *Journal of the Royal Society of New Zealand*, vol. 21, no. 3, 1991, pp. 237–47; Harry Morton, *The Whale's Wake*, Otago University Press, Dunedin, 1982, p. 164.

8 Generally, Ged Martin (ed.), *The Founding of Australia*, Hale & Iremonger, Sydney, 1978; Margaret Steven, *Trade, Tactics and Territory: Britain in the Pacific, 1783–1823*, Melbourne University Press, Melbourne, 1983, chapters 1–2; Harold B. Carter, *Sir Joseph Banks*, British Museum, London, 1988, pp. 213–14; Robert McNab (ed.), *Historical Records of New Zealand*, Government Printer, Wellington, 1908, vol. 1, pp. 36–37; Alan Frost, *The Global Reach of Empire: Britain's Maritime Expansion in the Indian and Pacific Oceans, 1764–1815*, Miegunyah Press, Melbourne, 2003, chapter 8; Alan Frost, *Botany Bay: The Real Story*, Black Inc., Melbourne, 2012, pp. 203–09.

9 The affair is discussed in Salmond, *Between Worlds*, chapter 9. For detail, see Carter, *Banks*, p. 218, and Lieutenant-Governor King to Nepean, April 1791; Dundas to Lords of Admiralty, 6 July 1791, McNab, *Historical Records*, vol. 1, pp. 126, 128.

10 Salmond, *Between Worlds*, pp. 227–33; for detail, King to Nepean 19 November 1793, in McNab, *Historical Records*, vol. 1, p. 180; King to Dundas 19 November 1793, in McNab, *Historical Records*, vol. 1, pp. 171–75; King to Dundas, 10 March 1794, in McNab, *Historical Records*, vol. 1, p. 192.

11 Much of the detail on the Britannia and Francis is in Robert McNab, *Murihiku and the Southern Islands*, William Smith, Invercargill, 1907, chapters 4–6; Salmond, *Between Worlds*, pp. 285–94; D.R. Hainsworth, 'Trade within the Colony', in G.J. Abbott & N.B. Nairn (eds), *Economic Growth of Australia 1788–1821*, Melbourne University Press, Carlton, 1969, p. 268; Steven, *Trade, Tactics, Territory*, p. 87.

12 George Bass to Capt Waterhouse 30 January 1802, in McNab, *Historical Records*, vol. 1, p. 225; George Bass to Capt. Waterhouse 5 January 1803, 2 February 1803, in McNab, *Historical Records*, vol. 1, pp. 240–42, 245; Bass to King 30 January 1803, in McNab, *Historical Records*, vol. 1, pp. 242–24.

13 Frost, *Global Reach*, pp. 314–20.

14 McNab, *Murihiku*, pp. 57–58; Salmond, *Between Worlds*, chapters 11, 16; R.P. Wigglesworth, 'The New Zealand Timber and Flax Trade 1769–1840', PhD thesis, Massey University, 1981, pp. 14, 259–61.

15 R.B. Joyce, 'Government Policy', in Abbott & Nairn, *Economic Growth of Australia*, p. 70; Frost, *Global Reach of Empire*, p. 308.

16 M.J.E. Steven, 'Exports other than Wool', in Abbott & Nairn, *Economic Growth of Australia*, p. 289; Abbott, 'Economic Growth', in ibid., pp. 147–48; see also Steven, *Merchant Campbell, 1769–1846: A Study of Colonial Trade*, Oxford University Press, Melbourne, 1965, and Hainsworth, 'Exploiting the Pacific Frontier'; details in 'Some Remarks on the Present State of the Colony of Sydney 4 June 1806', in McNab, *Historical Records*, vol. 1, pp. 271–74; Morton, *Whale's Wake*, p. 113; Rhys Richards, 'New Market Evidence on the Depletion of Southern Fur Seals: 1788–1833,' *New Zealand Journal of Zoology*, vol. 30, no.1, 2003, pp. 1–9.

17 Morton, *Whale's Wake*, pp. 90–94, 106, 121, 134, 144; Carter, *Banks*, pp. 433–44.
18 McNab, *Murihiku*, chapters 8–10 and 16, esp. pp. 113–14, 182.
19 McNab, *Murihiku*, p. 230 and, generally, chapters 12, 18, 23, quoted Atholl Anderson, *The Welcome of Strangers: An Ethnohistory of Southern Maori AD 1650–1850*, Otago University Press, Dunedin, 1998, p. 67.
20 Quoted in Anderson, *Welcome of Strangers*, p. 67.
21 McNab, *Murihiku*, p. 261.
22 Morton, *Whale's Wake*, chapter 15 and esp. pp. 51–53, 121–25, 131–33, 140–47. On whaling rights see Edward Shortland, *The Southern Districts of New Zealand*, reprinted, Capper Press, Christchurch, 1974, pp. 83–91.
23 Salmond, *Between Worlds*, chapters 17 and 18 and p. 339; Morton, *Whale's Wake*, p. 126; 'Rev. S. Marsden's Account of his First Visit to New Zealand', in McNab, *Historical Records*, vol. 1, pp. 343–48; Young, 'Australia's Pacific Frontier', p. 379; 'Governor King to Earl Camden 30 April 1805', in McNab, *Historical Records*, vol. 1, p. 255; Morton, *Whale's Wake*, p. 126; Salmond, *Between Worlds*, p. 339; 'King Papers, 2 January 1806' in McNab, *Historical Records*, vol. 1, pp. 263–66.
24 Anderson, *Welcome of Strangers*, p. 74.
25 Anderson, *Welcome of Strangers*, p. 75.
26 'Report on New Zealand Flax', by R. Williams, in McNab, *Historical Records*, vol. 1, pp. 463–64; Morton, *Whale's Wake*, p. 183; Joel Polack, *New Zealand*, Capper Press, Christchurch, 1974, vol. 1, p. 251; Richard A. Cruise, *Journal of a Ten Months' Residence in New Zealand*, Capper Press, Christchurch, 1974, pp. 38, 48; William Yate, *An Account of New Zealand*, Reed, Wellington, 1970, pp. 107–11.
27 On Marsden, see 'Rev. S. Marsden's Account of his First Visit to New Zealand', McNab, *Historical Records*, vol. 1, p. 358; and 'Letter from Marsden 20 September 1815', McNab, *Historical Records*, vol. 1, p. 403; Salmond, *Between Worlds*, p. 445. For Polack's dealings, see Polack, *New Zealand*, pp. 172–82. It must be emphasised that we have only Polack's account of this discussion: see, in general, Hazel Petrie & Hohipere Tarau, 'Māori Texts and Official Ventriloquism,' *New Zealand Journal of History*, vol. 46, no. 2, 2012, pp. 129–41.
28 Hazel Petrie, *Chiefs of Industry: Māori Tribal Enterprise in Early Colonial New Zealand*, Auckland University Press, Auckland, 2006, p. 28.
29 Anderson, *Welcome of Strangers*, p. 96.
30 Evelyn Stokes, *A History of Tauranga County*, Dunmore Press, Palmerston North, 1980, p. 55; Polack, *New Zealand*, p. 120.
31 Generally, Wigglesworth, *The New Zealand Timber and Flax Trade*, pp. 71–110 and Michael Roche, *History of New Zealand Forestry*, GP Books, Wellington, 1990, chapter 1. For detail, 'Robert Torrens to Under-Secretary Horton', 14 June 1827, in McNab, *Historical Records*, vol. 1, pp. 676–78; Augustus Earle, *A Narrative of a Nine Month's Residence in New Zealand in 1827*, Whitcombe & Tombs, Christchurch, 1909, pp. 66, 68, 82; Salmond, *Between Worlds*, p. 250; Cruise, *Journal*, p. 21, Ernst Dieffenbach, *Travels in New Zealand*, Capper Press, Christchurch, 1974, vol. 1, pp. 201–53; E.J. Wakefield, *Adventure in New Zealand*, John Murray, London, 1845, vol. 1, p. 412. The spar specifications are in Wigglesworth, *The New Zealand Timber and Flax Trade*, pp. 108–09.
32 For example, 'Robert Sugden to Earl Bathurst, 18 January 1821', in McNab, *Historical Records*, vol. 1, p. 516; 'Lt-Col Nicholls (Royal Marines) to Earl Bathurst 8 November 1823', McNab, *Historical Records*, vol. 1, pp. 599–603; 'Memorial to … Earl Bathurst, 24 April 1826', McNab, *Historical Records*, vol. 1, pp. 663–65.
33 The point about a new international division of labour is made by Erik Olssen, 'Wakefield and the Scottish Enlightenment', in Friends of the Turnbull Library (eds), *Edward Gibbon Wakefield and the Colonial Dream: A Reconsideration*, Government Printer, Wellington, 1997, p. 58. Descriptions are from Wakefield, *Adventure in New Zealand*, pp. 77, 92; Charles Heaphy, *Narrative of a Residence in Various Parts of New Zealand*, Capper Press, Christchurch, 1972, pp. 3, 22–23. The tart comment is David Monro, 'Notes of a Journey Through a Part of the Middle Island of New Zealand', Appendix C, in T.M. Hocken, *Contributions to the Early History of New Zealand*, Sampson Low, London, 1898, p. 231. On the New Zealand Company generally, see Philip Temple, *A Sort of Conscience: The Wakefields*, Auckland University Press, Auckland, 2002. Peter Holland, *Home in the Howling Wilderness: Settlers and the Environment in Southern New Zealand*, Auckland University Press, Auckland, 2013, chapter 2, discusses how settlers learnt to read the climate.
34 Newspaper quotes *Nelson Examiner*, 10 June 1843, p. 262; 8 March 1845, p. 2. On FitzRoy, see Dean Cowie, '"To Do All the Good I Can": Robert FitzRoy, Governor of New Zealand 1843–45', MA thesis, University of Auckland, 1994; on complexities of purchase, see Robyn Anderson & Keith Pickens, *Waitangi Tribunal Rangahaua Whanui Series: District 12 – Wellington District*, Waitangi Tribunal, Wellington, 1996.
35 Dieffenbach, *Travels in New Zealand*, vol. 1, pp. 15–19.

36 R.C.J. Stone, *Young Logan Campbell*, Auckland University Press, Auckland, 1982, quote at p. 86; generally, chapter 8; Petrie, *Chiefs of Industry*.
37 Fox to William Wakefield, 1 December 1843, *Fourteenth Report of the New Zealand Company*, Appendix 17; *Nelson Examiner*, 8 June 1844, p. 54; Ruth M. Allan, *Nelson: A History of Early Settlement*, Reed, Wellington, 1965, pp. 237–38; *Nelson Examiner*, 16 December 1843, p. 370; Edmund Bohan, *Edward Stafford*, Hazard Press, Christchurch, 1994, p. 23.
38 Donal P. McCracken, *Gardens of Empire: Botanical Institutions of the Victorian British Empire*, Leicester University Press, London, 1997 p. 85; utilitarian exploration is modelled by John Overton, 'A Theory of Exploration', *Journal of Historical Geography*, vol. 7, no. 1, 1981, pp. 53–70.
39 Brad Patterson, 'A Queer Cantankerous Lot', in David Hamer & Roberta Nicholls (eds), *The Making of Wellington, 1800–1914*, Victoria University Press, Wellington, 1990, esp. p. 65.
40 See generally A.G. Bagnall, *Wairarapa*, Masterton Trust Lands Trust, Masterton, 1976, chapters 2–4. James Kemp on prosperity quoted in Paul Goldsmith, *Waitangi Tribunal Rangahaua Whanui District 11A: Wairarapa*, Waitangi Tribunal, Wellington, 1996, p. 26. See also John C. Weaver, 'Frontiers into Assets: The Social Construction of Property in New Zealand 1840–1865', *Journal of Imperial and Commonwealth History*, vol. 27, no. 3, 1999, pp. 17–54.
41 Petrie, *Chiefs of Industry*, pp. 82, 88, 110.
42 Heaphy quoted in N.M. Taylor (ed.), *Early Travellers in New Zealand*, Clarendon Press, Oxford, 1959, p. 248; *Nelson Examiner*, 2 June 1860, p. 3; Julius Haast, *Report of a Topographical and Geological Exploration of the Western Districts of the Nelson Province*, Nelson Provincial Government, Nelson, 1861, pp. 43, 130–32. See also Mike Johnston and Sascha Nolden. *Travels of Hochstetter and Haast in New Zealand, 1858–60*, Nikau Press, Nelson, 2011.
43 Tuckett's Diary, in Hocken, *Contributions*, Appendix A, p. 207.
44 Monro in Hocken, *Contributions*, p. 258–60.
45 On Australia, see Sylvia Morrissey, 'The Pastoral Economy, 1821–1850' in James Griffen (ed.), *Essays in Economic History of Australia*, Jacaranda, Milton, 1967, pp. 51–112; Jill Ker, 'The Wool Industry in New South Wales 1803–30', *Business Archives and History*, vol. 2, no. 1, 1962, pp. 18–54; E.A. Beever, 'The Origin of the Wool Industry in New South Wales', *Business Archives and History*, vol. 5, no. 2, 1965, pp. 91–106; S. H. Roberts, *The Squatting Age in Australia*, Melbourne University Press, Carlton, 1970; Lynette Peel, *Rural Industry in the Port Phillip Region*, Melbourne University Press, Carlton, 1974; Tony Dingle, *The Victorians: Volume 2 – Settling*, Fairfax, Syme & Weldon Associates, McMahons Point, 1984.
46 Jim McAloon, 'Mobilising Capital and Trade' in Tom Brooking and Eric Pawson, *Seeds of Empire: The Environmental Transformation of New Zealand*, I.B. Tauris, London, 2011, pp. 94–115.
47 Robert Peden, *Making Sheep Country: Mt Peel Station and the Transformation of the Tussock Lands*, Auckland University Press, Auckland, 2011, chapter 1.
48 On the *Acheron*, see R. J. B. Knight, 'John Lort Stokes and the New Zealand Survey, 1848–1851', in Alan Frost & Jane Samson (eds), *Pacific Empires: Essays in Honour of Glyndwr Williams*, Melbourne University Press, Melbourne, 1999, p. 87–99; Sheila Natusch, *The Cruise of the Acheron*, Whitcoulls, Christchurch, 1978; on Thomson, see John Hall-Jones, *John Turnbull Thomson*, John McIndoe, Dunedin, 1992, esp. p. 83.
49 On the practice of surveying, see Giselle Byrnes, 'Surveying Space: Constructing the Colonial Landscape', in Bronwyn Dalley & Bronwyn Labrum (eds), *Fragments. New Zealand Social and Cultural History*, Auckland University Press, Auckland, 2000, pp. 54–75, and Janet Holm, *Caught Mapping. The Life and Times of New Zealand's Early Surveyors*, Hazard Press, Christchurch, 2005.

PART II COLONISING

5. Settlers transforming the open country

1 Peter Holland, Kevin O'Connor & Alexander Wearing, 'Remaking the Grasslands of the Open Country', in Eric Pawson & Tom Brooking (eds), *Environmental Histories of New Zealand*, Oxford University Press, Melbourne, 2002, p. 72.
2 B.P.J. Molloy, C.J. Burrows, J.E. Cox, J.A. Johnston & P. Wardle, 'Distribution of Subfossil Forest Remains in Eastern New Zealand', *New Zealand Journal of Botany*, vol. 1, no. 1, 1963, pp. 68–77; Matt S. McGlone & Janet M. Wilmshurst, 'Dating Initial Maori Environmental Impact in New Zealand', *Quaternary International*, vol. 59, no. 1, 1999, pp. 5–16.
3 Edward Shortland, *The Southern Districts of New Zealand: A Journal with Passing Notices of the Customs of the

Aborigines, Longman, Brown, Green & Longman, London, 1851.
4   Thomson W. Leys (ed.), *Brett's Colonist's Guide and Cyclopaedia of Useful Knowledge*, Auckland, Brett, 1883, reprinted Capper Press, Christchurch, 1980.
5   Peter Holland & Bill Mooney, 'Wind and Water: Environmental Learning in Early Colonial New Zealand', *New Zealand Geographer*, vol. 62, no. 1, 2006, pp. 39–49.
6   W.D. McIntyre, (ed.) *The Journal of Henry Sewell, 1853–1857*, 2 vols, Whitcoulls, Christchurch, 1980.
7   Robert Peden, *Making Sheep Country: Mt Peel and the Transformation of the Tussock Lands*, Auckland University Press, Auckland, 2011.
8   Sheila S. Crawford, *Sheep and Sheepmen of Canterbury 1850–1914*, Simpson & Williams, Christchurch, 1949; Andrew Hill Clark, *The Invasion of New Zealand by People Plants and Animals*, Rutgers University Press, New Brunswick, 1949, p. 181; R.P. Hargreaves, 'Speed the Plough: An Historical Geography of New Zealand Farming Before the Introduction of Refrigeration', PhD thesis, Otago, 1966, p. 458.
9   L.G.D. Acland, *The Early Canterbury Runs*, Whitcombe & Tombs, Christchurch, complete edition, 1951, p. 260; Kenneth B. Cumberland, *Landmarks*, Reader's Digest, Surry Hills, NSW, 1981, p. 108.
10 Frederick Weld, *Hints for Intending Sheep-Farmers*, Trelawney Saunders, London, 1851, p.10; Robert L. Peden, 'Sheep Farming Practice in Colonial Canterbury 1843 to 1882: The Origin and Diffusion of Ideas, Skills, Techniques and Technology in the Creation of the Pastoral System', MA thesis, University of Canterbury, 2002.
11 Peter J. Bowler, *The Invention of Progress: The Victorians and the Past*, Blackwell, Oxford, 1989, pp. 193–95; L.E. Lochhead, 'Preserving the Brownie's Portion: A History of Voluntary Nature Conservation Organisations in New Zealand, 1888–1935', PhD thesis, Lincoln University, 1994, pp. 16–28.
12 John C. Weaver, *The Great Land Rush and the Making of the Modern World, 1650–1900*, McGill-Queens University Press, Montreal, 2003, pp. 81–84; Donald Worster, *Nature's Economy: A History of Ecological Ideas*, Cambridge University Press, Cambridge, 1977, pp. 178–79.
13 E.J.T. Collins, 'Introduction', *The Agrarian History of England and Wales, Volume VII, 1840–1915*, Cambridge University Press, Cambridge, 2000, p. 11.
14 Ibid., p.11; Vaughan Wood & Eric Pawson, 'Flows of Agricultural Information', in Tom Brooking & Eric Pawson, *Seeds of Empire: The Environmental Transformation of New Zealand*, I.B. Tauris, London, 2011, pp. 139–58.
15 Jim McAloon, 'The Colonial Wealthy in Canterbury and Otago: No Idle Rich', *New Zealand Journal of History*, vol. 30, no. 1, 1996, p. 52.
16 Samuel Butler, *A First Year in the Canterbury Settlement*, A.C. Field, London, 1914, p. 50.
17 Wood & Pawson, 'Flows of Agricultural Information'.
18 Peden, 'Sheep Farming Practice', pp. 143–45; Eric Pawson & Neil C. Quigley, 'The Circulation of Information and Frontier Development: Canterbury 1850–1890', *New Zealand Geographer*, vol. 32, no. 2, pp. 65–76.
19 Peden, 'Sheep Farming Practice', pp. 140–43; Wood & Pawson, 'Flows of Agricultural Information'.
20 Wood & Pawson, 'Flows of Agricultural Information'; Peden, 'Sheep Farming Practice', pp. 145–47.
21 Peden, 'Sheep Farming Practice', pp. 142–43.
22 Peden, 'Sheep Farming Practice', pp. 148–51; Wood & Pawson, 'Flows of Agricultural Information'.
23 Miles Fairburn, 'The Rural Myth and the New Urban Frontier: An Approach to New Zealand Social History, 1870–1940', *New Zealand Journal of History*, vol. 9, no. 1, 1975, p. 11; Vaughan Wood & Tom Brooking, 'Canterbury Farming Intensifies', in Garth Cant & Russell Kirkpatrick (eds), *Rural Canterbury. Celebrating its History*, Daphne Brasell, Wellington, 2001, pp. 81–99.
24 Tom Brooking, *Lands for the People? The Highland Clearances and the Colonisation of New Zealand: A Biography of John McKenzie*, University of Otago Press, Dunedin, 1996, p. 260; McAloon, 'The Colonial Wealthy', p. 141.
25 Peden, *Making Sheep Country*, p. 204.
26 Ibid., pp. 216–17.
27 Holland et al., 'Remaking the Grasslands of the Open Country', p. 71.
28 Herbert Guthrie-Smith, *Tutira: The Story of a New Zealand Sheep Station*, 4th edn, Random House, Auckland, 1999, pp. 165–68.
29 A.H. Cockayne, 'The Natural Pastures of New Zealand 1: The Effect of Burning on Tussock Country', *Journal of the Department of Agriculture*, vol. 1, no. 1, 1910, pp. 12–15; V.D. Zotov, 'Survey of the Tussock Grasslands of the South Island of New Zealand', *Department of Scientific and Industrial Research Bulletin*, 1938, pp. 212A, 228A; Kenneth B. Cumberland, 'A Century's Change: Natural to Cultural Vegetation in New Zealand', *Geographical Review*, vol. 31, no. 4, 1941, p. 536.

30 B.P.J. Molloy, 'The Fire History', in C.J. Burrows (ed.) *Cass: History and Science in the Cass District, Canterbury, New Zealand*, University of Canterbury, 1977, p. 163; M.S. McGlone, 'The Origin of the Indigenous Grasslands of Southeastern South Island in Relation to Pre-human Woody Ecosystems', in *New Zealand Journal of Ecology*, vol. 25, no. 1, 2001, p.11; Ian E. Whitehouse, 'Erosion in the Eastern South Island High Country: A Changing Perspective', *Tussock Grasslands and Mountain Lands Institute Review* no. 42, 1984, p. 17; M.J. McSaveney & Ian E. Whitehouse, 'Anthropic Erosion of Mountain Land in Canterbury', *New Zealand Journal of Ecology* 12 (supplement 1989), p. 159; I.E. Whitehouse & A.J. Pearce, 'Shaping the Mountains of New Zealand', in J.M. Soons & M.J. Selby (eds), *Landforms of New Zealand*, Longman Paul, Auckland, 2nd edn, 1992, p. 151.
31 James F. Hoy & Thomas D. Isern, 'Bluestem and Tussock: Fire and Pastoralism in the Flint Hills of Kansas and the Tussock Grasslands of New Zealand', *Great Plains Quarterly*, Summer 1995, pp. 173–74; P.G. Holland & R.P. Hargreaves, 'The Trivial Round, The Common Task: Work and Leisure on a Canterbury Hill Country Run in the 1860s and 1870s', *New Zealand Geographer*, vol. 47, no. 1, 1991, p. 22.
32 Robert Peden, '"The Exceeding Joy of Burning": Pastoralists and the Lucifer Match. Burning the Rangelands of the South Island of New Zealand in the Nineteenth Century, 1850 to 1890', *Agricultural History*, vol. 80, no. 1, 2006, pp. 17–34.
33 S. Grant & J.S. Foster, *New Zealand: A Report on its Agricultural Conditions and Prospects*, G. Street, London, 1880; Peter Holland, Paul Star & Vaughan Wood, 'Pioneer Grassland Farming: Pragmatism, Innovation and Experimentation', in Brooking & Pawson, *Seeds of Empire*, pp. 51–77.
34 Crawford, *Sheep and Sheep Men of Canterbury*, p. 68.
35 A.H. Cockayne, 'The Grasslands of New Zealand: Component Species', *New Zealand Journal of Agriculture* 17, 1918, pp. 210–20.
36 Paul Star & Tom Brooking, 'Fescue to the Rescue: Chewing's Fescue, Paspalum, and the Application of Non-British Experience to Pastoral Practice in New Zealand, 1880–1920', *Agricultural History*, vol. 80, no. 3, pp. 312–35; Vaughan Wood & Eric Pawson, 'The Banks Peninsula Forests and Akaroa Cocksfoot: Explaining a New Zealand Forest Transition, *Environment and History*, vol. 14, no. 4, 2008, pp. 449–68.
37 Anon., *New Zealand Farmer*, 19, 1899, p. 334.
38 R.A. Loughnan, *The New Zealand Settlers' Handbook*, Government Printer, Wellington, 1911.
39 E.B. Levy, 'The Grasslands of New Zealand: Principles of Pasture Establishment', *New Zealand Journal of Agriculture*, 23, 1921, pp. 259–65 and 321–30.
40 Holland, Star & Wood, 'Pioneer Grassland Farming'.
41 Wood & Pawson, 'Flows of Agricultural Information'.
42 Cockayne, 'The Grasslands of New Zealand: Component Species'.
43 *Lyttelton Times*, 11 June 1862; 27 December 1864.
44 Peden, *Making Sheep Country*, pp. 180–92.
45 Eric Pawson & Peter Holland, 'People and the Land', in Michael Winterbourn, George Knox, Colin Burrows & Islay Marsden (eds), *The Natural History of Canterbury*, 3rd edn, Canterbury University Press, Christchurch, 2008, pp. 37–64.
46 B.L. Evans, (compiler), *Agricultural and Pastoral Statistics of New Zealand 1861–1954*, Government Printer, Wellington, 1956.
47 Peden, *Making Sheep Country*, pp. 176–77; Jim McAloon, 'Mobilising Capital and Trade', in Brooking & Pawson, *Seeds of Empire*, pp. 97–101.
48 *Appendix to the Journal of the House of Representatives* (hereafter *AJHR*), 1895, H–23; 1905, H–28; 1912, H–23B.
49 In 1897 the Waihemo County Council in East Otago bought 15 bushels of wheat dressed with phosphorus to eliminate small birds that uprooted grain seedlings and attacked developing seedhead. This was repacked into 7- or 8-pound lots and made available to ratepayers across the county. One farmer requested a 1-ounce bottle of strychnine to formulate his own poisoned bait, and the Council agreed to provide it. Similar policies were in place elsewhere in eastern and southern Otago where grain cultivation was compromised by the predations of small birds. Waihemo County Council Letterbook, 1895 to 1901, Hocken Collections, AG-624-003/004.
50 *AJHR*, 1887, C-17, pp. 1–2.
51 Joint Committee of Canterbury Chamber of Commerce and Canterbury Agricultural and Pastoral Association, 'Report on Rabbit Pest', *New Zealand Country Journal*, vol. 12, no. 1, 1888, p. 23.
52 *New Zealand Country Journal*, vol. 12, no.1, January 1888, p. 23.
53 Peden, *Making Sheep Country*, pp. 81–84.
54 Peter Newton, *High Country Journey*, A.H. & A.W. Reed, Auckland, 1952, p. 65.
55 University of Otago, Hocken Collection, MS-1271/088.

56 Tony Nightingale, 'Government and Agriculture – Formation of the Department of Agriculture', *Te Ara – the Encyclopedia of New Zealand*, updated 3-Jul-2012: www.TeAra.govt.nz/en/government-and-agriculture/4

6. Mining the quarry

1 For brief surveys of mining in New Zealand, see the *New Zealand Official Yearbooks*. G.J. Williams, *Economic Geology of New Zealand*, Office of the Congress and of the Australasian Institute of Mining and Metallurgy, Melbourne, 1965, includes a brief survey of the history of mining in New Zealand. The history of mining for minerals other than gold and coal remains largely untold, although M.R. Johnston, in *High Hopes: The History of the Nelson Mineral Belt and New Zealand's First Railway*, Nikau Press, Nelson, 1987, offers an account of Nelson's mineral belt of Dun Mountain–Red Hills.
2 W.N. Blair, 'Artificial Earth Sculpture', *Zealandia*, vol. 1, no. 8, 1890, pp. 474–81. Reprinted, and with an introduction by R.P. Hargreaves, in *Earth Science Journal*, vol. 2, no. 2, 1968, pp. 175–81.
3 The same is largely true for Australia. See, for example, J.M. Powell, *Environmental Management in Australia, 1788–1914. Guardians, Improvers and Profit: An Introductory Survey*, Oxford University Press, Melbourne, 1976; William J. Lines, *Taming the Great South Land: A History of the Conquest of Nature in Australia*, Allen & Unwin, North Sydney, 1991, p. 91; Neil Barr & John Cary, *Greening a Brown Land: The Australian Search for Sustainable Land Use*, Macmillan, South Melbourne, 1992, pp. 54–55; Stephen Dovers, 'The History of Natural Resource Use in Rural Australia: Practicalities and Ideologies', in Geoffrey Lawrence, Frank Vanclay & Brian Furze (eds), *Agriculture, Environment and Society: Contemporary Issues for Australia*, Macmillan, South Melbourne, 1992, pp. 1–18.
4 A.M. Honore, 'Ownership', in A.G. Guest (ed.), *Oxford Essays in Jurisprudence*, Oxford University Press, London, 1961, pp. 107–47.
5 See, for example, Votes & Proceedings, Otago Provincial Council, Session 27, 1870, Appendix, p. 94; *Mount Ida Chronicle*, 7 February 1873; *Dunstan Times*, 21 February 1873.
6 *Appendices to the Journal of the House of Representatives* (hereafter *AJHR*) G18, 1871, p. 4; G4, 1872, p. x.
7 *Borton v. Howe* and *Glassford v. Read*. See New Zealand Jurist 2, 1874–75, pp. 97–100; New Zealand Court of Appeal 3, 1875–77, pp. 17–18.
8 See Land Act 1877; Land Act 1877 Amendment Act 1882; Mining Act 1886; Mining Act 1886 Amendment Act 1887 (No. 2); Mining Act 1886 Amendment Act 1888; Mining Act 1891; Mining Act Amendment Act 1892; Mining Act 1898.
9 Victorian Parliamentary Papers, *Report of the Board Appointed to Inquire into the Sludge Question 1887*, pp. xv–xvi. See also R. Southern, 'Mining Wastes, Bucket Dredging, and Public Policy: The Sludge Abatement Board in Victoria', unpublished paper; Roger Southern, 'John Montgomery Coane (1848–1923), Surveyor and Consulting Engineer', *Victorian Historical Journal*, vol. 49, no. 2, 1978, pp. 119–28.
10 See Charles W. McCurdy, 'Stephen J. Field and Public Land Law Development in California 1850–1866: A Case Study of Judicial Resource Allocation in Nineteenth Century America', *Law and Society Review*, vol. 10, no. 2, 1976, pp. 235–66. See also Robert L. Kelley, *Gold vs Grain: The Hydraulic Mining Controversy in California's Sacramento Valley*, A.H. Clark Co., Glendale, California, 1959; Clarke C. Spence, 'The Golden Age of Dredging: The Development of an Industry and its Environmental Impact', *Western Historical Quarterly*, vol. 11, no. 4, 1980, pp. 401–14.
11 See *New Zealand Parliamentary Debates* (hereafter *NZPD*), vol. 134, 1905, p. 240; *NZPD*, vol. 135, 1905, p. 315; *AJHR*, I4, 1905, p. 4; *Bruce Herald*, 23 October 1905.
12 *AJHR*, H21, 1901, p. 12.
13 *AJHR*, C14 and C14A, 1907.
14 For a brief history of the field, see J.H.M. Salmon, *A History of Gold Mining in New Zealand*, Government Printer, Wellington, 1963; J.B. McAra, *Gold Mining at Waihi, 1878–1950*, Waihi Historical Society, Waihi, 1978; Laurie Barber, *No Easy Riches: A History of Ohinemuri County, Paeroa and Waihi: 1885–1985*, Ray Richards Publisher, Auckland/Ohinemuri County Council, Paeroa, 1985. A more recent essay, dealing largely with post 1970 mining, is Jennifer Dixon, 'Coromandel Gold: Conquest and Conservation', in John Connell & Richard Howitt (eds), *Mining and Indigenous Peoples in Australasia*, Sydney University Press, Sydney, 1991, pp. 169–82. See also Graham Watton, *Taming the Waihou: The Story of the Waihou Valley Catchment Flood Protection and Erosion Control Scheme*, Waikato Regional Council, Hamilton, 1995.
15 National Archives, Mines Department file, MD 1910/402.
16 See *AJHR*, I4, 1903, p. 3; I4, 1904, p. 5; I4, S.2, 1906, pp. 1–3; I4A, 1907, pp. 16–18; MD 1910/402.
17 *NZPD*, vol. 142, 1907, pp. 813–18.

18 *Waihi Daily Telegraph*, 1 July 1907; *New Zealand Herald*, 24 April 1909.
19 The following sections are based on *AJHR*, C14, 1910.
20 See *AJHR*, Session 2, 1921–22, D6F.
21 *AJHR*, D6, 1920, pp. 1–2.
22 *AJHR*, D6B, 1920, pp. 5, 7, 15.
23 *AJHR*, D6D, 1920, p. 13.
24 *Mount Ida Chronicle*, 17 June 1870. Concern that that principle was not being observed led the Alexandra Miners' Association to urge the classification of all lands within the goldfields and the strict reservation of all potentially auriferous land from settlement, a demand endorsed by the Otago Mining Conferences in 1873 and 1874. See National Archives, Legislative Department file, Le 1, 1873, p. 5; *AJHR*, C6, S2, 1873, pp. 3–4; A8, 1874, p. 3.
25 *Dunstan Times*, 8 July 1881.
26 T.J. Hearn, 'After the Gold Rush: Economic Change and Resource Use Conflict in Central Otago', in Geoff Kearsley & Blair Fitzharris (eds), *Southern Landscapes: Essays in Honour of Bill Brockie and Ray Hargreaves*, Department of Geography, University of Otago, Dunedin, 1990, pp. 55–76.
27 See, for example, *NZPD*, vol. 43, 1882, pp. 690–91.
28 *New Zealand Times*, 28 March 1876.
29 See Terry Hearn, 'Mining and Land: A Conflict Over Use 1858–1953', *New Zealand Law Journal*, August 1983, pp. 235–38.
30 *Evening Star*, 30 April 1871. The issue was raised in the Otago Provincial Council in 1874, but no action ensued. See Votes & Proceedings, Otago Provincial Council, Session 33, 1874, p. 4.
31 *Otago Witness*, 17 June 1903; *Evening Star*, 17 April 1903; *NZPD*, vol. 133, 1905, p. 58.
32 *Otago Witness*, 17 June 1903. See also *NZPD*, vol. 134, 1905, pp. 615–16.
33 Victorian Parliamentary Papers, Report of the Dredging and Sluicing Inquiry Board, 1914, p. 18; Spence, 1980.
34 *Dunstan Times*, 28 April 1903; *Otago Witness*, 17 June, 1 July 1903; *Evening Star*, 17 April 1903; *Tuapeka Times*, 29 September 1906; Mines Department file, MD N12/123, National Archives, Wellington.
35 Mines Department file, MD N12/123.
36 Len Richardson, *Coal, Class and Community: The United Mineworkers of New Zealand 1880–1960*, Auckland University Press, Auckland, 1995, pp. 24, 92.
37 *AJHR*, Session II, C3, 1891; C4, 1901; *AJHR*, Session I, C4, 1912; H44A, 1919.
38 Clyde Warden's Court, Leases for coal-mining purposes, 1873–1904; Cromwell Warden's Court, Coal-mining leases, 1881–1942; Cromwell Warden's Court, Lignite licences, Cromwell, 1899–1920; Queenstown Warden's Court, Mineral licences, Queenstown.
39 When the Nelson Provincial Council, in 1865, granted a lease for the first mine on the Brunner field, the company concerned was required to build a railway to Greymouth. In fact, it merely constructed a horse track along the northern bank of the Grey River, with coal barges being towed upstream. A report in 1891 indicated that 'enormous damage' followed on the destruction of the banks and the widening of the river channel to 'a great shingle-bed, and thus supplying the shingle-bars that are continually on the move downstream, and add to the cost of the [Greymouth] harbour-works'. Failure to meet that condition resulted in the cancellation of the lease. See *AJHR*, Session II, C3, 1891, p. 25.
40 On the Nightcaps–Ohai field, overburden was removed by hand shovels, skips, horses and drays, traction engines, and hydrosluicing, local streams being used as sludge channels. See John S. Thomson, *Pasture, Coal Seam, and Settlement: A Centennial History of Nightcaps and District 1880–1980*, Nightcaps and District Centennial Committee, Nightcaps, 1979, p. 225; also D.C.W. Muir, *Mataura: City of the Falls*, Mataura Historical Society, Gore, 1991.
41 See *AJHR*, C3B, 1892, pp. 18–24; C3B, 1896, pp. 20–26; C4, 1901, pp. 28–33; C3A, 1906, pp. 32–44; C3A, 1911, pp. 31–39; C2, 1916, pp. 75–6; C2, 1922, pp. 46–50.
42 In the former Ashburton County, for example, 2-hectare gravel reserves sited some 3 kilometres apart were set aside from 1868, while extensive quantities of gravel were also removed from local riverbeds. See Rosemary Britten, *Between the Wind and the Water: Ashburton County Council, 1876–1989*, Ashburton District Council, Ashburton, 1991, p. 50.
43 See *AJHR*, C2, each year from 1912.
44 Ian McBride, *The Paparua County: A Concise History*, Canterbury Public Library, Christchurch, 1990, p. 38; Greg G. Smith, *Divine Rock: The Quarrymen's Gift*. Christchurch City Council, Christchurch, 1993; Ian Church, *The Stratford Inheritance: A History of Stratford and Whangamomona Counties*, Heritage Press, Waikanae, 1990, pp. 160–71.

45 James Watson, *Along the Hills: A History of the Heathcote Road Board and the Heathcote County Council 1864–1989*, Heathcote County Council, Christchurch, 1989, pp. 73–74.
46 See Auckland Planning Association, *Auckland's Unique Heritage: 63 Wonderful Volcanic Cones and Craters*, Auckland, 1925. The association insisted that 'Our heritage in these hills must be guarded', their scenic value, geological interest, and 'their close association with and evidence of the work of the Maori people', necessitating measures to prohibit further quarrying. See also George M. Fowlds, 'The Volcanic Cones of Auckland', *Journal of the Auckland Historical Society*, vol. 1, no. 2, 1963, pp. 25–28.

7. Destruction under the guise of improvement? The forest, 1840–1920

1 Frank Sargeson, *Sargeson*, Penguin Books, Auckland, 1981, pp. 53–56.
2 George Jobberns, 'Life and Landscape: 50 years in Retrospect', in *New Zealand: Inventory and Prospect*, New Zealand Geographical Society, Wellington, 1956, p. 3. See also Kenneth B. Cumberland, 'A Century's Change: Natural to Cultural Vegetation in New Zealand', *Geographical Review*, vol. 31, no. 4, 1941, pp. 529–54.
3 Guy H. Scholefield, *New Zealand in Evolution: Industrial, Economic and Political*, Fisher Unwin, London, 1909, pp. 45–59.
4 There is an enormous international literature in environmental history and historical geography: Richard H. Grove, *Green Imperialism: Colonial Expansion, Tropical Island Edens and the Origins of Environmentalism. 1600–1860*, Cambridge University Press, Cambridge, 1995; Alfred W. Crosby, *Ecological Imperialism: The Biological Expansion of Europe, 900–1900*, Cambridge University Press, Cambridge, 1986; Gordon G. Whitney, *From Coastal Wilderness to Fruited Plain: A History of Environmental Change in Temperate North America from 1500 to the Present*, Cambridge University Press, Cambridge, 1994; William Cronon, *Changes in the Land: Indians, Colonists and the Ecology of New England*, Hill & Wang, New York, 1983; Gregory Barton, *Empire Forestry and the Origins of Environmentalism*, Cambridge University Press, Cambridge & New York, 2002.
5 Thomas R. Dunlap, *Nature and the English Diaspora: Environment and History in the United States, Canada, Australia, and New Zealand*, Cambridge University Press, Cambridge & New York, 1999; S. Ravi Rajan, *Modernizing Nature: Forestry and Imperial Eco-Development 1800–1950*, Oxford University Press, New York, 2006; Rollo Arnold, *Settler Kaponga 1881–1914: A Frontier Fragment of the Western World*, Victoria University Press, Wellington, 1997; James Beattie, 'W.L. Lindsay, Scottish Environmentalism, and the "Improvement" of Nineteenth-century New Zealand', in Tony Ballantyne & Judith A. Bennett (eds), *Landscape/Community: Perspectives from New Zealand History*, University of Otago Press, Dunedin, 2005, pp. 43–56; and Peter Holland, *Home in the Howling Wilderness: Settlers and the Environment in Southern New Zealand*, Auckland University Press, Auckland, 2013.
6 Cook and Banks quoted by Geoff Park, *Ngā Uruora: The Groves of Life: Ecology and History in a New Zealand Landscape*, Victoria University Press, Wellington, 1995, pp. 30, 34; 'Steep and Irregular Hills', from George French Angas, *Savage Life and Scenes in Australia and New Zealand*, Smith Elder and Co, London, 1847, reprinted A.H. & A.W. Reed, Wellington, 1967, vol. 1, p. 233.
7 G.C. Petersen, 'Pioneering the North Island Bush', in R.F. Watters (ed.), *Land and Society in New Zealand*, A.H & A.W. Reed, Wellington, 1965, pp. 66–79.
8 This picture is simplified. The range of southern beech forests extends from the Coromandel Ranges to Southland; in more northern areas they are generally confined to the most barren slopes, with mixed conifer–broad-leaved forest occupying the flats and valleys.
9 This discussion draws from Park, *Ngā Uruora*; Nic Bishop, *Natural History of New Zealand*, Hodder & Stoughton, Auckland, 1992, pp. 43–72; L. Cockayne, *New Zealand Plants and Their Story*, 4th edn, Government Printer, Wellington, 1967; R.G. Robbins, 'The Podocarp-Broadleaf Forests of New Zealand', *Transactions of the Royal Society of New Zealand, Botany*, vol. 1, no. 5, 1962, pp. 34–75; P. Wardle, 'Facets of the Distribution of Forest Vegetation in New Zealand', *New Zealand Journal of Botany*, vol. 2, no. 4, 1964, pp. 352–66.
10 For Māori environmental change see Atholl Anderson & Matt McGlone, 'Living on the Edge – Prehistoric Land and People in New Zealand', in John Dodson (ed.), *The Naïve Lands*, Longman Cheshire, Melbourne, 1992, pp. 199–219; M.S. McGlone, 'The Polynesian Settlement of New Zealand in Relation to Environmental and Biotic Changes', *New Zealand Journal of Ecology*, vol. 12 (supplement), 1989, pp. 115–29; M.S. McGlone, 'Polynesian Deforestation of New Zealand: A Preliminary Synthesis', *Archaeology of Oceania*, vol. 18, no. 1, 1983, pp. 11–25.

11 Sarah Harris quoted in Raewyn Dalziel, 'Popular Protest in Early New Plymouth: Why Did it Occur?', *New Zealand Journal of History*, vol. 20, no. 1, 1986, p. 12; Schnackenberg quoted in Park, *Ngā Uruora*, pp. 134–35; F.W. Strum, 'Timber and Firewood', *Transactions of the New Zealand Institute* (hereafter *TNZI*), vol. 11, 1878, pp. 568–70.
12 For a general commentary on the challenges of interpretation here see John M. MacKenzie, 'Empire and The Ecological Apocalypse: The Historiography of the Imperial Environment', in Tom Griffiths & Libby Robin (eds), *Ecology and Empire: Environmental History of Settler Societies*, Keele University Press, Edinburgh, 1997, pp. 215–28. For the wooden world see the New Zealand Historic Places Trust statement on the Auckland Timber Company Building (Former) 104 Fanshawe Street, Auckland at: www.historic.org.nz/TheRegister/RegisterSearch/RegisterResults.aspx?RID=9583. For a broad commentary on environmental anxieties in New Zealand see James Beattie, 'Environmental Anxiety in New Zealand, 1840–1941: Climate Change, Soil Erosion, Sand Drift, Flooding and Forest Conservation', *Environment and History*, vol. 9, no. 4, 2003, pp. 379–92; and also Paul Star, 'From Acclimatisation to Preservation: Colonists and the Natural World in Southern New Zealand, 1860–1894', PhD thesis, University of Otago, 1997.
13 W.J. Butler of Kokiri Mills, Greymouth, quoted in Thomas E. Simpson, *Kauri to Radiata: Origin and Expansion of the Timber Industry of New Zealand*, Hodder & Stoughton, Auckland, 1973, p. 249.
14 Vaughan Wood & Eric Pawson, 'The Banks Peninsula Forests and Akaroa Cocksfoot: Explaining a New Zealand Forest Transition', *Environment and History*, vol. 14, no. 4, 2008, pp. 449–68; see also P.J. Grant, *Hawke's Bay Forests of Yesterday: A Description and Interpretation*, self-published, Havelock North, 1996.
15 Ferdinand von Hochstetter, *New Zealand: Its Physical Geography, Geology, and Natural History*, Cotta, Stuttgart, 1867, p. 141; Hector, *Appendices to the Journals of the House of Representatives* (hereafter *AJHR*), 1870, both cited by Simpson, *Kauri to Radiata*, pp. 224, 226.
16 T. Kirk, *AJHR*, C3, 1886, p. 5; Rollo Arnold, 'The Virgin Forest Harvest and the Development of Colonial New Zealand', *New Zealand Geographer*, vol. 32, no. 2, 1976, pp. 105–26; G.A. Wilson, 'The Urge to Clear the "Bush": A Study of the Nature, Pace and Causes of Native Forest Clearance on Farms in the Catlins District (SE South Island, New Zealand)', PhD thesis, University of Otago, 1991; and Geoff A. Wilson, 'The Pace of Indigenous Forest Clearance on Farms in the Catlins District, South Island, New Zealand, 1861–1991', *New Zealand Geographer*, vol. 49, no. 1, April 1993, pp. 15–25.
17 M.M. Roche, 'From Forest to Pasture', plate 47 in Malcolm McKinnon (ed.), *New Zealand Historical Atlas*, David Bateman, Auckland with Department of Internal Affairs, Wellington, 1997; W. B. Johnston, 'Pioneering the Bushland of Lowland Taranaki: A Case Study', *New Zealand Geographer*, vol. 17, no. 1, April 1961, pp. 1–18; H.C.D. Somerset, *Littledene: A New Zealand Rural Community*, New Zealand Council for Educational Research, Auckland, 1938, Auckland. I am indebted to James Belich, *Making Peoples: A History of the New Zealanders from Polynesian Settlement to the End of the Nineteenth Century*, Penguin Books, Auckland, 1996, p. 372, for this point.
18 Timber merchant W. Leyland of Auckland, 1896, cited in Simpson, *Kauri to Radiata*, p. 252; G.S. Perrin, Report upon the Conservation of New Zealand Forests, *AJHR*, C8, 1897; Evelyn Stokes, 'Kauri and White Pine: A Comparison of New Zealand and American Lumbering', *Annals of the Association of American Geographers*, vol. 56, no. 3, 1966, pp. 440–50.
19 See Tom Brooking & Eric Pawson, 'The Contours of Transformation', in Tom Brooking & Eric Pawson, *Seeds of Empire: The Environmental Transformation of New Zealand*, I.B. Tauris, London, 2011, pp. 13–33.
20 Tregear cited from the Wellington *Evening Post*, 14 January 1886, by Rollo Arnold, *New Zealand's Burning: The Settlers' World in the Mid 1880s*, Victoria University Press, Wellington, 1994, p. 158; Petersen, 'Pioneering', pp. 66–79.
21 Catherine Knight, 'The Paradox of Discourse Concerning Deforestation in New Zealand: A Historical Survey,' *Environment and History*, vol. 15, no. 3, 2009, pp. 323–42.
22 Tregear, 14 January 1886; Petersen, 'Pioneering'; 'rough bush' in a letter from immigrant settler John King, 27 October, 1875, cited in Rollo Arnold, *The Farthest Promised Land*, Victoria University Press, Wellington, 1981, p. 300; Roche, 'Forest to Pasture'.
23 James Inglis, *Our New Zealand Cousins*, Sampson, Lowe Marston, Searle & Rivington, London, 1887, pp. 103, 108.
24 Simpson, Kauri to Radiata, p. 264; D.E. Hutchins, *Forestry in New Zealand. Part 1: Kauri Forests and Forests of the North and Forest Management*, Government Printer, Wellington, 1919; Arnold, *New Zealand's Burning*, p. 84; Inglis, *Cousins*, p. 104; Stephen J. Pyne, *Vestal Fire: An Environmental History, Told Through Fire, of Europe and Europe's Encounter with the World*, University of Washington Press, Seattle, 1997.
25 Julius Vogel, *New Zealand Parliamentary Debates* (hereafter *NZPD*), vol. 4, 1868, p. 190; Graeme Wynn,

'Pioneers, Politicians and the Conservation of Forests in Early New Zealand', *Journal of Historical Geography*, vol. 5, no. 2, 1979, pp. 171–88.
26 For Potts, see Paul Star, 'Potts, Thomas Henry, 1824–1888', from *The Dictionary of New Zealand Biography. Te Ara – the Encyclopedia of New Zealand*, updated 30-Oct-2012: www.teara.govt.nz/en/biographies/2p27/potts-1thomas-henry
27 Graeme Wynn, 'Conservation and Society in Late Nineteenth-Century New Zealand', *New Zealand Journal of History*, vol. 11, no. 2, 1977, pp. 124–36; M. M. Roche, *Forest Policy in New Zealand: An Historical Geography 1840–1919*, Dunmore Press, Palmerston North, 1987.
28 Simpson, *Kauri to Radiata*, p. 245–46; *AJHR*, H24, 1896; *AJHR*, C8, 1897.
29 Anon, 'Forestry and Land Settlement – A New Zealand View', *The Australian Forestry Journal*, vol. 2, 1919, p. 58.
30 *AJHR*, C12, 1913.
31 Roche, *Forest Policy*.
32 I. Campbell Walker, 'State Forestry: Its Aim and Object', *TNZI*, vol. 9, 1876, pp. 187–203; I. Campbell Walker, 'The Climatic and Financial Aspects of Forest Conservancy as Applicable to New Zealand', *TNZI*, vol. 9, 1877, pp. xxvii–xxix; L. Brown & A.D. McKinnon, *Captain Inches Campbell Walker, New Zealand's First Conservator of Forests*, NZFS Information Series, No. 54, 1966; Roche, *Forest Policy*, pp. 76–92; *New Zealand Times*, 15 December 1885.
33 L. Brown, *The Forestry Era of Professor Thomas Kirk, FLS*, NZFS Information Series, No. 56, 1968; T. Kirk to Commissioner of State Forests, 1 May 1886, National Archives, LS 53/15, p. 113; T.W. Hickson to T. Kirk, 1 November 1886, National Archives, LS 53/4 F & A 87–59; T. Hickson to Chief Conservator of State Forests, 4 July 1887, National Archives, LS 53/6 F&A 87–339.
34 Roche, *Forest Policy*; Tom Brooking, *Lands for the People: The Highland Clearances and the Colonisation of New Zealand*, Otago University Press, Dunedin, 1996.
35 Simpson, *Kauri to Radiata*, pp. 272–73; H. Guthrie-Smith, *Mutton Birds and Other Birds*, Whitcombe & Tombs, Christchurch, 1914, and his retrospective *Sorrows and Joys of a New Zealand Naturalist*, A.H. & A.W. Reed, Dunedin, 1936; E. Phillips Turner, 'A Retrospect of Forestry in New Zealand, 1894–1931', *Empire Forestry Journal*, vol. 11, no. 1, 1932, pp. 198–212. Paul Star, 'Native Forest and the Rise of Preservation in New Zealand (1903–1913)', *Environment and History*, vol. 8, no. 3, 2002, pp. 275–94.
36 The debate on the Forests Bill is in *NZPD*, vol. 16, 1874, pp. 79–94, 350–81, 399–426. For discussion, see Wynn, 'Conservation and Society', and 'Pioneers, Politicians and the Conservation of Forests'; Roche, *Forest Policy*, pp. 76–93.
37 Michael Roche has done much in recent years to chart the global circulation and influences of these debates; see his 'Sir David Hutchins and Kauri in New Zealand', in John Dargavel (ed.), *Araucarian Forests*, Australian Forest History Society Inc, Kingston, 2005, pp. 33–40; and 'Edward Phillips Turner: The Development of a "Forest Consciousness" in New Zealand 1890s to 1930s', *Proceedings* of the 6th National Australian Forest History Conference, Augusta, WA, 12–17 September 2004, pp. 143–53.
38 George Perkins Marsh, *Man and Nature; or, Physical Geography as Modified by Human Action*, Charles Scribner, New York, 1864.
39 *AJHR*, C3, 1877; J.P. Grossman, *The Evils of Deforestation*, Brett & Co., Auckland, 1909; Sargeson, *Sargeson*, pp. 55–57.
40 Herbert Guthrie-Smith, *Tutira: The Story of a New Zealand Sheep Station*, 4th edn, Godwit, Auckland/University of Washington Press, Seattle, 1999, p. xxiii.

PART III WILD PLACES
8. Children of the burnt bush: New Zealanders and the indigenous remnant, 1880–1930

1 'Burnt Bush', in B.E. Baughan, *Shingle-Short and Other Verses*, Whitcombe & Tombs, Christchurch, 1908, pp. 63–67.
2 M.M. Roche, *Forest Policy in New Zealand: An Historical Geography, 1840–1919*, Dunmore Press, Palmerston North, 1987, pp. 96–99; Tom Brooking, *Lands for the People? The Highland Clearances and the Colonisation of New Zealand*, Otago University Press, Dunedin, 1996, pp. 131–56.
3 Judith Johnston, 'The New Zealand Bush: Early Assessments of Vegetation', *New Zealand Geographer*, vol. 37, no. 1, 1981, pp. 19–24; Paul Star, 'From Acclimatisation to Preservation: Colonists and the Natural World in Southern New Zealand, 1860–1894', PhD thesis, University of Otago, 1997.

4   Francis Pound, *Frames on the Land: Early Landscape Painting in New Zealand*, Collins, Auckland, 1983.
5   Lynne Lochhead, 'Preserving the Brownie's Portion: A History of Voluntary Nature Conservation Organisations in New Zealand, 1888–1935', PhD thesis, Lincoln University, 1994, pp. 10–46; Paul Star, 'Plants, Birds and Displacement Theory in New Zealand, 1840–1900', *British Review of New Zealand Studies*, no. 10, 1997, pp. 5–21.
6   Otago Institute Minutes, Hocken Library, 11 January 1870; *New Zealand Parliamentary Debates* (hereafter *NZPD*), vol. 48, 1884, p. 340.
7   *NZPD*, vol. 16, 1874, p. 93; Graeme Wynn, 'Conservation and Society in Late Nineteenth Century New Zealand', *New Zealand Journal of History*, vol. 11, 1977, pp. 124–36; Graeme Wynn, 'Pioneers, Politicians and the Conservation of Forests in Early New Zealand', *Journal of Historical Geography*, vol. 5, no. 2, 1979, pp. 179–85; George Perkins Marsh, *Man and Nature*, 1864, ed. David Lowenthal, Harvard University Press, Cambridge, Mass., 1965.
8   Susanne and John Hill, *Richard Henry of Resolution Island*, John McIndoe, Dunedin, 1987; Ross Galbreath, *Walter Buller: The Reluctant Conservationist*, GP Books, Wellington, 1989; Chris Maclean, *Kapiti*, Whitcombe Press, Wellington, 1999; T.H. Potts, 'Help Us Save Our Birds', *Nature*, 2 May 1872, p. 6.
9   W.T.L. Travers, 'On the Changes Effected in the Natural Features of a New Country by the Introduction of Civilized Races', *Transactions of the New Zealand Institute* (hereafter *TNZI*), vol. 2, 1869, pp. 299–313, 313–30; vol. 3, 1870, pp. 326–36; W.W. Smith, 'On the Birds of the Lake Brunner District', *TNZI*, vol. 20, 1887, pp. 205–24; T.H. Potts, 'Stock in the Chathams', *New Zealand Country Journal*, vol. 12, 1888, pp. 370–75.
10 W.H. Guthrie-Smith, *Tutira: The Story of a New Zealand Sheep Station*, Reed, Dunedin, 1921; G.M. Thomson, *The Naturalisation of Animals and Plants in New Zealand*, Cambridge University Press, Cambridge, 1922.
11 *NZPD*, vol. 13, 1872, p. 204.
12 *NZPD*, vol. 152, 1910, p. 326; vol. 113, 1900, pp. 25–26, 34.
13 W.W. Harris, 'Three Parks: An Analysis of the Origins and Evolution of the New Zealand National Park Movement', MA thesis, University of Canterbury, 1974; David Thom, *Heritage: The Parks of the People*, Lansdowne Press, Auckland, 1987.
14 *NZPD*, vol. 86, 1894, p. 678; *NZPD*, vol. 57, 1887, p. 401.
15 Lochhead, 'Preserving the Brownie's Portion'; *Taranaki Herald*, 5 May 1891.
16 Ross Galbreath, *Working for Wildlife: A History of the New Zealand Wildlife Service*, Bridget Williams Books/Historical Branch, Department of Internal Affairs, Wellington, 1993.
17 Geoff Park, *Ngā Uruora: The Groves of Life: Ecology and History in a New Zealand Landscape*, Victoria University Press, Wellington, 1995, p. 318.
18 Lynne Lochhead, 'The Battle for the Rai (1898)', *Environment and Nature in New Zealand*, vol. 7, 2012, pp. 25–40.
19 *Nelson Evening Mail*, 23 February 1898.
20 *Appendices to the Journals of the House of Representatives* (hereafter *AJHR*), I5B, 1898, pp. 6, 16, 27.
21 *Evening Post*, 10 August 1898; M.M. Roche, *History of New Zealand Forestry*, GP Books, Wellington, 1990, pp. 405–14.
22 Kynan Gentry, 'Associations Make Identities: The Origins and Evolution of Historic Preservation in New Zealand, 1870–1954', PhD thesis, University of Melbourne, 2009, pp. 106, 114–15.
23 M.M. Roche, 'The Origins and Evolution of Scenic Reserves in New Zealand', MA thesis, University of Canterbury, 1979.
24 Tom Brooking, *Richard John Seddon*, Penguin Books, Auckland, 2014, ch. 14.
25 Paul Dingwall, 'Harry Ell's Vision in Nature Conservation', *Landscape*, vol. 10, 1981, pp. 23–27; Lochhead, 'Preserving the Brownie's Portion', pp. 181–91.
26 *AJHR*, H2, 1902, p. 21, H2, 1903, pp. iii–iv, C6, 1909, p. 2.
27 *AJHR*, C12, 1913, pp. xiv, xxii.
28 T.F. Cheeseman, 'The Naturalized Plants of the Auckland Provincial District', *TNZI*, vol. 15, 1882, pp. 268–74, *Otago Witness*, 30 June 1883, p. 17.
29 *AJHR*, C11, 1908, pp. 2–3.
30 *AJHR*, C8, 1907, pp. 15, 4; C11, 1908, p. 31.
31 Lochhead, 'Preserving the Brownie's Portion', pp. 192–295; Robin Hodge, 'Seizing the Day: Pérrine Moncrieff and Nature Conservation in New Zealand', *Environment and History*, vol. 9, 2003, pp. 407–17.
32 Kirstie Ross, *Going Bush: New Zealanders and Nature in the Twentieth Century*, Auckland University Press,

2008, pp. 26–92; Ross Galbreath, *Scholars and Gentlemen Both*, Royal Society of New Zealand, Wellington, 2002, p. 278; Chris Maclean, *Tararua: The Story of a Mountain Range*, Whitcombe Press, Wellington, 1994; Kate Hunter, *Hunting: A New Zealand History*, Random House, Auckland, 2009.
33 *AJHR*, C8, 1897, p. 43.
34 Keith Sinclair, *A Destiny Apart: New Zealand's Search for National Identity*, Allen & Unwin/Port Nicholson Press, Wellington, 1986; Ross Galbreath, 'Colonisation, Science, and Conservation: The Development of Colonial Attitudes toward the Native Life of New Zealand', PhD thesis, University of Waikato, 1989; Peter Gibbons, '"Going Native": A Case Study of Cultural Appropriation in a Settler Society', PhD thesis, University of Waikato, 1992; Claudia Bell, *Inventing New Zealand: Everyday Myths of Pakeha Identity*, Penguin, Auckland, 1996.
35 *NZPD*, vol. 122, 1902, p. 26.
36 Paul Star, 'Native Bird Protection, National Identity and the Rise of Preservation in New Zealand to 1914', *New Zealand Journal of History*, vol. 36, no. 2, 2002, pp. 123–36; Paul Star, 'Native Forest and the Rise of Preservation in New Zealand (1903–1913)', *Environment and History*, vol. 8, no. 3, 2002, pp. 275–94.
37 *Otago Daily Times*, 29 June 1900, 28 June 1901, 2 October 1902.
38 *Birds*, No. 6 [c. 1924]; *New Zealand Life*, vol. 1, 1925, p. 3; vol. 5, 1925, p. 7.
39 *AJHR*, C12, 1913, p. 16; Leonard Cockayne, *The Cultivation of New Zealand Plants*, Whitcombe & Tombs, Christchurch, 1923, p. 8.
40 Galbreath, 'Colonisation, Science, and Conservation', p. 350.
41 Paul Hamer, 'Nature and Natives: Transforming and Saving the Indigenous in New Zealand', MA thesis, Victoria University of Wellington, 1992, pp. 11, 105–07.
42 James Muir, 'Changing the Forest: Ecological Colonialism, Statute Law and the New Zealand Bush, 1840–1940', MPhil thesis, University of Waikato, 1995, pp. ii, 172, 174.
43 Geoff Park, 'Going Between Goddesses', in Klaus Neumann, Nicholas Thomas & Hilary Ericksen (eds), *Quicksands: Foundational Histories in Australia and Aotearoa New Zealand*, UNSW Press, Sydney, 1999, pp. 177, 189–90; Geoff Park, *Effective Exclusion? An Exploratory Overview of Crown Actions and Maori Responses Concerning the Indigenous Flora and Fauna, 1912–1983*, Waitangi Tribunal, Wellington, 2001; Cathy Marr, Robin Hodge & Ben White, *Crown Laws, Policies and Practices in Relation to Flora and Fauna, 1840–1912*, Waitangi Tribunal, Wellington, 2001, pp. 259–90.
44 Thomas R. Dunlap, *Nature and the English Diaspora: Environment and History in the United States, Canada, Australia, and New Zealand*, Cambridge University Press, Cambridge, 1999, pp. 71–163; Ian Tyrrell, *True Gardens of the Gods: Californian-Australian Environmental Reform, 1860–1930*, University of California Press, Berkeley, 1999, p. 15.
45 James Beattie, *Empire and Environmental Anxiety: Health, Science, Art and Conservation in South Asia and Australasia, 1800–1920*, Palgrave Macmillan, Basingstoke, 2011, pp. 72–99.
46 Muriel Williams, *Charles Blomfield: His Life and Times*, Hodder & Stoughton, Auckland, 1979, pp. 35, 168.

9. The meanings of mountains

1 It was very helpful to be able to discuss the updating of this chapter with James Beattie and Ted Catton.
2 Laurie Franks, *All the Stamps of New Zealand*, A. H. & A. W. Reed, Wellington, 1977, pp. 26–28. Mountain scenes also feature prominently on early banknotes, particularly those of the Bank of New Zealand.
3 Jon Mathieu, 'The Sacralization of Mountains during the Modern Age', *Mountain Research and Development*, vol. 26, no. 4, 2006, pp. 343–49.
4 John Ruskin, *Modern Painters*, Smith, Elder, London, new edn, 1873; Marjorie Hope Nicolson, *Mountain Gloom and Mountain Glory: The Development of the Aesthetics of the Infinite*, Cornell University Press, New York, 1959; Simon Schama, *Landscape and Memory*, Harper Collins, London, 1995.
5 Paul Shepard, *English Reaction to the New Zealand Landscape before 1850*, Pacific Viewpoint Monograph no. 4, Wellington, 1969, p. 42. See also James Beattie, 'Wilderness Found, Lost and Restored: the Sublime and Picturesque in New Zealand', in Mick Abbott & Richard Reeve (eds), *Wild Heart. The Possibility of Wilderness in Aotearoa New Zealand*, Otago University Press, Dunedin, 2011, pp. 89–103.
6 Hong-key Yoon, *Maori Mind, Maori Land*, Peter Lang, Berne, 1986; Ernst Dieffenbach, *Travels in New Zealand*, John Murray, London, vol. 1, 1843; Chris Maclean, *Tararua: The Story of a Mountain Range*, Whitcombe Press, Wellington, 1994.
7 Maclean, *Tararua*; John Pascoe, *Land Uplifted High*, Whitcombe & Tombs, Christchurch, 1952; Waitangi Tribunal, *The Ngāi Tahu Report 1991* (Wai 27), Brooker & Friend Ltd, Wellington, 1991, pp. 186–98.

8. *The Ngāi Tahu Report*, pp. 186–88; G.G.M. Mitchell, *Maori Place Names in Buller County*, A.H. & A.W. Reed, Wellington, 1948, p. 19; John Pascoe (ed.), *Over the Whitcombe Pass: The Narrative of Jakob Lauper*, reprinted from the *Canterbury Gazette*, July 1863, Whitcombe & Tombs, Christchurch, 1960.
9. Maclean, *Tararua*, pp. 80–81; John Overton, 'A Theory of Exploration', *Journal of Historical Geography*, vol. 7, no. 1, 1981, pp. 53–70.
10. Cheryll Sotheran, 'The Later Paintings of William Fox', *Art New Zealand*, no. 11, 1978, p. 45; Julie King, 'Facing up to Fox: The Colonial Watercolours of William Fox', *Art New Zealand*, no. 95, 2000, pp. 84–87, 99. See also Beattie, 'Wilderness Found, Lost and Restored'.
11. Jill Trevelyan, *Picturing Paradise: The Colonial Watercolours of William Fox*, National Library of New Zealand, Wellington, 2000; John H. Angus, *Aspiring Settlers: European Settlement in the Hawea and Wanaka Region to 1914*, John McIndoe, Dunedin, 1981, p. 52; Melvin D. Day, *Nicholas Chevalier Artist: His Life and Work with Special Reference to His Career in New Zealand and Australia*, Millwood Press, Wellington, 1981; Neil Roberts, *Nicholas Chevalier: An Artist's Journey through Canterbury in 1866*, Robert McDougall Art Gallery, Christchurch, 1992.
12. *Illustrated New Zealand Herald*, 21 September 1878.
13. *Illustrated New Zealand Herald*, 12 March 1874; *Otago Witness*, 28 March 1874; J. Hingston, 'The New Zealand Sounds', *The Victorian Review*, 1 September 1883, pp. 622–36; *Maoriland: An Illustrated Handbook of New Zealand*, issued by the Union Steam Ship Company of New Zealand, George Robertson & Co., Melbourne, 1884, pp. 136–37.
14. Anthony Trollope, *New Zealand*, Chapman & Hall, London, 1874, pp. 43, 50; William Spotswood Green, *The High Alps of New Zealand*, Macmillan, London, 1883, p. 303.
15. *Maoriland*, p. 116.
16. Blanche E. Baughan, *The Finest Walk in the World*, Whitcombe & Tombs, Christchurch, 1917; *Overland to Milford Sound*, Department of Tourist and Health Resorts, 1903, pp. 25, 46.
17. Annual Report of the Tourist and Health Resorts Department, *Appendices to the Journals of the House of Representatives* (hereafter *AJHR*), H2, 1906, pp. 10–11; 1907, p. 10, 1909, p. 13; 1914, p. 4; Annual Report of Department of Lands and Survey, *AJHR*, C1, 1903, pp. 153–54; Maclean, *Tararua*, p. 112.
18. 'High Places in the Southern Alps', *The New Zealand Railways Magazine*, vol. 1, no. 2, 1926, p. 35; see also Lee Davidson, 'Publicising Peaks: Early Promotion of Mountain Tourism', *New Zealand Geographic*, no. 117, September–October, 2012, pp. 58–72.
19. Verse from an advertisement for the 'Great Alpine Carnival, Mount Cook' in *The Press*, 15 July 1923.
20. 'High Places in the Southern Alps', p. 35.
21. 'Franz Josef Glacier, New Zealand', Poster, Publicity Branch, NZ Railways, n.d., Poster collection, Hocken Collections, University of Otago.
22. 'World-famed Otira Walk', n.d., Pamphlet collection, Hocken Collections, University of Otago; Kirstie Ross, *Going Bush: New Zealanders and Nature in the Twentieth Century*, Auckland University Press, Auckland, 2008, pp. 69–82.
23. *AJHR*, D1, 1936, p. xi; Charlotte Macdonald, *Strong, Beautiful and Modern: National Fitness in Britain, New Zealand, Australia and Canada, 1935–60*, Bridget Williams Books, Wellington, 2011; Ross, *Going Bush*, pp. 86–92.
24. *The New Zealand Motorists' Road Guide (North Island)*, Andrews, Baty & Co., Christchurch, 1935, p. 244.
25. Public Works Statement, *AJHR*, D1, 1938, pp. 187–205.
26. Alfred Runte, 'The West: Wealth, Wonderland and Wilderness', in J. Wreford Watson & Timothy O'Riordan (eds), *The American Environment: Perceptions and Policies*, John Wiley & Sons, London, 1976, pp. 47–62; Tom Brooking, 'So Different from the USA: Our Late Appreciation of Wilderness', in Abbott & Reeve, *Wild Heart*, p. 128. The quote is from Les Molloy, 'Wilderness Recreation: the New Zealand Experience', in Leslie F. Molloy (ed.), *Wilderness Recreation in New Zealand*, Federated Mountain Clubs of New Zealand, Wellington, 1983, p. 6.
27. M.H. Holcroft, *Discovered Isles: A Trilogy*, The Caxton Press, Christchurch, 1950, pp. 22–23.
28. Kate Hunter, *Hunting, A New Zealand History*, Random House, Auckland, 2009.
29. L. Cockayne, 'A Glimpse into the Alps of Canterbury', in *Canterbury Old and New 1850–1900*, New Zealand Natives Association, Christchurch, 1900, p. 215.
30. Molloy, 'Wilderness Recreation', p. 7.
31. Green, *The High Alps*, p. 4.
32. A.P. Harper, 'Climbing in the Alps of Switzerland and New Zealand', *New Zealand Alpine Journal*, vol. 1, no. 3, 1893, pp. 134–42.
33. Graham Langton, 'A History of Mountain Climbing in New Zealand to 1953', PhD thesis, University of Canterbury, 1996, p. 116.

34 Maclean, *Tararua*, from which much of the material on the TTC is drawn, along with Ross, *Going Bush*; W.H. Field & F.W. Vosseler, 'Message for the Founders', in B.D.A. Greig (ed.), *Tararua Story: Jubilee of a Mountain Club*, Tararua Tramping Club, Wellington, 1946, p. 4; G.E. Mannering, 'The Ascent of Mount Rolleston', *New Zealand Alpine Journal*, vol. 1, no. 2, 1892, pp. 98–112.
35 Ross, *Going Bush*, p. 65.
36 Annual Report of the Tourist and Health Resorts Department, *AJHR*, H2, 1913, p. 5; Freda du Faur, *The Conquest of Mount Cook and Other Climbs*, George Allen & Unwin, London, 1915. In 1914 Miss du Faur had moved to England. 'She never climbed again … she was entirely deterred by the fact that women were not allowed to join the English Alpine Club': Bee Dawson, *Lady Travellers: The Tourists of Early New Zealand*, Penguin Books, Auckland, 2001, p. 194.
37 Ross, *Going Bush*, p. 65.
38 Pascoe, *Land Uplifted High*, p. 217.
39 Margaret Johnston & Eric Pawson, 'Challenge and Danger in the Development of Mountain Recreation in New Zealand, 1890–1940', *Journal of Historical Geography*, vol. 20, no. 2, 1994, pp. 175–86; R.J. Keen (ed.), *Outdoors: the Official Journal of the Otago Tramping and Mountaineering Club*, Fiftieth Anniversary Issue, 1973; Hunter, *Hunting*, p. 214.
40 Johnston & Pawson, 'Challenge and Danger'.
41 Ibid., pp. 179–80.
42 *New Zealand Parliamentary Debates* (hereafter *NZPD*), vol. 249, 1937, p. 415; vol. 299, 1953, p. 189.
43 Annual Report, Department of Lands, *AJHR*, C6, 1907, pp. 2–3.
44 The relevant clause in the National Parks Act 1952, which was carried forward into the succeeding (and still current) Act in 1980, stemmed originally from policy developed by the FMC in 1938: 'The Federation and National Parks', *F.M.C. Bulletin*, no. 27, 1967, pp. 1–2. On 'naturalness', see David N. Cole & Laurie Yung (eds), *Beyond Naturalness. Rethinking Park and Wilderness Stewardship in an Era of Rapid Change*, Island Press, Washington DC, 2010.
45 *NZPD*, vol. 100, 1897, pp. 155–56.
46 A. Maud Moreland, 'Through South Westland. A Journey to the Haast and Mount Aspiring, New Zealand', Whitcombe & Tombs, Christchurch, 1911, p. 56; Public Works Statement, *AJHR*, D1, 1937; *The New Zealand Motorists' Road Guide* (South Island), 1938, p. 259; *NZPD*, vol. 271, 1945, p. 233.
47 Maclean, *Tararua*, who mentions the hydro potential. In fact the Public Works Statement of 1912 refers to investigations for this purpose on the Hutt River and the Waiohine and Tauherenikau, both major eastern Tararua rivers: *AJHR*, D1, 1912, p. 85.
48 Pascoe, *Land Uplifted High*, pp. 32–33; *Centennial Memorial Park Golden Jubilee Edition 1940–90: Conserving the Waitakere Ranges*, Auckland Regional Council Parks Department, Auckland, 1990; E. Earle Vaile, *Waitakere National Park, with a Short History of the Ranges*, The Waitakere National Centennial Park Citizens' Association, Auckland, 1939, p. 4.
49 Including, ironically, that of 'naturalness': see note 44 above.
50 'Landscape and Nature Conservation', in L.F. Molloy (comp.), *Land Alone Endures: Land Use and the Role of Research*, Discussion Paper 3, Department of Scientific and Industrial Research, Wellington, p. 67.
51 Roberta McIntyre, *Whose High Country? A History of the South Island High Country of New Zealand*, Penguin Books, Auckland, 2008, pp. 284–304; Jonathan West, 'Running Wild: What Path Will We Walk Through the Wilderness', in Abbott & Reeve, *Wild Heart*, pp. 40–50.
52 West, 'Running Wild'; Beattie, 'Wilderness Lost, Found and Restored'; William Cronon, 'The Trouble with Wilderness; or, Getting Back to the Wrong Nature', in William Cronon (ed.), *Uncommon Ground: Toward Reinventing Nature*, W.W. Norton, New York, 1995, pp. 69–90.

10. Swamp drainage and its impact on the indigenous

1 J.C. Beaglehole (ed.), *The 'Endeavour' Journal of Joseph Banks, 1768–1771*, 2 vols, 2nd edn, Public Library of New South Wales/Angus & Robertson, Sydney, 1962.
2 Charles Hursthouse, *New Zealand or Zealandia, the Britain of the South*, Edward Stanford, London, 1857, p. 69.
3 Clarence J. Glacken, *Traces on the Rhodian Shore: Nature and Culture in Western Thought from Ancient Times to the End of the Eighteenth Century*, University of California Press, Berkeley, 1967, p. 348.
4 Figures from Ministry for the Environment, *The State of New Zealand's Environment*, Wellington, 1997, pp. 7.24, 7.62.
5 Alfred W. Crosby, *Ecological Imperialism: The Biological Expansion of Europe, 900–1900*, Cambridge University

Press, Cambridge, 1986.
6   Figures from Edward B. Barbier, Joanne C. Burgess & Carl Folke, *Paradise Lost? The Ecological Economics of Biodiversity*, Earthscan, London, 1994.
7   Analysis by Landcare Research commissioned for Ministry for the Environment, *The State of New Zealand's Environment*.
8   Crosby, *Ecological Imperialism*.
9   Beaglehole, *The 'Endeavour' Journal*, September 1769, p. 252.
10 Patricia Fara, 'Presidential Portraits: Joseph Banks in the National Library', *National Library of Australia News*, December 1998.
11 The nineteenth century Sanitary Movement began in Britain and spread to its colonies. It was based on the utopian and educative impulse that good drainage meant good morals. See Rod Giblett, *Postmodern Wetlands: Culture, History, Ecology*, Edinburgh University Press, Edinburgh, 1996, pp. 120–25.
12 Hursthouse, *New Zealand or Zealandia*, p. 69.
13 Beaglehole, *The 'Endeavour' Journal*.
14 J.C. Beaglehole (ed.), *The Journals of Captain James Cook on His Voyages of Discovery, Vol. 1: The Voyage of the Endeavour, 1768–1771*, Cambridge University Press for the Hakluyt Society, Cambridge, 1955, p. 207.
15 Robert Silvester, 'Medieval Reclamation of Marsh and Fen', in Hadrian Cook & Tom Williamson, *Water Management in the English Landscape: Field, Marsh and Meadow*, Cambridge University Press, Cambridge, 1998, p. 127.
16 Christopher Taylor, 'Post-medieval Drainage of Marsh and Fen', in Cook & Williamson, *Water Management*, pp. 142, 149.
17 Giblett, *Postmodern Wetlands*, pp. 11, 116.
18 Geoff Park, *Nga Uruora; The Groves of Life. Ecology and History in a New Zealand Landscape*, Victoria University Press, Wellington, 1995, p. 38.
19 Simon Best, 'Oruarangi Pa: Past and Present Investigations', *New Zealand Journal of Archaeology*, vol. 2, 1980, pp. 56–71, cited in Park, *Ngā Uruora*, p. 49; Wendy Pond, *The Land with All Woods and Waters*, Waitangi Tribunal Rangahaua Whanui Series, Wellington, 1997.
20 Hursthouse, *New Zealand or Zealandia*, p. 33.
21 For example, Paul Carter, *The Road to Botany Bay: An Essay in Spatial History*, Faber, London, 1987.
22 W.J.T. Mitchell, 'Holy Landscape: Israel, Palestine, and the American Wilderness', *Critical Inquiry*, vol. 26, no. 2, 2000, p. 194.
23 Paul Shepard, *Nature and Madness*, University of Georgia Press, Athens, 1998, p. 83.
24 George J. Murray, *The Story of the Rangitaiki*, Presbyterian Bookroom, Christchurch, 1968.
25 Chairman (Assistant Director General of Lands) to the Director General of Lands, ABWN 6095 W5021/421 15/11 p.1–S.A.L.D.–Rangitaiki Drainage, National Archives, Auckland.
26 Rufus Tye, *Hauraki Plains Story*, Thames Valley News, Paeroa, 1974.
27 Arthur Young, 1799, quoted in Giblett, *Postmodern Wetlands*, p. 111.
28 Giblett, *Postmodern Wetlands*, pp. 114–15.
29 Ibid., p. 219.
30 Edward Gibbon Wakefield, *The British Colonization of New Zealand*, John W. Parker, London, 1840, p. 25.
31 The New Zealand Colonisation Company to be Incorporated by Charter, or Act of Parliament, New Zealand Company Papers, 325.931.77099, Alexander Turnbull Library, Wellington.
32 Charles Heaphy, 'Notes on Port Nicholson and the Natives in 1839', *Transactions of the New Zealand Institute* (hereafter *TNZI*), vol. 12, 1879, pp. 32–39.
33 Charles McGurk, *Aurora* letter, in 'Extracts from Colonists' Letters', *New Zealand Journal*, vol. 1, 1840.
34 William Wakefield, *Journal*, No. 2, 1839–1842, Typescript, New Zealand Company Papers 131/9, Alexander Turnbull Library, Wellington.
35 Wakefield, *Journal*.
36 Ruth France, 'Return Journey', in Allen Curnow (ed.), *The Penguin Book of New Zealand Verse*, Blackwood & Janet Paul, Auckland, 1966.
37 William Pember Reeves, 'An Old Chum on New Zealand Scenery', in George Phipps Williams & W.P. Reeves, *Colonial Couplets: Being Poems in Partnership*, Simpson & Williams, Christchurch, 1889, pp. 1–4.
38 Giblett, *Postmodern Wetlands*.
39 William Pember Reeves, *The Long White Cloud: Ao Tea Roa*, Horace Marshall & Son, London, 1898.
40 Modified, considerably, from Paul A. Colinvaux, *The Fates of Nations: A Biological Theory of History*, Simon & Schuster, New York, 1980.
41 John Locke, *Two Treatises of Government*, Awnsham Churchill, London, 1690.

42 'Is it not the will of God that the earth should be replenished and subdued, that the desert should give rise to the fruitful field, the frantic war-cry to the hymn of praise, and the frightful depository of the unburied dead to the country steeple and the village school?' said 'MA of Trinity College Cambridge', 'Exceptional Laws in Favour of the Natives of New Zealand', Appendix A in Wakefield, *The British Colonization of New Zealand*, p. 417.
43 William Swainson, *New Zealand and its Colonization,* Smith Elder & Co, London, 1859.
44 Geo. Gipps, 7 March 1841 (letter incl. Clarke's Instructions), Governor's Despatches, vol. 35, A 1224, CY POS p. 657. Mitchell Library, Sydney.
45 George Clarke, Report on the River Thames in New Zealand, from Extracts from the Final Report of the Chief Protector of Aborigines in New Zealand, CY POS, pp. 679–92, Mitchell Library, Sydney.
46 Best, 'Oruarangi Pa'.
47 Clarke, *Report*.
48 Pond, *The Land with All Woods and Waters*, p. 129.
49 Park, *Nga Uruora*, p. 54.
50 Pond, *The Land with All Woods and Waters*.
51 Pond, citing R.M. McDowall, *Gamekeepers of the Nation: The Story of New Zealand's Acclimatisation Societies, 1861–1990*, Canterbury University Press, Christchurch, 1994.
52 The US ecologist Daniel Janzen quoted in Gary Paul Nabhan, *Cultures of Habitat: On Nature, Culture and Story*, Counterpoint Press, Washington, DC, 1997.
53 *Ngai Tahu Report*, vol. 1, Waitangi Tribunal, Wellington, 1991, p. 158.
54 Ben White, *Inland Waterways: Lakes,* Waitangi Tribunal Rangahaua Whanui Series, Wellington, 1998, p. 1.
55 White, *Inland Waterways*.
56 Ibid.
57 Ibid., esp. chapter 2, Wairarapa lakes.
58 Recommendations, Report of the Royal Commission on Forestry, *Appendices to the Journals of the House of Representatives*, C12, 1913.
59 United Nations Environmental Programme and Wetlands International – Asia Pacific, *Wetlands and Integrated River Basin Management: Experiences in Asia and the Pacific*, Kuala Lumpur, 1997.
60 *A Wetlands Guideline*, National Water and Soil Conservation Organisation, Wellington, 1982.
61 Geoff Park, 'Effective Exclusion? An Exploratory Overview of Crown Actions and Māori Responses Concerning the Indigenous Flora and Fauna, 1912–1983', Waitangi Tribunal, Wellington, 2001.
62 William Massey in parliamentary debate on Bush and Swamp Crown Land Bill, *New Zealand Parliamentary Debates* (hereafter *NZPD*), vol. 124, 1903, p. 491.
63 Ibid.
64 Hone Heke Rankin, quoted in Evaan Aramakutu, 'Colonist and Colonials: Animals Protection Legislation in New Zealand from 1861–1910', MA thesis, Massey University, 1997; see also Geoff Park, 'Going between Goddesses', in Klaus Neumann, Nicholas Thomas & Hilary Ericksen (eds), *Quicksands: Foundational Histories in Australia and Aotearoa New Zealand*, UNSW Press, Sydney, 1999, pp. 176–97.
65 Hone Heke Rankin, *NZPD*, vol. 113, 1900, pp. 32–33.
66 Ibid.
67 Alfred K. Newman, 'A study of the causes leading to the extinction of the Maori', *TNZI*, vol. 14, 1881, pp. 459–77.
68 Hursthouse, *New Zealand or Zealandia*, p. 71.

PART IV MODERNISING
11. The grasslands revolution reconsidered

1 P.W. Smallfield, *The Grasslands Revolution in New Zealand*, Hodder & Stoughton, Auckland, 1970, p. 97. For further development of this view of the 'grasslands revolution' see Tom Brooking & Eric Pawson, *Seeds of Empire: The Environmental Transformation of New Zealand*, I.B. Tauris, London, 2011. On Darwin's metaphor, see John Stenhouse, 'The Battle between Science and Religion over Evolution in Nineteenth Century New Zealand', PhD thesis, Massey University, 1985, p. 11.
2 See R.J. Bremer, 'Federated Farmers and the State', in Brian Roper & Chris Rudd (eds), *State and Economy in New Zealand*, Oxford University Press, Auckland, 1993, p. 121. Figures on economic production and fertiliser use are from *New Zealand Official Yearbooks* (hereafter *NZOYB*), 1920–90. International figures come from

John McNeill, *Something New Under the Sun: An Environmental History of the Twentieth-Century World*, W.W. Norton, New York, 2000, p. 25.
3   On this and economic history, see Tom Brooking, 'Economic Transformation' and Gary Hawke, 'Economic Trends and Economic Policy' in Geoffrey W. Rice (ed.), *Oxford History of New Zealand*, 2nd edn, Oxford University Press, Auckland, 1992, pp. 230–53, 414–21; John Gould, *The Rake's Progress*, Hodder & Stoughton, Auckland, 1982; C.B. Schedvin, 'Staples and Regions of Pax Britannica', *Economic History Review*, vol. 43, no. 4, 1990, pp. 533–59; Brian Easton, *In Stormy Seas: The Post War New Zealand Economy*, University of Otago Press, Dunedin, 1997; Geoff Bertram, 'The New Zealand Economy, 1900–2000', in Giselle Byrnes, (ed.), *The New Oxford History of New Zealand*, Oxford University Press, Melbourne, 2009, pp. 537–72.
4   B.L. Evans, *A History of Agricultural Production and Marketing in New Zealand*, Keeling & Mundy, Palmerston North, 1969, and 'Grasslands Research' in *NZOYB*, 1960, pp. 1248–64; F.R. Callaghan, 'Science and Agriculture' in F.R. Callaghan (ed.), *Science in New Zealand*, Reed, Wellington, 1957, pp. 137–53.
5   Michael M. Roche, 'Empire, Duty and Land: Soldier Settlement in New Zealand 1915–1924' in Lindsay J. Proudfoot & Michael M. Roche (eds), *(Dis)placing Empire: Renegotiating British Colonial Geographies*, Ashgate, Aldershot, England, 2005, pp. 135–54.
6   This summary of the writing up of the 'grasslands revolution' by both participants and observers is based on: A.H. Cockayne, 'Grasslands of New Zealand: A Grasslands Philosophy', *Proceedings of the Seventh International Grasslands Conference*, 1956, pp. 585–95; Kenneth B. Cumberland, *Landmarks*, Reader's Digest, Auckland, 1981, pp. 152–207; Kenneth B. Cumberland, *Soil Erosion in New Zealand: A Geographic Reconnaissance*, Soil Conservation and Rivers Control Council, Wellington, 1944; Ross Galbreath, 'A Grassland Utopia? Pastoral Farming and Grassland Research in New Zealand', in *DSIR: Making Science Work for New Zealand, 1926–1992*, Victoria University Press/Historical Branch, Wellington, 1998, pp. 58–79; E. Bruce Levy, *The Grasslands of New Zealand*, Government Printer, Wellington, three editions, 1923, 1951 and 1973; L.W. McCaskill, *Hold This Land: A History of Soil Conservation in New Zealand*, Reed, Wellington, 1973; K.J. O'Connor, 'The Improvement and Utilization of Tussock Grasslands: A Scientist's Viewpoint', *Proceedings of the New Zealand Grasslands Association*, vol. 28, 1966, pp. 59–78; Smallfield, *The Grasslands Revolution*; essays in *New Zealand Agricultural Science*, vol. 8, no. 3, 1974, especially J.K. Syers, 'Effect of Phosphate Fertilizers on Agriculture and the Environment', pp. 149–64.
7   See J.D. Gould, *The Grass Roots of New Zealand History: Pasture Formation and Improvement, 1871–1911*, Massey University Occasional Publications, Palmerston North, 1974.
8   D.A. Campbell, 'Developments in Aerial Top-dressing During 1949', *New Zealand Journal of Science and Technology*, vol. 33A, no. 3, 1951, pp. 1–12.
9   *The State of New Zealand's Environment 1997*, Ministry for the Environment, Wellington, 1997, chapter 8, p. 9.
10 See Eric Pawson, 'Kenneth Cumberland 1913–2011', in Hayden Lorimer & Charles W. J. Withers (eds), *Geographers Biobibliographical Studies*, Vol. 31, Commission on the History of Geography of the International Geographical Union and the International Union of the History and Philosophy of Science, Continuum International Publishing Group, London, 2012, pp. 137–60.
11 Levy, *Grasslands*, 2nd edn, pp. 285–304; 3rd edn, pp. 238–70. On the loss of agricultural land to sprawling Auckland and Christchurch, see *New Zealand Agricultural Science*, vol. 8, no. 4, 1974; on agricultural chemicals, see Cumberland, *Landmarks*, pp. 191, 295; Smallfield, *Grasslands Revolution*, p. 130.
12 L.W. McCaskill, *Molesworth*, A.H. & A.W. Reed, Wellington, 1969, pp. 257–70.
13 Maslyn Williams & Barrie Macdonald, *The Phosphateers: A History of the British Phosphate Commissioners and the Christmas Island Phosphate Commissioners*, Melbourne University Press, Carlton, Victoria, 1985; T.W. Walker, 'Superphosphate and Biological Nitrogen Fixation' in Dennis Hogen & Bryce Williamson (eds), *New Zealand is Different: Chemical Milestones in New Zealand History*, New Zealand Institute of Chemistry, Christchurch, 1999, pp. 37–39. Also see Damen Salesa, 'New Zealand's Pacific' in Giselle Byrnes (ed.), *The New Oxford History of New Zealand*, Oxford University Press, Melbourne, 2009, pp. 162–63.
14 Paul Star & Tom Brooking, 'Fescue to the Rescue: Chewings Fescue, Paspalum, and the Application of Non-British Experience to Pastoral Practice in New Zealand, 1880–1920', *Agricultural History*, vol. 80, no. 3, 2006, pp. 312–35; and Paul Star & Tom Brooking, 'The Farmer, Science and the State in New Zealand', and Tom Brooking & Paul Star, 'Remaking the Grasslands: the 1920s and 1930s' in Brooking & Pawson, *Seeds of Empire*, pp. 161–66, 169–73 and 183–97.
15 Brooking & Star, 'Remaking the Grasslands', pp. 195–97.
16 Gordon M. Winder, 'Grasslands Revolutions in New Zealand: Disaggregating a National Story', *New Zealand Geographer*, vol. 65, no. 3, 2009, pp. 187–200.

17 The 'Southern Pastoral Lands Commission', *Appendices to the Journals of the House of Representatives* (hereafter *AJHR*), C15, 1920; L. McCaskill, 'Fertilisers in New Zealand, 1867–1929', M.Ag. thesis, University of Otago, 1929; 'Royal Commission To Inquire Into And Report Upon The Sheep Farming Industry', *AJHR*, H46A, 1949; H. Belshaw, F. B. Stephens and D. O. Williams et al., *Agricultural Organisation in New Zealand*, Melbourne University Press/OUP, Melbourne, 1936, pp. 386–87; A. N. Tait, 'The Growth of Fertiliser Use in New Zealand', *New Zealand Journal of Agriculture* (hereafter *NZJA*), vol. 95, no. 2, 1957, pp. 186–97; *NZOYB*, 1920–90. See W. J. Gardner, *The Amuri: A County History*, 2nd edn, Amuri County Council, Culverden, 1983, pp. 448–58, on irrigation. Figures on sheep and cattle numbers are based on the *NZOYB*, 1920, p. 224; 1949, p. 914; and 1980, p. 391.
18 Eric Pawson & Vaughan Wood, 'The Grass Seed Trade', and Brooking & Star, 'Remaking the Grasslands' in Brooking & Pawson, *Seeds of Empire*, pp. 117–38 and 188–99.
19 Smallfield, *Grasslands Revolution*, pp. 49, 99, 87.
20 Neil Clayton, 'Weeds, People and Contested Places: Selected Themes from the History of New Zealanders and their Weeds', PhD thesis, University of Otago, 2007.
21 See Helge Kragh, *An Introduction to the Historiography of Science*, Cambridge University Press, Cambridge, 1989, p. 180.
22 For example, Labour's Minister of Agriculture, Ben Roberts, stressed New Zealand's role in 'winning the peace', *NZJA*, vol. 73, no. 6, 1946, pp. 345–47. National's Minister of Agriculture, K.J. Holyoake, talked of New Zealand's role in feeding a hungry world, vol. 80, no. 1, 1950, p. 3.
23 On disasters resulting from noble intentions, see Nancy Langstone, *Forest Dreams Forest Nightmares: The Paradox of Old Growth in the Inland West*, University of Washington Press, Seattle, 1995. Cumberland claimed 'the lessons of the failures after 1918 had been thoroughly learned', *Landmarks*, p. 166. 'Sheep Industry Commission', *AJHR*, H46A, 1949, p. 88, warns of 'the tragedy of abandoned lands'.
24 Edward Cullen, Labour Minister of Agriculture, urged farmers to increase food production to head off totalitarianism, *NZJA*, vol. 78, no. 3 (March 1949), p. 227. Similarly, Cumberland wrote in *Landmarks*, p. 295: 'Food and Agricultural Organisation researchers predict substantially increased world demand for New Zealand's main export foodstuffs'.
25 On 'ruralism', see Tom Brooking, 'Use it or Lose it: Unravelling the Land Debate in Late Nineteenth-Century New Zealand', *New Zealand Journal of History*, vol. 30, no. 2, 1996, pp. 145–47; Miles Fairburn, *The Ideal Society and its Enemies: The Origins of Modern New Zealand Society*, Auckland University Press, 1988, pp. 19–73; Rodney B. Lyon, 'The Principles of New Zealand Liberal Thinking in the Late Nineteenth Century', PhD thesis, University of Auckland, 1982, p. 355; on the 'rural problem', see David B. Danbom, *Born in the Country. A History of Rural America*, The Johns Hopkins University Press, Baltimore, 1995, pp. 185–232.
26 Brooking & Star, 'Remaking the Grasslands', pp. 188–92.
27 Lord Bledisloe, *Grasslands: The Main Source of the Nation's True Wealth*, Third Conference of the New Zealand Grasslands Association, Wellington, 1934, pp. 20–25, 35.
28 Also see Harry Evison, *Te Wai Pounamu, the Greenstone Island: A History of the Southern Maori During the European Colonization of New Zealand*, Aoraki Press in association with the Ngāi Tahu Maori Trust Board & Te Runanganui o Tahu, Christchurch, 1993.
29 Libby Robin, 'Ecology: A Science of Empire?', in Tom Griffiths & Libby Robin (eds), *Ecology and Empire: Environmental History of Settler Societies*, Melbourne University Press, Melbourne, 1997, pp. 63–75.
30 Robin, 'Ecology: A Science of Empire?'
31 Libby Robin, *How a Continent Created A Nation*, University of New South Wales Press, Sydney, 2007, especially pp. 56–74 where she shows that 'sown pastures' increased from 4.5 million hectares in 1946 to 19.1 million hectares by 1965 and 48 million hectares by 1998.
32 *AJHR*, H 46A, 1949. The 'Problems' section, p. 42, blamed erosion on the nature of New Zealand soils, low fertility, rivers, and rabbits, and concluded, 'The hue and cry on erosion is, therefore, misleading.' Similarly, in 1985 economist Dr Alistair McArthur in *NZJA*, vol. 150, no. 5, p. 14, argued that there was no proof of the link between soil conservation and flooding.
33 On urbanisation and the dwindling political power of farmers, see Graeme Dunstall, 'The Social Pattern', in Geoffrey W. Rice (ed.), *The Oxford History of New Zealand*, pp. 455–58; Bremer, 'Federated Farmers and the State', p. 116.
34 Michael Roche, '"The Land We Have We Must Hold": Soil Erosion and Conservation in the Late Nineteenth- and Twentieth-Century New Zealand', *Journal of Historical Geography*, vol. 23, no. 4, 1997, pp. 447–58.
35 Herbert Guthrie-Smith, *Tutira: The Story of a New Zealand Sheep Station*, Godwit, Auckland/University of Washington Press, Seattle, reprint, 1999; 'The Changing Land' in *Making New Zealand*, II, No. 30,

Department of Internal Affairs, Wellington, 1940, pp. 12, 19–20.
36 Vaughan Wood, 'Soil Fertility Management in Nineteenth Century New Zealand', PhD thesis, University of Otago, 2003, pp. 398–403.
37 Rachel Carson, *Silent Spring;* Houghton, Mifflin Co. Boston, 1962, pp. 75–78, 204, 213.
38 McCaskill, *Hold This Land*, p. 260; Margaret Ritchie & Hugh Campbell, 'Historical Development of the Organic Agriculture Movement in New Zealand', in Hugh Campbell, *Recent Developments in Organic Food Production in New Zealand*, Department of Anthropology, University of Otago, Dunedin, 1996, pp. 10–18; Anthony Haystead, 'Organic Farming Survey', *Soil and Health*, Spring 1987, pp. 8–12; Frank Bradshaw, 'Disposing of Farm Wastes', *NZJA*, vol. 132, no. 3, 1975, p. 25; *Report on the State of New Zealand's Environment*, Ministry for the Environment, Wellington, 1997, chapter 6, p. 17; Robin Hodge, 'Nature's Trustee: Perrine Moncrieff and Nature Conservation in New Zealand 1920–1950', PhD thesis, Massey University, 1999, pp. 260–96.
39 Carson: *NZJA*, vol. 110, no.1, 1965, p. 9; 2,4,5-T: vol. 140, no. 4, 1980, pp. 28–34; vol. 146, no. 1, 1983, p. 29; DDT: vol. 140, no. 2, 1980, p. 5; *State of New Zealand's Environment*, chapter 8, pp. 66–67.
40 QE II Trust: *State of New Zealand's Environment*, chapter 9, p. 139; Agrichemicals: *State of New Zealand's Environment*, chapter 8, p. 36, Bruce Wildblood-Crawford, 'Environmental (In)justice and "Expert Knowledge": the Discursive Construction of Dioxins, 2,4,5-T and Human Health in New Zealand, 1940–2007', PhD thesis, University of Canterbury, 2008.
41 Jenny Philips, editor of the section entitled 'Women's World', championed legislation to protect the environment and recycling, *Straight Furrow*, 12 December 1974, p. 18.
42 *State of New Zealand's Environment*, chapter 7, p. 33.
43 'Effluent disposal': *NZJA*, vol. 120, no. 3, March 1970, pp. 25–37; 'Dairy Effluent Disposal': vol. 130, no. 2, 1975, p. 14; 'Farm Forestry': vol. 100, no. 4, 1960; vol. 110, no. 6, 1965, pp. 34–37 advocated the planting of 'radiata'. Planting of native trees was not advocated until vol. 153, no. 1, 1988, p. 50. *Straight Furrow* advocated harmonious landscape and planting more trees, including natives, from 5 February 1982, p. 13. Two months later, however, it expressed concern at good farming country being lost to forest planting: 14 May 1982, p. 4; landcare: *Landlink*, Summer 1994–99; stocking rates: *Straight Furrow*, 14 May 1982, p. 19.
44 Quin: *NZJA*, vol. 146, no. 3, 1983, pp. 4–6; Stephenson: vol. 148, no. 2, 1984, pp. 7–8; Haynes & Swift: vol. 148, no. 2, 1984, pp. 9–11.
45 Crowder, *NZJA*, vol. 147, no. 4, 1983, pp. 6–7; K.F. O'Connor, 'The Sustainability of Pastoralism', *Proceedings of the 1987 Hill and High Country Seminar*, 1987, Tussock Grasslands and Mountain Lands Institute Special Publication No. 30, pp. 161–88.
46 Ritchie & Campbell, 'Historical Development of the Organic Agriculture Movement in New Zealand'.
47 *State of New Zealand's Environment*, chapter 8, p. 33.
48 R.H.M. Langer (ed.), *Pastures: Their Ecology and Management*, Oxford University Press, Auckland, 1990: especially W. Harris, 'Pasture as Ecosystem', pp. 75–131; K.F. O'Connor, 'Pasture and Soil Fertility', pp. 157–96 R.J. Field & G.T. Daly, 'Weed Biology and Management', pp. 409–47.
49 www.fonterra.com/nz/en and communication from Dr Chris Rosin, Director of the Centre for Sustainability/ Kā Rakahau o te Ao Tūroa, University of Otago, 20 March 2013.
50 Ritchie & Campbell, 'Historical Development of the Organic Agriculture Movement in New Zealand'; and Hugh Campbell, Rob Burton, Mark Cooper, Matthew Henry, Erena Le Heron, Richard Le Heron, Nick Lewis, Eric Pawson, Harvey Perkins, Mike Roche, Chris Rosin & Toni White, 'From Agricultural Science to "Biological Economies"?', *New Zealand Journal of Agricultural Research*, vol. 52, no. 1, 2009, pp. 91–97.
51 As Brooking & Pawson, *Seeds of Empire*, passim and especially the conclusion, pp. 200–10, makes clear.

12. An interventionist state: 'wise use' forestry and soil conservation

1 M.M. Roche, *History of New Zealand Forestry*, New Zealand Forestry Corporation/GP Books, Wellington, 1990; M.M. Roche, *Land and Water. Water and Soil Conservation and Central Government in New Zealand, 1941–1988*, Historical Branch, Department of Internal Affairs, Wellington, 1994.
2 E.g. Donald Worster, 'Transformations of the Earth: Towards an Agroecological Perspective in History', *Journal of American History*, vol. 76, no. 4, 1990, pp. 1087–1106; Michael Williams, 'The End of Modern History', *The Geographical Review*, vol. 88, no. 2, 1998, pp. 275–300.
3 Alan Lester, 'Imperial Circuits and Networks: Geographies of the British Empire', *History Compass*, vol. 4, no. 1, 2006, pp. 124–41; see also Michael Roche, 'Forestry as Imperial Careering: New Zealand as the End and Edge of Empire in the 1920s–40s', *New Zealand Geographer*, vol. 68, no. 1, 2012, pp. 201–10.

4   These early estimates were of merchantable timber remaining. In 1905 this was put at 43 billion superficial feet, or seventy years' duration. The Royal Commission on Forestry's 1913 figure of 30 years was therefore notably less: *Appendices to the Journals of the House of Representatives* (hereafter *AJHR*), C4, 1905; C12, 1913.
5   Alfred Cockayne, 'The Monterey Pine, The Great Timber-Tree of the Future', *New Zealand Journal of Agriculture*, vol. 8, no. 1, 1914, pp. 1–26.
6   William Schlich, 'Forestry in the Dominion of New Zealand', *Quarterly Journal of Forestry*, vol. 1, no. 1, 1918, p. 23.
7   David Hutchins, *A Discussion of Australian Forestry with Special References to Forestry in Western Australia*, Perth, Government Printer, 1916, p. 391.
8   First articulated by Sir Francis Bell but taken up by Ellis and the State Forest Service (SFS): 'Afforestation', *Hawera & Normanby Star* 2 August, 1924; Commissioner of State Forests, 'A Forestry Policy for New Zealand' *New Zealand Journal of Agriculture*, vol. 18, no. 5. pp. 313–18.
9   Michael Roche, 'Colonial Forestry at its Limits: The Latter Day Career of Sir David Hutchins in New Zealand 1915-1920', *Environment and History* vol. 16, no. 4, 2010, pp. 431–54.
10 Michael Roche, 'Latter Day "Imperial Careering": L.M. Ellis – A Canadian Forester in Australia and New Zealand, 1920–1941', *Environment and Nature in New Zealand*, vol. 4, no. 1, 2009, pp. 58–77.
11 Phillips Turner was a surveyor by background and an accomplished amateur botanist. He was Inspector of Scenic Reserves from 1906 to 1912 and the secretary of the Royal Commission on Forestry in 1913. He became convinced of the need to establish a Forests Department headed by a professionally qualified forester and was involved in much behind-the-scenes advocacy. See Michael Roche, 'Edward Phillips Turner: The Development a 'Forest Consciousness' in New Zealand, 1890s to 1930s', *Proceedings of the 6th National Australian Forest History Conference*, Mill Press Rotterdam, 2005, pp. 143–53.
12 *AJHR*, C3, 1922, p. 2.
13 *AJHR*, C3, 1921, p. 7.
14 L. McIntosh Ellis, 'The Empire Forestry Conference: Some Impressions', *New Zealand Life and Forest Magazine*, vol. 3, no. 1, 1923, p. 6.
15 Michael Roche and John Dargavel, 'Imperial Ethos, Dominions Reality Forestry Education in New Zealand and Australia, 1910–1965', *Environment and History*, vol. 14, no. 4, 2008, pp. 523–43.
16 *AJHR*, C3, 1922, p. 2.
17 *AJHR*, C3, 1925, p. 7.
18 *AJHR* , C3A, 1920, p. 28; *AJHR* C3, 1925, p. 5.
19 Cecil M. Smith, 'The Use of Exotic Species in New Zealand Forestry', *New Zealand Journal of Forestry*, vol. 4, no. 2, 1937, p. 63.
20 'N.Z. Forests', *Evening Post*, 24 October 1928.
21 *New Zealand Official Year Book*, Government Printer, Wellington, 1936, pp. 375, 378. The SFS took on over 1000 men during the winter planting season in 1929. Others were employed under the Unemployment Board's Scheme No. 5, and during the winters from 1932 to 1935 over 1000 were employed on tree planting, but the decision to expand the exotic forest estate had been taken much earlier, in 1925.
22 Alexander Entrican & V. Rapson, 'The Establishment of a Local Pulp and Paper Industry in New Zealand' PR 347, 1943, Scion Library, Rotorua. Entrican in many ways typified the expert natural resource manager: Alex. R. Entrican, 'The State and The Development of Natural Resources', *New Zealand Journal of Public Administration*, vol. 21, no. 1, 1958, pp. 29–36.
23 Roche & Dargavel, 'Imperial Ethos'; J.M. Powell, 'Dominion over Palm and Pine: the British Empire Forestry Conferences, 1920–1947', *Journal of Historical Geography*, vol. 33, no. 4, 2007, pp. 852–77.
24 The early years of Fletchers are discussed in Neil Robinson, *James Fletcher: Builder*, Hodder & Stoughton, Auckland, 1970, and those of Tasman in Selwyn Parker, *Made in New Zealand: The Story of Jim Fletcher*, Hodder & Stoughton, Auckland, 1994.
25 Astrid Baker, 'Governments, Firms, and National Wealth: A New Pulp and Paper Industry in Post-war New Zealand', *Enterprise and Society*, vol. 5, no. 4, 2004, pp. 669–90.
26 J. Bain, 'Forest Monocultures: How Safe Are They?' *New Zealand Journal of Forestry*, vol. 26, no. 1, 1981, pp. 37–42; C.K. Chou, 'Monoculture, Species Diversification and Disease Hazards in Forestry', *New Zealand Journal of Forestry*, vol. 26, no. 1, 1981, pp. 20–36.
27 A. Priestly Thomson, *Problems of Sustained Yield Forestry in New Zealand. Paper Prepared for the Empire Forestry Conference, Canada, 1952*, Government Printer, Wellington, 1952, p. 6.
28 *AJHR*, C3, 1956, p. 28.
29 Alexander Entrican, Hugo Hinds & J. Stanley Reid, *Forest Trees and Timbers of New Zealand*, New Zealand

Forest Service Bulletin, No. 12, Government Printer, Wellington, 1957, p. 69.
30 *AJHR*, C3, 1961, p. 15.
31 Cecil M. Smith, 'The Use of Exotic Species in New Zealand Forestry', *New Zealand Journal of Forestry*, vol. 4, no. 2, 1937, pp. 67, 63.
32 Roche, *History of New Zealand Forestry*, p. 409.
33 'Editorial Notes', *New Zealand Journal of Forestry*, vol. 5, no. 3, 1946, p. 175.
34 *Evening Post*, 3 January 1945.
35 Roche, *History of New Zealand Forestry*, pp. 405–14.
36 Kenneth B. Cumberland, 'A Century's Change: Natural to Cultural Vegetation in New Zealand', *Geographical Review*, vol. 31, no. 4, 1941, pp. 529–54; James Beattie, 'Environmental Anxiety in New Zealand, 1840–1941: Climate Change, Soil Erosion, Sand Drift, Flooding and Forest Conservation', *Environment and History*, vol. 9, no. 4, 2003, pp. 379–92.
37 Graeme Wynn, 'Remapping Tutira. Contours of the Environmental History of New Zealand', *Journal of Historical Geography*, vol. 23, no. 4, 1997, pp. 418–46; Brad Coombes, 'The Historicity of Institutional Trust and the Alienation of Maori Land for Catchment Control at Mangatu, New Zealand', *Environment and History*, vol. 9, no. 3, 2003, pp. 333–59.
38 Roche, *Land and Water*, pp. 40–44.
39 Kenneth B. Cumberland, 'Burning Tussock Grassland: A Geographic Survey', *New Zealand Geographer*, vol. 1, no. 2, pp. 149–64.
40 Norman Taylor, 'Land Deterioration in the Heavier Rainfall Districts of New Zealand', *New Zealand Journal of Science and Technology*, vol. 19, no. 11, 1989, pp. 657–81; V.D. Zotov, 'Survey of the Tussock Grasslands of the South Island, New Zealand', *New Zealand Journal of Science and Technology*, vol. 20, no. 4A, 1938, pp. 212–44. For a revisionist view about the South Island high country, see Robert Peden, *Making Sheep Country: Mount Peel Station and the Transformation of the Tussock Lands*, Auckland University Press, Auckland, 2011.
41 R.P. Connell, 'Land Deterioration as a Grave Phase of Our National Economy', *New Zealand Journal of Agriculture*, vol. 59, no. 1, 1939, pp. 4–12.
42 E. Bruce Levy, *Grasslands of New Zealand*, Government Printer, Wellington, 1951.
43 Commission of Inquiry, *Maintenance of Vegetation Cover in New Zealand with Special Reference to Land Erosion*, Department of Scientific and Industrial Research Bulletin No. 77, Government Printer, Wellington, 1939.
44 Michael Roche, '"The Land We Have We Must Hold": Soil Erosion and Soil Conservation in Late Nineteenth- and Twentieth-century New Zealand', *Journal of Historical Geography*, vol. 23, no. 4, 1997, pp. 447–58.
45 Roche, *Land and Water*; Eric Pawson, 'Creating Public Spaces for Geography in New Zealand: Towards an Assessment of the Contributions of Kenneth Cumberland', *New Zealand Geographer*, vol. 67, no. 3, 2011, pp. 102–15.
46 Frederick Furkert, 'The Control of Rivers', *New Zealand Journal of Science and Technology*, vol. 10, 1928, pp. 434–47.
47 Roche, *Land and Water*, pp. 56–57.
48 *AJHR*, D1A, 1963, p. 10.
49 Michael Roche, 'Campbell, Douglas Archibald', from *The Dictionary of New Zealand Biography. Te Ara – the Encyclopedia of New Zealand*, updated 30-October–2012: www.teara.govt.nz/en/biographies/5c5/campbell-douglas-archibald
50 A.R. Acheson, *River Control and Drainage in New Zealand*, Ministry of Works, Wellington, 1968, p. 22.
51 Coombes, 'The Historicity of Institutional Trust'.
52 M.P.K. Sorrenson, 'Towards A Radical Reinterpretation of New Zealand History: The Role of the Waitangi Tribunal', in I.H. Kawharu (ed.), *Waitangi. Maori and Pakeha Perspectives of the Treaty of Waitangi*, Oxford University Press, Auckland, 1989, pp. 158–78; Paul Temm, *The Waitangi Tribunal: The Conscience of the Nation*, Random Century, Auckland, 1990.
53 Worster, 'Transformations of the Earth'.

13. On the edge: making urban places

1 I appreciate the comments of Ben Schrader and Simon Kingham in helping me to update this chapter.
2 Anon, *Auckland, The Gateway to New Zealand*, Wilson & Horton, Auckland, 1931, p. 1; Cilla McQueen quoted in David Eggleton, 'Introduction', *Here on Earth: The Landscape in New Zealand Literature*, Craig

Potton Publishing, Nelson, 1999, p. 14. The census definition of 'urban' in New Zealand includes all places with more than 1000 people, compared with 2500 in the United States. Many 'urban-dwellers' therefore lived in very small places.
3   William Cronon, 'Modes of Prophecy and Production. Placing History in Nature', *Journal of American History*, vol. 76, no. 4, 1990, p. 1131. A number of environmental histories about specific American cities have been published in the last decade, including: Matthew Klingle, *Emerald City: An Environmental History of Seattle*, Yale University Press, Newhaven, 2007; Michael F. Logan, *Desert Cities: The Environmental History of Phoenix and Tucson*, University of Pittsburgh Press, Pittsburgh, 2006; Anthony N. Penna & Conrad Edick Wright (eds), *Remaking Boston: An Environmental History of the City and Its Surroundings*, University of Pittsburgh Press, Pittsburgh, 2009.
4   Stephen Dovers (ed.), *Environmental History and Policy: Still Settling Australia*, Oxford University Press, Melbourne, 2000, p. viii; Jock Phillips, *A Man's Country? The Image of the Pakeha Male: A History*, Penguin Books, Auckland, 1987, p. 39.
5   The rural focus is clearly reflected in sources such as the collection edited by Carolyn Merchant, *American Environmental History: An Introduction*, Columbia University Press, New York, 2007. Amongst historical geographies of New Zealand, see, for example: Andrew Hill Clark, *The Invasion of New Zealand by People, Plants and Animals, South Island*, Rutgers University Press, New Brunswick, 1949; Kenneth B. Cumberland, *Landmarks*, Reader's Digest, Surry Hills, 1981.
6   J.W. Reps, *Cities of the American West: A History of Frontier Urban Planning*, Princeton University Press, Princeton NJ, 1979, and for New Zealand: G. Anderson, 'Wakefield Towns', in The Friends of the Turnbull Library, *Edward Gibbon Wakefield and the Colonial Dream: A Reconsideration*, GP Publications, Wellington, 1997, pp. 143–58; David Hamer, *New Towns in the New World: Images and Perceptions of the Nineteenth Century Urban Frontier*, Columbia University Press, New York, 1990; David Hamer, 'Wellington on the Urban Frontier', in David Hamer & Roberta Nicholls (eds), *The Making of Wellington 1800–1914*, Victoria University Press, Wellington, 1990, pp. 227–54.
7   Tim Low, *The New Nature. Winners and Losers in Wild Australia*, Penguin Books, Camberwell; Peter Newman & Isabella Jennings, *Cities as Sustainable Ecosystems: Principles and Practices*, Island Press, Washington DC, 2008; Marc Cioc & Char Miller, 'Interview: Joel Tarr', *Environmental History*, vol. 16, no. 1, 2011, pp. 121–36. Earlier critiques include Christine Meisner Rosen & Joel Arthur Tarr, 'The Importance of an Urban Perspective in Environmental History', *Journal of Urban History*, vol. 20, no. 3, 1994, pp. 299–310; and Maureen A. Flanagan, 'Environmental Justice in the City: A Theme for Urban Environmental History', *Environmental History*, vol. 5, no. 2, 2000, pp. 159–64.
8   Graeme Davison, 'Cities and Ceremonies: Nationalism and Civic Ritual in Three New Lands', *New Zealand Journal of History*, vol. 24, no. 2, 1990, pp. 97–117; Peter Holland, *Home in the Howling Wilderness: Settlers and Environment in Southern New Zealand*, Auckland University Press, Auckland, 2013.
9   James E. Vance, *The Merchant's World: The Geography of Wholesaling*, Prentice Hall, Englewood Cliffs, 1970; Jean Gottmann, *Megalopolis: The Urbanized Northeastern Seaboard of the United States*, Twentieth Century Fund, New York, 1961; N.C. Quigley, 'Patterns and Processes in the Development of Frontier Canterbury, 1850–1890', MA thesis, University of Canterbury, 1980; Eric Pawson & Neil C. Quigley, 'The Circulation of Information and Frontier Development: Canterbury 1850–1890', *New Zealand Geographer*, vol. 38, no. 2, 1982, pp. 65–76; William Cronon, *Nature's Metropolis: Chicago and the Great West*, W.W. Norton, New York, 1991.
10 Andre Siegfried, *Democracy in New Zealand*, repr. Victoria University Press / Price Milburn, Wellington, 1982 (first published 1914), pp. 252–53.
11 James Belich, *The New Zealand Wars and the Victorian Interpretation of Racial Conflict*, Auckland University Press, Auckland, 1986; *Making Peoples: A History of the New Zealanders, from Polynesian Settlement to the End of the Nineteenth Century*, Penguin Books, Auckland, 1996.
12 Tom Griffiths, 'We Have Still Not Lived Long Enough', *Inside Story*, 16 February 2009: inside.org.au/we-have-still-not-lived-long-enough (accessed 27 October 2012); Eric Pawson, 'Environmental Hazards and Natural Disasters', *New Zealand Geographer*, vol. 67, no. 3, 2011, pp. 143–47.
13 Andrew Hurley, 'Common Fields: An Introduction', in Andrew Hurley (ed.), *Common Fields: An Environmental History of St Louis*, Missouri Historical Society Press, St Louis, 1997, p. 10; see also Peter Newman, Timothy Beatley & Heather Boyer, *Resilient Cities. Responding to Peak Oil and Climate Change*, Island Press, Washington DC, 2009.
14 Dominic Alessio, 'Travel, Tourism and Booster Literature: New Zealand's Towns and Cities at the Turn of the Twentieth Century', *Studies in Travel Writing*, vol. 14, no. 4, 2010, pp. 383–96.

15 James Cowan, *New Zealand Cities*, Auckland, Whitcombe & Tombs, Auckland, 1917, pp. 10–11; John Barr, *The City of Auckland, New Zealand, 1840–1920*, Whitcombe & Tombs, Auckland, 1922, p. 199.
16 *Auckland, The Gateway to New Zealand*, p. 4.
17 Robert McGregor, *The Art Deco City: Napier, New Zealand*, Art Deco Trust, Napier, 1998; Patrick van Daele & Roy Lumby, *A Spirit of Progress: Art Deco Architecture in Australia*, Craftsman House, North Ryde, 1997.
18 Anon, *Auckland City and Province, Gateway to New Zealand*, Auckland, c. 1950, pages unnumbered.
19 *Auckland, The Gateway to New Zealand*, pages unnumbered; Julia Gatley (ed.), *Long Live the Modern. New Zealand's New Architecture 1904–1984*, Auckland University Press, Auckland, 2008, p. 12.
20 James Cowan, *Official Record of the New Zealand International Exhibition of Arts and Industries, Held at Christchurch 1906–7. A Descriptive and Historical Account*, Government Printer, Wellington, 1910, pp. 12, 21; see also John Mansfield Thomson (ed.), *Farewell Colonialism: The New Zealand International Exhibition, Christchurch 1906–07*, Dunmore Press, Palmerston North 1998.
21 S. Percy Smith, *Taranaki: The Garden of New Zealand*, W.G. Malone, New Plymouth, 1906; *New Plymouth: Health, Sunshine, Happiness*, New Plymouth Tourist and Expansion League, 1922; *Guide and Map of Gisborne and the Picturesque East Coast*, Gisborne Publishing Co., 1942; *Come to Sunny Gisborne on Sea: The California of New Zealand*, Gisborne 30,000 Club, 1946; see also D. Alessio, 'Colonial Views: Images of the New Zealand City and Town, 1880–1930', PhD thesis, Victoria University of Wellington, 1992.
22 *The Future's in Rotorua: A Complimentary Brochure*, City of Rotorua Publications Office, 1962; Wellington Early Settlers' and Historical Association Inc., *Pioneers Progress*, Wellington, 1960, p. 19; Gatley, *Long Live the Modern*.
23 Gatley, *Long Live the Modern*; Douglas Lloyd Jenkins, *At Home: A Century of New Zealand Design*, Random House, Glenfield, 2004.
24 Frank Whitford, 'The City in Painting', in Edward Timms & David Kelley (eds), *Unreal City: Urban Experience in Modern European Literature and Art*, Manchester University Press, Manchester, 1985, p. 57.
25 Mike Davis, *Ecology of Fear: Los Angeles and the Imagination of Disaster*, Metropolitan Books, New York, 1998. For an application of this argument to New Zealand, see Pawson, 'Environmental Hazards'.
26 Pawson, 'Environmental Hazards'; Rodney Grapes, *Magnitude Eight Plus: New Zealand's Biggest Earthquake*, Victoria University Press, Wellington, 2000; Rodney Grapes, *The Visitation: The Earthquakes of 1848 and the Destruction of Wellington*, Victoria University Press, Wellington, 2011.
27 N.J. Ericksen, *Creating Flood Disasters? New Zealand's Need for a New Approach to Urban Flood Hazard*, Water and Soil Miscellaneous Publication No. 77, National Water and Soil Conservation Authority, Wellington, 1986, p. 46.
28 Report on the Waimakariri River by James M. Balfour, Marine Engineer, 9 April 1866, MS CP 631a/1, National Archives, Christchurch.
29 Geo Nelson, *Report on the Waimakariri River (New Zealand)*, Cook, Hammond and Kell, London, 1928, p. iv.
30 F.W. Maclean, J.M. Willis & J.G. Alexander, *Water of Leith and Lindsay's Creek*, 'The Engineers' Commission', 1931, Hoc Bliss and RUS MacL, Hocken Collections, University of Otago; W.J. Watt, *Dunedin's Historical Background*, Dunedin City Council, Dunedin, 1972.
31 Pamela Wood, *Dirt, Filth and Decay in a New World Arcadia*, Auckland University Press, Auckland, 2005, p. 2.
32 William Pratt, *Canterbury and Incidentally New Zealand, AD 1900*, Christchurch Press, Christchurch, 1900, pp. 93–94; Wood, *Dirt, Filth and Decay*, p. 6.
33 Eric Pawson, 'Confronting Nature', in John Cookson & Graeme Dunstall (eds), *Southern Capital Christchurch: Towards a City Biography*, Canterbury University Press, Christchurch, 2000, pp. 60–84.
34 Report of the Commission of Inquiry into Certain Alleged Nuisances in Districts Within or Contiguous to the Southern Auckland Metropolitan Area (Fumes Inquiry) August 1955, *Appendices to the Journals of the House of Representatives*, H31A, 1955; C.J. Sparrow, 'Some Geographic Aspects of Air Pollution in Auckland', in *Auckland Air Pollution Symposium*, Royal Society of Health, Auckland, 1968; G.W.A. Bush, *Moving Against the Tide: The Brown's Island Drainage Controversy*, Dunmore Press, Palmerston North, 1980; *Finding of the Waitangi Tribunal on the Manukau Claim*, (Wai 8), Government Printer, Wellington, 1985.
35 Pawson, 'Confronting Nature'.
36 David Hamer, 'Centralization and Nationalism (1891–1912)', in Keith Sinclair (ed.), *The Oxford Illustrated History of New Zealand*, Oxford University Press, Oxford, 1990, p. 125–52; Miles Fairburn, 'The Rural Myth and the New Urban Frontier. An Approach to New Zealand Social History, 1870–1940', *New Zealand Journal of History*, vol. 9, no. 1, 1975, pp. 3–21.
37 Siegfried, *Democracy*, p. 247; Cowan, *New Zealand Cities*, p. 42; Eggleton, 'Introduction', p. 21.
38 The advertisements referred to in this section are from the archives of H.G. Livingstone Ltd, Christchurch, and

the Hocken Collections, University of Otago. See also Caroline Miller, 'Images and Reality: Selling Suburbia', *Proceedings of the Twentieth New Zealand Geography Conference*, New Zealand Geographical Society, Hamilton, 1999, pp. 223–26.
39 Ben Schrader, *We Call It Home: A History of State Housing in New Zealand*, Reed Publishing, Birkenhead, 2005; Gatley, *Long Live the Modern*.
40 D.C. Thorns, 'Owner Occupation, the State and Class Relations', in Chris Wilkes & Ian Shirley (eds), *In the Public Interest: Health, Work and Housing in New Zealand Society*, Benton Ross, Auckland, 1984, pp. 213–30; Eric Pawson, 'The Social Production of Urban Space', *New Zealand Geographer*, vol. 43, no. 3, 1987, pp. 123–29; Fairburn, 'The Rural Myth', p. 16.
41 *Auckland, The Gateway to New Zealand*, p. 10; Samuel P. Hays, *Beauty, Health and Permanence: Environmental Politics in the United States 1955–1985*, Cambridge University Press, Cambridge, 1987.
42 Penrose Industrial Progress Association Inc., *Penrose Progress: The Growth of a Great Industrial Centre*, Penrose, 1946, p. 5.
43 Michael Gunder, 'Auckland's Motorway System: A New Zealand Genealogy of Imposed Automotive Progress', *Urban Policy and Research*, vol. 20, no. 2, 2002, pp. 129–42; Christopher E. Harris, 'Slow Train Coming: The New Zealand State Changes its Mind about Auckland Transit', *Urban Policy and Research*, vol. 23, no. 1, 2005, pp. 37–55.
44 Ministry of Works, *The New Zealand Environment. National Report to the United Nations Preparatory Committee for the 1972 Conference on the Human Environment*, Town and Country Planning Division, Wellington, 1972, p. 7.
45 Ericksen, *Creating Flood Disasters?*, pp. 64–69.
46 'Sunny Timaru Welcomes You', *Shopping Week, September 27 to October 2*, Timaru Post Co., Timaru, 1926, p. 7.
47 Rob Shields, *Places on the Margin: Alternative Geographies of Modernity*, Routledge, London, 1991. Shields uses Brighton in Sussex as the focus of his argument.
48 'Sunny Timaru Welcomes You', p. 35.
49 Neill Atkinson, '"Call of the Beaches": Rail Travel and the Democratisation of Holidays in Interwar New Zealand', *Journal of Transport History*, vol. 33, no. 1, 2012, pp. 1–20.
50 *Timaru Herald*, 18 December 1922.

14. The empire of the rhododendron: reorienting New Zealand garden history

1 This research was made possible by a University of Waikato Vice-Chancellor's Research Award. The author thanks the many people who contributed to the chapter.
2 William Beinhart & Karen Middleton, 'Plant Transfers in Historical Perspective: A Review Article', *Environment and History*, vol 10, no. 1, 2004, pp. 3–29; James Beattie, *Empire and Environmental Anxiety: Health, Science, Art and Conservation in South Asia and Australasia*, Palgrave Macmillan, Houndsmills, 2011; Tom Brooking & Eric Pawson, *Seeds of Empire: The Environmental Transformation of New Zealand*, I.B. Tauris, London, 2011.
3 Jodi Frawley, 'Making Mangoes Move', *Transforming Cultures eJournal*, vol. 3, no. 1, 2008, pp. 165–84: http://epress.lib.uts.edu.au/journals/TfC (accessed 13 January 2012); Vaughan Wood & Eric Pawson, 'The Banks Peninsula Forests and Akaroa Cocksfoot: Explaining a New Zealand Forest Transition', *Environment and History*, vol. 14, no. 4, 2008, pp. 449–68; Eric Pawson, 'Plants, Mobilities and Landscapes: Environmental Histories of Botanical Exchange', *Geography Compass*, vol. 2, 2008, pp. 1464–77.
4 Alfred W. Crosby, *Ecological Imperialism: The Biological Expansion of Europe, 900–1900*, 2nd edn, Cambridge University Press, Cambridge; New York, 2004. On recent work, James Beattie, 'Making Home, Making Identity: Asian Garden Making in New Zealand, 1850s–1930s', *Studies in the History of Gardens & Designed Landscapes*, vol. 31, no. 2, 2011, pp. 139–59; James Beattie, J.M. Heinzen & John P. Adam, 'Japanese Gardens and Plants in New Zealand, 1850–1950: Transculturation and Transmission', *Studies in the History of Gardens & Designed Landscapes*, vol. 28, no. 2, 2008, pp. 219–36.
5 Paxton is quoted in Maggie Campbell-Culver, *The Origins of Plants: The People and Plants that have Shaped Britain's Garden History*, Eden Project Books, London, 2004, p. 352; Jane Kilpatrick, *Gifts from the Gardens of China*, Frances Lincoln, London, 2007, p. 140; Fa-ti Fan, *British Naturalists in Qing China: Science, Empire, and Cultural Encounter*, Harvard University Press, Cambridge, 2004, p. 249.
6 Kilpatrick, *Gifts from the Gardens of China*, p. 140; Fan, *British Naturalists in Qing China*; S. Heriz-Smith, 'The Veitch Nurseries of Killerton and Exeter c. 1780 to 1863: Part 1', *Garden History*, vol. 16, no. 1, 1988, pp. 41–57.

7   Richard H. Drayton, *Nature's Government: Science, Imperial Britain, and the 'Improvement' of the World*, Yale University Press, New Haven, 2000.
8   J. Livingstone, 'Observations on the Difficulties Which Have Existed in the Transportation of Plants from China to England, and Suggestions for Obviating Them', *Transactions of the Horticultural Society of London*, vol. III, 1819, pp. 421–29.
9   Campbell-Culver, *The Origins of Plants*, p. 350.
10 S. Challenger, 'Pioneer Nurserymen of Canterbury, New Zealand (1850–65)', *Garden History*, vol. 7, no. 1, 1979, pp. 25–64; Winsome Shepherd, *Wellington's Heritage: Plants, Gardens and Landscape*, Te Papa Press, Wellington, 2000; Louise Shaw, *A History of the Dunedin Horticultural Society, 1851–2001*, Dunedin Horticultural Society, Dunedin, 2000.
11 James Beattie, 'Colonial Geographies of Settlement: Vegetation, Towns, Disease and Well-Being in Aotearoa/New Zealand, 1830s–1930s', *Environment and History*, vol. 14, no. 4, 2008, pp. 583–610; Franklin Ginn, 'Colonial Transformations: Nature, Progress and Science in the Christchurch Botanic Gardens', *New Zealand Geographer*, vol. 65, no. 1, 2009, pp. 35–47; Matt Morris, 'A History of Christchurch Home Gardening from Colonisation to the Queen's Visit of Christchurch', PhD thesis, University of Canterbury, 2006.
12 Matthew Bradbury (ed.), *A History of the Garden in New Zealand*, Viking, Auckland, 1995; Katherine Raine, 'Domesticating the Land: Colonial Women's Gardening', in Bronwyn Dalley & Bronwyn Labrum (eds), *Fragments: New Zealand Social and Cultural History*, Auckland University Press, Auckland, 2000, pp. 76–96.
13 James Beattie & John Stenhouse, 'Empire, Environment and Religion: God and Nature in Nineteenth-century New Zealand', *Environment and History*, vol. 13, no. 4, 2007, pp. 413–46.
14 D.R. Given, E.G. Brockerhoff & J. Palmer, 'Nationally Networked Plant Collections Are a Necessity', *New Zealand Garden Journal*, vol. 9, no. 1, 2006, p. 15. In Given's article the incorrect botanical name is given for Chinese Hill Cherry. Records at the Allan Herbarium, Landcare Research, confirm Given collected Chinese Hill Cherry at the former site of McDonnell's garden: Ines Schonberger, pers. comm.
15 Stevan Eldred-Grigg, *A Southern Gentry: New Zealanders Who Inherited the Earth*, Reed, Wellington, 1980; Jim McAloon, *No Idle Rich: The Wealthy in Canterbury and Otago 1840–1914*, Otago University Press, Dunedin, 2002.
16 D.D. Muter to Editor, *Lyttelton Times*, 12 November 1853.
17 Sir John Cracroft Wilson, Diary/Recollections, 1854, of Canterbury, typed transcript by Ron Chapman, 1989, ARC1989. 80, Canterbury Museum, pp. 3, 23.
18 Wilson, Diary/Recollections, p. 23.
19 Thora Parker, *And Not to Yield: The Story of a New Zealand Family, 1840–1940*, David Bateman, Auckland, 1987, p. 26. On their popularity in India, see Eugenia W. Herbert, *Flora's Empire: British Gardens in India*, University of Pennsylvania State Press, Philadelphia, 2011.
20 *Catalogue of Plants Cultivated for Sale by William Martin & Son, Nurseryman and Seedsman, "Fairfield"*, Mills, Dick and Co., Dunedin, 1880, p. 10; Wm. Martin, 'Early History of Fairfield', November 1963, typewritten, in Martin DC-0320, Toitū Otago Settlers Museum; 'W. Martin, 'Some Notes on the Rhododendrons at Fairfield', Garden History, Box 5 Nurseries, Toitū Otago Settlers Museum; William Martin, 'Mr William Martin, a Pioneer Horticulturist of Otago', handwritten notes by Wm. Martin (grandson) 3 February 1953, Toitū Otago Settlers Museum, Dunedin; *Otago Witness*, 24 August 1893.
21 H. Mason, 'Thomas Mason (1818–1903)', in *The Family of Thomas and Jane Mason of Taita*, compiled by R. Evans and A. Evans, Evagean Publishing, Auckland, 1994; Thomas Mason to Aunt, Taita, 26 November 1871, Thomas Mason, MS Papers 54, Alexander Turnbull Library, Wellington; Shepherd, *Wellington's Heritage*, p. 43.
22 Thomas Mason to Uncle, 8 June 1841; Mason to Aunt, 28 May 1883; Mason to Aunt, 9 December 1883, Wellington, Alexander Turnbull Library; Jane Brown, *Tales of the Rose Tree: Ravishing Rhododendrons and Their Travels Around the World*, Harper Collins, London, 2004; Winsome Shepherd, 'The Finest Gardens in the Southern Hemisphere "The Gums", Taita, Wellington', in *The Family of Thomas and Jane Mason*, Evagean Publishing, Auckland, 1994, p. 51.
23 A. Ludlam, 'Essay on the Cultivation and Acclimatization of Trees and Plants', *Transactions and Proceedings of the New Zealand Institute*, vol. 1, 1868, pp. 1, 9–11, 14–16.
24 Shepherd, *Wellington's Heritage*; Shaw, *Southern Gardening*; Thomas R. Dunlap, *Nature and the English Diaspora: Environment and History in the United States, Canada, Australia, and New Zealand*, Cambridge University Press, New York, 1999.
25 Keith Stewart, *Rosa Antipodes: The History of Roses in New Zealand*, David Bateman, Auckland, 1994, pp. 63, 91, 96–97.

26 *Hay's Annual Garden Book Containing Descriptive Priced Catalogue of Ornamental Trees and Shrubs ... And Calendar of Monthly Operations*, Geo. T. Chapman, Auckland, 1867, pp. 56–58; *Hay's Annual Garden Book*, Geo. T. Chapman, Auckland, 1892, pp. 39–43; *Otago Witness*, 3 August 1904.
27 'Murphy's *Handbook of New Zealand Gardening* (1885) devoted almost 70 pages (out of 172) to florists' flowers'. Raine, 'Domesticating the Land', p. 95; *Press*, 15 May 1873; Shaw, *Southern Gardening*, pp. 57–58; *Otago Witness*, 8 April 1882.
28 Beattie, Heinzen & Adam, 'Japanese Gardens and Plants'; A.R. Ferguson, '1904 – The Year that Kiwifruit (*Actinidia deliciosa*) came to New Zealand', *New Zealand Journal of Crop and Horticultural Science*, vol. 32, no. 1, 2004, pp. 3–27; E.H. Wilson to David Tannock, Arnold Arboretum, Harvard University, 3 April 1922, 'Superintendent of Reserves: Correspondence', Series 3/14, 1914–1925, 1927, 1933–1937, Dunedin City Council Archives, Dunedin; Morris, 'A History of Christchurch Home Gardening', p. 142.
29 Mrs William, *Facts: Or, The Experiences of a Recent Colonist in New Zealand by a Lady 1882*, self-published, Yalding, 1883, p. 24.
30 James Ng, *Windows on a Chinese Past*, vol. 3, Otago Heritage Books, Dunedin, 1993, p. 55, note 35b.
31 For excellent overviews see Lily Lee & Ruth Lam, *Sons of the Soil: Chinese Market Gardeners in New Zealand*, Dominion Federation of New Zealand Chinese Commercial Growers, Pukekohe, 2012; and Nigel Murphy, *Success Through Adversity: A History of the Dominion Federation of New Zealand Chinese Commercial Growers*, Dominion Federation of New Zealand Chinese Commercial Growers, Pukekohe, 2012. *Otago Witness*, 18 December 1880; *New Zealand Parliamentary Debates*, vol. 92, 1896, p. 254.
32 Jessica Heine, 'Colonial Anxieties and the Construction of Identities: The Employment of Maori Women in Chinese Market Gardens, Auckland, 1929', MA thesis, University of Waikato, Hamilton, 2006, p. 26; *Evening Post*, 19 October 1891; 'Interim Report (No II) of the Chinese Immigration Committees', *Appendices to the Journal of the House of Representatives* (hereafter *AJHR*), H–5a, 1871, p. 8; Lily Lee & Ruth Lam, 'Chan Dah Chee (1851–1930): Pioneer Chinese Market Gardener and Auckland Businessman', *ENNZ: Environment and Nature in New Zealand*, vol. 6, no. 1, June 2011, pp. 24–54; Julia Bradshaw, *Golden Prospects: Chinese on the West Coast of New Zealand*, West Coast Historical & Mechanical Society, Greymouth, 2009, p. 70.
33 Robert B. Marks, *Tigers, Rice, Silk, & Silt: Environment and Economy in Late Imperial South China*, Cambridge University Press, New York, 1998; Warwick Frost, 'Migrants and Technological Transfer: Chinese Farming in Australia, 1850–1920', *Australian Economic History Review*, vol. 42, no. 2, July 2002, pp. 113–31; James Gerber & Lei Guang (eds), *Agriculture and Rural Connections in the Pacific, 1500–1900*, Ashgate, Aldershot, 2006; Sucheng Chan, *This Bitter Sweet Soil: The Chinese in California Agriculture, 1860–1910*, University of California Press, Berkeley and Los Angeles, 1986; H.P.M., 'Remunerative Farming for Otago', *North Otago Times*, 2 September 1870.
34 'Interim Report (No II) of the Chinese Immigration Committees', *AJHR*, H–5a, 1871, p. 8 (Owhiro Valley); *Evening Post*, 4 October 1887 (Merbett); H.J. Hawkins, 'The Kitchen Garden, Fruit and Flower Garden …', in *Chapman's Settler's Hand-Book to the Farm and Garden, Arranged for the Seasons & Climate of New Zealand*, G.T. Chapman, Auckland, date unknown, pp. 67–68.
35 *Wanganui Herald*, 25 August 1897; *Auckland Star*, 23 April 1887; *Evening Post*, 26 January 1923; J.B.J. Lee, *Jade Taniwha: Māori-Chinese Identity and Schooling in Aotearoa*, Rautaki Ltd, Auckland, 2007.
36 *Evening Post*, 26 September 1887; Brian Moloughney & John Stenhouse, '"Drug-besotten, Sin-begotten Fiends of Filth": New Zealanders and the Oriental Other, 1850–1920', *New Zealand Journal of History*, vol. 33, no. 1, 1999, pp. 43–64.
37 *Otago Witness*, 18 March 1871; *Otago Witness*, 23 March 1872. See also Shaw, *Southern Gardening*, pp. 43, 59–60. *Ashburton Guardian*, 18 February 1905 (Tinwald).
38 Don, *Nineteenth Inland Otago Tour, 1905–1906*, p. 19; *Southland Times*, 25 May 1887.
39 Lee & Lam, *Sons of the Soil*.
40 Ng, *Windows on a Chinese Past*, vol. 3, pp. 304–12.
41 See James Beattie, 'Chinese Environmental Networks: Otago and Canton, 1860s–1910s', in James Beattie, Edward Melillo & Emily O'Gorman (eds), *Networks of Nature in the British Empire: New Views on Imperial Environmental History*, Continuum/Bloomsbury, New York and London, 2014.
42 Duncan Campbell, 'Transplanted Gardens: Aspects of the Design of the Garden of Beneficence, Wellington, New Zealand', *Studies in the History of Gardens and Designed Landscapes*, vol. 31, no. 2, 2011, pp. 160–66; James Beattie, 'Growing Chinese Influences in New Zealand: Chinese Gardens, Identity and Meaning', *New Zealand Journal of Asian Studies*, vol. 9, no. 1, 2007, pp. 38–60; James Beattie (ed.), *Lan Yuan. The Garden of Enlightenment: Essays on the Intellectual, Cultural, and Architectural Background to the Dunedin Chinese Gardens*, Dunedin Chinese Gardens Trust, Hamilton, 2008.

PART V PERSPECTIVES
15. Postcolonial environments

1. I am grateful to Eric Pawson for so many useful suggestions. Garth Cant and Mike Roche provided information and Eve McCosker constantly alerted me to the importance of the senses.
2. See Giselle Byrnes & Catharine Coleborne, 'Editorial Introduction: The Utility and Futility of "the Nation" in Histories of Aotearoa New Zealand', *New Zealand Journal of History*, vol. 45, no. 1, 2011, pp. 1–14, 7–8.
3. See Andrew Hill Clark, *The Invasion of New Zealand by People, Plants and Animals: The South Island*, Rutgers University Press, New Brunswick, 1949; Kenneth B. Cumberland, *Landmarks*, Reader's Digest, Surry Hills, NSW, 1981.
4. Eric Pawson, 'Postcolonial New Zealand?', in Kay Anderson & Faye Gayle (eds), *Cultural Geographies*, Addison Wesley Longman, Australia, 1999, p. 26. See also Tom Brooking & Eric Pawson, *Seeds of Empire: The Environmental Transformation of New Zealand*, I.B. Tauris, London, 2011; James Beattie, 'Recent Themes in the Environmental History of the British Empire', *History Compass*, vol. 10, no. 2, 2012, pp. 129–39; James Beattie, *Empire and Environmental Anxiety: Health, Science, Art and Conservation in South Asia and Australasia 1800–1920*, Palgrave, Basingstoke, 2011; Graeme Wynn, 'Remapping Tutira: Contours in the Environmental History of New Zealand', *Journal of Historical Geography*, vol. 23, no. 4, 1997, pp. 418–46.
5. Tom Brooking, *The History of New Zealand*, Greenwood Press, Westport CT, 2004, p. 3.
6. Mark M. Smith, *Sensory History*, Berg, Oxford and New York, 2007, p. 1.
7. Elizabeth A. Foyster & Christopher A. Whatley (eds), *A History of Everyday Life in Scotland 1600–1800*, Edinburgh, Edinburgh University Press, 2010, p. 217.
8. See G.H. Roeder, 'Coming to Our Senses', *Journal of American History*, vol. 81, no. 3, 1994, pp. 1112–22; Constance Classen, *Worlds of Sense: Exploring the Senses in History and Across Cultures*, Routledge, London and New York, 1993; Constance Classen, David Howes & Anthony Synnott, *Aroma: The Cultural History of Smell*, Routledge, London and New York, 1994; Constance Classen (ed.), *The Book of Touch: A Collection of Essays on the History, Sociology and Anthropology of Touch*, Berg, Oxford and New York, 2005; Constance Classen, *The Deepest Sense: A Cultural History of Touch*, University of Illinois Press, Illinois, 2012.
9. Michel Foucault, *The Order of Things: An Archaeology of the Human Sciences*, New York, Pantheon Books, 1970, p. 17.
10. Robert D. Sack, *Conceptions of Space in Social Thought: A Geographic Perspective*, University of Minnesota Press, Minneapolis, 1980, p. 9.
11. Sack, *Conceptions of Space*, pp. 12, 15.
12. Keith Thomas, *Religion and the Decline of Magic*, Charles Scribner's Sons, New York, 1971, p. 168.
13. J. Douglas Porteous, *Landscapes of the Mind: Words of Sense and Metaphor*, University of Toronto Press, Toronto, 1990, p. xiv.
14. See Joy Parr's call 'Notes for a More Sensuous History of Twentieth Century Canada: The Timely, the Tacit and the Material Body,' *Canadian Historical Review*, vol. 82, no. 4, 2001, pp. 720–45.
15. Noel Castree, David Demeritt, Diana Liverman & Bruce Rhoads (eds), *A Companion to Environmental Geography*, Wiley-Blackwell, Oxford, 2009; Bruce Braun & Noel Castree (eds), *Remaking Reality: Nature at the Millennium*, Routledge, London and New York, 1998.
16. Christine Dann, 'Ecofeminism, Women and Nature', in Rosemary du Plessis with Phillida Bunkle, Kathie Irwin, Alison Laurie & Sue Middleton (eds), *Feminist Voices: Women's Studies Texts for Aotearoa/New Zealand*, Oxford University Press, Auckland, 1992, pp. 338–53.
17. Carolyn Merchant, *The Death of Nature: Women, Ecology and the Scientific Revolution*, Harper and Row, San Francisco, 1980, p. 1.
18. Annette Kolodny, *The Lay of the Land: Metaphor as Experience and History in American Life and Letters*, University of North Carolina Press, Chapel Hill, 1975; Annette Kolodny, *The Land Before Her: Fantasy and Experience of the American Frontiers, 1630–1860*, University of North Carolina Press, Chapel Hill, 1984.
19. Joy Parr, *Sensing Changes: Technologies, Environments and the Everyday, 1953–2003*, UBC Press, Vancouver, 2009, p. 4.
20. Graeme Wynn, 'Foreword' in *Sensing Changes*, p. xiii.
21. Rollo Arnold, *New Zealand's Burning: The Settlers' World in the Mid 1880s*, Victoria University Press, Wellington, 1994. p. 11.
22. Te Maire Tau, 'Ngai Tahu and the Canterbury Landscape' in John Cookson & Graeme Dunstall (eds), *Southern Capital: Christchurch: Towards a City Biography*, Canterbury University Press, Christchurch, 2000, p.

42.
23 Tau, 'Ngai Tahu and the Canterbury Landscape', p. 43.
24 See M.M. Roche, 'Charles Torlesse: A Pioneer Recorder of Canterbury Weather 1849–1858', *Weather and Climate*, vol. 4, 1984, pp. 66–69.
25 Peter Holland, *At Home in the Howling Wilderness: Settlers and the Environment in Southern New Zealand*, Auckland University Press, Auckland, 2013, p. 197.
26 Tau, 'Ngai Tahu and the Canterbury Landscape', p. 42.
27 'Climate', *Cyclopedia of New Zealand (Canterbury Provincial District)*, The Cyclopedia Company Ltd, Christchurch, 1903, http://nzetc.victoria.ac.nz/tm/scholarly/tei-Cyc03Cycl-t1-body1-d1-d21.html
28 James Cowan, 'Jessie Mackay: Poet, Idealist and Celtic Patriot', *The New Zealand Railways Magazine*, vol. 11, no. 10, 1 January 1937, pp. 13–14.
29 Katie Pickles, 'Colonial Sainthood in Australasia', *National Identities*, vol. 7, no. 4, 2005, pp. 389–408.
30 Quoted in Rosemary Britten, *Lake Coleridge: The Power, the People, the Land*, Hazard Press, Christchurch, 2000, pp. 48–49.
31 Interview 1, by author, June 2010.
32 Interview 2, by author, June 2010.
33 http://en.wikipedia.org/wiki/Nor'west_arch and www.neilcherry.com
34 See Eric Pawson, 'Confronting Nature', in Cookson & Dunstall, *Southern Capital*, pp. 72–75; Holland, *At Home in the Howling Wilderness*, pp. 71–72.
35 http://www.norwestercafe.co.nz/ http://www.duxbrew.co.nz
36 Larry W. Price, 'Hedges and Shelterbelts on the Canterbury Plains, New Zealand: Transformation of an Antipodean Landscape', *Annals of the Association of American Geographers*, vol. 83, no. 1, 1993, pp. 119–40.
37 James Belich, *Paradise Reforged: A History of the New Zealanders from the 1880s to the Year 2000*, Allen Lane: Penguin, Auckland, 2001, p. 391.
38 Robert McGregor, *The Hawke's Bay Earthquake: New Zealand's Greatest Natural Disaster*, 2nd edn, Art Deco Trust, Napier, 2002, p. 20.
39 Malcolm McKinnon (ed.), *New Zealand Historical Atlas*, David Bateman, Auckland and the Department of Internal Affairs, Wellington, 1997, Plate 87: Fire, Flood and Quake; McGregor, *The Hawke's Bay Earthquake*, p. 44.
40 Bob Brockie (ed.), *The Penguin Eyewitness History of New Zealand: Dramatic First-Hand Accounts from New Zealand's History*, Penguin, Auckland, 1998, 2002, p. 160.
41 www.redcross.org.nz/canterbury
42 McGregor, *The Hawke's Bay Earthquake*, p. 34.
43 Matthew Wright, *Hawke's Bay: The History of the Province*, Palmerston North, Dunmore Press, 1994, p. 159.
44 Tina Law, 'Christchurch Staggers Forward as Work Gets Underway', *The Press*, 22 August 2011.
45 Fiona Farrell, *The Broken Book*, Auckland University Press, Auckland, 2011.
46 McGregor, *The Hawke's Bay Earthquake*, pp. 31–34.
47 *New Zealand Parliamentary Debates*, vol. 229, 18 September 1931, p. 698.
48 Wright, *Hawke's Bay*, p. 161; McGregor, *The Hawke's Bay Earthquake*, p. 34.
49 McGregor, *The Hawke's Bay Earthquake*, pp. 38, 43; McKinnon, *New Zealand Historical Atlas*, Plate 66.
50 McKinnon, *New Zealand Historical Atlas*, Plate 67.
51 Eric Pawson, 'Environmental Hazards and Natural Disasters', *New Zealand Geographer*, vol. 67, no. 3, 2011, pp. 143–47.
52 Statistics New Zealand, Census of Population and Dwellings, 2006.
53 Julie Cupples & Kevin Glynn (eds), 'Counter-cartographies: New (Zealand) Cultural Studies/Geographies and the City', Special issue of the *New Zealand Geographer*, vol. 65, no. 1, 2009, pp. 1–89.

16. An updated history of New Zealand environmental law

1 The author wishes to thank Royden Somerville, QC, Associate Professor Ken Palmer, and Dr Benjamin Richardson for their contributions to early research for the first edition of this chapter. This time, she would like to add Associate Professor Janine Hayward and Ms Ceri Warnock.
2 'Environmental law includes judicial decisions, legislation, and administrative regulation directly concerning the use, management, and protection of the physical and biological elements of the biosphere and the effects of human and natural interactions within and between these physical and biological elements': David A.R. Williams (ed.), *Environmental and Resource Management Law*, 2nd edn, Butterworths, Wellington, 1997, p.

7. See also Klaus Bosselmann, David Grinlinton & Prue Taylor (eds), *Environmental Law for a Sustainable Society*, 2nd edn, Monograph Series, vol. 1, New Zealand Centre for Environmental Law, Auckland, 2013.
3   See *Wi Parata v. Bishop of Wellington* (1877) 3 New Zealand Jurist Reports (New Series) Supreme Court, p. 77, per C.J. Prendergast, and the discussion in E.T. Durie, 'F.W. Guest Memorial Lecture 1996: Will the Settlers Settle? Cultural Conciliation and Law', *Otago Law Review*, vol. 8, 1996, pp. 449–65.
4   Even before the abolition of the provinces in 1875, local government had entered a period of 'variegated fragmentism' (Graham W.A. Bush, *Local Government and Politics in New Zealand*, George Allen & Unwin, Auckland, 1980, p. 20), which persisted until the Local Government Act 1974 was enacted. The 'tinkering' approach is invited by the outdated view of nature as a 'perfectly intelligible clockwork … [whose] moving parts were automata or mechanisms in miniature': J. Baird Callicott, 'La Nature est Morte, Vive la Nature!', *Hastings Center Report*, vol. 22, no. 5, 1992, pp. 17–18.
5   Ton Bührs & Robert V. Bartlett, *Environmental Policy in New Zealand: The Politics of Clean and Green?*, Oxford University Press, Auckland, 1993, p. 92.
6   The obvious exception being sea fisheries, where legislation has protected Māori fishing rights since 1877.
7   'The Thermal Springs Districts Act, 1881', *Bay of Plenty Times*, 22 December 1881.
8   The presumption of protection has merits, but also 'has the strange effect of not specifying by name [the wildlife which is] protected. For instance, nowhere in the statutes of New Zealand will one find a provision which specifically protects the kiwi …' N. Wells, 'Protection of Wildlife in the Environment' *Auckland University Law Review*, vol. 4, 1980–83, p. 392. The kiwi is New Zealand's national bird.
9   Neville Peat, *Manapouri Saved! New Zealand's First Great Conservation Success Story*, Longacre Press, Dunedin, 1993, p. 1.
10 Peat, *Manapouri Saved!*, p. 33.
11 Alan F. Mark, 'Foreword' in Peat, *Manapouri Saved!*, p. vii.
12 *Royal Forest and Bird Protection Society of New Zealand Inc. v. Bay of Plenty Regional Water Board* (1978) 6 New Zealand Town Planning Appeals, p. 361. The New Zealand waterfowl affected were the grey and blue ducks and the black and brown teal. The association of all four in one habitat was described as nationally unique (p. 365).
13 T. Black, 'Defending the Environment', *New Zealand Law Journal*, 1978, p. 153.
14 Roger Wilson, *From Manapouri to Aramoana: The Battle for New Zealand's Environment*, Earthworks Press, Auckland, 1982, p. 20.
15 Town and Country Planning Act Review Committee, *Report to Government*, September 1973.
16 Wilson, *From Manapouri to Aramoana*, p. 63. For a detailed and critical description of events leading to the damming of the Clutha River, see Paul Powell, *Who Killed the Clutha?*, J. McIndoe, Dunedin, 1978.
17 See F.M. Brookfield, 'High Courts, High Dam, High Policy: The Clutha River and the Constitution', *Recent Law*, 1983, pp. 62–72.
18 See, for example, *CREEDNZ Inc. v. Governor-General* [1981] 1 New Zealand Law Reports, p. 172; *Environmental Defence Society v. South Pacific Aluminium* [1981] 1 New Zealand Law Reports, p. 146, *(No. 2)* [1981] 1 New Zealand Law Reports, p. 153, *(No. 3)* [1981] 1 New Zealand Law Reports, p. 216; *Re An Application by NZ Synthetic Fuels Corporation Ltd* [1981] 8 New Zealand Town Planning Appeals, p. 138.
19 Ton Bührs, 'Environmental Policy', in Raymond Miller (ed.), *New Zealand Politics in Transition*, Oxford University Press, Auckland, 1997, p. 289.
20 Owen Furuseth & Chris Cocklin, 'An Institutional Framework for Sustainable Resource Management: The New Zealand Model', *Natural Resources Journal*, vol. 35, 1995, p. 255.
21 Bührs & Bartlett, *Environmental Policy in New Zealand*, p. 94. On Labour's programmes of reform of environmental administration and local government, see Steve Britton, Richard Le Heron & Eric Pawson (eds), *Changing Places in New Zealand: A Geography of Restructuring*, New Zealand Geographical Society, Christchurch, 1992.
22 Furuseth & Cocklin, 'An Institutional Framework for Sustainable Resource Management', p. 254.
23 Mark N.H. Seabrook-Davison, Weihong, J. Ji & Dianne H. Brunton, 'Survey of New Zealand Department of Conservation Staff Involved in the Management and Recovery of Threatened Species', *Biological Conservation*, vol. 143, 2012, pp. 212–19; Kenneth F.D. Hughey, Geoffrey Kerr, Ross Cullen & Andrew J. Cook, *Perceptions of Conservation and the Department of Conservation: Interim Findings from the 2008 Environmental Perceptions Survey*, Lincoln University Environment Society & Design Division, Christchurch, 2008.
24 Furuseth & Cocklin, 'An Institutional Framework for Sustainable Resource Management', p. 255.
25 Pavan Sharma, 'A Productive Anomaly: New Zealand's Parliamentary Commissioner for the Environment', University of Tasmania, School of Law Research Paper, 2010/2011, pp. 13, 19.

26 Nicola Wheen & Jacinta Ruru, 'The Environmental Reports', in Janine Hayward & Nicola R. Wheen (eds), *The Waitangi Tribunal/Te Roopu Whakamana i te Tiriti o Waitangi*, Bridget Williams Books, Wellington, 2004, p. 98.
27 Waitangi Tribunal, *Ko Aotearoa Tenei: A Report into Claims Concerning New Zealand Law and Policy Affecting Māori Culture and Identity*, Wai 262, Legislation Direct, Wellington, 2011.
28 For example, *Te Runanga o Taumarere v. Northland Regional Council* [1996] New Zealand Resource Management Appeals, p. 77 and *TV3 Network Services Ltd* v. *Waikato District Council* [1997] New Zealand Resource Management Appeals, p. 539.
29 Nicola R. Wheen & Janine Hayward, 'The Meaning of Treaty Settlements and the Evolution of the Treaty Settlement Process', in Nicola R. Wheen & Janine Hayward (eds), *Treaty of Waitangi Settlements*, Bridget Williams Books, Wellington, 2012, p. 14.
30 Ministry for the Environment, *Introducing the Resource Management Bill*, Ministry for the Environment, Wellington, 1989, p. 1. For a description and assessment of the making of the Resource Management Act, see Rt Hon. Sir Geoffrey Palmer, 'Sustainability – New Zealand's Resource Management Legislation', in Monique Ross & J. Owen Saunders (eds), *Growing Demands on a Shrinking Heritage: Managing Resource-Use Conflicts, Essays from the Fifth Institute Conference on Natural Resources Law*, 1992, pp. 408–28. For a history of New Zealand water law, culminating in the Resource Management Act, see Nicola R. Wheen, 'A Natural Flow: A History of Water Law in New Zealand', *Otago Law Review*, vol. 9, 1997, pp. 71–110.
31 Furuseth & Cocklin, 'An Institutional Framework for Sustainable Resource Management', p. 259.
32 Ministry for the Environment, *Environment New Zealand 2007*, Ministry for the Environment, Wellington, 2007.
33 Hon. David Carter, Minister for Local Government, 'Environment Canterbury (Temporary Commissioners and Improved Water Management) Amendment Bill, First Reading', *New Zealand Parliamentary Debates*, vol. 684, 2012, p. 5301.
34 These are the words of Dr Nick Smith, Minister for the Environment, reported in Paul Gorman 'ECan Councillors Sacked', *The Press*, 30 March 2010.
35 Hon. David Carter Minister for Local Government, 'Environment Canterbury (Temporary Commissioners and Improved Water Management) Amendment Bill,' p. 5301.
36 Ministry for the Environment, 'Chapter 10: Freshwater', *Environment New Zealand 2007*, p. 304.
37 Office of the Auditor-General, Managing Freshwater Quality: Challenges for Regional Councils, Office of the Auditor-General, Wellington, 2011, at 3.72, 5.22, 5.3, 5.4 and 5.5.
38 Stephen Tromans, 'High Talk and Low Cunning: Putting Environmental Principles into Legal Practice', *Journal of Planning & Environmental Law*, vol. 2, 1995, p. 791.
39 The concept may be indirectly incorporated into other Acts: for example the purpose of the Climate Change Response Act 2002, which houses the emissions trading scheme for greenhouse gases, includes enabling New Zealand to meet its international obligations under the United Nations Framework Convention on Climate Change, 1992 and its Kyoto Protocol, 1998, both of which promote sustainable development.
40 *New Zealand Rail Ltd v. Marlborough District Council* [1994] New Zealand Resource Management Appeals, p. 86.
41 *North Shore City Council v. Auckland Regional Council* [1997] New Zealand Resource Management Appeals, p. 94.
42 Geoffrey Palmer, *Environment: The International Challenge*, Victoria University Press, Wellington, 1995, pp. 172–73.
43 See note 12 above and accompanying text.
44 *New Zealand Recreational Fishing Council Inc v. Sanford Ltd* [2009] New Zealand Supreme Court p. 54; [2009] 3 New Zealand Law Reports, p. 438, at 39.
45 *Canterbury Regional Council v. Selwyn District Council* [1996] Environmental Law Reports of New Zealand, vol. 2, p. 395 at 407; [1997] New Zealand Resource Management Appeals, p. 25 at 35. For a critical review on uncertainty and the Resource Management Act, see Nicola R. Wheen, 'The Resource Management Act 1991: A "Greener" Law for Water?' *New Zealand Journal of Environmental Law*, vol. 1, 1997, pp. 165–98; I.H. Williams, 'The Resource Management Act 1991 – Well Meant but Hardly Done', *Otago Law Review*, vol. 9, 2000, pp. 673–95.
46 D.F. Dugdale, 'Framing Statutes in an Age of Judicial Supremacism', *Otago Law Review*, vol. 9, 2000, p. 604.
47 Warwick Gullett, 'The Precautionary Principle in Australia: Policy, Law & Potential Precautionary EIAs', *Risk*, vol. 11, 2000, p. 95.
48 Alexander Gillespie, 'Precautionary New Zealand', *New Zealand Universities Law Review*, vol. 24, 2011, pp.

364–85. The Fisheries Act, s 10 in particular has been criticised. Dara Modeste, 'The Precautionary Principle and the Fisheries Act', *New Zealand Law Journal*, 2011, pp. 179–84.
49 Tromans, 'High Talk and Low Cunning', p. 780.
50 Bruce Pardy, 'Abstraction, Precedent, and Articulate Consistency: Making Environmental Decisions', *California Western Law Review*, vol. 34, 1998, p. 429. Bruce Pardy has written a lot about discretion and the failure to secure ecological bottom lines in environmental decision-making, see for example, 'In Search of the Holy Grail of Environmental Law: A Rule to Solve the Problem', *International Journal of Sustainable Development Law and Policy*, vol. 1, 2005, pp. 29–57.

17. Ngāi Tahu and the 'nature' of Māori modernity

1 Although this epigraph is drawn from O'Regan's critique of Te Rūnanga o Ngai Tahu's commercial investment strategies since 1998, it can arguably be extended out to include its approach to the environment over the same time. Paul Diamond, *A Fire in Your Belly: Māori Leaders Speak*, Huia Wellington, 2003, pp. 38–39.
2 Michael King, *Penguin History of New Zealand*, Penguin Books, Auckland, 2003, pp. 61–62, 77.
3 Jacinta Ruru, 'Indigenous Restitution in Settling Water Claims: The Developing Cultural and Commercial Redress Opportunities in Aotearoa, New Zealand', *Pacific Rim Law & Policy Journal*, vol. 22, no. 2, 2013, p. 314; Te Maire Tau & Atholl Anderson (eds), *Ngāi Tahu: A Migration History: The Carrington Text*, Bridget Williams Books, in association with Te Rūnanga o Ngāi Tahu (TRoNT), Wellington, 2008, p. 114.
4 Kāi Tahu Whānui, or Ngāi Tahu Whānui, most commonly known as Ngāi Tahu, is the iwi that holds mana whenua over Rakiura/Stewart Island and the majority of Te Wai Pounamu/the South Island of New Zealand: www.ngaitahu.iwi.nz/About-Ngai-Tahu
5 Angela Wanhalla, *In/visible Sight: The Mixed Descent Families of Southern New Zealand*, Bridget Williams Books, Wellington, 2009, pp. 4–9; Bill Dacker, *The People of the Place: Mahika Kai*, New Zealand 1990 Commission, Wellington, 1990.
6 Tony Ballantyne, 'Economic Systems, Colonization and the Production of Difference: Thinking Through Southern New Zealand', paper presented in the 'Paucity & Plenty' lecture series, Eisenberg Institute for Historical Studies, University of Michigan, Ann Arbor, 2 December 2010, p. 20.
7 Ann Parsonson, 'Ngāi Tahu – The Whale That Awoke: From Claim to Settlement (1960–1998)', in John Cookson & Graeme Dunstall (eds), *Southern Capital: Christchurch: Towards a City Biography 1850–2000*, Canterbury University Press, Christchurch, 2000, p. 259.
8 Te Maire Tau, 'Ngāi Tahu', in Jock Phillips (ed.), *Māori Peoples of New Zealand: Ngā Iwi o Aotearoa*, David Bateman, Auckland, 2006, pp. 129–31; Nicola Carrell, 'Innovation in Reconciliation – the Ngai Claims Settlement Act 1998', *New Zealand Journal of Environmental Law*, vol. 3, 1999, pp. 179–91; John Dawson, 'A Constitutional Property Settlement Between Ngai Tahu and the New Zealand Crown', in Janet McLean (ed.), *Property and the Constitution*, Hart Publishing, Portland, 1999, pp. 207–23; Michael Stevens, 'Settlements and "Taonga": A Ngāi Tahu Commentary', in Nicola R. Wheen & Janine Hayward (eds), *Treaty of Waitangi Settlements*, Bridget Williams Books, Wellington, 2012, 124–37; Parsonson, 'Ngāi Tahu', pp. 248–75.
9 Carrell, 'Innovation in Reconciliation', p. 179.
10 Liesl Johnstone, 'From the Outside', *Te Karaka*, no. 40, Kana/Spring 2008, pp. 33–37; Diamond, *A Fire in Your Belly*, p. 23.
11 Harry C. Evison, *The Long Dispute: Maori Land Rights and European Colonisation in Southern New Zealand*, Canterbury University Press, Christchurch, 1997, p. 335; Tipene O'Regan, 'Impact on Māori – a Ngāi Tahu perspective', in Jacinta Ruru (ed.), *In Good Faith: Symposium Proceedings Marking the 20th Anniversary of the Lands Case*, New Zealand Law Foundation, Wellington, 2008, pp. 48–49.
12 Matthew Littlewood, 'Mayors Agree with Critical Report, *Timaru Herald*, 20 February 2010; 'Govt Names ECan Commissioners', *Otago Daily Times*, 22 April 2010; Ministry for the Environment, 'Environment Canterbury Commissioners', www.mfe.govt.nz/rma/central/investigation/ecan/environment-canterbury.html
13 Environment Canterbury (Temporary Commissioners and Improved Water Management) Act 2010, Schedule 1, Part 1.
14 'Important Steps Taken in Ngāi Tahu, Environment Canterbury Partnership', 19 February 2013, http://ecan.govt.nz/news-and-notices/news/pages/ngai-tahu-environment-canterbury-partnership.aspx; Faumuinā Tafuna'i & Adrienne Rewi, 'Silver Linings', *Te Karaka*, no. 52, Raumati/Summer 2011, p. 50.
15 Howard Keene, 'Fresh Take', *Te Karaka*, no. 54, Makariri/Winter 2012, p. 16. Implicit in Solomon's comment, therefore, is a sense of guilt for declining freshwater mahinga kai not actually caused by her or Ngāi Tahu Whānui. This is a type of response shared by other indigenous people. In 2010, for instance, a delegation of

Winnemem Wintu from the McCloud River in California visited Ngāi Tahu. They performed ceremonial dances on the banks of the Rakaia River for chinook or quinalt salmon introduced to South Island rivers in the late 1800s from a hatchery on the McCloud River. Due to dam projects for irrigation and hydro-electricity, however, they are now endangered species in their native waters. The Winnemem Wintu, themselves severely affected by these forces of colonisation, nonetheless came to meet the South Island-based salmon to 'atone for our failure and protect the rivers so they could come home'. Ngāi Tahu who hosted the Winnemem Wintu saw parallels with long-finned eels which are endangered by dams on the Waitaki River: Ila Couch, 'Atonement', *Te Karaka*, no. 47, Makariri/Winter, 2010, pp. 14–19.
16 Mark Solomon, 'Ngai Tahu and Water', speech notes from Tikanga and Technology IV, Environmental Protection Agency, Huirapa Marae, Puketeraki, 24–26 October 2012, p. 10, at www.epa.govt.nz/te-hautu/wananga-and-hui/Pages/2012-13-Tikanga-and-Technology-IV.aspx; O'Regan, 'Impact on Māori', p. 42.
17 It is worth noting that an interim Deed of 'On-Account' Settlement entered into by Crown and Ngāi Tahu negotiators centred on two tribally significant water bodies. Although they did quite different things, the Ngāi Tahu (Pounamu Vesting Act) 1997 and the Ngāi Tahu (Tūtaepatu Lagoon) Vesting Act 1998 both sought to reconnect and strengthen the relationship between specific Ngāi Tahu communities and adjacent water-bodies: in one case the Arahura River on the South Island's West Coast, in the other case, a degraded but still ecologically and culturally significant lagoon close to Kaiapoi pā, north of Christchurch: Stevens, 'Settlements and "Taonga"', p. 125; Te Kōhaka o Tūhaitara Trust, 'About the Park', http://tuhaitarapark.org.nz/Tuhaitara_Park/About_the_Park.html
18 Te Rūnanga o Ngāi Tahu, 'Economic Security', http://www.ngaitahu.iwi.nz/About-Ngai-Tahu/Settlement/Settlement-Offer/Economic-Security.php
19 Diamond, *A Fire in Your Belly*, p. 28.
20 Waitangi Tribunal, 'Mahinga Kai Summary', *Ngai Tahu Land Report 1991*, 2.12, www.waitangi-tribunal.govt.nz/reports/viewchapter.asp?reportID=D5D84302-EB22-4A52-BE78-16AF39F71D91&chapter=23
21 Te Rūnanga o Ngāi Tahu, *Freshwater Policy*, Te Rūnanga o Ngāi Tahu, Christchurch, 1999, p. 8.
22 Te Rūnanga o Ngāi Tahu, *Ngāi Tahu 2025*, Te Rūnanga o Ngāi Tahu, Christchurch, 2001, pp. 9, 45.
23 *Ngāi Tahu 2025*, p. 11. *Ngāi Tahu 2025* describes State of the Takiwā reporting and monitoring as integrating 'Mātauranga Māori and Western science to gather information about the environment and to establish a baseline for the creation of policy and improvement of environmental health', as 'an alternative to conventional state of the environment reporting used by the Ministry for the Environment, that takes into account tangata whenua values': pp. 48–49; Te Rūnanga o Ngāi Tahu, 'Ki Uta ki Tai', www.ngaitahu.iwi.nz/Ngai-Tahu-Whanui/Natural-Environment/Environmental-Policy-Planning/Ki-Uta-Ki-Tai.php; also Halina Ogonowska-Coates & Felolini Maria Ifopo, 'In Hot Water', *Te Karaka*, no. 38, Kahuru/Autumn 2008, pp. 16–21.
24 Matthew Littlewood, 'Ngai Tahu Alarmed at Plans', *Timaru Herald*, 20 January 2010.
25 David Williams, 'Carter Opposes Cubicle-dairying', *The Press*, 26 January 2010.
26 Ibid.
27 David Williams, 'Govt "calls in" cubicle-farm proposals', *The Press*, 28 January 2010; 'Factory farm plan "called in" by Environment Minister', *New Zealand Herald*, 27 January 2010.
28 David Bruce, 'Dairying Plan Hits Snag', *Otago Daily Times*, 24 November 2011.
29 In 2010 the National Institute of Water and Atmospheric Research, a Crown Research Institute, rated Te Waihora at 6.9 on the trophic level index. By comparison, the North Island lakes Rotorua and Wairarapa that measure 80 km^2 and 77 km^2 respectively had trophic levels of 4.7 and 5.0: Minister for the Environment, 'Questions & Answers', 25 August 2011, p. 1, http://beehive.govt.nz/release/116-million-clean-plan-nz%E2%80%99s-most-polluted-lake; Tafuna'i & Rewi, *Silver Linings*, p. 48.
30 Jamie Morton, 'No Swimming: 52% Impure NZ Rivers', *New Zealand Herald*, 17 October 2012.
31 Quoted in Te Maire Tau, 'The Death of Knowledge – Ghosts on the Plains', *New Zealand Journal of History*, vol. 35, no. 2, 2001, p. 149.
32 S.G. Kitto, 'The Environmental History of Te Waihora – Lake Ellesmere', MSc thesis, University of Canterbury, 2010, pp. 25–26.
33 These three reports are the Ngai Tahu [Land] Report 1991, the Ngai Tahu Sea Fisheries Report 1992, and the Ngai Tahu Ancillary Claims Report 1995.
34 Around $250,000 per year since the Ngāi Tahu Settlement: Tafuna'i & Rewi, 'Silver Linings', p. 49. TRoNT planned to spend another $500,000 as part of the 2011 clean-up package: Minister for the Environment, 'Questions & Answers', p. 3.
35 Tafuna'i & Rewi, *Silver Linings*, p. 48. Nearby Te Waihora is Te Roto o Wairewa/Lake Forsyth, also an important source of mahinga kai that is likewise extremely degraded. It has a particular problem with

sedimentation and blue-green algae that produces a deadly cyanotoxin in summer. This poses a health hazard to humans and other mammals. It also limits the water oxygen needed for fish and shellfish to survive. However, Wairewa Rūnanga, another papatipu rūnanga of TRoNT, is part of an innovative and ambitious bioengineering project that seeks to restore the lake and provide new economic and amenity opportunities for Ngāi Tahu and the wider community: Adrienne Rewi, 'A New World', *Te Karaka*, no. 50, Makariri/Winter 2011, pp. 48–50.
36 Michael Wright, '$11.6m Not Enough To Clean Up NZ's Most Polluted Lake', *The Press*, 22 March 2012.
37 Central Plains Water, 'Technical Elements of the Scheme', www.cpwl.co.nz/technical/technical-elements.html; Ministry for Primary Industries, 'Government Funding for Central Plains Water Irrigation', www.mpi.govt.nz/news-resources/news/government-funding-for-central-plains-water-irriga
38 Central Plains Water, 'A New Irrigation Scheme For Central Canterbury – Do We Need One?' www.cpwl.co.nz/home/irrigation.html; Central Plains Water, 'Facts and Fictions', www.cpwl.co.nz/home/facts-fictions.html
39 Faumuinā Tafuna'i, 'New Pastures', *Te Karaka*, no. 49, Raumati/Summer 2010, pp. 34–36; *Te Manu Korihi News*, 'Ngai Tahu Defends Dairy Farming Plans', Radio New Zealand, 10 November 2011; Keene, 'Fresh Take', p. 19.
40 Maki Akiyama, 'Contemporary Kaitiakitanga: Case Study of the Hurunui River Catchment', Geography 420 Final Report, University of Canterbury, p. 23; David Williams, 'Ngai Tahu Withdraws from Irrigation Scheme', *The Press*, 23 October 2009.
41 Tony Chaston, 'Big Dairy Plans for Ngāi Tahu', *interest.co.nz*, 6 January 2011.
42 Mark Solomon, 'Leadership – Recovery and Rebuild', press release, 23 April 2013.
43 Alan Wood, 'Ngai Tahu Trials Dairy Conversions', *The Press*, 25 March 2013.
44 Greg Byrnes, 'Culture in Conflict', 22 April 2013; Chris Hutching, 'New Ngai Tahu Board Reflects Internal Wrangle', *National Business Review*, 28 October 2009.
45 Tafuna'i, 'New Pastures', pp. 34–35.
46 Ibid., pp. 35–36.
47 See regulations 17–29 (especially regulation 25) of the Fisheries (South Island Customary Fishing) Regulations 1999.
48 Tiny Metzger, pers. comm., April 2012.
49 Evison, *The Long Dispute*, p. 242.
50 Michael J. Stevens, 'Kāi Tahu me te Hopu Tītī ki Rakiura: An Exception to the Colonial Rule?', *Journal of Pacific History*, vol. 41, no. 3, 2006, pp. 273–91; Stevens, 'Settlements and "Taonga"', pp. 133–36.
51 Rob Tipa, 'Something in the Water', *Te Karaka*, no. 51, Kana/Spring 2011, pp. 29–32.
52 Alex Fensome, 'Estuaries in Decline, Study Finds, *Southland Times*, 10 November 2012.
53 Alex Fensome, 'Race on to Save Estuaries', *Southland Times*, 11 November 2012.
54 Environment Southland, 'Waituna: What's All the Fuss About?', p. 2, www.es.govt.nz/media/11944/waituna-whats-all-the-fuss-about.pdf
55 Rob Tipa, 'Last Chance', *Te Karaka* no. 53, Kahuru/Autumn 2012, p. 29.
56 Tipa, 'Last Chance'. The nearby Mataura catchment is another severely modified and degraded water system, which also negatively impacts upon a number of mahinga kai, especially kanakana (lamprey). Hokonui Rūnanga, another papatipu rūnanga of TRoNT, successfully applied for New Zealand's first freshwater mātaitai for the Mataura River (this covers 10 km of the river's 240 km). Rūnanga members have undertaken environmental monitoring and restoration activities in relation to this, including riparian planting: Ogonowska-Coates & Ifopo, 'In Hot Water', pp. 18–21. In late 2011, kanakana from the Mataura River and some other Southland rivers besides, were marked with hitherto unobserved red contusions on their bodies. It transpired that this was the result of a kind of bacterium that until then had never been detected in New Zealand: Environment Southland, 'Unusual Illness Affecting Southland Fish', *Envirosouth: Environment Southland News*, December 2011, p. 12.
57 Tau, 'Death of Knowledge', p. 149.
58 Editorial, 'Our Degraded Estuaries', *Southland Times*, 12 January 2013.
59 Tipa, 'Something in the Water', p. 31.
60 'Meat Firm Fined for Polluting Water', *Southland Times*, 18 April 2013.
61 Tahu Potiki, 'Ngai Tahu's New Boss Has Strong Whanau Focus', *The Press*, 24 August 2012.
62 Arihia Bennett, Chief Executive Officer, Te Rūnanga o Ngāi Tahu, *Te Karaka*, no. 57, Kahuru/Autumn 2013, p. 3.
63 *Te Manu Korihi News*, 'Ngai Tahu Defends Dairy Farming Plans'.

64 In 2005, 42 per cent of registered Ngāi Tahu lived within the tribal area compared to 45 per cent outside of it, and 7.5 per cent overseas: Stevens, 'Kāi Tahu me te Hopu Tītī ki Rakiura, p. 291.
65 Parsonson, 'Ngāi Tahu – The Whale That Awoke', p. 257.
66 www.whalewatch.co.nz
67 Tipene O'Regan, Guest Lecture, School of Business, University of Otago, 2002.
68 Te Maire Tau, 'Preface', in Tau & Anderson, *Ngāi Tahu: A Migration History*, p. 9.
69 Antoinette Burton, 'Introduction: The Unfinished Business of Colonial Modernities', in Antoinette Burton (ed.), *Gender, Sexuality and Colonial Modernities*, Routledge, London, 1999, p. 1.

18. Mastering the land: mapping and metrologies in Aotearoa New Zealand

1 For an analysis of the theoretical aspects of such notions of map making see John Pickles, *A History of Spaces. Cartographic Reason, Mapping and the Geo-coded World*, Routledge, London, 2004.
2 Atholl Anderson, 'A Fragile Plenty: Pre-European Maori and the New Zealand Environment', in Eric Pawson & Tom Brooking (eds), *Environmental Histories of New Zealand*, Oxford University Press, Melbourne, 2002, pp. 30–31.
3 Anderson, 'A Fragile Plenty', p. 33.
4 Evelyn Stokes, 'Contesting Resources: Māori, Pākehā, and a Tenurial Revolution', in Pawson and Brooking (eds), *Environmental Histories*, p. 36.
5 Stokes, 'Contesting Resources', p. 36.
6 Ibid., p. 35.
7 Jan Kelly, 'Maori Maps', *Cartographica*, vol. 36, no. 2, 1999, pp. 12–13.
8 Danny Keenan, 'Bound to the Land: Māori Retention and Assertion of Land and Identity', in in Pawson and Brooking (eds), *Environmental Histories*, p. 250.
9 Kelly, 'Maori Maps', p. 15.
10 Mere Roberts, 'Mind Maps of the Maori', *GeoJournal*, vol. 77, no. 6, 2012, pp 741–51.
11 While an unknown southern continent had been hypothesised among European scholars and included on world maps since antiquity, the first known map to show a coast at approximately the location of New Zealand is the Dieppe map drawn by Jean Rotz and presented to King Henry VIII of England c. 1535–42. Numerous later maps of the same tradition repeat and elaborate on the coastline in the southern ocean drawn by Rotz.
12 William Grent, *A New and Accurate Map of the World Drawne According to the Truest Descriptions, Latest Discoveries, and Best Observations, that Have Been Made by English, or Strangers*, London, 1625.
13 Jeppe Strandsbjerg, 'The Cartographic Reality of Space. Territory, Globalisation and International Relations', PhD thesis, University of Sussex, 2008.
14 Roger J.P. Kain & Elizabeth Baigent, *The Cadastral Map in the Service of the State: A History of Property Mapping*, University of Chicago Press, Chicago, 1992, pp. 318–25.
15 John Brian Harley, 'Deconstructing the Map', *Cartographica*, vol. 26, no. 2, 1989, pp. 1–20.
16 Brian Marshall, 'From Sextants to Satellites: A Cartographic Timeline for New Zealand', *New Zealand Map Society Journal*, no. 18, 2005, p. 104.
17 Phillip Lionel Barton, 'Maori Cartography and the European Encounter', in David Woodward & G. Malcolm Lewis (eds), *The History of Cartography*, vol. 2, book 3, University of Chicago Press, Chicago, 1998, pp. 493–532.
18 Kelly, 'Maori Maps', pp. 4, 11.
19 Bruno Latour, *Politics of Nature: How to Bring the Sciences into Democracy*, Harvard University Press, Cambridge, MA and London, 2004, pp. 15–18.
20 Latour, *Politics of Nature*, p. 10.
21 Kelly, 'Maori Maps', p. 15.
22 Stokes, 'Contesting Resources'; Keenan, 'Bound to the Land'.
23 Jim McAloon, 'Resource Frontiers, Environment and Settler Capitalism 1769–1860', in Pawson and Brooking (eds), *Environmental Histories*, p. 52.
24 Cited in Marshall, 'From Sextants to Satellites', p. 36.
25 Jeremy Black, *War in the Modern World Since 1815*, Routledge, London, 2003, p. 16.
26 Kelly, 'Maori Maps', pp. 1–2.
27 Geoff Conly, *Piets Eye in the Sky. The Story of New Zealand Aerial Mapping Ltd.*, Grantham House, Wellington, 1986, p. 71.

28 Conly, *Piets Eye*, p. 81.
29 Piet van Asch, in *Whites Aviation Magazine*, vol. 1, 1945, cited in Conly, *Piets Eye*, p. 95.
30 Conly, *Piets Eye*, p. 97.
31 Marshall, 'From Sextants to Satellites', p. 97.
32 Rafael Capurro & Birger Hjørland, 'The Concept of Information', *Annual Review of Information Science and Technology*, vol. 37, no. 1, 2003, pp. 343–411.
33 Chris Perkins, 'Cartography: Mapping Theory', *Progress in Human Geography*, vol. 27, no. 3, 2003, pp. 341–51.
34 Adam Smith, *An Inquiry into the Nature and Causes of the Wealth of Nations: A Selected Edition*, Oxford Paperbacks, Oxford, 2008 (first published 1776).
35 Land Information New Zealand Toitu Te Whenua, *Statement of Intent 2012–2015*, New Zealand Government, Wellington, 2012, p. 4.
36 McAloon, 'Resource Frontiers', p. 59.
37 Peter Swann, *The Economics of Metrology and Measurement*, Report for the National Measurement Office, Department of Business, Innovation and Skills, Wellington, 2009, pp. i–ix.

19. Epilogue

1   We are grateful to Peter Holland, Harvey Perkins and Vaughan Wood for discussions during the preparation of this Epilogue, although we alone are responsible for the final result.
2   C.P. Snow, *The Two Cultures*, Cambridge University Press, Cambridge, 1993. In this edition, Stefan Collini, in the 'Introduction', argues that the nature of academic disciplines and changes since the late 1950s 'have made any binary division into *two* cultures look more implausible than ever' (p. lv). Whether this makes it any easier for conversations to occur between those in the humanities and those in sciences is another matter.
3   Tom Griffiths, 'We Have Still Not Lived Long Enough', *Inside Story*, 16 February 2009, http://inside.org.au/we-have-still-not-lived-long-enough/; Eric Pawson, 'Environmental Hazards and Natural Disasters', *New Zealand Geographer*, vol. 67, no. 3, 2011, pp. 143–47.
4   A useful collection of essays that engages with this theme is Kimberly Coulter & Christof Mauch (eds), *The Future of Environmental History: Needs and Opportunities*, Rachel Carson Center Perspectives, issue 3, Munich, 2011.
5   Sverker Sörlin & Paul Warde, 'The Problem of the Problem of Environmental History: A Re-Reading of the Field', *Environmental History*, vol. 12, no. 1, 2007, p. 118.
6   B.B. Fitzharris, 'The 1992 Electricity Crisis and the Role of Climate and Hydrology', *New Zealand Geographer*, vol. 48, no. 2, 1992, pp. 79–83; Peter Holland & Bill Mooney, 'Wind and Water: Environmental Learning in Early Colonial New Zealand', *New Zealand Geographer*, vol. 62, no. 1, 2006, pp. 39–49.
7   E.T.J. Durie, *New Approaches to Maori Land in the 1980's with Particular Reference to Its Settlement and Resettlement in the Northern Half of the North Island*, Auckland Branch, New Zealand Geographical Society, Auckland, 1981; and Stephen Dovers, *Environmental History and Policy: Still Settling Australia*, Oxford University Press, Melbourne, 2000.
8   Land and Water Forum, *Third Report of the Land and Water Forum: Managing Water Quality and Allocating Water*, Land and Water Trust, 2012, www.landandwater.org.nz.
9   Paul James & Eric Pawson, 'Contested Places: The Significance of the Motunui-Waitara Claim to the Waitangi Tribunal', *Aboriginal History*, vol. 19, 1995, 111–25; Nicola R. Wheen & Jacinta Ruru, 'The Environmental Reports', in Janine Hayward & Nicola R. Wheen (eds), *The Waitangi Tribunal/Te Roopu Whakamana i te Tiriti o Waitangi*, Bridget Williams Books, Wellington, 2004, pp. 97–112.
10 Ministry for the Environment, 'Recreational Water Quality in New Zealand', *Indicator Update*, INFO 653, 2012.
11 Eric Pawson, 'Kenneth Cumberland 1913–2011', in Hayden Lorimer & Charles W.J. Withers (eds), *Geographers Biobibliographical Studies*, vol. 31, Commission on the History of Geography of the International Geographical Union and the International Union of the History and Philosophy of Science, Continuum International Publishing Group, London, 2012, pp. 137–60.
12 Marianne Elisabeth Lien & John Law, '"Emergent Aliens": On Salmon, Nature, and Their Enactment', *Ethnos, Journal of Anthropology*, vol. 76, no. 1, 2011, p. 83.
13 William Cronon, 'The Trouble with Wilderness; or, Getting Back to the Wrong Nature', in William Cronon (ed.), *Uncommon Ground: Toward Reinventing Nature*, W.W. Norton, New York, 1995, pp. 69–90; Mick Abbott & Richard Reeve (eds), *Wild Heart: The Possibility of Wilderness in Aotearoa New Zealand*, Otago University Press, Dunedin, 2011.

14 The question is posed in William Cronon, 'The Uses of Environmental History', *Environmental History Review*, vol. 17, no. 3, 1993, pp. 1–22; see also Christof Mauch, 'The Magic of Environmental History and Hopes for the Future', in Coulter & Mauch, *The Future of Environmental History*, pp. 60–63.
15 Rachael Selby, Pātaka Moore & Malcolm Mulholland, *Māori and the Environment: Kaitiaki*, Huia Publishers, Wellington, 2010; Mairi Jay, 'Remnants of the Waikato: Native Forest Survival in a Production Landscape', *New Zealand Geographer*, vol. 61, no. 1, 2005, pp. 14–28; Mairi Jay, 'The Political Economy of a Productivist Agriculture: New Zealand Dairy Discourses', *Food Policy*, vol. 32, no. 2, 2007, pp. 266–79; Grace Karskens, 'Water Dreams, Earthen Histories: Exploring Urban Environmental History at the Penrith Lakes Scheme and Castlereagh, Sydney', *Environment and History*, vol. 13, no. 2, 2007, pp. 115–54.
16 J.K. Gibson-Graham & Gerda Roelvink, 'An Economic Ethics for the Anthropocene', *Antipode*, vol. 41, no. S1, 2009, pp. 320–46; Mauch, 'The Magic of Environmental History'.
17 Franklin Ginn, 'Colonial Transformations: Nature, Progress and Science in the Christchurch Botanic Gardens', *New Zealand Geographer*, vol. 65, no. 1, 2009, pp. 35–47.
18 Rodney Grapes, *Magnitude Eight Plus: New Zealand's Biggest Earthquake*, Victoria University Press, Wellington, 2000.
19 Blair Fitzharris, 'How Vulnerable is New Zealand to the Impacts of Climate Change?', *New Zealand Geographer*, vol. 63, no. 3, 2007, pp. 160–68.
20 Patrick Maxwell, 'Extracurricular University Matters', *Cambridge Alumni Magazine*, no. 67, 2012, p. 41.
21 Donald Worster, 'Appendix: Doing Environmental History', in Donald Worster (ed.), *The Ends of the Earth. Perspectives on Modern Environmental History*, Cambridge University Press, Cambridge, 1988, p. 294; Eric Pawson & Stephen Dovers, 'Environmental History and the Challenge of Interdisciplinarity: An Antipodean Perspective', *Environment and History*, vol. 9, no. 1, 2003, 53–75.
22 David N. Livingstone, 'Science, Text and Space: Thoughts on the Geography of Reading', *Transactions of the Institute of British Geographers*, vol. 30, no. 4, 2005, pp. 391–401.
23 Charles S. Elton, *The Ecology of Invasions by Animals and Plants*, Methuen, London, 1958; Libby Robin, 'History for Global Anxiety', in Coulter & Mauch, *The Future of Environmental History*, pp. 41–44.
24 Ministry for the Environment, *The State of New Zealand's Environment*, GP Publications, Wellington, 1997, Part 10, p. 6. See also Part 9, 'The State of Our Biodiversity'; Jay, 'Remnants of the Waikato'.
25 Frank Uekoetter, *Consigning Environmentalism to History? Remarks on the Place of the Environmental Movement in Modern History*, Rachel Carson Center Perspectives 2011/7, Ludwig-Maximilians-Universität München/ Deutsches Museum, Munich, 2011, p. 4.
26 Worster, 'Appendix', p. 306.
27 Sörlin & Warde, 'The Problem of the Problem', make a powerful case for a closer engagement with social and political theory.
28 Vaughan Wood, 'Soil Fertility Management in Nineteenth Century New Zealand', PhD thesis, University of Otago, 2003; Vaughan Wood, Tom Brooking & Peter Perry, 'Pastoralism and Politics: Reinterpreting Contests for Territory in Auckland Province, New Zealand, 1853–1864', *Journal of Historical Geography*, vol. 34, no. 2, 2008, pp. 220–41.
29 J. Donald Hughes, *What is Environmental History?*, Polity Press, Cambridge, 2006, p. 1.
30 Mauch, 'The Magic of Environmental History', p. 60.
31 See for example Sörlin & Warde, 'The Problem of the Problem'.
32 Snow, *The Two Cultures*, p. 50.

Glossary of Māori terms

Michael J. Stevens

ahi kā literally 'keeping the fires burning on the land'; a Māori system of claiming rights to land through sustained occupation
arohatanga 'aroha' means love or compassion, and 'arohatanga' suggests relationships based on love and empathy
atua Māori deities
hau wind; or the vitality of man; vital essence of the land
hapū group of extended families or clan
harakeke (*Phormium tenax*) New Zealand flax; actually a type of lily
heke migration
Hineahuone mythological first woman created out of the earth
huia (*Heteralocha acutirostris*) native wattlebird with hooked beak, much prized by Māori for its feathers; it became extinct by the early twentieth century because of destruction of its forest habitat, and settler hunting for sale to collectors
iwi tribal federation of hapū
kahikatea (*Dacrycarpus dacrydioides*) tall, soft-wooded native tree of the podocarp family, which dominated the wetlands but was heavily milled to make butter-boxes, among other things
kāinga unfortified village
kaitiakitanga stewardship
kaiwhakahaere chairperson; director
kākā (*Nestor meridionalis*) large parrot with red, brown and green plumage

karaka (*Corynocarpus laevigatus*) native tree with broad green leaves; the fruit is edible once it has been treated for toxicity
karakia prayer or incantation
kānuka (*Kunzea ericoides*) hardy, second-growth native plant that grows to 10 metres tall
kaumātua adult, generally elderly, man or woman
kaupapa principle; philosophy
kawanatanga governance
kiore (*Rattus exulans*) small Pacific rat
kauri (*Agathis australis*) a native conifer tree, among the largest in the world, that produces excellent timber suitable for house- and boat-building, and resin that can be used in industrial processes such as varnish manufacture
kererū (*Hemiphaga novaeseelandiae*) native woodpigeon
kia kaha! be strong!
Kīngitanga Māori King Movement
kōkako (*Callaeas cinerea wilsoni*) large green flightless parrot, now highly endangered
kōura (*Paranephrops planifrons* and *Paranephrops zealandicus*) native freshwater crayfish
kōwhai (*Sophora* spp.) native tree famous for its yellow flower
kūmara (*Ipomoea batatas*) sweet potato brought to New Zealand from Polynesia, originally from Latin America

kurī native dog brought by Māori from Polynesia

māhoe (*Melicytus ramiflorus*) small native tree

mahinga kai traditional resource-gathering area, food source

mana prestige, status, authority

mana whenua right of an iwi, hapū or whānau to claim land through genealogy, occupation and use, or conquest

manaakitanga generosity, caregiving, compassion

mangā (*Thyrsites atun*; makā in Southern dialect) barracouta

mānuka (*Leptospermum scoparium*) shrub with small leaves and usually white flowers, though coloured forms are widely cultivated

māpou (*Myrsine australis*) small native tree sometimes cultivated for its reddish stems and foliage

marae courtyard area in front of wharenui; frequently applied to the complex of buildings surrounding a marae

matai (*Prumnopitys taxifolia*; black pine) native tree of the podocarp family

mātaitai a form of reserve designed to give effect to non-commercial Māori fishing rights

matuku (*Botaurus poiciloptilus*; Australasian bittern) bittern, a swamp bird

mauri life force or vital essence of an individual, species or place

miro (*Prumnopitys ferruginea*) a native tree of the podocarp family

moa (*Dinornithiformes* spp.) native ratites (flightless birds), now extinct, which ranged from small to even larger than modern-day ostrich and emu

Moriori separate tribe of Māori who lived on the Chatham Islands from the fifteenth century; not an earlier non-Polynesian stratum of people in New Zealand as imagined by the early twentieth-century ethnographers S. Percy Smith and Elsdon Best

Murihiku southern New Zealand

Muriwhenua the Far North of New Zealand with which Te Rārawa, Te Aupōuri, Ngāti Kahu, Ngāti Kurī and Ngāti Takoto are associated

Ngāti Toa an iwi grouping from the Kawhia area which migrated south to Kapiti Island and Otaki in the 1820s under the leadership of the chief Te Rauparaha; they also migrated to the top of the South Island in the early 1830s and clashed with the major South Island tribe of Ngāi Tahu as far south as Canterbury

nīkau (*Rhopalostylis sapida*) native palm

paepae ground-lying horizontal beam that separates a wharenui from a marae; the area of a marae where speeches are made

Pai Mārire 'good and peaceful' religion founded in the early 1860s by Te Ua Haumene in Taranaki and revived by Princess Te Puea of Tainui and the Kīngitanga in the 1920s

Pākehā Non-Māori New Zealanders of mainly British descent

Papatūānuku Earth Mother

pāua (*Haliotis* spp.) black-lipped, iridescent-green, univalve shellfish

pōhā a receptacle made out of bull kelp used to store and transport preserved food

pōhuehue (*Muehlenbeckia* spp.) native groundcover and climbing plant that grows in coastal areas

pōhutukawa (*Metrosideros excelsa*) native tree that grows in coastal areas, famous for its crimson flowers that blossom around Christmas time

rāhui prohibition on harvesting of food supplies, normally enforced in the interests of resource management and conservation

Rakiura Stewart Island

rangatiratanga full chiefly authority

Ranginui Sky Father

raupō (*Typha orientalis*) native bulrush used for construction of dwellings and as food

raupatu conquer; confiscate

rimu (*Dacrydium cupressinum*; red pine) native podocarp that produces top-quality timber

rimurapa (*Durvillaea antarctica*) bull kelp

rohe territorial area claimed by an iwi, hapū or whānau, including all resource rights

Rūaumoko deity of earthquakes and volcanoes

rūnanga, papatipu rūnanga tribal council, local level council

ruru (*Ninox novaeseelandiae*; morepork) native owl

tai ('*Ki uta, ki tai*') sea; coast

taonga treasures, including fisheries, forests and properties

taonga tuku iho something valued handed down; heirloom

Tāne Mahuta deity of the forest

Tangaroa deity of the sea

tangata whenua people of the land; those who hold mana whenua

taniwha water monster, often a guardian; from the Pacific tiger shark, *Galeocerdo cuvier*

tapu set apart, restricted, sacred

taro (*Colocasia* spp.) edible root plant brought from Polynesia and grown in the northern parts of the North Island

tawa (*Beilschmiedia tawa*) native tree

tāwhara edible flower bracts of kiekie (*Freycinetia baueriana* spp. *banksii*), a native palm much prized by Māori

Tāwhirimatea deity of wind and atmospheric events

Te Ao the world of light

Te Hiku o Te Ika the tail of the fish of Māui; Northland

Te Ika a Māui the captive fish of the demigod Māui; the North Island

Te Mauru/Te Hau Kai Tangata nor'west wind; or the wind that devours humankind

Te Pō primeval darkness before creation

Te Wai Pounamu the South Island

tī (*Cordyline australis*) native cabbage tree; an imported variant (*Cordyline terminalis*) was also cultivated

tītī (*Puffinus griseus*; muttonbird; sooty shearwater) member of petrel family of seabirds

toatoa (*Phyllocladus toatoa*), native conifer that grows in the northern half of the North Island

toetoe (*Cortaderia* spp.) robust New Zealand tussock grass, closely related to the pampas grasses of South America

tōtara (*Podocarpus totara* and *P. hallii*) tall native podocarp trees; the wood is valued for carving, and was once used for fenceposts

tuatara (*Sphenodon punctatus*) ancient and long-lived native reptile related to the dinosaurs through the order Rhynchocephalia

tūī (*Prosthemadera novaeseelandiae*); also called parson bird because of the white feathers growing at its throat

uta inland

utu reciprocity; redressing an imbalance in relationships to maintain harmony; more complex than simple 'revenge'

waimāori freshwater (as opposed to waitai – seawater)

Waitaha early South Island grouping of people

weka (*Gallirallus* spp.) small flightless native rail with brown and green plumage

whakapapa genealogy

whānau extended family

whānaungatanga the primacy of kinship bonds in determining action and in settling rights and status

whānui broad; wide

Index

Page numbers in **bold** refer to illustrations.

acclimatisation societies 102, 245, 248, 250
Acland, John Barton 82, 95
aerial topdressing 193, 195, 196, 197, 198, 200, 204, 221, **222**, 225
afforestation 206, 211, 214–15, 216, 221–22, 224, 225
Agrichemical Users' Code of Practice 205
Agricultural and Pastoral (A&P) associations 94
agricultural chemicals 205–06; *see also* fertilisers; herbicides; pesticides
agricultural science 93, 94, 104, 194, 195, 197, 200, 202–03, 204, 205, 207
agricultural societies and shows 94
agriculture: 'best use' of land 210; colonial 79, 92, 93; and conservation movement 149, 153; deforestation 19, 21, **21**, 105, 122, 127, 131–34, **132**, 135, 138, 153, 205, 209, 210, 218, 334, **334**; impacts on land 104–05; intensive farming 99, 100–01, 206, 207, 219, 289, 290, 298–99, 306–07, 329; knowledge dissemination 93–94, 98, 99; land reclamation 19, 79, 96–97, 99, 104, 174, 176, 185, **186**, 186–87, 188, 189 (*see also* deforestation); loss of farming land to urbanisation 237; Māori 76–77, 302–04, 306–08; mining conflicting with 115–18; organic farming 205, 206–07; pastoralism 80–82, 84, 89–90, 92, 93, 94, 95, 97, 99–100; run-off 22, 206, 290, 301, 329; small farms 94, 95; subsidies 194, 198, 221, 223, 224; waste disposal 22, 206, 298–99; *see also* cattle; dairy farming and industry; meat industry; pasture; sheep; wool
Ah Chee (Chan Dah Chee) 253, 255, 256
Ah Sam **253**
air pollution 233–24, 274, 282
aluminium smelting 279, 283, **284**, 285
Andersen, Johannes 154, 155

animals: economic use 72, 74, 75–77; extinctions 35, 36, 44–46; introduced 54, 71, 72, 73, 84, 89, 123; Māori names 41; protection by legislation 278; species, Māori settlement period 42; *see also* pests, introduced
Animals Protection Acts 144–45, 155, 188
Aoraki/Mt Cook 146, 158, 163, 166–67, 170, 171, 172, **331**; *see also* Hermitage, Mt Cook
archaeological data 35, 38, 39, 40, 43, 44, 45, 46, 160, 177
Arnold, Rollo 23, 264
Arthur's Pass 162, 163
Arthur's Pass National Park 146–47, 166, 171, 172
Auckland 47, 64, 79, 94, 120, 128, 129, 146, 147, 171, 226, 228, 229–30, **230**, 231, 234, **234**, 235, 237, 252
Auckland Islands 40, 152
Australia: agricultural science 203; British settlement 72; colonial trade 57, 58, 59, 60, 70, 72–74, **73**, 75, 76, 78, 84, 128; gold mining 106, 107–08, 109, 117, 121; influence on New Zealand pastoralism 81, 82, 92, 93, 99; plants from 244, 245, 248, 249, 250, 251
avifauna *see* birds
Banks, Joseph 71, 72, 75, 174, 175–76, 177, 183, 244
Banks Peninsula 45, 47, 91, 98, 128, 133
Barnes, Felicity 23
Barnicoat, John 160
Barton, John Saxon 271
Bass, George 74
Baughan, Blanche 141, 162, 166
Bay of Islands 54, 55, 56, 57, 58, 59, 61, 62, 64, 76, 78, 94
Bay of Plenty 47, 59, 62, 66, 174, 179, 281
Beattie, Herries 300, 306
beech, southern (*Nothofagus* sp.) 47, 125–26, 215, 282

381

Belich, James 22, 269
Bell, Sir Francis 211
Bennett, Arihia 307
Binney, Judith 56
biodiversity loss 333
birds: conservation 144–45, 147, 149, 154, 155; extinctions 35, 36, 44–46, 50, 311; introduced 71, 144, 150; pest birds 102; species 42, 48, 144–45; wetland habitat 175, 183, 184–85, 187, 188; *see also* moa
Blenheim 231
Blomfield, Charles 156, **157**
botanical gardens 154, 245
bracken fern *(Pteridium esculentum)* 43, 46, 47, 48, 50, 54, 79, 311
Bragge, James **18**, 18–19
Brown, William 79
Brownlee, Gerry 271
Brunner, Thomas 81, 161
Buller, James 60
burning: accidental fires 133; attitudes to 19; of forests and bush 19, 20, **21**, 35–36, 39–40, 44, 46–48, 61, 91, 132–33, 311, 334, **334**; for land reclamation 19, 79, 96–97, 99; sensory engagement of settlers 264; tussock grassland 46, 47, 91, 92, 96, 97, 99, 218
Busby, James 60, 61
Butler, Samuel 82, 93

Californian gold mining 106, 109–10, 117, 121
Callaghan, F.R. 194
Campbell, Douglas 193, 195, 196, 197, 221, 222
Campbell Island 152
Campbell, John Logan 79
Campbell, Lachlan Bain 271
Campbell, Robert 82
Canterbury 82–83, **83**, 161, 166, 171, 227; agriculture 82, 92, 93, 95, 101, 196, 197, 290, 301–02, 329; deforestation 91, 133; earthquakes 261, **264**, 268, 269–75, **273**, 328, 332; exploration 161; forests and forest management 42, 134; freshwater management 295; government intervention in resource management 289; *Hills and Plains, Waikari 1956* (Sutton) **267**; moa 45; nor'west wind 265–69, **268**; pasture 97, 98; roads 165; sheep farming 81; soils and soil erosion 47, 91, 205, 219, 221; tussock burning 97; *see also* Ngāi Tahu; and individual place names, e.g. Christchurch
Canterbury Water Management Strategy (CWMS) 295, 296
capitalism: expansionist policies 17, 70, 82, 84; integration of Māori 72; moral economy 76–78, 81
Carson, Rachel 205

catchment boards 220–21, 223, 330
Catholic mission 56, **57**
cats 71
cattle 80–81, 99, 185, 198
Caughley, Graeme 47
census, Bay of Islands–Hokianga 60, 61
Chan Dah Chee (Ah Chee) 253, 255, 256
Chapman, Valentine 217
Chateau Tongariro 165
Chatham Islands 40, 58, 63, 150
Cherry, Neil 267
Chevalier, Nicholas 161
Chew Chong 255–56
Chinese: European colonist interactions 254; gold miners 251–52, **252**; horticulture 251–56, **253**, 257; plant introductions **242**, 243–45, 246, 247, 248–49, 250–52, 254–55, 256–57
Christchurch 91, 94, 120, 137, 146, 147, 152, 229, 230, 235, 237, 238, 247–48; air pollution 233–34, 282; colonial settlement 273–74; earthquakes 231, 268, 269–75, **273**, 302, 328, 332; ethnic groups 275; flooding 232, **233**
Church Missionary Society (CMS) 56, **57**, 59
Clark, Andrew Hill 20
Clarke, George 183
Clarke, Jack 166–67
Clifford, Charles 92
climate 78, 91–92; impact on Māori settlement of New Zealand 41–42; windward–leeward cline 42, **43**, 45, 47, 48, 49, **50**
climate change 35, 46, 48, 328, 332; and deforestation 136, 143; emissions trading scheme 288
Clutha Development (Clyde Dam Empowering) Act 1982 284–85
Clutha River 110, 114, 224, 282–85, **283**
coal mining 81, 106, 118–19, **119**
Cockayne, A.H. (Alfred) 98, 99, 193, 198, 211
Cockayne, Leonard 150–52, 154, 166, 169, 171, 217
Colenso, William 80
colonisation: British 78–84, 92; 'ecological imperialism' 175–79, 242–43, 332; environmental appropriation 155–56; European colonists' perceptions 18–21, 23–24, 54, 92–93, 136–37; global webs and networks 23; imperial setting for forest and soil conservation 210; by Māori 38–41; metaphor of a garden 246; sensory engagement with the environment 264; *see also* Māori, and Pākehā colonisation; postcolonialism; settler economy
Commission for the Environment 281
Connell, R.P. 220
conservation: ecological motives 150–52; economic motives 143; environmental appropriation 155–56; imperial setting 210; and increasing scarcity of species 149–50; management

after 1914 152–54; Māori ethic 277; and nationhood 154–55; scenery preservation 145–49; state conservation 206, 209, 224–25, 278; wetlands 179, 187; *see also* environmental law; resource management; soil conservation; water availability and conservation; and also under individual subjects, e.g. forests – conservation
Conservation Act 1987 285
conservation estate 27, 172–73, 277, 285, 287, 288, 316
Constitution Act 1952 (UK) 64
Cook, James 59, 70–71, 72, 84, 176, 181, 183, 313
Coromandel Peninsula 60, 61, 128
Couch, Donald 296
Cowan, James 154, 155, 228, 229, 235, 266
Cronon, William 21, 22, 23, 173, 226, 227, 330
Crosby, Alfred 21, 174, 242–43
Crowder, Bob 206
Cruise, Richard 59
Cumberland, Kenneth 20–21, 35, 127, 193, 195, 196, 200, 202, 203, 220, 330
cyanide 110, 111, 112, 113

Dacre, Ranulph 60
dairy farming and industry 94, 98, 99, 132, 185, 197, 198, 199, 207, 289, 290, 329, 332; Mackenzie Basin 298–99, 303; Ngāi Tahu Property 302–04, 307–08; Waituna Lagoon 306–07
Dann, Christine 263
Darwin, Charles 137, 193, 332
Davies, William 201
DDT 205
deer 89, 153, 166
deforestation: for agriculture 19, 21, **21**, 105, 122, 127, 131–34, **132**, 135, 138, 153, 205, 209, 210, 218, 334, **334**; consequences 47–48, 129, 133–34, 136–37, 143, 150, 153, 203, 205, 218, 329; for gardens 20; by Māori 35–37, 38–40, 44, 46–48, **49**, 50, 91, 126, 311; for timber 57, **58**, 59–61, 127–30, 133–35, 136, 210; *see also* afforestation; timber industry
Department of Agriculture 104, 196, 197, 198, 219, 220, 221; Grasslands Research Station 198
Department of Conservation 153, 285
Department of Internal Affairs 153
Department of Lands 135, 170
Department of Lands and Survey 148, 149, 153, 210, 285, 324; Forestry Branch 211
Department of Scientific and Industrial Research (DSIR) 203, 219–20, 321
Department of Tourist and Health Resorts 146, 162
developmentalism 282–85, **284**, 288–90, 291–92
Dieffenbach, Ernst 61, 79, 160
displacement theory 150
DNA analysis 40
dogs 71; Māori (kurī) 42, 44, 45, 50

Don, Alexander 251–52, **252**
Donne, Thomas 148–49, 153
Du Faur, Freda 167–68
Dugdale, D.F. 291
Dunedin 128, 137, 146, 147, 148, 162, 229, 232, 235, 236, **236**, 238, 248, 251, 252
Dunlap, Thomas 21

Earthquake and War Damage Commission 271
earthquakes 229, 231, 261, 263, **264**, 268, 269–75, **273**, 332
East Coast, North Island 66, 71, 126, 198, 203, 206, 218, 237
East India Company 75, 244
ecofeminism 263
ecology: Cockayne's botanical conservation work 150–52; 'ecological imperialism' 175–79, 242–43, 332; as 'science of Empire' 203
econationalism 330
Egmont National Park 146–47, 158, 170, 171
Egmont Reserve 148
electricity developments *see* hydroelectricity developments
Ell, Harry 145, 148, 149, 154
Ellis, L.M. 211, 213–15, 217
Ellison, Edward 299
embodiment theories 263–64
Empire Forestry Conferences 213, 214, 215
Entrican, Alex 214, 215, **216**, 217, 218
Environment Canterbury 289, 295, 296, 298, 302; co-governance agreement with Ngāi Tahu 296
environmental appropriation 155–56
environmental awareness, historical perspective 18–21
Environmental Council 281
environmental degradation: environmentalists' perception 285; from forest and bush clearance 20–21, 47, 136, 203, 205, 207, 218, **219**; management 28, 136, 207, 327; from mining 114–15, 116–18, **117**; from overstocking 21, 105, 203; from pest invasions 21; from rabbit burrowing and grazing 21, 102, 105, 114; *see also* soil erosion; water quality
environmental history: applied 330; colonial transformations 18–21, 23, 331; contemporary historiography and practice 21–23; definition 18; and future actions, 18, 332, 334, 335; historical writing and representation 23–24; interdisciplinarity 332–34; themes for further research 329–32
environmental law: and developmentalism 282–85, **284**, 288–90, 291–92; enabling and facilitating resource exploitation 277–78, 281, 282, 288; hydroelectricity developments 154, 224, 279–81, 282–85, **283**, 289, **289**; legal change following European settlement 277–79; and

sustainability 287–88, 290–92; *see also* resource management; and individual Acts, e.g. Fisheries Act 1996
Environmental Protection Authority 289
erosion *see* environmental degradation; soil erosion
Evans, B.L. 194
exploration, by Europeans 70–72, 79–80, 81, 82–84, 313–14
exports: edible fungus *(Auricularia polytricha)* 255, **256**; meat 94, 95–96, 101, 105; minerals 106; primary sector generally 194, 199, 202, 219, 329, 331; rabbit skins 103; timber 60, 75, 76, 78, 128, 137, 211; wool 81–82, 93, 95, 101, 200; *see also* trade, colonial
extinctions 35, 36, 44–46
Eyrewell, Ngāi Tahu dairying 302–03, 307–08

Farm Forestry Association 206
farming *see* agriculture
fauna *see* animals
Federated Farmers 200, 205, 206, 207, 298; Women's Division 206
Federated Mountain Clubs (FMC) 169
feminist embodiment 263–64
fens, England 175–76, 177, 179, 180, 182
fernlands 46, 47, 48, 79, **90**, 91, 96, **126**
fertilisers 98, 138, 154, 197, 200, 333; aerial topdressing 193, 195, 196, 197, 198, 200, 204, 221, **222**, 225; land carrying capacity and productivity 193–95, **194**, 196, 197–99, **199**, 221; natural fertilisers 205; 'urine and dung showers' 195
Field, W.H. 167, 168
Fiordland 144, 146, 153, 154, 162, 171, 280
Fiordland National Park 170, 171
fire *see* burning
fish and fisheries: conservation 143, 278, 287; and cyanide disposal 112, 113; early explorers and traders 71, 72, 74, 76; exports 76; Māori 50, 54, 60, 71, 112, 113, 114, 184, 185, 287, 299, 301, 304–06, 307; overfishing 288
Fish Protection Act 1877 143
Fisheries Act 1996 287, 290–91
Fisheries Conservation Act 1884 143
FitzRoy, Robert 64, 79
Flannery, Tim 17
flax *(Phormium tenax)* industry 53, 54, **55**, 57, **58**, 59, 61, 63, 71, 72–73, 75, 76, 78, 188, 314
Fletcher Holdings/Construction Company 215, 271
Fletcher, James 215
flooding 17, 92, 199, 253, 268; coastal subdivisions 329; control 114, 121, 153, 171, 177, 179, 186, 203, 218, 220, 221, 223, 232, **233**, 278; resulting from deforestation 143, 150, 153, 203, 205, 218, 220; river siltation from mine waste 26, 108, 111–13, 114, 121; urban

flooding 28, 228, 231–32, **233**, 234, 238, 268
flora *see* plants
food resources: introduced by Cook 71–72, 84; Māori 42–43, 48, 50, 54, 60, 71, 76–77, 114, 155, 184, 185, 188, 293, 296, 297–98, 299, 301, 305–06, 307, 311; New Zealand's role as food producer 200, 201, 202; *see also* fish and fisheries; kūmara; pigs; potatoes; wheat
food safety and purity 205, 207
forest reserves 134–35, 147, 210, 214
forests: conservation 134–35, 137, 143, 147, 149–50, 209, 210–14, 215, 216–18, 224, 282; distribution **49**, **58**, **126**, **131**; early European impressions 123–24, **124**; exotic plantings 134, 150, 206, 211, **212**, 214–15, 217; forest cover, 1840s **126**, 127; forest cover, 1880-1910 **131**, 141, 210; indigenous species 42, 47, 123, 124–26, 216–18, 224, 225, 287–88; management 134–37, 213, 215, 216–17, 225, 278, 282, 287–88; Māori forest 213, 215, 222; perceptions of 142–43; state forests 135, 209, 210, 211–17, **212**, 224, 278, 282, 287–88; structure 124, **125**; wetlands 176–77, 183, 186–87; 'wise use' 225; *see also* afforestation; deforestation; kahikatea; kauri; pulp and paper industry; timber industry
Forests Acts 135, 136, 143, 213, 218, 287–88, 290
Forests Department 211, 213
Forster, George 71–72
Forty Mile Bush 18–19; *see also* Seventy Mile Bush
Foucault, Michel 262
Fox Glacier 171
Fox, William 24, **24**, 81, 145, 161
freshwater management 295, 297, 299–302, 303–04; *see also* irrigation
Friends of the Earth 282
fungus, edible *(Auricularia polytricha)* 255, **256**
Fyfe, Tom 166–67

Galbreath, Ross 146, 155
gardens: as antithesis of wilderness 230; Chinese plant introductions **242**, 243–45, 246, 247, 248–49, 250–52, 254–55, 256–57; concepts of plant distribution 241–43; early New Zealand plant networks **242**, 243–45; European colonists' perceptions 20; flower gardening and shows 250–51; home gardens 24, **24**, 237, 245–51, **247**, **249**; nurseries 250; organic practice 205; public gardens 154, 245, 257; William Fox 24, **24**; *see also* horticulture; plants
Gisborne 230
global economy: and environmental change 70; New Zealand, pre-1860 72, **73**, 74, 75–76, 84; *see also* resource economy
Godley, John Robert 272, **273**, 275
Gold Fields Act Amendment Act 1875 (No 1) 108–09

gold mining 83, 106–07, **107**, 246, 277; Chinese miners 251–52, **252**; preservation of agricultural land 116–18, **117**; and private property rights 108, 116; waste discharge and dumping 108–16, 121
gorse 89, 102, 105, 196, 202, 207
Graham, George 166–67
grasses *see* pasture
grasslands: burning 46, 47, 91, 92, 96, 97, 99; conservation 152, 220; distribution **49**, **90**, 91–92, **125**
'grasslands revolution' 193–207, **194**
Green, W.S. 162, 166
Grey, George 64
Greymouth 232
Gully, John 161–62, 172
Guthrie-Smith, Herbert 21, 138, 144, 204–05, 218, 330

Haast, Julius von 38, 81, 160
Hall, William 56
Hamer, Paul 155
Hamilton 237
Hanson, James 53
Harper, Leonard 161
Hastings 269, 270, 271
Hauraki Gulf 62, 83
Hauraki Plains 176–77, 183; drainage **175**, 176, 177, **178**, 179–80, **180**, 181, 183
Hawker, Tony 301
Hawke's Bay 47, 49, 64, 80, 81, 96, 131, 132, 138, 205, 220, 237, 261; earthquake, 1931 229, 261, 269–72, 275; *see also* individual place names, e.g. Napier
Haynes, R.J. 206
hazards, environmental 231–35, 238, 263, 268
Heaphy, Charles 78, 81, 160, 161, 181
heather 154
Hector, James 127–28
Henry, Richard 148
herbicides 196, 197, 200, **204**, 205, 206
Hermitage, Mt Cook 163, **164**, 167, 168, 169
Hickson, Thomas 135
historic places 148
Hochstetter, Ferdinand von 128
Hokianga 54, 59, 60–62, 126, 246
Hokitika 163, 232
Holcroft, M.H. 166
Holdaway, Richard 44
Holloway, John 35
home ownership 235–36
Hongi Hika 56, 61
Horeke 246, **247**
horticulture: Chinese 205, 251–56, **253**; Māori 17, 36, 42, 46, 48, 76, 84, 245, 246, 311–12; organic practices 207

Hughey, Ken 301
hunting 153, 166, 167, 169
Hursthouse, Charles 174, 175–76, 177, 189
Hurunui Water Project (HWP) 302, 303
Hutchins, David 211
Hutt Valley 64, 78, 231, 232, 237, 248–50
hydroelectricity developments 154, 224, 279–81, 282–85, **283**, 289, **289**

indigenous peoples, rights *versus* conservation movement 155–56
Inglis, James 133, 138
invasion biology 332–33
Invercargill 93, 128, 238, 305, 306
irrigation 198, 298–99, 301–02, 303, 307, 332
island reserves and sanctuaries **142**, 144–45, 147, 148, 154

Kaharoa ash deposits 40
kahikatea 176–77, 183, 186–87, 211, 217
Kaingaroa Forest 211, 215
kaitiakitanga (guardianship) 20, 295–96
Kaituna claim 1984 330
Kapiti Island 63, 144, 150, 152, 153, 154
kauri: conservation 149, 211, 216; distribution **55**, 125; logging 57, 59–61, 129–30, **130**, 133, 149, 216, 314, 327; Waipoua Forest 148, 150, 152, 216, 217–18
kauri gum 130, 195
Kawharu, Sir Hugh 67
Kehu 81, 161
Kelly, Jan 319
Kendall, Thomas 56
Kennedy's Bush **149**
Kermadec Islands 40
King Country 122, 131, 218
King, Michael 22
King, Philip Gidley 53–54, 73, **74**, 75, 76
Kīngitanga movement 64, 66
Kirk, Norman 280, 282
Kirk, Thomas 128, 135, 144
kiwi (*Apteryx* sp.) 144, 145
Kolodny, Annette 263
Korokoro 56
kūmara 54, 76

Labour government environmental policy 280, 282, 283, 285–86, 287, 288, 289, 290
Lake Wanaka Preservation Act 1973 282
land: commodification 84; forms of land tenure 115; loss of farming land to urbanisation 237; mining rights, and land 108, 115–18, 120–21; *see also* Māori land
Land Acts 118, 135, 141, 145
land degradation *see* environmental degradation
Land Drainage Act 1893 186

Land Information New Zealand (LINZ) 324, 325
Landcare Research 206
Lands Department 135
Langer, R.H.M. 207
law *see* environmental law; Māori – customary law
leeward–windward cline 42, **43**, 45, 47, 48, 49, **50**
Lester, Alan 210
Levy, Sir Bruce 193, 195, 196, 197, 199, 200, 201, 202, 207, 220
libertarian ideology 285, 286–87
Lincoln Agricultural College 94, 196, 203, 206, 207, 221
Lindis Pass **165**
Little Barrier Island 144, 155
Livingstone, John 244
Ludlam, Fanny and Alfred **249**, 249–50
Lyttelton 128

Mackenzie Country 92, 102, 103, 279, 297–99, **300**, 303
Mackenzie, Thomas 146, 153
Manapouri controversy **279**, 279–81, **280**; environmental gains 281–82; subsequent resurgence of developmentalism 282–85, **284**
Manawatu 64, 98, 131, 132, 133, 290
Mannering, Guy 167
Mantell, Walter 35, 38, 160
Māori: and Chinese 254, 256–57; chronology of Māori colonisation 38–41; cosmogony 52; customary law 53, 63, 64, 66, 69, 277, 278; deforestation 35–37, 38–40, 44, 46–48, **49**, 50, 91, 126, 311; fisheries 50, 54, 60, 71, 112, 113, 114, 184, 185, 287, 299, 301, 304–06, 307; food resources 42–43, 48, 50, 54, 60, 71, 76–77, 114, 155, 184, 185, 188, 293, 296, 297–98, 299, 301, 305–06, 307, 311; forests 213, 215, 222; founding population 40–41; horticulture 17, 36, 42, 46, 48, 76, 84, 245, 246, 311–12; inter-tribal warfare 62–63, 188; kaitiakitanga (guardianship) 20, 295–96; language 61; and nor'west wind, Canterbury 265; pre-European environmental impact on Māori 41–43; pre-European environmental impacts of Māori 17, 35–37, 38–40, 44–50, **50**, 155; relationship to environment 22, 52–53; renaissance 286; resource management 20, 44–48, 52, 277, 278, 297–98, 307–09, 311; society 53, 54, 61–63, 72, 77–78, 188–89, 311–12; spatial knowledge 311–13, 314, 325; traditional valuations of landscapes 22; treaty claims 36, 64, 66, 69, 261, 286, 287, 294, 297, 301, 329–30; understandings of mountains 160–61, 172; urbanisation 237, 275; and water (*see also* Māori – fisheries, above) 23, 296, 297–98, 299–302, **300**, 304–07, 329–30; *see also* Ngāi Tahu; Treaty of Waitangi

Māori, and Pākehā colonisation: early encounters 53–54, **55**; environmental appropriation by Pākehā 155–56; impacts on Māori society 61–63, **62**, 72, 77–78, 188–89; Māori trade with Pākehā 54, 57, 61, 71, 72, 75–78, 79, 183, 314; missionaries 56, **57**; Pākehā reliance on Māori cooperation in resource exploitation 57–58, 59–61, 62, 78; *see also* resource economy
Māori land: alienation in North Island, 1860–1939 **68**; confiscation **65**, 66, 203; Crown acquisition 64, **65**, 81; customary tenure system 52–53, 63, 67, 312, 315, 325; European views on Māori land use 20, 23, 78–79, 81, 202, 246; extinguishment of native title 63–64, **65**, 66–67; leases 80, 81; Māori resistance to sales 64, 66; Native Land Court policies 66–67, **68**, 69, 187, 315; Ngāi Tahu 293–94; place names 17, 23, 41, 91; purchases by colonists **57**, 64, 67, 78–79, 81, 82, 84, 203; riparian land 113; taken under Public Works Act and other legislation 179, 185–86, 187; whakapapa whenua 312, 314
mapping *see* surveying and mapping
Marine Mammals Protection Act 1978 282
marine reserves 282
Marion du Fresne, Marc-Joseph 71
Mark, Alan 281
market gardening *see* horticulture
Marlborough 63, 72, 81, 82, 92, 97, 102, 106, 128, 196
Marsden, Ernest 220
Marsden, Samuel 54, 56, 59, 76, 77
Marsh, George Perkins 143, 150
Martin, William 248
Mason, Thomas and Jane 248–49
Massey Agricultural College 203, 207
Massey, William 188, 211
mātaitai 304–05
McCaskill, Lance 193, 196, 198, 200, 203, 220
McDonnell, Thomas 246, **247**
McGavock, A.D. 214
McGlone, Matt 40
McGregor, William 217
McKay, Jessie 266
McLean, Donald 81
meat industry 94, 95–96, 98, 99, 100, 101, 105, 185, 199
Merchant, Carolyn 263
Meridian Energy 289
metrologies 311, 325, **326**, 327
Milford Track 162–63, 167
mining: coal 81, 106, 118–19, **119**; conflicting with land settlement and agriculture 115–18; environmental impact 106, 108–17, **117**, 118–19, 120–21; quarrying 106, 120; *see also* gold mining
Mining Act 1891 116
Mining Amendment Acts 117

Ministry for the Environment 285, 287, 299, 330
Ministry of Agriculture and Fisheries 206, 207
Ministry of Works (and Development) 165, 221, 285
missionaries 56, **57**, 59, 60, 61, 63, 193, 246, 251
moa (Dinornithiformes): biomass 45; distribution 42, **43**; extinction 35, 44, 45–46, 47, 48; hunting 36, 39, 41, **43**, 46, 48, 91, 311
Moncrieff, Perrine 152–53
Moore, George 82
Motu River 281
Motunui–Waitara claim 1983 330
Mount Cook Company 163
mountaineering 166–68, **168**, 169; accidents and fatalities 169–70, **170**
mountains: in art 160, 161–62, 172; cultural meanings 159–61, 172; location **159**; Māori understandings 160–61, 172; and national identity 165–66; in national parks 170–72; significance 158; sport 166–70; tourism 161–65, 166
Mt Cook 146, 158, 163, 166–67, 170, 171, 172, **331**
Mt Taranaki (Egmont) **132**, 146, 158, 160, 163, 172
Muir, James 155
mustelids 104, 144
Myers, J.G. 154

Napier 128, 228, 230, 269, 270, 271, **272**
National Development Act 1979 285, 288
National Forest Inventory 213
National government policy 215, 279, 283–85, 286, 289
national identity 154–55, 165–66
national parks 141, **142**, 145–46, 147–48, 155, 172–73, 210, 218, 278; mountains 170–72
National Parks Act 1952 170–71, 172
National Parks Authority 172
National Water and Soil Conservation Authority 206, 224
Native Bird Protection Society 152–53, 154
Native Forests Action Council 282
Native Land Court policies 66–67, **68**, 69, 187, 315
Native Lands Act 1862 66, 69
Native Plants Protection Act 1934 145, 278–79
Natives' Association 154
nature: nature/culture divide 22, 330; perceptions of nature 22–23, 70, 142–43, 262–63, **264**
Nature Conservation Council 281
Nauru Island 193, 195, 197
Nelson 24, **24**, 42, 63, 79, 81, 82, 128, 137, 146, 161, 176
New Plymouth **80**, 163, 227, 229, 230, 255
New Zealand Aerial Mapping (NZAM) 321, **322**
New Zealand Agricultural Chemical Trust 205
New Zealand Alpine Club (NZAC) 158, 166, 167, 169

New Zealand Company 64, 78, 79, 80, 84, 161, 181, 183
New Zealand Deerstalkers Association 169
New Zealand Forest and Bird Protection Society (1914–ca.1918) 152
New Zealand Forest Service (NZFS) 216, 217, 218
New Zealand Forestry League 152, 153, 154
New Zealand Historical Atlas 21, 335
New Zealand Institute 152
New Zealand Institute of Agricultural Science 197
New Zealand Journal of Agriculture 99, 200, **201**, **202**, 205, 206
New Zealand Railways (NZR) 238, **331**
New Zealand Settlements Act 1863 65, 66
New Zealand Wars 64, 66, 315
Newman, Alfred 188–89
Ng, James 252
Ngahuruhuru 53, 73
Ngāi Tahu 29–30, 75, 82, 91, 161, 265, 293–96, 307–09; conflict between commercial and cultural interests 302–04, 307–09; and waimāori 297–98, 304–07, 329; *see also* Te Rūnanga o Ngāi Tahu (TRoNT)
Ngāi Tahu Property 302–04, 307–08
Ngāi Tahu Seafood 304–06
Ngāpuhi confederation 62, 76
Ngāti Tamaterā 112, 113
Ngāti Toa 63, 79
Ngāti Tuwharetoa 146, 171
Nicholas, John 54
Norfolk Island 40, 53
Northland (Te Tai Tokerau) 42, 46, 47, 188–89, 197, 210; *see also* individual place names
nor'west wind, Canterbury 265–69, **268**
Noxious Weeds Act 1901 148

O'Connor, Kevin 196, 206
Office of Treaty Settlements 286
Ohinemuri River 110–14
Omaui **304**, 305–06
open country 80–90, 104–05; in 1840 91–92; acres under cultivation **101**; distribution 49, **90**; *see also* agriculture; fernlands; grasslands; pastoralism; scrublands; wetlands
Opouri Valley 147–48
O'Regan, Sir Tipene 294, 297, 306
organic farming 205
Otago 42, 76, 81, 83, 84, 91, 93, 101, 102, 106, 108, 118, 134, 135, 161, 162, 227, 246, 248, 251, 252, 253, 305; Catlins district 47, 128; Central Otago 91, 92, 102, 103, 110, 115, 119, 165; East Otago 46; North Otago 35, 84, **95**, 114, 165; South Otago 83–84; *see also* individual place names, e.g. Dunedin
Otehake wilderness area 172
Otira Gorge 146, 158, 162, 163, 166, 171

pā (Māori forts) **43**, 48–49
Pai Mārire 66
Palmer, Geoffrey 290
Palmer, Henry Spencer 316–17
Parliamentary Commissioner for the Environment 285, 298–99
Parr, Joy 263
Pascoe, John 169, 171
pastoralism 80–82, 84, 89–90, 92, 93, 94, 95, 97, 99–100
pasture: exotic species 96, 97–98, 99, 104, 194, 195, 196, 197, 198, 201–02, 242; 'grasslands revolution' 193–207, **194**; improvement and maintenance 97, 98–99, 104, 105; land coverage 132; native pastures 96, 99, 201, 207
pāua fishery 304–06
Perrin, George 134, 153
pesticides 197, 205, 206
pests, introduced 71, 89, 101, 215; control 197, 205, 206, 278; integrated management schemes 206; *see also* cats; deer; dogs; mustelids; rabbits; rats, Pacific; rats, ship
Petchey, Fiona 45
Peter, Juliet 267, **268**
Phillip, Arthur 72
Phillips, Jock 226
photographic records 18–19, 23
pigs 54, 57, 71, 72, 73, 76, 79, 84, 166
place names, Māori 17, 23, 41, 91
plants: concepts of plant distribution 241–43; conservation 145, 150–52; distribution of New Zealand plants and seeds to Britain and beyond 80; early New Zealand networks **242**, 243–45; economic use 70–72, 76–77; food plants, Māori 42–43, 48; introduced 23, 71–72, 73, 76, 84, 89, 123, 243–45, 246, 247–52, 256–57; Māori names 41; protection by legislation 278–79; *see also* gardens; weeds
pōhā preparation 305, **305**, 306
Polack, Joel 77
pollen analysis 38–39, 40, 97
pollution *see* air pollution; soils – pollution; waste disposal; water quality
Polynesian technology 41
Poole, A.L. 214, 215
Porteous, Douglas 263
postcolonialism 261, 262, 264, 269, 272–75
Potatau Te Wherowhero 64
potatoes 54, 57, 60, 71–72, 73, 75, 76, 77, 79, 84, 245
Potts, Thomas 133, 144, 150
Poverty Bay 91, 98
Powell, Joe 18
preservation *see* conservation; scenery preservation societies
progress leagues 230

Project Aqua 289, **289**
public works 120, 165; *see also* railways; roads
Public Works Act 1876 120, 179, 185–86
Public Works Department 111, 114, 221, 321
pulp and paper industry 214, 215

Queen Elizabeth II National Trust 206
Quin, Bert 206
Quin, Joe **253**

rabbits 21, 89, 95, 101–03, **103**, 105, 114; control 103–04, 144
radiocarbon dating 35, 38, **39**, 40, 47
Rai Valley 147–48
railways 120, 128, 148–49, 162, 163, 210, 230, 237, 238
Rangitaiki River 281
Rangitaiki Swamp, drainage 179, 180, 181, 187
Rankin, Hone Heke 188
rats, Pacific *(Rattus exulans)* 38, 39, 44, 45, 50, 160
rats, ship 71
recreation, outdoor 153, 162–63, 166, **331**; *see also* hunting; mountaineering; skiing; tramping
Reeves, William Pember 182
reserves *see* forest reserves; island reserves and sanctuaries; marine reserves; mātaitai; national parks; scenic reserves
Reserves and Other Lands Disposal and Public Bodies and Empowering Act 1915 120
Resolution Island 144, 148, 155–56
resource economy 70–76, 84; and environmental change 19, 70, 72, 76–78, 84, 123; Pākehā reliance on Māori cooperation 57–58, 59–61, 62, 78; *see also* agriculture; fish and fisheries; flax *(Phormium tenax)* industry; food resources; global economy; mining; seals and sealing; settler economy; timber industry; trade, colonial; whaling
resource management: adaptations to environmental variability 332; government intervention 288–90; by Māori 20, 44–48, 52, 277, 278, 297–98, 307–09, 311; state responsibility 224–25, 286; 'wise use' of resources 224, 225; *see also* conservation; environmental law; sustainability
Resource Management Act 1991 21, 286–87, 288–90, 296
Resource Management (Waitaki Catchment) Amendment Act 2004 289
Resumption of Land for Mining Purposes Act 1873 116
Riccarton Bush 147, 152, 153
riparian rights 108–09, 110
Rivers Board Act 1884 186
Rivers Commission 110, 114–15, 121
roads 162, 165, **165**, 198, 210, 237, 278
Robin, Libby 203

Rolleston, William 272–73, 274
Ronga Valley 147–48
Rotorua 230, 237
Royal Commission on Forestry 1913 134, 150, 186, 210
Royal Commission on the Sheep-Farming Industry 1949 200, 203, 206, 220–21
Royal Forest and Bird Protection Society 152–53, 154, 218, 219
Royal Society of New Zealand 152, 153, 154, 220
Ruatara 56, 76, 77
run-off, agricultural 22, 206, 290, 301, 329
'ruralism' 200, **202**, 235

Sack, Robert D. 262
Sang, Mike 303
Sargeson, Frank 122, 123, 137
Sauer, Carl 20
Savage, John 54
Save Manapouri campaign 280–81
Scenery Preservation Act 1903 148, 209
scenery preservation societies 146–48, 156
scenic reserves 141, **142**, 145–46, 148–49, 152, 210, 278
Schlich, William 211, 213
Scott, Robert Falcon 273, 275
scrublands 47, 48, **49**, 54, **90**, 196
seals and sealing 36, 42, 46, 48, 72, 73, 74, 75, 76, 181, 311, 314
Sears, P.D. 197
Seddon, Richard John 134, 148
sensory history 262–72, 275
settler economy 78–84, 89, 92–96; and conservation 152, 210; and environmental change 69, 84, 89–90, 104–05, 120–21, 123, 127–34, 136–38, 143, 207; mining or land settlement 115–18; *see also* agriculture; resource economy
Seventy Mile Bush 132
Sewell, Henry 20, 22, 91, 265
Sewell, Tony 303
Sharpe, Alfred 156
sheep farming 80–82, 84, 92, 93, 99, 122, 132, 185, 199; acres under cultivation and sheep numbers **101**; breeds for open country environments 99–102, **100**, 104; and landscape values 173; and rabbit irruption 102–03, 105; sheep numbers and fertiliser use 194, 198; *see also* meat industry; wool
sheep stations 80, 82, 93, **95**, 95–96, **96**, **100**, 102–03, 196
Shortland, Edward 83
Siegfried, André 227, 235
skiing 165
Smallfield, P.W. 193, 196, 200
Smith, C.M. 214, 217

Smith, Nick 299, 301
Smith, William Mein 80
smog 233–34, 274
soil conservation 196–97, 206, 218–24, **222**, **223**, 224, 332
Soil Conservation and Rivers Control Act 1941 209, 220
Soil Conservation and Rivers Control Council (SCRCC) 218, 220–21, **222**, 223
soil erosion 97, 102, 105, 117, 153, 198–99, 329; control 196, 206, 219–24, 278; resulting from agricultural practices 117, 153, 196–97, 203, 206; resulting from deforestation 20–21, 47, 136, 150, 195, 203, 205, 209, 218–19, **219**, 329
soils: fertility 98, 104, 105, 138, 194, 195, 219; open country 91; pollution 205; restoration in areas once mined 116–18; trace element deficiencies 104, 105, 195
Solomon, Mark 296, 302, 303, 307
Solomon, Raewyn 296
Southern Alps 158, **159**, 160, 161, 166, 167, 172
Southern Pastoral Lands Committee 218
Southland 30, 47, 75–76, **83**, 92, 97, 98, 102, 128, 135, 196, 197, 205, 290, **304**, 305, 306–07
sport, in mountain regions 166–69; accidents and fatalities 169–70, **170**; *see also* hunting; mountaineering; skiing; tramping
Stapledon, George 200–01
State Forest Service (SFS) 153, 213–15, 216, 217, 218
State Forests Act 1885 135, 141
Stephenson, Ian 206
Stewart Island (Rakiura) 75, **83**, 150
Stoddart, Mark 266–67
subsidies, agricultural 194, 198, 221, 223, 224
suburbs 228, **234**, 235–38, 274, 275
Summit Road Association 148
superphosphate *see* fertilisers
surveying and mapping 310–11, 325, 327; aerial surveys 317, 321, **322**, 323; cadastral maps 316–17, **318**, 320, 325; digital maps 325; early colonists' use of maps as authoritative evidence 314–15, 320, 325; for environmental management 320, **323**, 323–24; European explorers 313; key sequences 319, **320**, 321; Māori spatial knowledge 311–13, 314, 325; military maps 317, **318**, 319, 320, 321; reform, privatisation and the information economy 324–25; surveying and mapping for settlement 69, 80, 81, 83–84, 161, 273, **315**, 316–17, **318**, 319, 320; topographical maps 317, **318**, 319, 320, 325
Surville, Jean François de 72
sustainability 143, 197, 207, 277, 287–88, 291–92; balancing approach 290–91; precautionary principle 291, 292

Sutton, Bill 266, **267**
Sutton, Roger 271
Swainson, William 183
Swamp Drainage Act 1915 187
swamps *see* wetlands
Swift, R.S. 206
Syers, J.K. 197

Tannock, David 251
Tapsell, Philip 62, 63, 77
Taranaki 63, 64, 66, 126, 127, 131–32, **132**, 146, 156, 176, 196, 219, 230; *see also* Mt Taranaki
Tararua Ranges 160, 161, 163, 167, 168–69, 171, 172
Tararua Tramping Club (TTC) 167, 168–69, 171
Tasman, Abel 313
Tasman Pulp and Paper Ltd 215, 216
Tau, Te Maire 265–66, 306, 308
Taupo and district 66, 79, 80, 158, 165, 237
Taupo eruption 46, 183
Tāwhiao 64
Taylor, Norman 220
Taylor, Richard 79–80
Te Anau controversy *see* Manapouri controversy
Te Heuheu Tukino IV 146, 156
Te Ika a Māui 17, 41
Te Kooti 66
Te Rauparaha 63, 77
Te Rūnanga o Ngāi Tahu (TRoNT) 294, **294**, **295**, 296, 297, 301, 302, 303, 304–05, 306, 307, 308; Freshwater Policy Statement 297, 299; *Ngāi Tahu 2025* 297–98, 299; *see also* Ngāi Tahu
Te Ua Haumene 66
Te Waihora/Lake Ellesmere 296, 299–302, **300**
Te Wharekorari 160
Thames district 59, 63, 75, 112, 113, 114, 183; *see also* Hauraki Plains
Thermal Springs Districts Act 1881 145, 278
'Think Big' policy 285
Thomas, Keith 262–63
Thomson, A.P. 216
Thomson, G.M. 144, 152, 153, 154
Thomson, John Turnbull 83–84, 161
Timaru 93, 163, 167, 171, 238–40, **239**
timber industry: depletion of native timber resources 133–34, 136, 149, 150, 209, 210, 211, 214, 224; exotic timber 134, 150, 211, **212**, 214–15, **216**, 217; indigenous timber felling and milling, Forests Amendment Act 1993 287–88; kauri logging 57, 59–61, 129–30, **130**, 133, 149, 216, 314, 327; pre-1840 trade 57, **58**, 59–60, 61, 71, 72, 74, 75, 76, 78, 126, 314, 327; regulation 134–35, 213; sawmills 60, 122, 128–29, **129**; and scenery preservation supporters 147–48, 209; selective logging of indigenous species 216–17; sustainability 287–88; *see also* forests; pulp and paper industry

Titokowaru 66
Tongariro National Park 146, 148, 150–52, **151**, 156, 158, 163, 170, 171
tourism 145–46, 148–49, 150, 153, 300; in mountain regions 161–65, 166, 171
town and country planning legislation 282, 286
trade, colonial 56, 57–58, **58**, 59, 60, 70, 72–76, **73**, 81, 84; Māori trade with Pākehā 54, 57, 61, 71, 72, 75–78, 79, 183, 314; New Zealand produce treated as Australian 75
'tragedy of the commons' 48
tramping 162–63, 165, 167, 168–69
Treaty of Waitangi: contraventions 69, 112, 286; and European world view 20; guarantee of Māori rights 278; principles 286, 287; signing 63; Treaty Settlement Process 286; and world view of Europeans 20; *see also* Waitangi Tribunal
Tregear, Edward 131–32
Tripp, Charles 82
Trollope, Anthony 162
tūī 145
Tuki Tahua 53, 73
Turner, E. Phillips 148, 213, 214
tussock grassland *see* grasslands
Tussock Grasslands and Mountain Lands Institute 196
Tutira (Guthrie-Smith) 21, 205, 218
2,4–D 205
2,4,5–T 196, 205

Union Steam Ship Company 146, 162
urban areas: architecture and design 229, 230–31, 238, 269, 271–72, 274; counterpoint of outdoor recreation to urban routine 153, 173; environmental historians' neglect 226–27; government investment 230–31, 237; progressive towns on the edge of wilderness 227, 228–31; quarries 120; scenery preservation 146–47; suburbs 228, **234**, 235–38, 274, 275; superiority of country over urban living 200, 235; Timaru case study 238–40, **239**; towns vulnerable to environmental hazards 227–28, 231–35, 238, 261, 269–75; urbanisation 203–04

Victorian gold mining 106, 107–08, 109, 117, 121
viticulture 207
Vogel, Julius 136, 143, 150
volcanic activity 40, 46, 126, 183
volcanic cones, quarrying 120
Volcanic Plateau 152, 195, 214
Vosseler, F.W. 167, 169

Waaka-Home, Mandy 298
Waihou–Ohinemuri goldfield dispute 110–14, **111**
Waikato 47, 62–63, 64, 66, 118, 183, 196; *see also* individual place names, e.g. Hamilton
Waimakariri River 232, **233**, 234

Waipoua Forest 148, 150, 152, 216, 217–18
Waipoua Forest Preservation Committee 217–18
Wairarapa 49, 64, 80–81, 92, 102, 161, 271, **323**
Wairau Affray 64, 81, 315
Wairau Valley 82, 162
Waitaha Nation myth 36
Waitakere Ranges 147, 171, 172
Waitaki River and Valley 45, 102, 160, 282, 289, **289**, 298–99, **300**
Waitangi Tribunal 36, 52, 64, 66, 69, 160, 185, 286, 294, 297, 301
Waitomo Caves 148
Waituna Lagoon 306–07
Wakefield, Edward Gibbon 20, 23, 78, 81, 177, 226
Wakefield, William 78, 80, 181
Walker, Inches Campbell 135, 137
Wanaka hydro development proposal 282
Ward, Alan 67, 68
Warming, Eugene 150
waste disposal: agriculture 22, 206, 298–99, 306–07; gold mining 108–16, 121; household waste 234; industrial developments, North Taranaki 330; Māori concerns 224; motivation for early legislation 278; sawmills 143; sewage 232, 233, 234, 237, 301, 330
Waste Lands Committee 147
Water and Soil Conservation Act 1967 206, 223, 281, 286
water availability and conservation 219, 223–24, 281, 287, 297, 329–30; *see also* hydroelectricity developments; irrigation
water quality 224, 288, 329–30; and agriculture 22–23, 195, 290, 298–99, 300, 301, 306–07, 329; freshwater management 295, 297, 299–302, 303–04; gold mine waste disposal 108–16, 121; legislation 278; New River Estuary 305–06, 307; non-point and diffuse pollution 290; sawmill waste disposal 143; Te Waihora/Lake Ellesmere 296, 299–302, **300**; urban areas 232–33; waterways and wetlands 22–23

weeds: control 148, 196, 200, **204**, 205, 206, 278; growth stimulated by fertiliser run-off 206; introduction and spread 89, 102, 105, 154; Māori land 202
Weld, Frederick 92
Wellington 49, 79, 128, 132, 137, 146, 147, 154, 161, 163, 181, 229, 230–31, 232, 237, 252, 253, 254, 261, 329, 332; *see also* Hutt Valley
Wesleyan mission 56, **57**, 60
West Coast 81, 82, 91, 106, 118, 128, 161, 162, 165, 171, 210, 252, 265, 288
Westport 231
wetlands: cattle grazing 99; conservation and restoration 179, 187, 206; distribution **90**; drainage 79, 174–76, **175**, 177, **178**, 179–81, **180**, 182–83, 185–88, **186**; forests 176–77, 183, 186–87; Māori perceptions and use 22, 160, 177, 179, 181, 183, 184, 185, 186, 188–89; Pākehā perceptions 22, 175–79, 181, 182–83; pollution 22–23
Whale Watch Kaikoura 308
whaling 72, 314; ships 56, 57, 58, 59, 60–61, 75, 76; shore stations 54, **55**, 57, 76, 79, **83**, 143, 181–82
Whanganui and district 80, 172, 195, 218, 227, 254, **318**
Wheao River 281
wheat 81, 97, 245
Whitcombe, John Henry 161
White, Ben 185
wilderness areas 166, 172, 173, 227, 330, 335
Wildlife Service 145, 187
Williams, Henry 59
Wilmshurst, Janet 40
Wilson, E.H. 251
Wilson, Sir John Cracroft 247
Wilson, William 247–48
windward–leeward cline 42, **43**, 45, 47, 48, 49, **50**
wool 81–82, 93, 94, 95, 100, 101, 199, 200
Worthy, Trevor 44